A FIRST COURSE IN APPLIED COMPLEX VARIABLES

LESTER A. RUBENFELD

Department of Mathematics

Rensselaer Polytechnic Institute

JOHN WILEY & SONS

New York · Chichester · Brisbane · Toronto · Singapore

WITH LOVE

TO MY PARENTS
who started the process

TO MY FAMILY
who provided the support and motivation

TO MY HERITAGE
which puts all of this in its proper perspective

Library of Congress Cataloging in Publication Data:

Rubenfeld, Lester A.
A first course in applied complex variables.

Includes index.
1. Functions of complex variables. I. Title.
QA331.R83 1984 515.9 84-11840
ISBN 0-471-09843-4

Printed in the United States of America

10 9 8 7 6 5 4 3 2 1

PREFACE

The subject of complex function theory is one of the most beautiful and elegant in all of mathematics and has proved to be extremely useful in many applied fields. Many students of engineering and science presently have mathematics requirements well beyond the first two years of calculus, linear algebra, and differential equations, and a choice of a course in complex variables is becoming more and more commonplace.

This book was written after many years of teaching this material to students at Rensselaer Polytechnic Institute and was designed as a first course on the subject. The book can be used as a one-semester text for either upper-level undergraduate mathematics majors or upper-level undergraduate or beginning graduate engineering and science majors, although it is recognized that the dominant audience will be students who are studying engineering and science. My objective is to present the material in a manner that is both intuitive and as immediately applicable as possible. To this end, very early in the discussion (in Chapter 3) some applications are made to the fields of fluid flow and electrostatics. The discussion is continued and extended in Chapter 4 and in assorted examples and problems in Chapters 5 and 6.

The beauty of complex function theory rests not only in its underlying mathematical structure but in its many applications. It is to this end that the text is structured. Many examples are worked out which illustrate all the topics discussed in the main body of the text. The numerous problems at the end of each section are balanced between those that are straightforward illustrations of the material, those that extend the material, and those that are of a more theoretical nature. Answers are given to all problems. Except for several sections, for instance, 3.2, 4.2, 5.6, and

6.8, which are well suited for more mathematically oriented students, an attempt has been made to dispense with mathematical rigor when it would detract from the main purpose of the text. Some of this rigor is presented when necessary, and some of it is left to the problems. However, all theorems and definitions are stated in precise mathematical language.

The text was written with the assumption that readers have had the usual two or three semesters of calculus. Although a course in multidimensional advanced calculus would be beneficial, it is by no means necessary. The material covered is, for the most part, the standard subject matter presented at this level together with as many applications as possible, again at this level. The stress is on the computational aspects of the subject. The sections in a chapter are numbered; for example, Section 2.1 is the first section in Chapter 2. The most important equations, examples, definitions, and theorems are numbered consecutively within each section. Thus, (2.5.6) refers to the sixth such item in Section 2.5, whereas Figure 5.6.2 refers to the second figure in Section 5.6.

I am grateful to the many people who have aided me in one way or another in the preparation of this book. In particular, my thanks go to Mrs. Theresa Testo and the staff of the Mathematical Sciences Department at Rensselaer Polytechnic Institute for typing the manuscript, and to Mr. Thomas Head and Mr. Craig Gleason for checking the answers to many of the problems. Special thanks go to my wife Alice, who helped to proofread the manuscript. Finally, I acknowledge with pleasure the cooperation of Andrew Ford, Gary Ostedt, and the editorial and production staff of John Wiley.

Lester A. Rubenfeld

TO THE INSTRUCTOR

With the exception of some of the more advanced topics in Chapters 4, 5, and 6, it should be possible to complete this text in a normal 15-week semester. A word is needed, however, concerning the order of the material. You will note that the material on conformal mapping and its applications occurs early on (in Chapter 4). Usually, this subject is presented after integration and infinite series. The reasons for this are very much of a mathematical nature, since integration and infinite series are needed to prove many of the important results about conformal mapping. However, the natural place for this material is soon after the elementary complex functions are introduced (in Chapter 2). This is the approach that I have taken, and it has worked quite well at Rensselaer. However, the material in Chapter 4 is quite self-contained and can be included either before or after Chapters 5 and 6.

The topic of the evaluation of real integrals is first introduced in Section 5.4, right after the Cauchy integral formula. Usually, this is delayed until after residue theory (in Section 6.6). I believe that since many of the integrals which are evaluated by residues can just as well be evaluated using the Cauchy integral formula, it is advantageous to introduce this topic as early as possible. Indeed, the topic of the evaluation of real integrals is again taken up, unified, and extended in Section 6.7. However, if the instructor wishes, the material in Section 5.4 can be presented, with some minor modification, after Section 6.6.

The table that follows is a recommendation as to which material should be covered each week during a 15-week course using this book. It is based on class testing of the material over many semesters at Rensselaer Polytechnic Institute.

This, of course, is subject to change depending on the makeup of the class. For example, a more mathematically oriented group of students should cover the beginning material more quickly, and thus there will be time for the inclusion of the more theoretical subject matter in Sections 3.2, 4.2, 5.6, and 6.8. Also, depending on the class makeup, a choice might have to be made as to which of the sections devoted to applications should be covered.

WEEK	*SECTIONS COVERED*
1	1.1, 1.2, about one half of 2.1
2	Finish 2.1, 2.2, about one half of 2.3
3	Finish 2.3, 2.4, about one third of 2.5
4	Finish 2.5, 2.6, start 2.7
5	Finish 2.7, 3.1 (briefly), 3.3
6	Exam. No. 1, 3.4 or about one half of 3.5
7	Finish 3.4 or 3.5, 4.1 (no equivalent curves)
8	4.2 (very briefly; just the basic ideas), 4.3 (emphasize results), 4.4, 4.5 (heat equation)
9	4.5 and 4.6 (applications depend on class), 5.1, start 5.2
10	Exam. No. 2, finish 5.2, start 5.3
11	Finish 5.3, 5.4
12	6.1 (briefly) 6.2 (no uniform convergence), 6.3, start 6.4
13	Finish 6.4, 6.5
14	Exam. No. 3, 6.6, 6.7 } (see note)
15	Finish 6.7, review } (see note)

Note: If time is left, try to cover some of the material on the argument principle and root location in Section 6.8, or Section 5.5 on the Schwarz–Christoffel transformation.

CONTENTS

1

COMPLEX NUMBERS

1.1 COMPLEX NUMBERS; SETS OF POINTS

As mathematics evolved out of the Dark Ages, it was realized that certain well-accepted concepts had to be extended and generalized. Thus, in addition to the integers, one had to introduce rational numbers (perhaps to barter for a half loaf of bread?). Mathematicians then recognized that though a quadratic equation such as $x^2 = \frac{1}{4}$ has rational solutions, the equation $x^2 = 2$ does not. Thus, irrational numbers were introduced, and the real number system was completed. This was sufficient until one considered the problem of finding (real) solutions to the quadratic equation $x^2 = -1$ (or perhaps the less obvious $x^2 + x + 1 = 0$). Problems such as these presented a formidable roadblock and led to the development of complex numbers.

What is done is rather simple. Entities, called **complex numbers**, are introduced together with formal rules of combination. Thus, a member of the **set of complex numbers**, denoted by the letter z, is represented symbolically in the form

(1.1.1) $$z = x + iy$$

where x and y are real numbers satisfying the usual rules of addition, multiplication, and so on, and the symbol i, called the **imaginary unit**, formally has the property

(1.1.2) $$i^2 = -1$$

In (1.1.1), the real numbers x and y are the **real part** and **imaginary part** of z, respectively, and are denoted by

(1.1.3) $x = \text{Re}(z), \quad y = \text{Im}(z)$

We say that **z is real** if $y = 0$, while it is **purely imaginary** if $x = 0$.

(1.1.4) ***Example.*** The complex number $z = 3 + 2i$ has real part 3 and imaginary part 2 (not $2i$), while the real number 5 can be viewed as the complex number $z = 5$ having real part 5 and imaginary part 0.

■

The above example shows that the set of real numbers, with which we are all familiar, can be considered as a subset of the complex numbers, namely, those complex numbers with zero imaginary part.

Geometrically, *complex numbers can be represented as points in the plane.* Thus, $z_0 = x_0 + iy_0$ can be interpreted as that point in the xy-plane whose coordinates are (x_0, y_0), as shown in Figure 1.1.1. (In Problem 6, we show how **complex numbers can also be interpreted as vectors in the plane**).

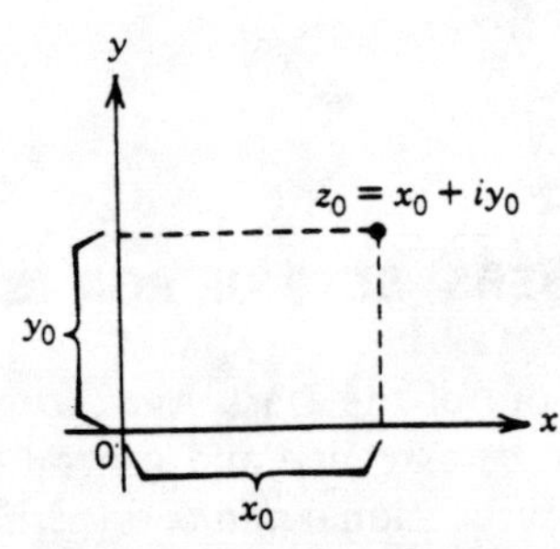

Figure 1.1.1 The z-plane.

We will call the xy-plane, when viewed in this manner, the **z-plane**, or **complex plane**, with the x-axis designated as the **real axis** and the y-axis as the **imaginary axis.** In addition, we will call the set of points in the z-plane having positive y-values the **upper half-plane** and those with positive x-values the **right half-plane**, with the rest of the z-plane similarly designated.

With complex numbers viewed as points in the z-plane, we designate the complex number **zero** as the origin. Thus,

(1.1.5) $x + iy = 0 \quad \text{means} \quad x = y = 0$

In addition, since two points in the xy-plane are the same if and only if both their x- and y-coordinates agree, we can define **equality of two complex numbers** by

(1.1.6) $x_1 + iy_1 = x_2 + iy_2 \quad \text{means} \quad x_1 = x_2 \text{ and } y_1 = y_2$

Thus, we see that a single equation between complex quantities really contains two real equations.

We must now provide rules which tell us how complex numbers are to be combined.

(1.1.7) ***Definition.*** Let $z_1 = x_1 + iy_1$ and $z_2 = x_2 + iy_2$. Then we define

(a) $z_1 \pm z_2 = (x_1 \pm x_2) + i(y_1 \pm y_2)$

(b) $z_1 z_2 = (x_1 x_2 - y_1 y_2) + i(x_1 y_2 + x_2 y_1)$

(c) for $z_2 \neq 0$, z_1/z_2 is that complex number w such that $z_1 = z_2 w$

■

In Problems 4 and 5, the reader is asked to show that

$$\frac{z_1}{z_2} = \frac{(x_1 x_2 + y_1 y_2) + i(x_2 y_1 - x_1 y_2)}{x_2^2 + y_2^2} \tag{1.1.8}$$

and that the usual rules of associativity and commutativity for addition and multiplication, which hold for real numbers, are also true for complex numbers. Also note that although the product of two complex numbers is *defined* as in (1.1.7b), it can be derived by *formal multiplication* using $i^2 = -1$.

(1.1.9) ***Example.*** Show that if $z_1 z_2 = 0$, then either $z_1 = 0$ or $z_2 = 0$.

Solution. If we write $z_1 = x_1 + iy_1$ and $z_2 = x_2 + iy_2$, then by (1.1.7b) and (1.1.5), the equation $z_1 z_2 = 0$ leads to the two *real* equations

$$x_1 x_2 - y_1 y_2 = 0 \quad \text{and} \quad x_1 y_2 + x_2 y_1 = 0 \tag{1.1.10}$$

Multiplying the first by y_2, the second by x_2, and subtracting leads to

$$y_1(x_2^2 + y_2^2) = 0$$

This implies, since the x's and y's are real, that either $y_1 = 0$ or $x_2^2 + y_2^2 = 0$. If $y_1 = 0$, then from (1.1.10) either $x_1 = 0$ (and hence $z_1 = 0$) or $x_2 = y_2 = 0$ (and hence $z_2 = 0$). If $x_2^2 + y_2^2 = 0$, then $x_2 = y_2 = 0$, and hence $z_2 = 0$.

■

(1.1.11) ***Example.*** Solve the quadratic equation $z^2 + z + 2 = 0$ without using the quadratic formula.

Solution. Writing $z = x + iy$, we get

$$x^2 - y^2 + x + 2 = 0 \quad \text{and} \quad 2xy + y = 0 \tag{1.1.12}$$

The second equation yields $y = 0$ or $x = -\frac{1}{2}$. If $y = 0$, then the first equation above would imply that $x^2 + x + 2 = 0$. However, since x is real, this has no real solutions (plot the curve $y = x^2 + x + 2$), and hence we cannot have $y = 0$. Thus, $x = -\frac{1}{2}$, which by the first equation in (1.1.12) yields $y = \pm\sqrt{7}/2$. Hence, the roots of the quadratic are

$$z = x + iy = \tfrac{1}{2}(-1 \pm i\sqrt{7})$$

■

We would remark here that one cannot define a concept of inequality between complex numbers which is consistent with the usual ordering of real numbers.

Finally, we have the following two definitions.

(1.1.13) ***Definition.***

(a) The **complex conjugate** $\bar{z}$ of $z = x + iy$ is that complex number defined by

$$\bar{z} = x - iy$$

(b) The **magnitude** $|z|$ (or **modulus**) of z is that real, nonnegative number given by

$$|z| = (x^2 + y^2)^{1/2}$$

■

Note that geometrically, as shown in Figure 1.1.2*a*, $\bar{z}$ is simply the reflection of z in the real axis, while $|z|$ is the distance of z from the origin. Also, as in Figure 1.1.2*b*, if $z_1 = x_1 + iy_1$ and $z_2 = x_2 + iy_2$, then

(1.1.14)
$$|z_1 - z_2| = [(x_1 - x_2)^2 + (y_1 - y_2)^2]^{1/2}$$

which is the **distance between z_1 and z_2.**

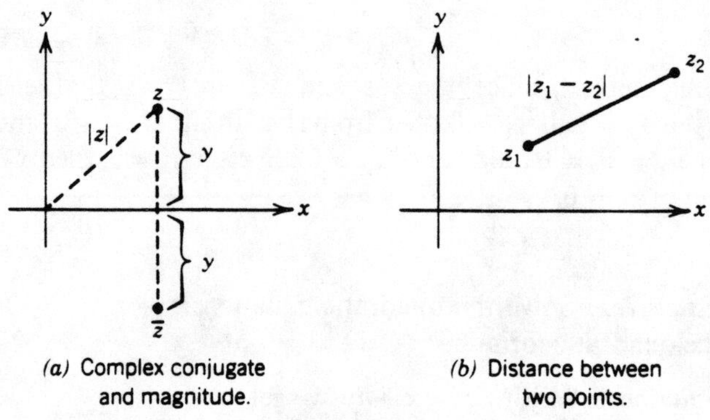

(a) Complex conjugate and magnitude.

(b) Distance between two points.

Figure 1.1.2

(1.1.15) ***Example.*** From the definition of $\bar{z}$ we have

(a) $x = \frac{1}{2}(z + \bar{z})$ and $y = \frac{1}{2i}(z - \bar{z})$

(1.1.16) **(b)** z is real means $z = \bar{z}$

(c) z is imaginary means $z = -\bar{z}$

■

In Problem 12*a*, the reader is asked to verify the following.

(1.1.17) ***Properties of Complex Numbers***

(a) $\overline{(\bar{z})} = z$
(b) $\overline{z_1 \pm z_2} = \bar{z}_1 \pm \bar{z}_2$
(c) $\overline{z_1 z_2} = \bar{z}_1 \bar{z}_2$ and, if $z_2 \neq 0$, $\overline{(z_1/z_2)} = \bar{z}_1/\bar{z}_2$
(d) $|\bar{z}| = |z|$
(e) $z = 0$ if and only if $|z| = 0$
(f) $z\bar{z} = |z|^2$
(g) $|z_1 z_2| = |z_1||z_2|$ and $\left|\dfrac{1}{z}\right| = \dfrac{1}{|z|}$ if $z \neq 0$

■

(1.1.18) ***Example.*** Show that

$$\left|\frac{z - i}{z + i}\right| < 1$$

when $\text{Im}(z) > 0$.

Solution. From (1.1.14), $|z - i|$ is the distance from z to i while $|z + i|$ is the distance from z to $-i$. Hence, for z in the upper half-plane, $|z - i| < |z + i|$, and the result follows.

■

(1.1.19) ***Example.*** Write the complex number

$$\frac{1 + i}{1 + 2i}$$

in the form $x + iy$.

Solution. Instead of using (1.1.8), we will proceed differently. If we multiply the numerator and denominator by the complex conjugate of the denominator, we get

$$\frac{1 + i}{1 + 2i} = \frac{(1 + i)(1 - 2i)}{|1 + 2i|^2} = \frac{3 - i}{5} = \frac{3}{5} - \frac{i}{5} \tag{1.1.20}$$

■

Sets of Points; Triangle Inequality. The concepts and definitions introduced above can be used to describe certain sets of points in the plane, as the following examples illustrate.

(1.1.21) ***Example.*** By definition, a **circle of radius R_0 with center at the point (x_0, y_0)** is the set of all points (x, y) at distance R_0 from (x_0, y_0). Thus, if $z = x + iy$ and $z_0 = x_0 + iy_0$, this circle can be described by the complex equation

$$|z - z_0| = R_0 \tag{1.1.22}$$

Also, the set of points lying in the **disk** bounded by this circle satisfies the inequality

$$|z - z_0| < R_0$$

■

As a matter of terminology, we define the **unit circle** by $|z| = 1$ and the **unit disk** by $|z| < 1$.

(1.1.23) ***Example.*** Show that the complex equation

$$|z - 1| + |z - 2| = 2 \tag{1.1.24}$$

describes an ellipse, and write its equation in standard form.

Solution. An ellipse has the geometrical property that the sum of the distances from an arbitrary point on it to two given points (the foci) is constant. The two magnitudes on the left side of (1.1.24) represent the distances from z to the points 1 and 2, respectively. In addition, their sum ($= 2$) is greater than the distance between these two foci, and hence we have a real curve, which must be the ellipse in question. To write its equation in standard form, we write (1.1.24) as

$$[(x - 1)^2 + y^2]^{1/2} + [(x - 2)^2 + y^2]^{1/2} = 2$$

Bringing $[(x - 1)^2 + y^2]^{1/2}$ to the right-hand side and squaring yields $4[(x - 1)^2 + y^2]^{1/2} = 2x + 1$. Squaring again, we get $12x^2 - 36x + 16y^2 + 15 = 0$ which, after completing squares in x, yields the standard equation of an ellipse

$$\left(x - \frac{3}{2}\right)^2 + \frac{4}{3}y^2 = 1 \tag{1.1.25}$$

■

We now state two important inequalities.

(1.1.26) ***Theorem.*** If z_1 and z_2 are two complex numbers, then

(1.1.27)
(a) $|z_1 + z_2| \leq |z_1| + |z_2|$ **(triangle inequality)**
(b) $|z_1 + z_2| \geq ||z_1| - |z_2||$

Proof. To prove **(a)**, first note that since $x^2 \leq x^2 + y^2$, we have

$$\text{Re}(z) \leq |z| \qquad \text{(for any } z\text{)}$$

Now (see Problem 12)

$$|z_1 + z_2|^2 = |z_1|^2 + |z_2|^2 + 2\,\text{Re}(z_1\bar{z}_2)$$

Combining this with the above yields

$$\begin{aligned} |z_1 + z_2|^2 &\leq |z_1|^2 + |z_2|^2 + 2|z_1 z_2| \\ &= (|z_1| + |z_2|)^2 \end{aligned}$$

Taking square roots yields the triangle inequality. The reader is asked to prove part (b) in Problem 23.

■

The inequality in (a) above is called the **triangle inequality** because, as in Figure 1.1.3, it is simply a statement of the fact that the length of one side of a triangle is less than the sum of the lengths of the other two sides.

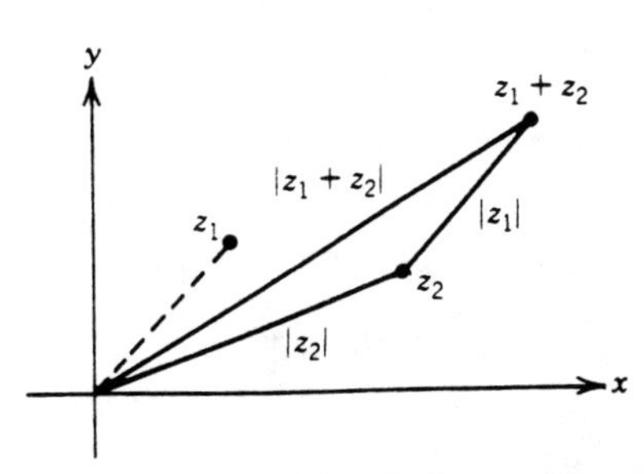

Figure 1.1.3 Triangle inequality.

(1.1.28) ***Example.*** If z is any point in the annulus $2 < |z| < 3$, show that

$$\frac{1}{10} < \left|\frac{z+i}{1+z^2}\right| < \frac{4}{3}$$

Solution. To begin with,

$$\left|\frac{z+i}{1+z^2}\right| = \frac{|z+i|}{|1+z^2|} \le \frac{|z|+1}{|1+z^2|} < \frac{4}{|1+z^2|} \tag{1.1.29}$$

where we have used the triangle inequality on the numerator together with the upper bound on $|z|$. Also, by (1.1.27b),

$$\begin{aligned} |1+z^2| &\ge |1-|z|^2| \\ &= |z|^2 - 1 \qquad \text{(since } |z| > 2 > 1\text{)} \\ &> 3 \end{aligned}$$

Thus, $1/|1+z^2| < 1/3$, and the result follows from (1.1.29). The reader is asked to show that

$$\left|\frac{z+i}{1+z^2}\right| > \frac{1}{10}$$

■

(1.1.30) ***Example.*** Show that the cubic equation

$$z^3 + 3z^2 + 5z + 10 = 0$$

cannot have a root lying inside the unit circle.

Solution. From the equation, if z_0 is a root, then

$$10 = |z_0^3 + 3z_0^2 + 5z_0| \tag{1.1.31}$$

If we assume that $|z_0| < 1$, then, from the triangle inequality applied twice, we have

$$\begin{aligned} 10 = |z_0^3 + 3z_0^2 + 5z_0| &\le |z_0^3 + 3z_0^2| + 5|z_0| \\ &\le |z_0|^3 + 3|z_0|^2 + 5|z_0| \\ &< 9 \qquad (\text{since } |z_0| < 1) \end{aligned}$$

Since this is a contradiction, we must conclude that there are no roots inside of the unit circle.

■

Topology; Point at Infinity. We will have need later for some topological properties of sets of complex numbers. These concepts are basic to theorem proving and will be used as the need arises.

A **δ-neighborhood** (or simply a **neighborhood**) **of a point z_0** is the set of all points z, as shown in Figure 1.1.4*a*, lying within a circle of radius δ centered at z_0; that is,

$$|z - z_0| < \delta \qquad (\delta\text{-neighborhood}) \tag{1.1.32}$$

A point is an **interior point** of a set D if there is some neighborhood of it completely contained in the set (see Figure 1.1.4*b*). The collection of all interior points is the **interior of the set**. A set is **open** if each point is an interior point. Thus, the set of points lying inside the ellipse $2x^2 + 7y^2 = 1$, that is, those points satisfying

$$2x^2 + 7y^2 < 1$$

is open, while the points lying on the ellipse is not open.

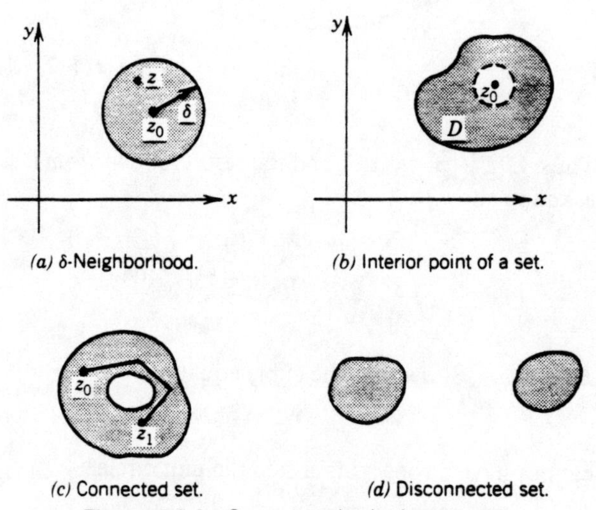

(a) δ-Neighborhood. *(b)* Interior point of a set.

(c) Connected set. *(d)* Disconnected set.

Figure 1.1.4 Some topological concepts.

A set of points is **bounded** if it lies inside of some neighborhood of the origin; otherwise it is **unbounded**. Thus, the set of points lying on the hyperbola

$$x^2 - 2y^2 = 1 \tag{1.1.33}$$

is unbounded.

A **boundary point** of a set is a point having the property that every neighborhood of it contains points in the set and points not in the set. Thus, the set of points satisfying

$$0 < |z - 1| \leq 1 \tag{1.1.34}$$

has as its boundary points the circle $|z - 1| = 1$ and the point $z = 1$. Note that boundary points need not belong to the set, as evidenced here by the point $z = 1$. The set described by (1.1.34) is called a **punctured disk**, or **deleted neighborhood**, of $z = 1$, since its center $z = 1$ is not in the set.

A set is **connected** (or **polygon connected**) if every pair of points in it can be joined by a broken line lying in the set; otherwise it is **disconnected**. For example, the region shown shaded in Figure 1.1.4*c* is connected, while that shown in Figure 1.1.4*d* is not. Loosely speaking, a connected set is one which consists of only one piece. A **domain** is a non-empty, open, connected set, while a **region** is a set consisting of a domain plus all, some or none of its boundary points.

So far, in discussing the z-plane, we have meant the finite plane. That is, every point in the **finite z-plane** is at some finite distance from the origin. However, we will have need to discuss the **point at infinity** which, together with the finite z-plane, constitutes what is called the **extended z-plane**. Unlike real calculus, where there are basically two directions on the axis, and thus we *can* talk about $+\infty$ and $-\infty$, in the complex plane there are many ways to go off to infinity. In the extended z-plane the point $z = \infty$ is that point which does not lie within any finite neighborhood of the origin. Perhaps one of the best ways to visualize this is to use a **stereographic projection**, as illustrated in Figure 1.1.5. Here, we draw the usual z-plane with the real

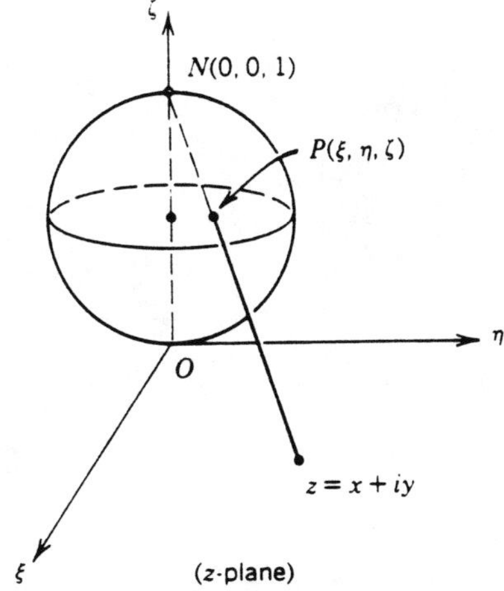

Figure 1.1.5 Stereographic projection onto the Riemann sphere.

and imaginary axes designated ξ and η, respectively, together with a third axis, designated ζ, and a sphere of radius $\frac{1}{2}$ with center on this axis and lying $\frac{1}{2}$ unit above the z-plane. Then, if we identify the origin with the south pole of the sphere, any other point z in the finite z-plane has a unique counterpart P on the sphere given by the intersection of the sphere and a line joining z and the north pole N of the sphere (see Figure 1.1.5). In the same manner, any point P on the sphere has a unique image in the z-plane. Clearly from the figure, the point $z = \infty$ in the extended z-plane would be identified with the north pole of the sphere. This sphere is called the **Riemann sphere**. What points in the z-plane correspond to points on the northern hemisphere of this sphere? On the southern hemisphere?

PROBLEMS (The answers are given on page 417.)

1. Find the magnitudes and the real and imaginary parts of the following complex numbers:

(a) $\dfrac{(1+i)(3+4i)}{(2+i)(1+2i)}$

(b) $\dfrac{(1+i)^2(4+i)}{(3-i)}$

(c) $\dfrac{1+\alpha}{1-|\alpha|\bar{\alpha}}$ where $\alpha = i$

(d) $\left(\dfrac{1-i}{1+i}\right)^7$

(e) $(1+i\sqrt{2})^3$

(f) $\left(\dfrac{-i}{3+i}\right)^4$

2. (a) If $n = 0, 1, 2, \ldots$, show that $i^n = \begin{cases} (-1)^{n/2}, & n \text{ even.} \\ i(-1)^{(n-1)/2}, & n \text{ odd.} \end{cases}$

(b) Compute $(1+i)^n$, $n = 0, 1, 2, \ldots$.

(c) Compute $(1-i)^n$, $n = 0, 1, 2, \ldots$.

3. If $z_1 = x_1 + iy_1$ and $z_2 = x_2 + iy_2$, show that

$$(z_1 + z_2)^2 = z_1^2 + 2z_1z_2 + z_2^2$$

4. Show that, if $z_{1,2} = x_{1,2} + iy_{1,2}$, then [see (1.1.8)]

$$\frac{z_1}{z_2} = \frac{(x_1x_2 + y_1y_2) + i(x_2y_1 - x_1y_2)}{x_2^2 + y_2^2} \qquad z_2 \neq 0$$

5. *Prove:*

(a) $z_1z_2 = z_2z_1$ (**commutativity** under multiplication).

(b) $(z_1 + z_2)z_3 = z_1z_3 + z_2z_3$.

6. Any real vector $\mathbf{v} = a\mathbf{i} + b\mathbf{j}$ in the plane can be associated with a complex number $z = a + ib$.
   (a) Show that definition (1.1.7a) is equivalent to the usual definition of vector addition. Show the same equivalence for multiplication by a real number.
   (b) If $\mathbf{v}_1$ and $\mathbf{v}_2$ are two vectors having associated complex numbers z_1 and z_2, express, in terms of z_1 and z_2, the fact that $\mathbf{v}_1$ and $\mathbf{v}_2$ are (i) parallel, (ii) perpendicular.
   (c) What conditions are sufficient in order that the four points z_1, z_2, z_3 and z_4 be the vertices of a parallelogram? What is the area of this parallelogram?
   (d) What is the area of a triangle having vertices at z_1, z_2, and z_3.
7. Consider the cubic equation $z^3 + az^2 + b = 0$.
   (a) Find *complex values* of a and b so that $2 + i$ is a root of the cubic.
   (b) Find *real values* of a and b so that $2 + i$ solves the equation.
8. Consider the quadratic equation $az^2 + bz + c = 0$, where $a(\neq 0)$, b, and c are *real*. Write $z = x + iy$ and show that there are nonreal roots only when $b^2 - 4ac < 0$, and these are given by $(1/2a)[-b \pm i(4ac - b^2)^{1/2}]$.
9. Define the **square root of** z ($\sqrt{z}$) to be those complex numbers w so that $w^2 = z$. Show that there are two square roots if $z \neq 0$, one being the negative of the other. [*Hint:* Let $z = x + iy$ and $w = u + iv$, and then solve for u and v.]
10. Consider the quadratic equation $az^2 + bz + c = 0$, where $a(\neq 0)$, b, and c are *complex*.
   (a) If $w = z + k$, find a value for the constant k so that the quadratic equation for w does not have a linear term (a term of the form αw).
   (b) Use the result in (a), and the definition of the square root in Problem 9, to show that the solutions of the quadratic are

$$z = \frac{-b \pm (b^2 - 4ac)^{1/2}}{2a}$$

11. Write $z = x + iy$ and compute the roots of the following polynomial equations:
   (a) $z^2 = i$
   (b) $z^2 = i + 1$
   (c) $z^3 = i$
   (d) $z^n = 1 \quad n = 3, 4$
12. (a) Prove properties (1.1.17).
   (b) Use induction to prove (i) $\overline{z_1 \cdot z_2 \cdots z_n} = \bar{z}_1 \cdot \bar{z}_2 \cdots \bar{z}_n$; (ii) $|z_1 \cdot z_2 \cdots z_n| = |z_1| \cdot |z_2| \cdots |z_n|$.
   (c) Prove that $|z_1 + z_2|^2 = |z_1|^2 + |z_2|^2 + 2\,\mathrm{Re}(z_1\bar{z}_2)$.
   (d) Prove the **Parallelogram law** $|z_1 + z_2|^2 + |z_1 - z_2|^2 = 2(|z_1|^2 + |z_2|^2)$. Why is this called a parallelogram law?
13. (a) Describe the location of $-z$ relative to z.
   (b) Do the same for $-\bar{z}$.
14. (a) Where is the point $1/\bar{z}$ located if z lies outside, on, or inside the unit circle? This point is called the **inversion of z in the unit circle**.

(b) Compare the signs of $\text{Re}(z)$ and $\text{Re}(1/z)$. Do the same with $\text{Im}(z)$ and $\text{Im}(1/z)$.

15. Consider the polynomial $p(z) = \sum_{k=0}^{n} a_k z^k$.

(a) If $z = x + iy$, show that $p(z) = p_1(x, y) + ip_2(x, y)$, where p_1 and p_2 are real polynomials in x and y of degree n.

(b) If the coefficients $\{a_k\}$ are real, show that $p(\bar{z}) = p_1 - ip_2$.

16. For which z do the following hold?

(a) $\left|\dfrac{z-i}{z+i}\right| < 1.$

(b) $\left|\dfrac{z-\alpha}{z+\alpha}\right| < 1 (\alpha = \alpha_1 + i\alpha_2).$

(c) $\text{Im}\left(\dfrac{1}{z}\right) < 1.$

(d) $z^2 = \bar{z}^2.$

(e) $z^3 = \bar{z}^3.$

(f) $|z-1| = |z-2|$ and $0 \le \text{Re}(iz) \le 1.$

(g) $\text{Re}(z+1) = |z|.$

17. Show that three distinct points z_1, z_2, and z_3 are colinear (lie on the same line) if and only if there is a *real* constant $k \ne 0$ so that $z_2 - z_1 = k(z_3 - z_2)$.

18. Find a point z on the line segment joining z_1 and z_2 so that the distances from z to z_1 and z to z_2 are in the ratio b/a, where a and b are positive. [*Hint:* See Problem 17.]

19. Describe in terms of z the following regions in the xy-plane:

(a) The upper half of the disk $(x - x_0)^2 + (y - y_0)^2 \le R_0^2$.

(b) The half-plane lying to the right of the line $y = 2 - x$.

(c) The region lying between the circles $x^2 + y^2 = 1$ and $(x-3)^2 + (y-1)^2 = 36$.

(d) The right-hand branch of the hyperbola $4x^2 - y^2 = 4$.

(e) The general **conic** $ax^2 + by^2 + cx + dy + e = 0$.

(f) The circle, centered at (x_0, y_0), which goes through the origin.

20. If k is real and A is complex, show that the equation $Az + \overline{Az} = k$ describes a **straight line** with slope $\text{Re}(A)/\text{Im}(A)$.

21. Describe the sets of points satisfying the following relationships:

(a) $|z - z_0| = k|z - z_1| \quad (k > 0)$

(b) $\left|\dfrac{z - z_0}{1 - \bar{z}_0 z}\right| < 1$

(c) $|z-1| + |z-2| = k (k > 0)$. Distinguish between the cases $k > 1$, $k = 1$, and $k < 1$. [*Hint:* view this geometrically.]

22. Use induction to prove the **general triangle inequality**

$$|z_1 + z_2 + \cdots + z_n| \le |z_1| + |z_2| + \cdots + |z_n|$$

23. (a) Prove inequality (1.1.27b). [*Hint:* Write $z_1 = (z_1 - z_2) + z_2$ and use the triangle inequality. Then, write $z_2 = (z_2 - z_1) + z_1$.]

(b) Use both parts of (1.2.27) to show that $\left|\frac{z - z_0}{z - z_1}\right| \leq \frac{|z| + |z_0|}{||z| - |z_1||}$.

(c) Prove $|z_1 + z_2 + z_3| \geq |z_1| - |z_2| - |z_3|$ if $|z_1| \geq |z_2 + z_3|$.

(d) Use parts (b) and (c) to prove that if z lies on a circle of radius R centered at the origin, then

$$\left|\frac{z^2 + iz + 2}{z^3 + z + 1}\right| \leq \frac{R^2 + R + 2}{R^3 - R - 1}$$

for "large enough" R.

(e) If z lies in the annular region $2 < |z - 1| < 3$, find a real positive M so that $\left|\frac{z - 1}{z - 2}\right| < M$.

(f) If $p(z)$ is a polynomial with positive coefficients, show that $|p(z)| \leq p(|z|)$.

(g) Let $p(z)$ and $q(z)$ be polynomials with degree $(p) = n$ and degree $(q) = m$, $m \geq n$. Show that if R is large enough, then, for z on the circle $|z| = R$, there is a constant k so that $|p/q| \leq k/R^{m-n}$ [*Hint:* Factor out the largest power from both p and q.]

24. (a) Describe the sets of points in Problem 21 using the words open, bounded, unbounded, connected, domain.

(b) Do the same as in (a) for the set of points satisfying $x^2 - 4y^2 > 1$.

25. (a) Refer to Figure 1.1.5 and show that the coordinates (ξ, η, ζ) of the stereographic projection of $z = x + iy$ satisfy

$$\xi + i\eta = \frac{z}{1 + |z|^2} \qquad \zeta = \frac{|z|^2}{1 + |z|^2}$$

(b) Use the above result to show that the stereographic images of lines in the z-plane are in general ellipses on the Riemann sphere.

1.2 POLAR FORM: POWERS AND ROOTS

As is well known, it is often convenient to use polar coordinates r and θ to represent points in the plane. These are as shown in Figure 1.2.1, and the relationship between x, y and r, θ is

(1.2.1) $$x = r\cos\theta \qquad y = r\sin\theta$$

Note that the radial distance r is inherently nonnegative and from the above is given by

(1.2.2) $$r = (x^2 + y^2)^{1/2} = |z|$$

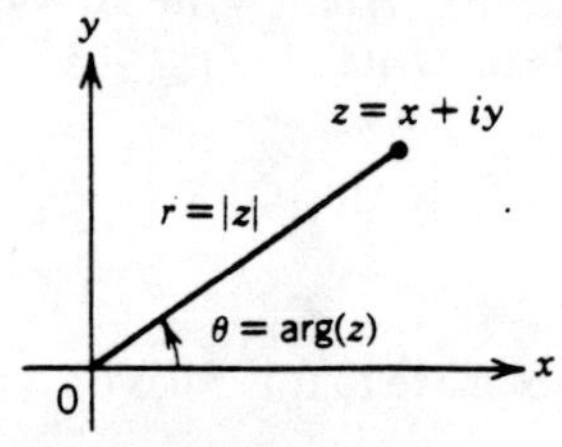

Figure 1.2.1 Polar coordinates.

Due to the periodicity of the trigonometric functions, the polar angle θ is not uniquely determined from x and y (any multiple of 2π can be added to it to yield the same x and y values) and in general is given by

$$\theta = \tan^{-1}(y/x)$$

where any branch of the inverse tangent function can be chosen.

From (1.2.1) and (1.2.2), with $z = x + iy$, we have

(1.2.3) $$z = |z|(\cos\theta + i\sin\theta)$$

In this **polar form for complex numbers**, the polar angle θ is called the **argument of z** (see Figure 1.2.1) and denoted by

(1.2.4) $$\theta = \arg(z)$$

As discussed above, $\arg(z)$ is not uniquely determined. Also, it is not defined at $z = 0$.

Note from (1.2.3) that points on the **unit circle** $|z| = 1$ have the polar form $\cos\theta + i\sin\theta$. Thus (as is easily verified),

(1.2.5) $$|\cos\theta + i\sin\theta| = 1 \qquad (\text{for any } \theta)$$

We can also define a **local polar angle** relative to a point $z_1 \neq 0$, as shown in Figure 1.2.2. In Problem 2 the reader is asked to show that the most general value of $\arg(z - z_1)$ is given by

(1.2.6) $$\arg(z - z_1) = \phi + 2\pi n \qquad 0 \leq \phi < 2\pi \qquad (n = 0, \pm 1, \ldots)$$

where ϕ is the angle shown in the figure.

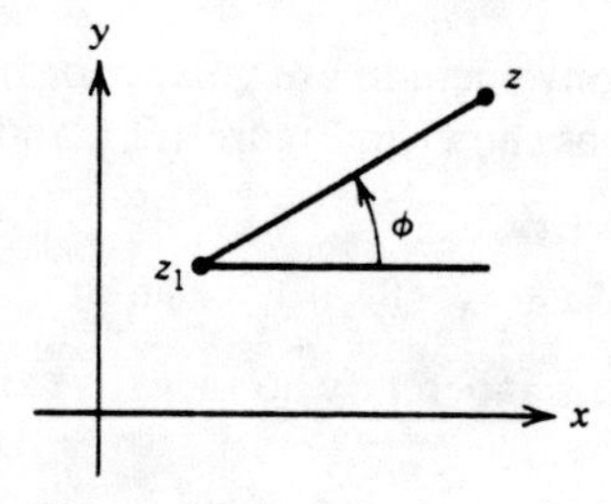

Figure 1.2.2 Local polar angle.

(1.2.7) ***Example.*** Write the most general polar form for the complex numbers $z_1 = 1 + i\sqrt{3}$ and $z_2 = -3$.

Solution. From Figure 1.2.3, $\theta_1 = \pi/3$. Hence, since $|z_1| = 2$,

$$z_1 = 2\left[\cos\left(\frac{\pi}{3} + 2\pi n\right) + i\sin\left(\frac{\pi}{3} + 2\pi n\right)\right]$$

Now, z_2 lies on the negative real axis, and $|z_2| = 3$. Thus,

$$z_2 = 3[\cos(\pi + 2\pi n) + i\sin(\pi + 2\pi n)]$$

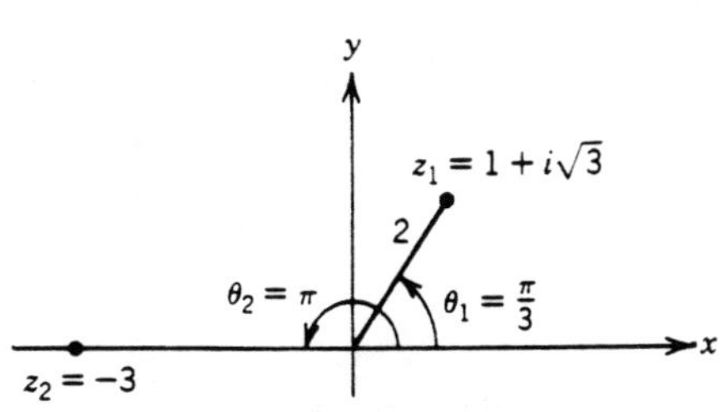

Figure 1.2.3 Example (1.2.7).

■

(1.2.8) ***Example.*** Compute all possible arguments of z_1, z_2, $z_1 z_2$, and z_1/z_2, where $z_1 = 1 + i$, $z_2 = 1 - i$.

Solution. Clearly, $z_1 z_2 = 2$ and $z_1/z_2 = i$. Thus,

$$\arg(z_1) = \frac{\pi}{4} + 2\pi n \qquad \arg(z_2) = -\frac{\pi}{4} + 2\pi k$$

$$\arg(z_1 z_2) = 2\pi l \qquad \arg\left(\frac{z_1}{z_2}\right) = \frac{\pi}{2} + 2\pi p$$

where n, k, l, and p are any positive or negative integers, or zero. Note that if we choose $n = l = 1$ and $k = 0$, then

$$\arg(z_1 z_2) = \arg(z_1) + \arg(z_2)$$

We will have more to say about this below.

■

We now state a rather obvious, but useful, result.

(1.2.9) ***Theorem.*** Let the polar forms of z_1 and z_2 be $z_1 = r_1(\cos\theta_1 + i\sin\theta_1)$, $z_2 = r_2(\cos\theta_2 + i\sin\theta_2)$. Then,

$$z_1 = z_2 \qquad \text{if and only if} \qquad r_1 = r_2 \qquad \text{and} \qquad \theta_1 = \theta_2 + 2\pi k$$

for some integer $k = 0, \pm 1, \ldots$.

■

We know that the polar coordinates of a point are uniquely determined if we restrict the polar angle to lie in some specified 2π range. We are thus led to the following standard (and really arbitrary) definition.

(1.2.10) ***Definition.*** The **principal argument** of the nonzero complex number z, which we denote by $\mathbf{Arg}(z)$ (*note the capital letter*), is that unique value of $\arg(z)$ lying in the range (see Figure 1.2.4),

(1.2.11)
$$-\pi < \mathrm{Arg}(z) \le \pi$$

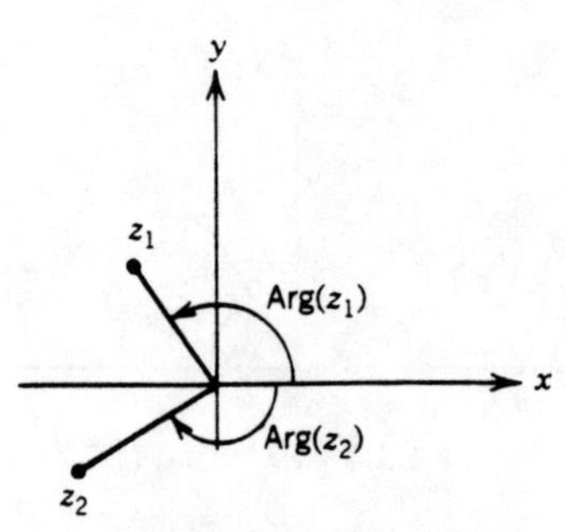

Figure 1.2.4 Principal argument.

■

Note that the above angle range is not the same as the usual range of the polar angle θ, namely, $0 \le \theta < 2\pi$.

(1.2.12) ***Example.*** The possible values of $\arg(1)$ are $2\pi n$, $n = 0, \pm 1, \ldots$. The only one of these lying in the principal argument range is by (1.2.11),

(1.2.13)
$$\mathrm{Arg}(1) = 0$$

Also, from Figure 1.2.5,

(1.2.14)
$$\mathrm{Arg}(-1 + i) = \frac{3\pi}{4} \qquad \text{and} \qquad \mathrm{Arg}(1 - i) = \frac{-\pi}{4}$$

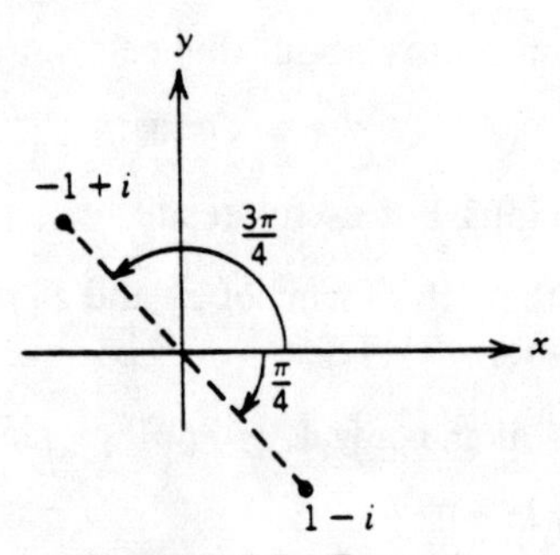

Figure 1.2.5 Example (1.2.12).

Finally, due to where the equality is in (1.2.11), we have

$$\text{Arg}(-1) = \pi$$

■

In Problem 3, the reader is asked to verify that the choice of principal argument is *not* equivalent to choosing the principal inverse tangent in the expression $\theta = \tan^{-1}(y/x)$. Also, since the difference between any two values of $\arg(z)$ must be a multiple of 2π, we have

(1.2.15) $$\arg(z) = \text{Arg}(z) + 2\pi n \qquad \text{for some} \qquad n = 0, \pm 1, \pm 2, \ldots$$

(1.2.16) *Theorem.* If values are assigned to any two of the arguments $\arg(z_1)$, $\arg(z_2)$, and $\arg(z_1 z_2)$, then there is a value of the third one for which

$$\arg(z_1 z_2) = \arg(z_1) + \arg(z_2)$$

Proof. Let $z_1 = |z_1|(\cos\theta_1 + i\sin\theta_1)$ and $z_2 = |z_2|(\cos\theta_2 + i\sin\theta_2)$. Then,

(1.2.17) $$\begin{aligned} z_1 z_2 &= |z_1| \cdot |z_2|[\cos\theta_1 \cos\theta_2 - \sin\theta_1 \sin\theta_2 \\ &\quad + i(\cos\theta_1 \sin\theta_2 + \cos\theta_2 \sin\theta_1)] \\ &= |z_1| \cdot |z_2|[\cos(\theta_1 + \theta_2) + i\sin(\theta_1 + \theta_2)] \end{aligned}$$

Hence, since $\theta_1 = \arg(z_1)$ and $\theta_2 = \arg(z_2)$, we have

$$\arg(z_1 z_2) = \arg(z_1) + \arg(z_2) + 2\pi n \qquad n = 0, \pm 1, \ldots$$

Now, since each argument is defined anyway only up to an additive multiple of 2π, the result is proved.

■

(1.2.18) *Example.* Show that in general

$$\text{Arg}(z_1 z_2) \neq \text{Arg}(z_1) + \text{Arg}(z_2)$$

For what set of points z_1 and z_2 can you be sure to have equality above?

Solution. Let $z_1 = -1 + i$ and $z_2 = i$. Then, $z_1 z_2 = -1 - i$, $\text{Arg}(z_1) = 3\pi/4$, $\text{Arg}(z_2) = \pi/2$ and $\text{Arg}(z_1 z_2) = -3\pi/4$. Clearly, $\text{Arg}(z_1 z_2) \neq \text{Arg}(z_1) + \text{Arg}(z_2)$. The reason for this is that $\text{Arg}(z_1) + \text{Arg}(z_2) = 5\pi/4$, which lies outside of the principal argument range.

Now, if z_1 and z_2 both lie either in the right half-plane or on the imaginary axis (with not both on the negative imaginary axis), then equality will always hold. (Why?) However, these are not the only possibilities. For example, we have equality when $z_1 = -1 + i$ and $z_2 = -1 - i$.

■

The reader is asked to show in Problem 4 that, in general,

(1.2.19) $$\operatorname{Arg}(z_1 z_2) = \operatorname{Arg}(z_1) + \operatorname{Arg}(z_2) + 2\pi n \qquad (n = 0, 1 \text{ or } -1)$$

Finally, from the general result (1.2.17), we have

(1.2.20) ***Theorem* (DeMoivre).** If n is a nonnegative integer, then

$$(\cos\theta + i\sin\theta)^n = \cos(n\theta) + i\sin(n\theta)$$

Proof. The proof follows from a repeated application of (1.2.17) with the choice $z_1 = z_2 = \cos\theta + i\sin\theta$.

■

From this result it easily follows by induction that, if $z = |z|(\cos\theta + i\sin\theta)$, then

(1.2.21) $$z^n = |z|^n[\cos(n\theta) + i\sin(n\theta)]$$

(1.2.22) ***Example.*** Write the complex number $(1 + i)^{25}$ in the form $x + iy$.

Solution. Certainly the reader will not want to multiply $1 + i$ by itself 25 times. Instead, let us first write $1 + i$ in polar form as

$$1 + i = \sqrt{2}\left[\cos\left(\frac{\pi}{4}\right) + i\sin\left(\frac{\pi}{4}\right)\right].$$

Then, from (1.2.21),

$$\begin{aligned}(1 + i)^{25} &= 2^{25/2}\left[\cos\left(\frac{25\pi}{4}\right) + i\sin\left(\frac{25\pi}{4}\right)\right].\\ &= 2^{25/2}\left[\cos\left(\frac{\pi}{4}\right) + i\sin\left(\frac{\pi}{4}\right)\right]\\ &= 4096(1 + i)\end{aligned}$$

■

(1.2.23) ***Example.*** Verify the trigonometric identities

$$\cos(3\theta) = \cos^3\theta - 3\cos\theta\sin^2\theta$$

$$\sin(3\theta) = 3\cos^2\theta\sin\theta - \sin^3\theta$$

Solution. From DeMoivre's Theorem with $n = 3$, we have

$$\begin{aligned}\cos(3\theta) + i\sin(3\theta) &= (\cos\theta + i\sin\theta)^3\\ &= \cos^3\theta + 3i\cos^2\theta\sin\theta - 3\cos\theta\sin^2\theta - i\sin^3\theta\end{aligned}$$

Equating real and imaginary parts yields the above results. Note that both identities are derived at the same time. Indeed, this is a rather painless way to derive many of the trigonometric identities which students are asked to simply *verify* in high school.

■

Roots of Complex Numbers. In writing the square root of $2^{1/2}$, what is usually meant is the positive square root of 2. However, there is also a negative square root of 2. In fact, if we set $x = \sqrt{2}$, then x satisfies the equation $x^2 = 2$, which clearly also has the solution $x = -\sqrt{2}$.

A more interesting case is the cube root of 2, that is, $2^{1/3}$. Again, if we set $x = 2^{1/3}$, then x satisfies $x^3 - 2 = 0$, which has only one real root. Thus, there is only one real cube root of 2. However, the cubic equation $z^3 - 2 = 0$ has three complex roots, one of which is our real root. The question is then how to find the remaining two complex roots. First, however, we must define what is meant by the nth roots of a complex number.

(1.2.24) ***Definition.*** Let n be a positive integer. Then by the **nth roots** of the complex number z_0, denoted by $z_0^{1/n}$, we mean all those z satisfying $z^n = z_0$, that is,

(1.2.25)
$$z = z_0^{1/n} \text{ means } z^n = z_0$$

■

Let us show by example how complex roots are found.

(1.2.26) ***Example.*** Find all of the roots $2^{1/3}$.

Solution. Let $z = 2^{1/3}$. Then, by (1.2.25), $z^3 = 2$. First write, in polar form,

(1.2.27)
$$z = |z|(\cos\theta + i\sin\theta) \qquad \text{and} \qquad 2 = 2(\cos 0 + i\sin 0)$$

By DeMoivre's Theorem, the equation $z^3 = 2$ becomes

$$|z|^3[\cos(3\theta) + i\sin(3\theta)] = 2(\cos\theta + i\sin\theta)$$

and by Theorem (1.2.9), we have

(1.2.28)
$$|z|^3 = 2 \qquad \text{and} \qquad 3\theta = 2\pi k \qquad k = 0, \pm 1, \dots$$

Hence, $|z| = 2^{1/3}$, by which we mean the unique and positive cube root of 2, and the complex roots are given by

(1.2.29)
$$z = 2^{1/3}\left[\cos\left(\frac{2\pi k}{3}\right) + i\sin\left(\frac{2\pi k}{3}\right)\right] \qquad k = 0, \pm 1, \dots$$

Observe that although there are an infinite number of different arguments above, there are really only three different values of z gotten by choosing, for example, $k = 0, 1, 2$. Hence, we have the three cube roots

(1.2.30)
$$z = 2^{1/3} \qquad 2^{-2/3}(-1 + i\sqrt{3}) \qquad 2^{-2/3}(-1 - i\sqrt{3})$$

Also, the nonreal roots occur in complex conjugate pairs and (see Figure 1.2.6) the three roots are evenly distributed on a circle of radius $2^{1/3}$. That this is a general result is shown in Problem 13*b*.

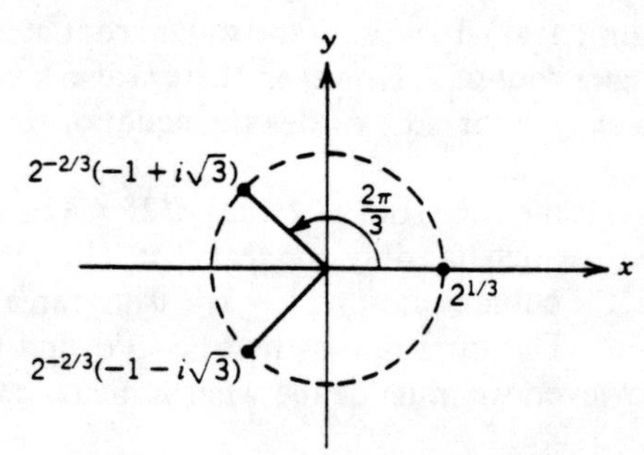

Figure 1.2.6 Example (1.2.26).

(1.2.31) ***Example.*** To find the cube roots $(1 + i)^{1/3}$, set $(1 + i)^{1/3} = r(\cos\theta + i\sin\theta)$. With $1 + i = 2^{1/2}\left[\cos\left(\frac{\pi}{4}\right) + i\sin\left(\frac{\pi}{4}\right)\right]$, we get $r^3 = 2^{1/2}$ and $3\theta = (\pi/4) + 2\pi k$. Hence, the cube roots have polar coordinates given by

$$r = 2^{1/6} \qquad \theta = \frac{\pi}{12} + \frac{2\pi k}{3} \qquad k = 0, 1, 2$$

Note that these roots are spaced evenly around a circle of radius $2^{1/6}$ (see Figure 1.2.7). However, none is a complex conjugate of any of the others. This is because the cubic polynomial $z^3 - (1 + i)$ has complex coefficients (see Problem 14).

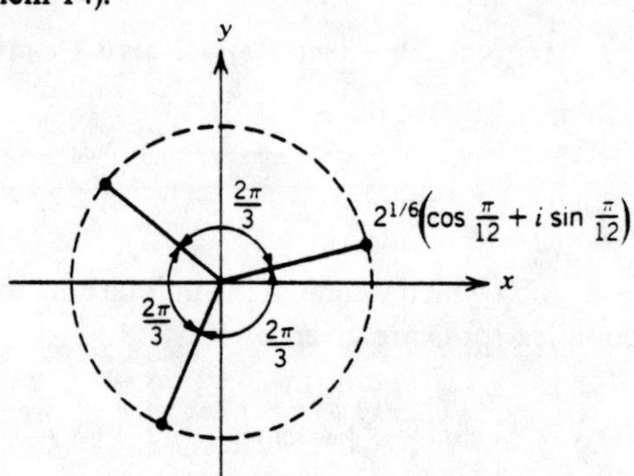

Figure 1.2.7 Example (1.2.31).

PROBLEMS (The answers are given on page 418.)

1. For each of the following complex numbers, compute the most general argument and the principal argument: (a) $27(1 + i)$, (b) $5(1 + i)/(1 - i)$, (c) $-2/(\sqrt{3} + i)$, (d) $(1 - i)^6$, (e) -2, (f) $\sqrt{3} - 3i$, (g) $2 + i$.

2. From Figure 1.2.2, verify that $\arg(z - z_1) = \phi + 2\pi n$, $n = 0, \pm 1, \ldots$.
3. The **principal inverse tangent function** Tan^{-1} is defined in calculus as that branch of the inverse tangent which takes values in the interval $(-\pi/2, \pi/2)$. If $z = x + iy$, show that

$$\mathrm{Arg}(z) = \begin{cases} \mathrm{Tan}^{-1}\left(\dfrac{y}{x}\right), \ \mathrm{Re}(z) > 0 \\ \mathrm{Tan}^{-1}\left(\dfrac{y}{x}\right) + \pi, \ \mathrm{Re}(z) < 0 \text{ and } \mathrm{Im}(z) > 0 \\ \mathrm{Tan}^{-1}\left(\dfrac{y}{x}\right) - \pi, \ \mathrm{Re}(z) < 0 \text{ and } \mathrm{Im}(z) < 0 \end{cases}$$

4. Prove that $\mathrm{Arg}(z_1 z_2) = \mathrm{Arg}(z_1) + \mathrm{Arg}(z_2) + 2\pi n$, $n = 0$, 1, or -1. [*Hint:* Consider all possible positions of z_1 and z_2, together with the principal argument range.]
5. *Prove:*
   (a) There are values of $\arg(z)$ and $\arg(1/z)$ so that $\arg(1/z) = -\arg(z)$.
   (b) There are values of $\arg(z_1)$, $\arg(z_2)$, and $\arg(z_1/z_2)$ so that $\arg(z_1/z_2) = \arg(z_1) - \arg(z_2)$.
6. If $k > 0$, prove that $\arg(kz) = \arg(z)$, with the same choice of argument on both sides of the equation.
7. For which z's does
   (a) $\mathrm{Arg}(\bar{z}) = -\mathrm{Arg}(z)$ (b) $\mathrm{Arg}(-z) = -\mathrm{Arg}(z)$
   (c) $\arg(\bar{z}) = -\arg(z)$ where $0 < \arg(z) \le 2\pi$
8. For which $k > 0$ are the two equations $|z - 1| = k$ and $\arg(z) = \pi/4$ satisfied? How many values of z are there?
9. Describe, or sketch, the sets of points defined by
   (a) $\mathrm{Arg}(z - z_0) = k$ (b) $\mathrm{Arg}(z) = \mathrm{Arg}(\bar{z})$
   (c) $0 < |z| < 2$ $\pi/4 < \arg(z) < \pi/2$
10. If $0 \le \mathrm{Arg}(z_1) < \mathrm{Arg}(z_2) < \pi$, show that the exterior angle at $z = z_1$ of the triangle with vertices at $z = 0$, $z = z_1$ and $z = z_2$ is $\mathrm{Arg}(z_2 - z_1) - \mathrm{Arg}(z_1)$. What are the exterior angles at $z = 0$ and at $z = z_2$?
11. If $|z_1| = |z_2|$, with z_1 and z_2 in the first quadrant, show that $\mathrm{Arg}(z_1 + z_2) = \frac{1}{2}[\mathrm{Arg}(z_1) + \mathrm{Arg}(z_2)]$. [*Hint:* Draw a parallelogram and its diagonal.]
12. (a) If $|z| = 1$, and $\theta = \arg(z)$, show that

$$\sum_{n=0}^{N} z^n = \frac{\left[\cos\left(\dfrac{N\theta}{2}\right) + i\sin\left(\dfrac{N\theta}{2}\right)\right]\sin\dfrac{1}{2}(N+1)\theta}{\sin\left(\dfrac{1}{2}\theta\right)}$$

[*Hint:* first show that $\sum_{n=0}^{N} z^n = \dfrac{1 - z^{N+1}}{1 - z}$, and then set $z = \cos\theta + i\sin\theta$.]

(b) Evaluate the sums

(i) $\sum_{n=0}^{N} \cos(n\theta)$ (ii) $\sum_{n=1}^{N} \sin(n\theta)$

13. (a) Prove that $|z^{1/n}| = |z|^{1/n}$, where the right-hand side denotes the real, positive nth root of $|z|$.

(b) Prove that the n-roots $z_0^{1/n}$ are distributed evenly on the circle $|z| = |z_0|^{1/n}$.

14. (a) Let $p(z)$ be the polynomial $p(z) = \sum_{k=0}^{n} a_k z^k$. Prove that if the coefficients $a_0, \ldots, a_N$ are real, then the roots of $p(z) = 0$ occur in complex conjugate pairs. See Example 1.2.31 for a case when the result is not true. [*Hint:* If $p(z_0) = 0$, what is $p(\bar{z}_0)$?]

(b) For the polynomial in (a), if $p(1 + i) = 3 + 2i$, what is $p[2/(1 + i)]$?

15. (a) If A is a complex constant, describe where the point $z - A$ is in relation to z. This is called a **translation.**

(b) Use polar coordinates to describe the location of the following points:

(i) αz, where α is a complex constant

(ii) $1/\bar{z}$ (this is called an **inversion**)

(iii) z^n, $n = 0, 1, 2, \ldots$

16. Compute the following:

(a) $(1 + i\sqrt{3})^{12}$ (b) $1/(1 + i)^{100}$ (c) $\text{Arg}[(1 - i)^{15}]$

(d) $(\sqrt{3} - i)^n$ $n = 0, 1, 2, \ldots$

17. Compute the following:

(a) $(4 + 4i)^{1/4}$ (b) $(-4)^{1/4}$ (c) $\left(\frac{1}{1 + i}\right)^{1/2}$

(d) $(1 - i)^{1/4}$ (e) $(1 - \sqrt{3}i)^{1/3}$

18. Solve the following polynomial equations:

(a) $z^4 + 16 = 0$ (b) $z^4 = 8iz$ (c) $(1/z^3) - i = 0$

(d) $z^2 + (1 - i)z - i = 0$

19. If m and n are positive integers, we define $z^{m/n}$ to be that complex w so that $z^m = w^n$. Compute

(a) $i^{2/3}$ (b) $(-1)^{2/3}$

20. (a) Verify the following trigonometric identities:

(i) $\cos(4\theta) = \cos^4\theta - 6\cos^2\theta\sin^2\theta + \sin^4\theta$

(ii) $\sin(4\theta) = 4\cos^3\theta\sin\theta - 4\cos\theta\sin^3\theta$

(iii) $\cos(5\theta) = \cos^5\theta - 10\cos^3\theta\sin^2\theta + 5\cos\theta\sin^4\theta$

(iv) $\sin(5\theta) = 5\sin\theta\cos^4\theta - 10\cos^2\theta\sin^3\theta + \sin^5\theta$

(b) If n is an arbitrary nonnegative integer, write an expression for $\cos(n\theta)$ and $\sin(n\theta)$ in terms of powers of $\sin\theta$ and $\cos\theta$.

21. Unlike in Example (1.2.23), we often need trigonometric identities which express powers of sine and cosine in terms of sines and cosines of multiple angles.

(a) Use complex variables to show that

(i) $\cos^2\theta = \frac{1}{2}(1 + \cos 2\theta)$ and $\sin^2\theta = \frac{1}{2}(1 - \cos 2\theta)$

(ii) $\cos^3\theta = \frac{1}{4}(\cos 3\theta + 3\cos\theta)$ and $\sin^3\theta = -\frac{1}{4}(\sin 3\theta - 3\sin\theta)$

[*Hint:* If $z = \cos\theta + i\sin\theta$, then $\cos\theta = \frac{1}{2}(z + \bar{z})$ and $\sin\theta = (1/2i)(z - \bar{z})$.]

(b) Use the hint in part (a), and the binomial theorem, to generate expressions for $\cos^n\theta$ and $\sin^n\theta$, where $n = 0, 1, 2, \ldots$.

22. Use complex variables to solve the following real trigonometric equations:

(a) $\sin(2\theta) = \sin\theta$ (b) $2\sin\theta + \cos\theta = 2$

(c) $2\sin\theta + \cos\theta = k$ (for which k are there solutions?)

[*Hint:* Express everything in terms of $z = \cos\theta + i\sin\theta$. Use $\bar{z} = 1/z$, and solve the resulting polynomial equation.]

23. The roots $1^{1/n}$ are all of the form $\cos(2\pi k/n) + i\sin(2\pi k/n)$, for $k = 0, 1, \ldots, n - 1$. The quantity $\omega_n = \cos(2\pi/n) + i\sin(2\pi/n)$ is called an **nth root of unity**.

(a) Show that the roots $1^{1/n}$ are of the form ω_n^k, $k = 0, 1, \ldots, n - 1$.

(b) Show that the roots $z^{1/n}$ are given by $\hat{z}\omega_n^k$, $k = 0, 1, \ldots, n - 1$, where $\hat{z}$ is *any* specific root of $z^{1/n}$.

(c) Prove that $\sum_{k=0}^{n-1} \omega_n^k = 0$ for $\omega_n \neq 1$. [*Hint:* multiply the sum by $1 - \omega_n$ and use the definition of ω_n.]

2

COMPLEX FUNCTIONS AND THEIR MAPPING PROPERTIES

2.1 FUNCTIONS, LIMITS, AND CONTINUITY

A **complex valued function f of a complex variable** z is a rule which assigns for each z in a specified region of the complex plane one or more complex numbers w, and is denoted by

$$w = f(z) \tag{2.1.1}$$

The **domain of definition of f** (or just the **domain** of f) is the set of those z's for which the function can be defined. If $f(z)$ is uniquely defined for each z in the domain of f, we say that f is a **single-valued function of z**. Otherwise, it is **multiple-valued**.

If $z = x + iy$ is in the domain of f, then $f(z)$, being complex valued, has a real part u and an imaginary part v. These real quantities u and v depend upon z which, in turn, depends upon the values of x and y. Hence, u and v must be real-valued functions of the two real variables x and y. Thus, we write

$$w = f(z) = u(x, y) + iv(x, y) \tag{2.1.2}$$

Since it is also possible to represent z in polar coordinates as $z = r(\cos\theta + i\sin\theta)$, the real and imaginary parts of $f(z)$ can likewise be viewed as real-valued functions of r and θ. Hence, we may also write

$$w = f(z) = \hat{u}(r, \theta) + i\hat{v}(r, \theta) \tag{2.1.3}$$

(2.1.4) ***Example.*** Consider the function $f(z) = z^2$. Since $z = x + iy$, we have

$$f(z) = x^2 - y^2 + 2ixy$$

Here, $u(x, y) = x^2 - y^2$ and $v(x, y) = 2xy$. In polar coordinates, we find from (2.1.3) that $\hat{u}(r, \theta) = r^2 \cos(2\theta)$ and $\hat{v}(r, \theta) = r^2 \sin(2\theta)$.

For the function $g(z) = |z|^2$, we have $u(x, y) = x^2 + y^2$ and $v(x, y) = 0$. Hence, $g(z)$ is always real and is a **real-valued function of a complex variable**.

Finally, consider

$$h(z) = \frac{1}{x} + iy^2$$

Here, $u(x, y) = 1/x$ and $v(x, y) = y^2$, and the domain of h is $x \neq 0$, that is, all z not on the imaginary axis. Now one might ask where the specific dependence upon z can be found in the above. Indeed, if $u(x, y)$ and $v(x, y)$ are *any* real-valued functions of x and y, then the complex function $u(x, y) + iv(x, y)$ can *always* be written explicitly in terms of z and $\bar{z}$, for we have $x = (1/2)(z + \bar{z})$ and $y = (1/2i)(z - \bar{z})$. Hence, substituting these into $h(z)$ above, we get

$$h(z) = \frac{2}{z + \bar{z}} - \frac{i}{4}(z^2 - 2|z|^2 + \bar{z}^2)$$

Note that f, g, and h are all single-valued functions.

■

(2.1.5) ***Example.*** Show that the function $w = \sqrt{z}$ is multiple-valued, and compute all of its possible real and imaginary parts.

Solution. By definition, we have that if $w = \sqrt{z}$, then $z = w^2$. Hence, $x + iy = (u + iv)^2$, which yields the two real equations

$$u^2 - v^2 = x, \qquad 2uv = y \tag{2.1.6}$$

The objective is to solve for u and v in terms of x and y. Now in the second equation, if $y \neq 0$, neither u nor v can vanish. Hence, for z not real, we can eliminate u in favor of v and substitute into the first equation to get

$$u = \frac{y}{2v} \quad \text{and} \quad v^4 + xv^2 - \frac{1}{4}y^2 = 0 \tag{2.1.7}$$

The second of these is quadratic in v^2 and yields $v^2 = \frac{1}{2}(-x \pm \sqrt{x^2 + y^2})$. Now, since $-x - \sqrt{x^2 + y^2}$ is always negative, and since v is to be real (hence $v^2 > 0$), we must eliminate the negative square root above. Thus, by (2.1.7), $u(x, y)$ and $v(x, y)$ are given, for nonreal z, by

$$u(x, y) = \frac{\pm y}{\{2[-x + (x^2 + y^2)^{1/2}]\}^{1/2}} \quad \text{and} \tag{2.1.8}$$

$$v(x, y) = \pm\left[\frac{1}{2}(-x + \sqrt{x^2 + y^2})\right]^{1/2}$$

In the above, the plus signs and the minus signs go together. Hence, as of course was expected, the function $\sqrt{z}$ is multiple-valued and takes on two values for each z, one being the negative of the other.

In Problem 5, the reader is asked to complete the example and determine $u(x, y)$ and $v(x, y)$ when z is real.

In order not to cause confusion, **from now on whenever the word function is used, we will mean single-valued function**. For example, we will say *the function* $f(z) = z^2$ while using the phrase *the multiple-valued function* $f(z) = \sqrt{z}$.

In calculus, we define the graph of a real-valued function $f(x)$ as the set of all points in the xy-plane of the form $(x, f(x))$, where x is in the domain of f. For functions of a complex variable, this approach is not practical since it would require the determination of a set of points in a four-dimensional $xyuv$-space. Instead, we attempt to describe a complex function by its **mapping properties**. To understand what is meant by this, observe from (2.1.2) that the function f maps a point z, with coordinates (x, y) in the xy-plane (the z-plane), into a point w with coordinates $(u(x, y), v(x, y))$ in a uv-plane (the w-plane). In this manner, a complex function gives rise to a **transformation (change of variables)** from (x, y) to (u, v) defined by

(2.1.9) $$u = u(x, y), \qquad v = v(x, y)$$

Hence, f can be viewed as mapping a region D_z in the z-plane into a region D_w in the w-plane, as depicted in Figure 2.1.1.

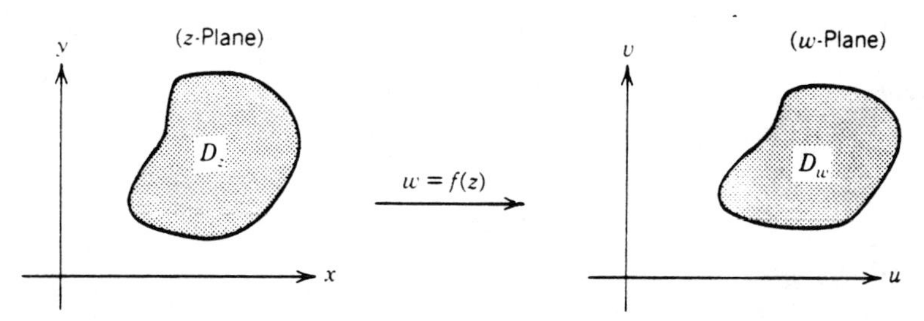

Figure 2.1.1 A complex function viewed as a mapping.

When we ask for the **mapping properties of a function $f(z)$**, we seek to determine the image in the w-plane of given regions in the z-plane. Below, we illustrate these ideas with some simple examples. In the later sections of this chapter, we define more complicated complex functions and follow each definition with a discussion of some of its mapping properties. In this manner, we will build up a table of mappings which we will in turn combine to produce more complicated ones.

(2.1.10) ***Example.*** Consider the function $w = z + b$, where b is the complex constant $b = b_1 + ib_2$. Then,

(2.1.11) $$u = x + b_1, \qquad v = y + b_2$$

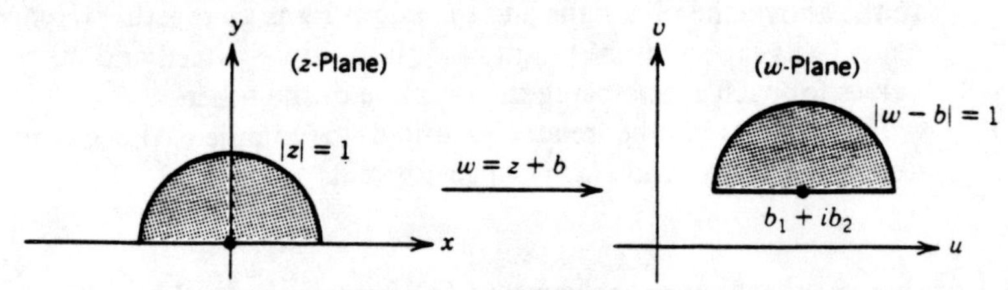

Figure 2.1.2 Example (2.1.10).

From calculus, we recognize this as a **translation** and thus expect that the image in the w-plane of any region in the z-plane is simply a translation of that region in the direction of the vector $\mathbf{b} = b_1\mathbf{i} + b_2\mathbf{j}$. In particular consider the half-disk shown in the z-plane in Figure 2.1.2. This can be described by $|z| < 1$ and $\operatorname{Im}(z) > 0$. Hence, its image in the w-plane is described by

$$|w - b| < 1 \qquad \text{and} \qquad \operatorname{Im}(w) > b_2$$

and is the translated half-disk shown in the figure.

Note that we can also view the half-disk as the intersection of the "full" disk $|z| < 1$ and the half-plane $\operatorname{Im}(z) > 0$. Its image is then the intersection of the images of these two regions, that is, the intersection of the disk $|w - b| < 1$ and the half-plane $\operatorname{Im}(w) > b_2$.

■

(2.1.12) ***Example.*** Under the mapping

(2.1.13)
$$w = x + iy^2$$

what is the image in the w-plane of

(a) The line $y = kx$, $0 \le x \le 2$ $\quad (k > 0)$?

(b) The region shown shaded in Figure 2.1.3?

(2.1.14) ***Solution.*** Since

$$u = x \qquad \text{and} \qquad v = y^2$$

the line in (a) transforms into the parabolic arc $v = k^2u^2$, $0 \le u \le 2$. A typical situation is shown dotted in Figure 2.1.3.

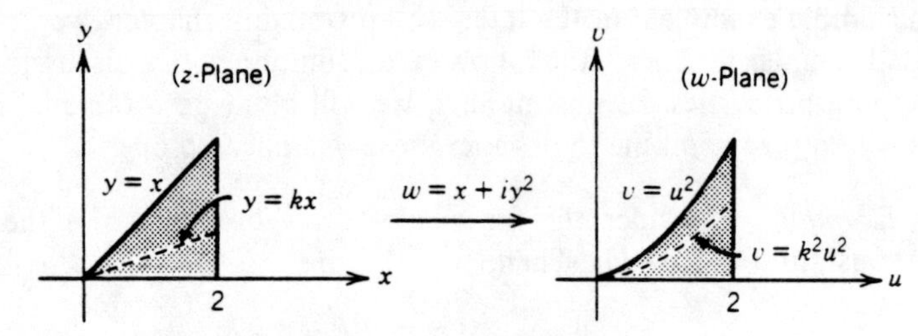

Figure 2.1.3 Example (2.1.12).

We will determine the image of the triangular region in the z-plane in two different ways. First of all, this region may be defined by the inequalities

(2.1.15)
$$0 \le y \le x, \qquad 0 \le x \le 2$$

Hence, from (2.1.14) its image in the uv-plane is described by the inequalities

$$0 \le v \le u^2, \qquad 0 \le u \le 2$$

and is the region bounded by the lines $u = 0$, $u = 2$, and $v = 0$, and the parabola $v = u^2$, as shown in the figure.

Another approach is to observe that the triangle in the z-plane can be generated by the lines $y = kx$, $0 \le x \le 2$, as k takes on values between 0 and 1, with a typical line shown dotted in the figure. Hence, the image of the triangle will be generated by the images of each of these dotted lines. These images are the parabolic arcs from part (a), and the image of the triangle is the region described above and shown in the figure.

The reader is urged to redo this problem by generating the triangle in the z-plane using horizontal and vertical line segments.

■

Limits and Continuity. As in elementary calculus, we have the following definitions:

(2.1.16) ***Definition.*** Let $f(z)$ be defined in some neighborhood of the point z_0, *but not necessarily at* z_0 *itself.* Then,

(a) $f(z)$ has a **limit L at $z = z_0$**, written

(2.1.17)
$$\lim_{z \to z_0} [f(z)] = L \qquad \text{or} \qquad f(z) \xrightarrow[z \to z_0]{} L$$

which means, loosely speaking, that $f(z)$ is "close to L" when z is "close to z_0" (but not equal to z_0). Analytically, we say that for every $\epsilon > 0$, there is a $\delta > 0$, so that

(2.1.18)
$$0 < |z - z_0| < \delta \Rightarrow |f(z) - L| < \epsilon$$

Note that the inequality $0 < |z - z_0|$ excludes the possibility that $z = z_0$.

(b) The **limit of $f(z)$ at z_0 is infinite**, that is,

(2.1.19)
$$\lim_{z \to z_0} [f(z)] = \infty$$

means that for any $R > 0$, there is a $\delta > 0$ so that

(2.1.20)
$$0 < |z - z_0| < \delta \Rightarrow |f(z)| > R$$

Hence, f is "large in magnitude" when z is "close to z_0."

(c) If $f(z)$ is defined at all points outside of a disk, we define

(2.1.21)
$$\lim_{z\to\infty} f(z) = \lim_{w\to 0} f\left(\frac{1}{w}\right)$$

if the limit on the right exists. Then, we say that f **has a limit at** $z = \infty$.

■

(2.1.22) ***Example.*** Show that

$$\lim_{z\to i} (z^2 + z) = i - 1$$

a not unexpected result.

Solution. We shall rigorously prove this using the ϵ–δ definition (2.1.18). Thus, we wish to find a δ for any given ϵ so that

(2.1.23)
$$0 < |z - i| < \delta \Rightarrow |z^2 + z - (i - 1)| < \epsilon$$

We first "fiddle around" with the quantity $|z^2 + z + 1 - i|$ (about which we wish to conclude something) and try to bound it from above. Since $z^2 + z + 1 - i = (z - i)(z + i + 1)$, we have

$$|z^2 + z + 1 - i| = |z - i| \cdot |z + 1 + i|$$

Note the occurence of the term $|z - i|$ above—this also occurs in (2.1.23). However, the other term $|z + 1 + i|$ is not of this form. Hence, we first write $z + 1 + i = (z - i) + 1 + 2i$ and then use the triangle inequality to get (since $|1 + 2i| = \sqrt{5}$)

(2.1.24)
$$|z^2 + z + 1 + i| \le |z - i|[|z - i| + \sqrt{5}]$$

Since we are demanding that $|z - i| < \delta$, the largest value of the right-hand side is $\delta(\delta + \sqrt{5})$. Hence, if we choose any δ so that $\delta(\delta + \sqrt{5}) < \epsilon$, which implies

$$\delta^2 + \sqrt{5}\delta - \epsilon < 0$$

the second inequality in (2.1.23) will be satisfied whenever $|z - i| < \delta$. Now a plot of the quadratic function $p(\delta) = \delta^2 + \sqrt{5}\delta - \epsilon$ (see Figure 2.1.4) shows that for $\delta > 0$, $p(\delta) < 0$ when

$$0 < \delta < \frac{1}{2}(-\sqrt{5} + \sqrt{5 + 4\epsilon})$$

It is now a straightforward matter to show that (2.1.23) follows from (2.1.24) for any δ satisfying the above inequality. Thus, $\lim_{z\to i} (z^2 + z) = i - 1$.

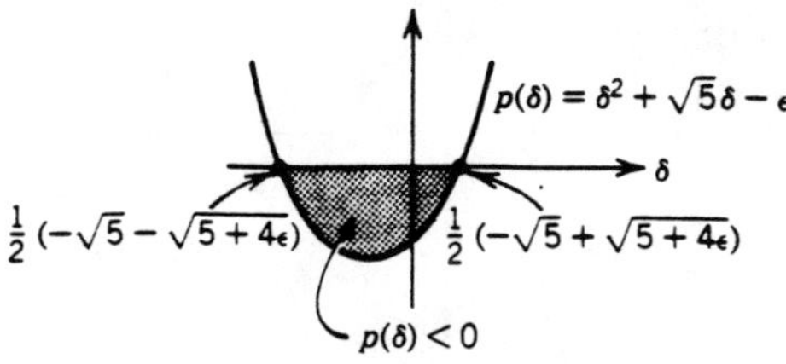

Figure 2.1.4 Example (2.1.22).

■

(2.1.25) ***Example.*** Consider $\lim_{z \to 1+2i} [x^2 + y^3 + i\sin(\pi y/2)]$. Since

$$z \to (1 + 2i) \Rightarrow x \to 1 \quad \text{and} \quad y \to 2,$$

we would expect the limit to be 9. The interested reader may prove this result using much the same type of argument as in the previous example.

Now consider

$$\lim_{z \to 0} \left(\frac{\bar{z}}{z}\right) \tag{2.1.26}$$

If we permit z to approach zero along the real axis, then $z = \bar{z} = x$ and the limit is 1. If z approaches 0 along the imaginary axis, then $z = iy$ while $\bar{z} = -iy$, and hence the limit is -1. Therefore, since we get different limiting values depending upon how z approaches 0, we would not expect the limit to exist. This in fact will be the result of Theorem (2.1.30a).

■

(2.1.27) ***Example.*** Compute $\lim_{z \to \infty} f(z)$, $\lim_{z \to 1} g(z)$, and $\lim_{z \to 1} h(z)$, where

$$f(z) = \frac{z}{2 + iz}, \qquad g(z) = \frac{z^2 - 1}{z - 1}, \qquad h(z) = \frac{1}{z^2 - 1}$$

Solution. Intuitively, for z large in magnitude, we expect that $f(z)$ would behave like $1/i = -i$, and hence that

$$\lim_{z \to \infty} f(z) = -i \tag{2.1.28}$$

To verify this, we use Definition (2.1.21) to get

$$f\left(\frac{1}{w}\right) = \frac{1/w}{2 + i/w} = \frac{1}{2w + i} \xrightarrow[w \to 0]{} -i$$

For $g(z)$, we note that the function $(z^2 - 1)/(z - 1)$ is not defined at $z = 1$. However, since $z \neq 1$ for the purposes of taking the limit, we can first write

$$\frac{(z^2 - 1)}{z - 1} = \frac{(z - 1)(z + 1)}{z - 1} = z + 1$$

and hence $\lim_{z \to 1} g(z) = 2$.

For $h(z)$, we first write

$$|h(z)| = \frac{1}{|z-1| \cdot |z+1|}$$

and observe that for z close to 1, $h(z)$ is large in magnitude. Hence,

(2.1.29)
$$\lim_{z \to 1} h(z) = \infty$$

The results of the next theorem closely parallel those from calculus. The proofs are left to the reader.

(2.1.30) ***Theorem.*** Let $f(z)$ and $g(z)$ be defined in some neighborhood of $z = z_0$ but not necessarily at $z = z_0$. Then,

(a) If $f(z)$ has a limit at z_0, the limit is unique.

(b) If $f(z) = u(x, y) + iv(x, y)$ and $z_0 = x_0 + iy_0$, then f has a limit at z_0 if and only if both real-valued functions u and v have limits at (x_0, y_0). Then, we have

(2.1.31)
$$\lim_{z \to z_0} [f(z)] = \lim_{(x,y) \to (x_0,y_0)} [u(x, y)] + i \lim_{(x,y) \to (x_0,y_0)} [v(x, y)]$$

(c) If f and g both have limits at z_0, and if α and β are complex constants, then (omitting the notation $z \to z_0$),

(i) $\lim(\alpha f + \beta g) = \alpha \lim f + \beta \lim g$
(ii) $\lim(fg) = (\lim f) \cdot (\lim g)$
(iii) $\lim(f/g) = \lim f / \lim g$.

where the last result is valid if $\lim g(z) \neq 0$.

■

(2.1.32) ***Example.*** Consider

$$\lim_{z \to 0} \left[\frac{1}{1 + e^{1/x}} + iy^2 \right]$$

The imaginary part y^2 clearly has limit zero as $y \to 0$. Hence, by part **(b)** of the above theorem, the limit will exist at $z = 0$ if and only if $(1 + e^{1/x})^{-1}$ has a limit at $(x, y) = (0, 0)$. Now, $1/x$ becomes large positively (negatively) if x approaches zero through positive (negative) values. Hence,

$$e^{1/x} \to \begin{cases} \infty, & \text{if } x \downarrow 0 \\ 0, & \text{if } x \uparrow 0 \end{cases}$$

Thus,

$$\frac{1}{1 + e^{1/x}} \to \begin{cases} 0, & \text{if } x \downarrow 0 \\ 1, & \text{if } x \uparrow 0 \end{cases}$$

Therefore, since these limiting values depend upon the manner in which $(x, y) \to (0, 0)$, the limit does not exist.

■

(2.1.33) ***Definition.*** Let $f(z)$ be defined in some neighborhood of $z = z_0$. Then f **is continuous at z_0** if the limit of f at z_0 exists and

$$\lim_{z \to z_0} f(z) = f(z_0)$$

while it is **continuous in a region D** if it is continuous at each point of D.

■

As in elementary calculus, we have the following results for continuous functions.

(2.1.34) ***Theorem.*** Let $f(z)$ and $g(z)$ be defined in a neighborhood N_0 of $z = z_0$.

(a) If $f(z) = u(x, y) + iv(x, y)$, and $z_0 = x_0 + iy_0$, then f is continuous at z_0 if and only if u and v are continuous at (x_0, y_0).

(b) If f and g are continuous at z_0, then so are $\alpha f + \beta g$, fg, and f/g [if $g(z_0) \neq 0$].

(c) Assume that for all values of z in some neighborhood of z_0, $g(z)$ is in N_0. Hence, the **composite function $f(g(z))$** is defined for z in some neighborhood of z_0. If f is continuous at $g(z_0)$ and g is continuous at z_0, then $f(g(z))$ is continuous at z_0.

Proof. Parts (a) and (b) follow from Theorem (2.1.30). To prove (c), we must show that for any ϵ, there is a δ so that

$$|z - z_0| < \delta \Rightarrow |f(g(z)) - f(g(z_0))| < \epsilon$$

From the assumed continuity of f at $g(z_0)$, we know that for any ϵ there is a $\hat{\delta}$ so that

$$|g(z) - g(z_0)| < \hat{\delta} \Rightarrow |f(g(z)) - f(g(z_0))| < \epsilon$$

Finally, from the continuity of g at z_0, we know that for the above $\hat{\delta}$ there is a δ so that

$$|z - z_0| < \delta \Rightarrow |g(z) - g(z_0)| < \hat{\delta}$$

Combining these last two results yields the continuity of $f(g(z))$ at z_0.

■

(2.1.35) ***Example.*** The principal argument function $f(z) = \text{Arg}(z)$ is continuous everywhere except at $z = 0$ (where it is not even defined) and for z on the negative real axis. The continuity follows either from the representation of $\text{Arg}(z)$ in terms of the inverse tangent (see Problem 3 of Section 1.2) or geometrically from Figure 2.1.5*a*. From the figure it is clear that for all z in a small enough neighborhood of z_0, $\text{Arg}(z)$ is "close to" $\text{Arg}(z_0)$.

For z_0 on the negative real axis, we have (see Figure 2.1.5*b*)

$$\lim_{z \to z_0} \text{Arg}(z) = \begin{cases} \pi, & \text{Im}(z) > 0 \\ -\pi & \text{Im}(z) < 0 \end{cases}$$

Thus, $\text{Arg}(z)$ is not continuous here, since the two limits are different.

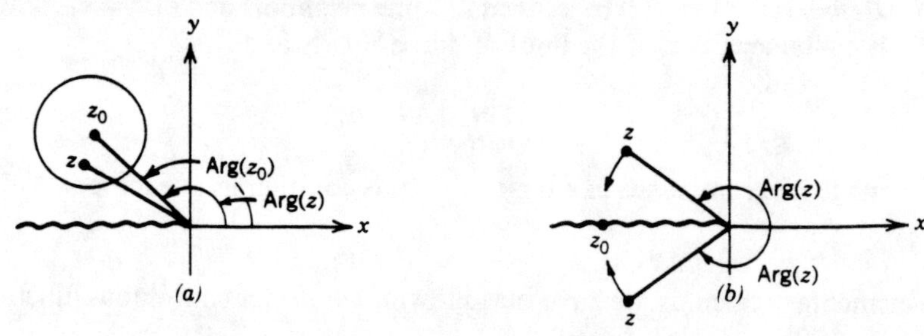

Figure 2.1.5 Continuity properties of Arg(z).

Finally, we have the following definition and results.

(2.1.36) ***Definition.*** A function $f(z)$ is **bounded in a region D** if there exists a positive number M so that

$$|f(z)| \leq M, \qquad \text{for all } z \text{ in } D$$

(2.1.37) ***Theorem.*** Let $f(z)$ be defined in some domain D and let z_0 be a point in D. Then,

(a) If f has a limit at z_0, it is bounded in some neighborhood of z_0.

(b) If f is continuous at z_0 and $f(z_0) \neq 0$, then $f(z) \neq 0$ for z in some neighborhood of z_0.

Proof. We will prove (a) and leave the proof of (b) to Problem 21. Since the limit exists at z_0, there is a number L such that, given ϵ, there is a δ so that

$$|z - z_0| < \delta \Rightarrow |f(z) - L| < \epsilon$$

Hence, for z in this δ-neighborhood of z_0, we have, by the above and the triangle inequality,

$$|f(z)| = |f(z) - L + L| \leq \epsilon + |L|$$

Thus, f is bounded.

PROBLEMS (The answers are given on page 420.)

(Problems with an asterisk are more theoretical.)

1. What is the domain of the following functions? Write each in Cartesian form $u(x, y) + iv(x, y)$.

(a) $\dfrac{z}{(z - i)^2}$ (b) $\dfrac{z^2 + 1}{z - 2i}$ (c) $\dfrac{z + 1}{\bar{z}}$

(d) $z^2 + \bar{z}^2$ (e) $\dfrac{z}{(z^2 + 4i)}$ (f) $\dfrac{z - iz}{z^2 + iz}$

(g) $z + \dfrac{1}{z}$ (h) $z[\text{Re}(\bar{z})]^2$ (i) $\dfrac{x^2 + iy}{y^2 - ix}$

2. Which of the following functions are single-valued? Multiple-valued?

(a) z^2 (b) $\bar{z}^2$ (c) $z^{1/3}$ (d) $\text{Arg}(z)$
(e) $\text{Arg}(\bar{z}^2 + 1)$ (f) $\arg(z)$

3. Write the following in terms of z and $\bar{z}$;

(a) $xy + ix^2$ (b) $r^2 \cos\theta \sin\theta + ir^3$

(c) $\dfrac{x + i}{x + y}$ (d) $r\cos(2\theta) + i\sin(2\theta)$

4. Find $f(2i)$ and $f(1 - i)$, where $f(z)$ is given by

(a) $\dfrac{z}{1 - \bar{z}}$ (b) $\dfrac{ax + 2iy}{y - ix}$ (c) $\text{Arg}\left(\dfrac{1}{\bar{z}}\right)$

(d) $\sqrt{x + y} + ixy$

5. (a) Complete Example (2.1.5) and find u and v for $\sqrt{z}$ when z is real.
(b) Redo Example (2.1.5) in polar coordinates. That is, if

$$w = \rho[\cos\phi + i\sin\phi] \quad \text{and} \quad z = r[\cos\theta + i\sin\theta]$$

find ρ and ϕ in terms of r and θ.

6. (a) Write all possible values of $\sqrt{z^2 + 1}$ in the form $u(x, y) + iv(x, y)$.
(b) Find all values of $(z^3)^{1/3}$ [*Hint:* Use polar coordinates.]

7. (a) Under the mapping $w = x + iy^2$, what is the image of
(i) The horizontal line segment $y = y_0$, $x_0 \le x \le x_1$.
(ii) The vertical line segment $x = x_0$, $y_0 \le y \le y_1$.
(b) Use either of the results in (a) to verify the mapping shown in Figure 2.1.3 [see Example (2.1.12)].

8. (a) Describe the image in the w-plane of any region D in the z-plane under the mapping $w = \bar{z}$.
(b) Do the same as part (a), but for $w = 1 + \bar{z}$.

9. Consider the half-disk $0 < |z| < R$, $\text{Im}(z) > 0$, and the function $w = f(z) = x + i(x^2 + y^2)$.
(a) Express x and y in terms of u and v, and use the inequalities describing the half-disk to compute its image under $f(z)$.
(b) What is the image of the semicircle $x^2 + y^2 = R_0^2$, $y > 0$ under $w = f(z)$?
(c) Use part (b) to rederive the result of part (a). [*Hint:* Generate the semidisk from the semicircles.]

10. Consider the function $w = f(z) = x + i(x + y^2)$.
(a) What is the image of the half-line $x = x_0$, $y > 0$?
(b) What is the image of the first quadrant?

11. Under the mapping $w = 1/z$, find the image of
   (a) The unit circle $|z| = 1$. (b) The real axis.
   (c) The upper half-plane. (d) The circle $|z - i| = 1$.
   (e) The quarter disk $|z| < 1$, $\text{Im}(z) > 0$, $\text{Re}(z) > 0$.
12. Find the image of the first quadrant under the mapping $w = (z - i)/z$.
13. Find the image of the unit circle under the mappings

   (a) $w = \dfrac{z - i}{z + i}$ (b) $w = z + \dfrac{1}{z}$

14. One might be tempted, in order to find the image of a region under a given mapping, to first find the image of its boundary. As shown below, this does not always work.
   (a) What is the image of the square region whose vertices are at $z = 1 \pm i$ and $-1 \pm i$ under the mapping $w = x^2 + y + i(x^2 - y)$.
   (b) Show that the image of the boundary of the square in (a) is not the boundary of the image.
15. Consider the function $w = z^2$.
   (a) Express the polar coordinates of w in terms of the polar coordinates of z.
   (b) Use the result of (a) to find the image under $w = z^2$ of
      (i) The first quadrant.
      (ii) A wedge with vertex at the origin, one edge on the positive real axis, and vertex angle $\theta_0 < \pi$.
      (iii) The unit disk.
      (iv) The annular sector $2 < |z| < 3$, $\pi/4 < \arg(z) < \pi/2$.
16. Find the image of the strip $\text{Re}(z) < 0$, $0 < \text{Im}(z) < \pi/2$ under the mapping $w = e^x \cos y + ie^x \sin y$. [*Hint:* What are the polar coordinates of w?]
17. Compute the following limits:

   (a) $\lim\limits_{z \to 1 - i} (z^2 + \bar{z}^2)$ (b) $\lim\limits_{z \to i} \left(xe^{xy} + i\dfrac{e^{xy}}{x + 1} \right)$

   (c) $\lim\limits_{z \to 0} [z \,\text{Arg}(z)]$ (d) $\lim\limits_{z \to i} \left[\dfrac{z^3 + (2 - i)z^2 + (1 - i)z - i}{z - i} \right]$

   (e) $\lim\limits_{z \to \infty} \left(e^{-x^2} + \dfrac{iy^2}{1 + y^2} \right)$ (f) $\lim\limits_{z \to 0} |z|$

   (g) $\lim\limits_{z \to \infty} \left(\dfrac{z}{z^2 + 1} \right)$ (h) $\lim\limits_{z \to \infty} \left(\dfrac{z^2}{z + 1} \right)$ (i) $\lim\limits_{z \to 0} \left(\dfrac{z^2}{|z|} \right)$

   (j) $\lim\limits_{z \to \infty} \left[\dfrac{p(z)}{q(z)} \right]$, where p and q are polynomials of order n and m, respectively, whose highest power has coefficient 1

   (k) $\lim\limits_{z \to 0} \left(\dfrac{z}{|z|} \right)$

18. Where are the following functions continuous?

(a) $\dfrac{z+1}{z^2+4z+5}$ (b) $\dfrac{p(z)}{q(z)}$ where p and q are polynomials

(c) $\begin{cases} \dfrac{x^2}{x^2+y^2} + 2i, & z \neq 0 \\ 2i \quad , & z = 0 \end{cases}$ (d) $\operatorname{Arg}(z^2+1)$

(e) $\begin{cases} \dfrac{x^4+y^4}{x^2+y^2} + xi, & z \neq 0 \\ 0 \quad , & z = 0 \end{cases}$ (f) $\operatorname{Arg}\left(\dfrac{z-i}{z+i}\right)$

(g) $(z^2+5z)^{12}$ (h) $|z|$

(i) $\begin{cases} \dfrac{\operatorname{Re}(z)}{z}, & z \neq 0 \\ 0 & z = 0 \end{cases}$

19.* (a) Prove that $f(z) \to L$ if and only if $|f(z) - L| \to 0$.
(b) Prove that if $f(z) \to L$, then $|f(z)| \to |L|$.
[*Hint:* Use (1.1.27b) to bound $||f| - |L||$.]
(c) Prove that if f is defined at z_0, then it is continuous at z_0 if and only if $|f(z) - f(z_0)| \xrightarrow[z \to z_0]{} 0$.

20. Compute the composite functions $f(g(z))$, where $f(z)$ and $g(z)$ are given by

(a) $f = z + 1, \quad g = z^2$ (b) $f = \dfrac{z}{z+i}, \quad g = \dfrac{z+i}{z}$

(c) $f = \dfrac{1}{x} + iy^2, \quad g = z^2$ (d) $f = x + 2iy, \quad g = f^2$

21.* (a) Prove Theorem (2.1.37b) [*Hint:* Express $|f(z)|$ in terms of u and v and use the corresponding result from calculus.]
(b) Prove that if $f(z)$ is continuous in a closed, bounded region of the z-plane, it is bounded there.

2.2 DERIVATIVES; CAUCHY–RIEMANN EQUATIONS; ANALYTICITY

As in elementary calculus, we have the following definition.

(2.2.1) ***Definition.*** Let $f(z)$ be defined in a neighborhood of $z = z_0$. Then the **derivative of f at $z = z_0$**, denoted by $f'(z_0)$ [or $(df(z_0)/dz)$], is given by

(2.2.2)
$$f'(z_0) = \lim_{z \to z_0} \left[\frac{f(z) - f(z_0)}{z - z_0} \right]$$

if the limit exists. We then say that f is **differentiable at z_0**. A function is **differentiable in a domain D** if it is differentiable at each point of D.

■

We can also define **higher-order derivatives** in the usual way.

(2.2.3) ***Example.*** Let $f(z) = z^{-2}$. Then, for $z_0 \neq 0$,

$$\frac{f(z) - f(z_0)}{z - z_0} = \frac{z_0^2 - z^2}{z_0^2 z^2 (z - z_0)} = -\frac{z + z_0}{z_0^2 \cdot z^2}$$

Hence, the limit in (2.2.2) exists for all $z_0 \neq 0$ and yields $f'(z_0) = -2z_0^{-3}$.

■

(2.2.4) ***Example.*** Let $f(z) = x + 2iy$. If $z_0 = x_0 + iy_0$, then

$$\frac{f(z) - f(z_0)}{z - z_0} = \frac{(x - x_0) + 2i(y - y_0)}{(x - x_0) + i(y - y_0)}$$

$$= \frac{(x - x_0)^2 + 2(y - y_0)^2 + i(x - x_0)(y - y_0)}{(x - x_0)^2 + (y - y_0)^2}$$

For f to be differentiable at z_0, both the real and the imaginary parts of the above must possess limits as $(x, y) \to (x_0, y_0)$. Consider the limit of the real part

$$\lim_{(x, y) \to (x_0, y_0)} \left[\frac{(x - x_0)^2 + 2(y - y_0)^2}{(x - x_0)^2 + (y - y_0)^2} \right]$$

If we permit (x, y) to approach (x_0, y_0) along the vertical line $x = x_0$, then this limit will be 2, while along the horizontal line $y = y_0$ it will be 1. Hence, the limit does not exist and $f(z)$ is differentiable nowhere.

■

The proofs of the following properties of differentiable functions follow along the exact same lines as in calculus.

(2.2.5) ***Theorem.*** Let $f(z)$ and $g(z)$ be differentiable at $z = z_0$. Then, if α and β are complex constants,

(a) $\alpha f + \beta g$, fg, and f/g [if $g(z) \neq 0$ in some neighborhood of z_0] are all differentiable at z_0, and

(i) $(\alpha f + \beta g)' = \alpha f' + \beta g'$

(ii) $(fg)' = f'g + fg'$

(iii) $\left(\dfrac{f}{g}\right)' = \dfrac{gf' - fg'}{g^2}$

(b) If f is differentiable at $g(z)$, the **composite function** $f(g(z))$ is differentiable at z and we have the **chain rule**

(2.2.6)
$$\frac{d}{dz}[f(g(z))] = f'(g(z)) \cdot g'(z)$$

■

Finally, we have that only continuous functions can be differentiable.

(2.2.7) ***Theorem.*** If $f(z)$ is differentiable at z_0, then it must be continuous there.

Proof. From Definition (2.2.1), since f is differentiable,

$$\lim_{z \to z_0} \left| \frac{f(z) - f(z_0)}{z - z_0} - f'(z_0) \right| = 0$$

Now, we can write

$$f(z) - f(z_0) = (z - z_0)\left[\frac{f(z) - f(z_0)}{z - z_0} - f'(z_0) + f'(z_0)\right]$$

and by the triangle inequality,

$$|f(z) - f(z_0)| \le |z - z_0|\left[\left|\frac{f(z) - f(z_0)}{z - z_0} - f'(z_0)\right| + |f'(z_0)|\right]$$

Hence, $\lim_{z \to z_0} |f(z) - f(z_0)| = 0$, and thus f is continuous at z_0.

■

Cauchy–Riemann Equations. Note from the Example (2.2.4) that even though the real-valued functions $u(x, y) = x$ and $v(x, y) = 2y$ possess continuous partial derivatives *of all orders*, the complex function $f(z) = u + iv$ is not differentiable anywhere. Hence, we must conclude that we *cannot simply piece together any two real-valued functions* u(x, y) *and* v(x, y) *and expect to get a differentiable complex-valued function* f(z) = u + iv. The two real-valued functions u and v must possess some additional structure. This structure, as we will see below, is brought about by virtue of the fact that the limit, in the definition of derivative, must not depend on how z is to approach z_0.

(2.2.8) ***Theorem.*** Let $u(x, y)$ and $v(x, y)$ be two real-valued functions defined in some neighborhood of the point (x_0, y_0). Consider the complex-valued function

$$f(z) = u(x, y) + iv(x, y)$$

(a) If f is differentiable at $z_0 = x_0 + iy_0$, then u and v possess first partial derivatives at (x_0, y_0) and satisfy the **Cauchy–Riemann equations**

(2.2.9)
$$u_x = v_y \quad \text{and} \quad u_y = -v_x$$

where the partial derivatives u_x, u_y, v_x, and v_y are evaluated at

(x_0, y_0). In addition, we have the following two expressions for the derivative

$$f'(z_0) = u_x + iv_x \\ = v_y - iu_y \tag{2.2.10}$$

(b) If u and v possess continuous first partial derivatives at (x_0, y_0) and satisfy the Cauchy–Riemann equations there, then $f(z)$ is differentiable at z_0.

Proof. We will prove only part (a) here and leave the proof of (b) to Problem 5. However, note that (b) is not strictly converse to (a). That is, if at (x_0, y_0), u and v possess first partial derivatives which are not continuous but which do satisfy the Cauchy–Riemann equations, then $f(z) = u + iv$ need not necessarily be differentiable at $z_0 = x_0 + iy_0$. (See Problem 9 for an example.)

To prove (a), we assume that f is differentiable and hence that the limit

$$f'(z_0) = \lim_{z \to z_0} \left[\frac{f(z) - f(z_0)}{z - z_0}\right] \\ = \lim_{z \to z_0} \frac{[u(x, y) - u(x_0, y_0)] + i[v(x, y) - v(x_0, y_0)]}{(x - x_0) + i(y - y_0)} \tag{2.2.11}$$

exists *independent of how z approaches* z_0. First, letting $z \to z_0$ along a horizontal line through z_0, for which $z = x + iy_0$, we get from the above

$$f'(z_0) = \lim_{x \to x_0} \left[\frac{u(x, y_0) - u(x_0, y_0)}{x - x_0} + i\frac{v(x, y_0) - v(x_0, y_0)}{x - x_0}\right]$$

Since this limit exists, so must the limits of the real and imaginary parts, which the reader will recognize as the partial derivatives $u_x(x_0, y_0)$ and $v_x(x_0, y_0)$. Hence,

$$f'(z_0) = u_x + iv_x \tag{2.2.12}$$

If we now take the limit in (2.2.11) along the vertical line $x = x_0$, then an argument similar to that above shows that u_y and v_y must both exist and

$$f'(z_0) = \frac{1}{i}(u_y + iv_y) \tag{2.2.13}$$

Equating these two results [which then verifies (2.2.10)] leads to the Cauchy–Riemann equations (2.2.9).

■

To reiterate, from part (a) of the theorem, we learn that the Cauchy–Riemann equations are necessary in order that a complex function be differentiable. Thus, at those points where the Cauchy–Riemann equations are not satisfied, $u + iv$ cannot be differentiable. However, from part (b) we learn that the Cauchy–Riemann equations (which are easy to check), plus continuous first partial derivatives, are sufficient for $f(z) = u + iv$ to be differentiable. The following examples illustrate these points.

(2.2.14) ***Example.*** Show that the function $f(z) = |z|^2$ is differentiable only at $z = 0$, and compute $f'(0)$.

Solution. Here, we have

$$u(x, y) = x^2 + y^2 \qquad \text{and} \qquad v(x, y) = 0 \tag{2.2.15}$$

Clearly, all partial derivatives of u and v are continuous everywhere, and the Cauchy–Riemann equations (2.2.9) yield

$$2x = 0 \qquad \text{and} \qquad 2y = 0$$

which are satisfied only at $z = 0$. Also, since $u_x(0,0) = v_x(0,0) = 0$, we have from (2.2.10) that $f'(0) = 0$.

■

(2.2.16) ***Example.*** Let $f(z) = e^{-y}\cos x + ie^{-y}\sin x$. Since $u(x, y) = e^{-y}\cos x$ and $v(x, y) = e^{-y}\sin x$ possess continuous partial derivatives, and

$$u_x = -e^{-y}\sin x = v_y$$

$$u_y = -e^{-y}\cos x = -v_x$$

$f(z)$ is differentiable everywhere. Hence, from (2.2.10),

$$\begin{aligned} f'(z) &= -e^{-y}\sin x + ie^{-y}\cos x \\ &= if(z) \end{aligned}$$

■

Sometimes, because of the form of the function $f(z)$, it is quite clumsy to verify that the Cauchy–Riemann equations are satisfied. For example, if

$$u(x, y) = \frac{x^2 - y^2}{(x^2 + y^2)^2} \qquad \text{and} \qquad v(x, y) = \frac{-2xy}{(x^2 + y^2)^2}$$

then, for $z \neq 0$, it can be shown that the Cauchy–Riemann equations are satisfied, but the calculations are somewhat tedious. Note, however, that in polar coordinates the above functions become the simpler-looking

$$u = \frac{\cos(2\theta)}{r^2}, \qquad v = -\frac{\sin(2\theta)}{r^2}$$

Thus, a polar coordinate version of the Cauchy–Riemann equations would clearly be helpful.

(2.2.17) ***Theorem.*** Let r and θ be polar coordinates and $\hat{u}(r, \theta)$, $\hat{v}(r, \theta)$ be two real-valued functions with continuous first partial derivatives in some

neighborhood of $r = r_0 > 0, \theta = \theta_0$. If at r_0 and θ_0,

(2.2.18)
$$\hat{u}_r = \frac{1}{r}\hat{v}_\theta \quad \text{and} \quad \hat{u}_\theta = -r\hat{v}_r$$

then $f(z) = \hat{u} + i\hat{v}$ is differentiable at $z_0 = r_0(\cos\theta_0 + i\sin\theta_0)$ and

(2.2.19)
$$f'(z_0) = (\hat{u}_r + i\hat{v}_r)(\cos\theta_0 - i\sin\theta_0)$$

Equations (2.2.18) are the **Cauchy–Riemann equations in polar coordinates.**

■

The proof of this result follows directly from the chain rule and the Cauchy–Riemann equations and is given in Problem 6.

(2.2.20) ***Example.*** With

$$u(x, y) = \frac{x^2 - y^2}{(x^2 + y^2)^2} \quad \text{and} \quad v(x, y) = \frac{-2xy}{(x^2 + y^2)^2}$$

show that $f(z) = u + iv$ is differentiable for $z \neq 0$ and $f'(z) = -2/z^3$.

Solution. In polar coordinates, u and v become

$$\hat{u}(r, \theta) = \frac{\cos(2\theta)}{r^2}, \qquad \hat{v}(r, \theta) = -\frac{\sin(2\theta)}{r^2}$$

Clearly, for $r \neq 0$, $\hat{u}$ and $\hat{v}$ possess continuous first partial derivatives with respect to r and θ, and it is easily verified that equations (2.2.18) are satisfied. Hence, f is differentiable for $z \neq 0$, and (2.2.19) yields

$$\begin{aligned} f'(z) &= -2r^{-3}[\cos(2\theta) - i\sin(2\theta)] \cdot [\cos\theta - i\sin\theta] \\ &= -2r^{-3}[\cos(3\theta) - i\sin(3\theta)] \\ &= -2\{r^3[\cos(3\theta) + i\sin(3\theta)]\}^{-1} \\ &= \frac{-2}{z^3} \end{aligned}$$

■

Analyticity. There exist functions such as $|z|^2$ in Example (2.2.14) which are differentiable only at isolated points. Intuitively, it seems that such functions may not be very useful. Thus, we are led to the concept of analyticity.

(2.2.21) ***Definition***

(a) A function $f(z)$ is said to be **analytic** (or **holomorphic**) **at a point** z_0 if it is differentiable not only at z_0 but at each point in some neighborhood of z_0.

(b) $f(z)$ is **analytic in a domain** if it is analytic at each point in the domain.

(c) $f(z)$ is **entire** if it is analytic in the whole finite z-plane.
(d) $f(z)$ is **analytic at $z = \infty$** if the function $f(1/z)$ is analytic at $z = 0$.

■

We shall see, as we proceed through this text, that the concept of analyticity is of fundamental importance in complex function theory.

The next result follows from Theorem (2.2.5).

(2.2.22) ***Theorem.*** Let f and g be analytic at z_0. Then,

(a) $\alpha f + \beta g$, fg, and f/g (if $g \neq 0$ in some neighborhood of z_0) are all analytic at z_0.

(b) If f is analytic at $g(z_0)$, the **composite function** $f(g(z))$ is analytic at $z = z_0$.

These results are also valid if the word entire is substituted for analytic.

■

(2.2.23) ***Example.*** The functions $f(z) = z^n$ $(n = 0, 1, 2, \ldots)$ and $g(z) = e^x \cos y + ie^x \sin y$ are differentiable everywhere and hence entire. However, $|z|^2$ is differentiable only at $z = 0$ and hence is analytic nowhere. The function

$$h(z) = e^{x^2 - y^2}[\cos(2xy) + i \sin(2xy)]$$

will be recognized as the composite $g(z^2)$ and hence is entire since z^2 and g are both entire.

The function $p(z) = z$ is not analytic at $z = \infty$ since $p(1/z) = 1/z$ is not analytic (nor even defined) at $z = 0$, while $q(z) = z/(i + z)$ *is* analytic at $z = \infty$. (Why?).

■

We will call points of nonanalyticity **singular points**; and as we shall see throughout this text, certain types of singular points play important roles both mathematically and physically.

PROBLEMS (The answers are given on page 421.)

(Problems with an asterisk are more theoretical.)

1. Use the definition of derivative to compute $f'(z)$ at those points where the derivative exists, where $f(z)$ is given by

   (a) $z^n \, (n = 0, 1, 2, \ldots)$ (b) $(1 + \bar{z})^{-1}$ (c) $x + iy^2$

   (d) $\dfrac{1}{z}$ (e) $\bar{z}$ (f) $\bar{z}^2$

2.* Show that Definition (2.2.1) is equivalent to

$$\lim_{\Delta z \to 0} \frac{1}{\Delta z}[f(z_0 + \Delta z) - f(z_0)] = f'(z_0)$$

3.* (a) Prove that if $f(z) = u + iv$ is continuously differentiable in a neighborhood of $z_0 = x_0 + iy_0$, then both u and v possess continuous first partial derivatives there.

(b) Prove that if $f''(z)$ exists and is continuous, then

(i) All second partial derivatives of u and v exist with $u_{xy} = u_{yx}$ and $v_{xy} = v_{yx}$.

(ii) $u_{xx} = v_{xy}$ and $u_{xy} = -v_{xx}$.

(iii) $f''(z) = u_{xx} + iv_{xx}$.

4.* Prove Theorem (2.2.5).

5.* Prove Theorem (2.2.8b) by verifying the following steps. First, with $z = x + iy$ and $z_0 = x_0 + iy_0$, write

$$\frac{f(z) - f(z_0)}{z - z_0} = \frac{[u(x, y) - u(x_0, y_0)] + i[v(x, y) - v(x_0, y_0)]}{z - z_0}$$

(a) Since u and v possess continuous first partial derivatives, verify that

$$u(x, y) - u(x_0, y_0) = (x - x_0)\,u_x(x_0, y_0) + (y - y_0)\,u_y(x_0, y_0) + \epsilon_1(x, y)$$

$$v(x, y) - v(x_0, y_0) = (x - x_0)\,v_x(x_0, y_0) + (y - y_0)\,v_y(x_0, y_0) + \epsilon_2(x, y)$$

where

$$\lim_{z \to z_0} \left[\frac{\epsilon_{1,2}(x, y)}{|z - z_0|}\right] = 0$$

[*Hint:* Use a result on differentials from advanced calculus.]

(b) $$\frac{f(z) - f(z_0)}{z - z_0} = g(z) + \frac{1}{z - z_0}\begin{Bmatrix}(x - x_0)u_x + (y - y_0)u_y \\ + i[(x - x_0)v_x + (y - y_0)v_y]\end{Bmatrix}$$

where the function $g(z)$ satisfies $\lim_{z \to z_0} [g(z)] = 0$.

(c) $$\lim_{z \to z_0} \left[\frac{f(z) - f(z_0)}{z - z_0}\right] = u_x + iv_x$$

[*Hint:* Use the Cauchy–Riemann equations in (b).]

6.* Prove Theorem (2.2.17). [*Hint:* With $x = r\cos\theta$ and $y = r\sin\theta$, the chain rule yields $u_r = u_x \cos\theta + u_y \sin\theta$. Now you finish.]

7. Indicate where the following functions are differentiable and compute the derivatives:

(a) $2x + iy^2$ (b) $\bar{z}^2$ (c) $2x^2 + y^2 + y - 1 + ix$

(d) $x^3 - 3xy^2 + i(3x^2y - y^3)$ (e) $x^2 + 3y^2 + 2ixy$

(f) $\dfrac{\bar{z}}{|z|^2}$ (g) $z\,\mathrm{Im}(z)$ (h) $z\,\mathrm{Re}(z)$ (i) $\mathrm{Arg}(z)$

(j) $z\,\mathrm{Arg}(z)$ (k) $\mathrm{Re}(z) \cdot \mathrm{Im}(z)$ (l) $z^2 + \bar{z}^2$ (m) $xe^{xy} + i\dfrac{e^{xy}}{x}$

(n) $e^{x^2 - y^2}[\cos(2xy) + i\sin(2xy)]$

(o) $|xy|$ (p) $e^x(\sin y - i\cos y)$ (q) $e^x(\sin y + i\cos y)$
(r) $\ln|z| + i\,\mathrm{Arg}(z)$ (s) $z\bar{z}^2$

8. Find values of a, b and c so that $f(z) = -x^2 + xy + y^2 + i(ax^2 + bxy + cy^2)$ is entire.

9. For the following function, show that the Cauchy–Riemann equations are satisfied at $z = 0$ but that the function is not differentiable there:

$$f(z) = \begin{cases} \dfrac{xy}{x^2 + y^2}, & z \neq 0 \\ 0, & z = 0 \end{cases}$$

Why doesn't this contradict Theorem (2.2.8b)?

10. (a) Show that all polynomials are entire.
(b) Where are rational functions analytic?
(c) Show that $f(z) = \begin{cases} \dfrac{z^2 - 1}{z - 1}, & z \neq 1, \\ 2, & z = 1 \end{cases}$ is entire.

11. Use the Cauchy–Riemann equations to show that $f(z) = z^n$, $n = 0, 1, 2, \ldots$ is entire and $f' = nz^{n-1}$.

12. (a) Which of the following are analytic at $z = \infty$?

(i) $\dfrac{1}{z}$

(ii) $\dfrac{1}{1 + z^2}$

(iii) z

(iv) $\dfrac{1}{1 + \bar{z}}$

(b) When is a rational function analytic at $z = \infty$?

13.* (a) Let $u(x, y)$ be a real valued continuously differentiable function whose first partial derivatives vanish in a domain D of the xy-plane. Prove that u is constant in D. [*Hint:* Since D is connected, any two points can be joined by a broken line. Write $u(x, y)$ as an integral along such a line.]
(b) Construct a counterexample to the result in (a) if D is not connected.
(c) Prove that if $f(z)$ is continuously differentiable in a domain D, with $f' \equiv 0$, then $f(z)$ is constant in D.

14. Use the result of Problem 13c to prove that if $f(z)$ is continuously differentiable, and hence analytic, in a domain D, then it must be a constant if any of the following hold:
(a) f is real-valued.
(b) f is purely imaginary.

(c) $|f|$ is constant.
(d) $f = u + iu^n (n = 0, 1, 2, \ldots)$ *where* $u = u(x, y)$ is continuously differentiable in D.
(e) $\bar{f}$ is also analytic.

15. Let $f(z) = u(x, y) + iv(x, y)$ be differentiable in a domain D. Show that $|f'(z)|^2$ is the magnitude of the **Jacobian of the mapping** $u = u(x, y)$, $v = v(x, y)$.

16. Any function $f(z) = u(x, y) + iv(x, y)$ can be viewed as a function $\hat{f}$ of both z and $\bar{z}$ defined by

$$f(z) = \hat{f}(z, \bar{z}) = u\left(\frac{1}{2}(z + \bar{z}), \frac{1}{2i}(z - \bar{z})\right) + iv\left(\frac{1}{2}(z + \bar{z}), \frac{1}{2i}(z - \bar{z})\right)$$

Assume that u and v are continuously differentiable functions of their arguments.

(a) Use the chain rule to compute $\partial\hat{f}/\partial z$ and $\partial\hat{f}/\partial\bar{z}$ in terms of the partial derivatives of u and v.
(b) Use (a) to show that f is differentiable if and only if $\partial\hat{f}/\partial\bar{z} = 0$, in which case $f' = \partial\hat{f}/\partial z$. Hence, **a differentiable function has no $\bar{z}$ dependence.**
(c) Use the result in (b) to show that if u and v have continuous second partial derivatives, then

$$\frac{\partial^2\hat{f}}{\partial z\,\partial\bar{z}} = \frac{1}{4}\left(\frac{\partial^2\hat{f}}{\partial x^2} + \frac{\partial^2\hat{f}}{\partial y^2}\right)$$

17. (a) If $f(z) = u + iv$ is analytic, and u and v possess continuous second partial derivatives, show that both u and v satisfy **Laplace's equation** $u_{xx} + u_{yy} = 0$.

(b) Use (a) to show that $\left(\frac{\partial^2}{\partial x^2} + \frac{\partial^2}{\partial y^2}\right)|f|^2 = 4|f'|^2$.

18. If $f(z)$ is entire and continuously differentiable, show that $g(z) = \overline{f(\bar{z})}$ is entire. [*Hint:* $f(\bar{z}) = u(x, -y) + iv(x, -y)$. Now verify that the Cauchy–Riemann equations are satisfied for $g(z)$.]

2.3 THE EXPONENTIAL FUNCTION

Here, we will attempt to define the complex analog of the exponential function e^x. There are several ways to motivate this, with the objective being to define a complex function, analytic in some neighborhood of the real axis, which reduces to e^x when z is real. Thus, in some sense we will have extended e^x into the complex plane. For example, the function

(2.3.1)
$$f(z) = e^x + iy$$

reduces to e^x when $y = 0$ but is differentiable only on the imaginary axis (why?), and hence analytic nowhere. Such a function would not serve to define the complex exponential.

(2.3.2) ***Example.*** Find an analytic function $f(z)$ satisfying the complex analog of the differential equation for the real exponential, that is,

(2.3.3)
$$\frac{df}{dz} = f$$

Solution. Since

(2.3.4)
$$f(z) = u(x, y) + iv(x, y)$$

is assumed analytic, we have $f' = u_x + iv_x$. Hence, using (2.3.3) and the Cauchy–Riemann equations, we get

(2.3.5)
$$\left.\begin{array}{ll}\textbf{(a)} & u_x = u \\ \textbf{(b)} & v_x = v\end{array}\right\} \text{(follows from } f' = f\text{)}$$
$$\left.\begin{array}{ll}\textbf{(c)} & u_x = v_y \\ \textbf{(d)} & u_y = -v_x\end{array}\right\} \text{(Cauchy–Riemann equations)}$$

Equations (a) and (b) can be integrated immediately to yield

(2.3.6)
$$u(x, y) = g(y)e^x \quad \text{and} \quad v(x, y) = h(y)e^x$$

where $g(y)$ and $h(y)$ are real-valued functions of the remaining variable y. With this, (2.3.5c, d) yield

(2.3.7)
$$h'(y) = g(y) \quad \text{and} \quad g'(y) = -h(y)$$

where the prime sign denotes derivatives with respect to y. If we eliminate g in favor of h, we get $h'' + h = 0$ whose solutions are

(2.3.8)
$$h(y) = c_1 \cos y + c_2 \sin y$$

where c_1 and c_2 are real constants. Then (2.3.7) yields

(2.3.9)
$$g(y) = -c_1 \sin y + c_2 \cos y$$

Finally, we find from (2.3.6) through (2.3.9) that

(2.3.10)
$$f(z) = ke^x(\cos y + i \sin y)$$

where we have defined the complex constant $k = c_2 + ic_1$.

∎

From the above we see that the only analytic function $f(z)$ satisfying $f' = f$ is that given by (2.3.10). If in addition we demand that $f(z)$ reduce to e^x when z is real ($y = 0$), then we must choose $k = 1$. Thus, we are led to the following definition.

(2.3.11) ***Definition.*** The **complex exponential function**, denoted by e^z, or $\mathbf{exp(z)}$, is defined by

(2.3.12)
$$e^z = e^{x+iy} = e^x(\cos y + i \sin y)$$

∎

(2.3.13) ***Theorem.*** The function e^z possesses the following properties:

(a) e^z is single-valued, entire, and satisfies

$$\frac{de^z}{dz} = e^z$$

(2.3.14) **(b)** (i) $e^{iy} = \cos y + i \sin y$ **(Euler's equation)**

(ii) $\cos y = \frac{1}{2}(e^{iy} + e^{-iy})$

(iii) $\sin y = \frac{1}{2i}(e^{iy} - e^{-iy})$

(c) $|e^z| = e^x \qquad \arg(e^z) = y + 2\pi n \qquad n = 0, \pm 1, \ldots$

(d) $e^{z_1} \cdot e^{z_2} = e^{z_1 + z_2}$

(e) $e^{-z} = \frac{1}{e^z}$

(f) $e^z \neq 0$ for any finite z

(g) e^z is periodic with imaginary periods $2\pi i n$, that is,

(2.3.15)
$$e^{z + 2\pi i n} = e^z \qquad n = 0, \pm 1, \pm 2, \ldots$$

Proof. Part (a) is immediate since the Cauchy–Riemann equations are satisfied everywhere, as is easily verified. To prove (b), let $x = 0$ in (2.3.12) to get (bi). From this, $e^{-iy} = \cos y - i \sin y$, and we can derive (bii, iii) by adding and subtracting this and (bi).

The remaining parts are easily derived from the definition of e^z and are left to the reader.

■

Note the interesting periodicity property (g). It shows that though the real exponential e^x is not periodic, its complex counterpart e^z is, but with *imaginary periods*.

(2.3.16) ***Example.*** Show that the only periods of e^z are those given in (2.3.15).

Solution. We wish to find all complex $p = p_1 + ip_2$ so that $e^{z+p} = e^z$ for all z. Hence, since the magnitude and argument of these two exponentials must be equal, we get from (2.3.13c)

(2.3.17)
$$e^{x+p_1} = e^x \quad \text{and} \quad y + p_2 = y + 2\pi n$$

The first of these yields $p_1 = 0$, while the second yields $p_2 = 2\pi n$. Thus, the periods are $p = 2\pi i n$.

■

(2.3.18) ***Example.*** From property (2.3.14), we have that

(a) $e^{2\pi i n} = 1, \qquad n = 0, \pm 1, \pm 2, \ldots$

(2.3.19) **(b)** $e^{\pi i n} = (-1)^n$

(c) $e^{(\pi i n)/2} = i^n$

(d) $|e^{iy}| = 1$

This last result shows that all points of the form e^{iy} lie on the unit circle.

Finally, again from (2.3.14), we can write the **polar form of z**, $z = |z|(\cos\theta + i\sin\theta)$, as

(2.3.20)
$$z = |z|e^{i\theta} \qquad \theta = \arg(z)$$

■

(2.3.21) ***Example.*** Find all values of z satisfying

$$e^z = 1 + i$$

Solution. Using (2.3.13c), we find

$$e^x = |1 + i| = \sqrt{2} \Rightarrow x = \frac{1}{2}\ln 2$$

while

$$y = \arg(1 + i) = \frac{\pi}{4} + 2\pi k \qquad k = 0, \pm 1, \ldots$$

Thus,

$$z = \frac{1}{2}\ln 2 + i\left(\frac{\pi}{4} + 2\pi i k\right) \qquad k = 0, \pm 1, \pm 2, \ldots$$

■

Using exactly the same technique as in the last example, the reader is asked to show that **e^z takes on all values, except zero, an infinite number of times.**

(2.3.22) ***Example.*** Discuss the limit $\lim_{z \to \infty} e^z$.

Solution. As pointed out in Section 1.1, there are many ways that z can go off to infinity in the complex plane. For example, z can approach infinity in the right half-plane along the horizontal line $y = y_0$. In this case, $x \to +\infty$, and we have from the definition of e^z that

(2.3.23)
$$\lim_{\substack{x \to +\infty \\ (y = y_0)}} (e^z) = \lim_{x \to +\infty} (e^{iy_0} \cdot e^x) = \infty$$

We also might have z approaching infinity in the left half-plane along horizontal lines, in which case $x \to -\infty$ and

(2.3.24)
$$\lim_{\substack{x \to -\infty \\ (y = y_0)}} (e^z) = 0$$

Hence, $\lim_{z \to \infty} (e^z)$ does not exist. (Why?) The reader is asked to apply Definition (2.1.21) to also show that the limit does not exist.

■

Mapping Properties. The following example will serve to illustrate an important mapping property of e^z.

(2.3.25) ***Example.*** Compute the image, under the mapping $w = e^z$, of the horizontal line segment $y = y_0, x_0 < x < x_1$, shown in Figure 2.3.1*a*. Use this result to find the image of the strips shown in parts (*b*) and (*c*) of the figure.

Solution. From $w = e^z$, we have

(2.3.26) $$|w| = e^x \quad \text{and} \quad \arg(w) = y + 2\pi n$$

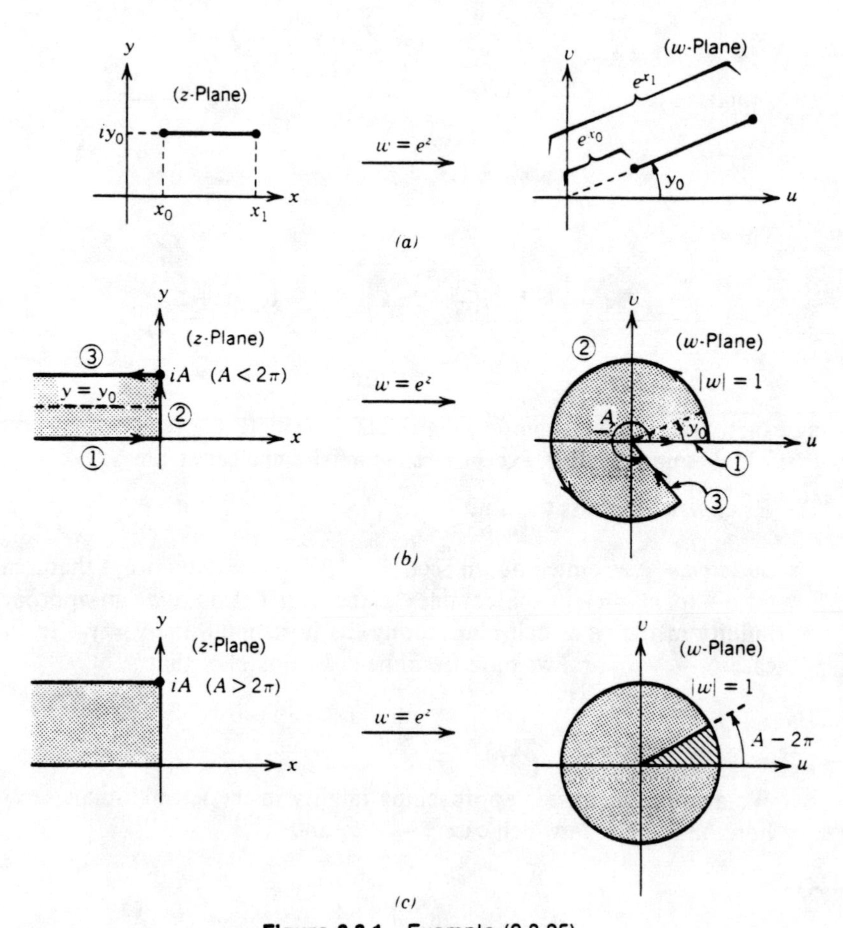

Figure 2.3.1 Example (2.3.25).

Hence, the line $y = y_0$ has as its image all, or a part of, the ray

$$\arg(w) = y_0$$

To determine what part of the ray is actually covered, we use the inequality $x_0 < x < x_1$ in the first of equations (2.3.26) to get

$$e^{x_0} < |w| < e^{x_1}$$

Hence, the image of the line segment is as shown in Figure 2.3.1*a*. Note that if $x_0 = -\infty$, then the ray segment in the w-plane would extend to the origin (but not include it); while if $x_1 = +\infty$, it would extend to infinity.

Now consider the semiinfinite strip of Figure 2.3.1*b*. This strip may be generated by moving the horizontal line $y = y_0$, $x < 0$, shown dotted in the diagram, up from $y_0 = 0$ to $y_0 = A$. Hence, the image of the strip will be generated by the images of these dotted lines, a typical image shown dotted in the w-plane of Figure 2.3.1*b*. Thus, as the dotted line segment in the z-plane moves up, the rays $\arg(w) = y_0$ rotate counterclockwise, and hence the image of the strip will be part, or all, of the punctured disk $0 < |w| < 1$. The final result will depend upon the height A of the strip. If $A < 2\pi$, then the pie-shaped region shown in the figure will be generated. Note that the image of the boundary of the strip is the boundary of the image with the corresponding pieces and directions shown numbered in the figure.

If $A > 2\pi$ (see Figure 2.3.1*c*), the whole interior of the disk $|w| < 1$, with the exception of the origin, will be covered, with some pieces covered more than once. This multiple covering is simply a reflection of the periodicity of e^z.

Another approach to finding the image of the strip is to first describe it by the inequalities

$$-\infty < x < 0, \qquad 0 < y < A$$

Then, from (2.3.26), we get

$$0 < |w| < 1, \qquad 0 < \arg(w) < A$$

which is a description of the region described above.

■

(2.3.27) ***Example.*** What is the image of the strip in the z-plane of Figure 2.3.2 under the mapping $w = e^{\bar{z}}$.

Solution. We will use this simple example to illustrate an important approach to mapping problems. We observe that w is composed of two operations—first conjugation of z and then exponentiation of the result. Hence, if we define an intermediate variable w_1 by

$$w_1 = \bar{z} \tag{2.3.28}$$

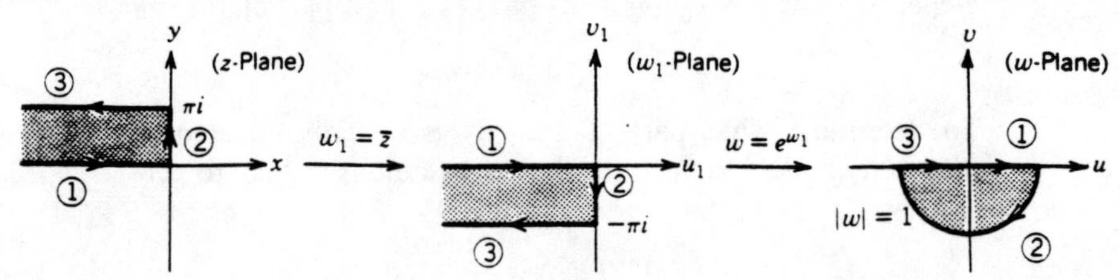

Figure 2.3.2 Example (2.3.27).

then,

(2.3.29)
$$w = e^{w_1}$$

Now we first map the strip in the z-plane into the w_1-plane using (2.3.28). The image is shown in the first part of the figure and is simply the reflection of the strip in the real axis. We then map this "w_1-strip" into the (final) w-plane using (2.3.29). This is a minor modification of the previous example and produces the half-disk shown in the w-plane. Hence, the **composite map** $e^{\bar{z}}$ takes the strip in the z-plane onto this half-disk. Note the direction of the arrows on the corresponding pieces of the boundaries. Compare this to the previous example.

■

PROBLEMS (The answers are given on page 422.)

1. Here, we show how we might be led to the definition of e^z using some well-known properties of real exponentials, namely,

$$e^x = \sum_{n=0}^{\infty} x^n/n! \quad \text{and} \quad e^{x_1+x_2} = e^{x_1} \cdot e^{x_2}.$$

   (a) "Formally" use the above infinite series to show that $e^{iy} = \cos y + i \sin y$.
   (b) "Formally" use (a) and the second property above to motivate the definition of e^z in (2.3.12).

2. (a) Is e^{iz} periodic in z? If so, what are the periods?
   (b) Answer the same question as in (a) for e^{z^2}.
   (c) Show that $e^{\bar{z}} = \overline{e^z}$.
   (d) Show that $(e^z)^n = e^{nz}$, $n = 0, \pm 1, \pm 2, \ldots$.

3. Write the following in the form $x + iy$.

   (a) e^i (b) e^{1+i} (c) $\dfrac{\exp(1 + 3\pi i)}{\exp[-1 + (\pi i/2)]}$ (d) e^{e^i}

4. Prove parts (c)–(g) of Theorem (2.3.13).

5. Compute the following limits, or indicate why they don't exist:

(a) $\lim_{z \to 1-i} (e^{\bar{z}^2})$ (b) $\lim_{z \to \infty} e^{-z^2}$ (c) $\lim_{z \to \infty} \left(\frac{e^z}{z}\right)$

(d) $\lim_{z \to \pi i/2} (e^z)$ (e) $\lim_{z \to e^i} (e^{\bar{z}})$ (f) $\lim_{z \to 0} \left(\frac{e^z - 1}{z}\right)$

[*Hint:* Find the real and imaginary parts of $z^{-1}(e^z - 1)$, and then use multidimensional Taylor series from advanced calculus.]

6. Find all values of z which satisfy the following:

(a) $e^z = i$ (b) $e^{z^2} = \frac{1}{\sqrt{2}}(1 + i)$ (c) $e^{1/z} = 1 + i$

(d) $e^z = 1 - i\sqrt{3}$ (e) $e^{3z} + i = 0$ (f) $e^{3z} + ie^z = 0$

(g) $e^{2z} + e^z + 1 = 0$ (h) $e^{\bar{z}} = \sqrt{3} + i$ (i) $e^z + e^{-z} = 0$

(j) $e^{e^z} = i$

7. (a) Where is e^{z^2} (i) imaginary? (ii) Real?
(b) Where is $|e^z|$ (i) <1? (ii) $= 1$? (iii) >1?

8. Show that in the strip $-\infty < x < \infty$, $y_0 < y \le y_0 + 2\pi$, the equation $e^z = w_0$ has a unique solution for any finite w_0.

9. In terms of the exponential, how can you represent points lying on a circle of radius R_0 centered at z_0.

10. Where are the following functions differentiable? Analytic? Compute the derivatives. (You may find it helpful to use the result of Problem 16b of Section 2.2.)

(a) e^{z^2} (b) $\bar{z}e^z$ (c) $\bar{z}e^{z^2}$

(d) $\exp(e^z)$ (e) $\frac{z}{e^z - 1}$ (f) $e^{1/z}$

(g) $\exp[(z-1)^{-2}]$ (h) $e^{\bar{z}}$ (i) $e^{\bar{z}^2}$

11. (a) Show that the **nth roots of unity** (see Problem 23 of Section 1.2) are of the form $e^{2\pi ik/n}$, $k = 0, 1, \ldots, n - 1$.
(b) If θ_0 is an argument of z_0, show that

$$z_0^{1/n} = |z_0|^{1/n} e^{i\theta_0/n} e^{2\pi ik/n} \qquad k = 0, 1, \ldots, n - 1$$

12. Expand the real and imaginary parts of e^z through quadratic terms in their **Taylor Series** about $x = 0$ and $y = 0$. Use these results to show that it is reasonable to expect that for $|z|$ small, $e^z \approx 1 + z + (z^2/2)$. (This subject will be taken up in great detail in Chapter 6.)

13. Use the inequalities (1.1.27) to show that

(a) $|1 + e^z| \le 1 + e^x$

(b) $$|(e^z + 1)^{-1}| \le \begin{cases} (e^x - 1)^{-1} & \text{Re}(z) > 0 \\ (1 - e^x)^{-1} & \text{Re}(z) < 0 \end{cases}$$

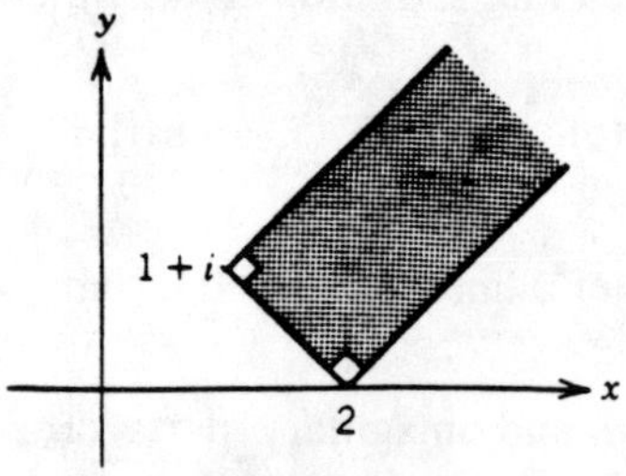

Figure 2.3.3

(c) $|z(e^z + 1)| \leq 1 + e,$ for z inside the unit disk

(d) $\left|\dfrac{z}{e^z + 4}\right| \leq \dfrac{1}{4 - e},$ for $|z| < 1$

14. (a) What is the image of the vertical line segment $x = x_0$, $y_0 < y < y$, under the mapping $w = e^z$?

(b) Use the result in (a) to generate the image of the strips of Example (2.3.25).

15. Find the images of the following sets of points under the given mapping:

(a) The strip $1 < x < (\pi/2) + 1$, $0 < y < \infty$ under $w = e^{i(z-1)}$.

(b) The rectangle $0 < x < x_0$, $0 < y < y_0 < 2\pi$ under $w = e^z$.

(c) The first quadrant under $w = e^z$.

(d) The half-line $y = x$, $x \geq 0$ under $w = e^z$.

(e) The half-strip shown in Figure 2.3.3 under $w = \exp[(1 + i)(z - 2)/\sqrt{2}]$.

2.4 LINEAR FUNCTIONS AND POSITIVE INTEGER POWERS

Here, we study the mapping properties of the **linear function**

(2.4.1) $$w = az + b \qquad a \neq 0 \qquad (a, b \text{ constant})$$

Note that w is entire. Also, if we write a in polar form,

$$a = |a|e^{i\theta_a} \qquad |a| > 0 \qquad [\theta_a = \arg(a)]$$

we can rewrite (2.4.1) as

$$w = e^{i\theta_a}|a|\left(z + \frac{b}{a}\right)$$

Hence, the general linear map is a composition of the three mappings

$$w_1 = z + \frac{b}{a}, \qquad w_2 = |a|w_1, \qquad w = e^{i\theta_a}w_2.$$

The first of these is a **translation** and has already been discussed in Example (2.1.10) (the reader is urged to reread this). The other two are discussed below.

(2.4.2) ***Example.*** Consider the special linear map $w = kz$, $k > 0$. If $z = |z|e^{i\theta}$, then

$$|w| = k|z| \quad \text{and} \quad \arg(w) = \theta \tag{2.4.3}$$

Thus, under the mapping, a point in the z-plane is moved *radially* to a new position $k|z|$ units from the origin, as shown in Figure 2.4.1a. Also, as in part (b) of the figure, which treats the case $k > 1$, two points z_0 and z_1, which are $|z_1 - z_0|$ units apart, are moved to the points kz_0 and kz_1, which are $k|z_1 - z_0|$ units apart. Hence, this mapping, which we will call a **magnification**, is either a **stretching** (if $k > 1$) or a **contraction** (if $k < 1$).

Now consider the line L_z: $y = c_1x + c_2$ shown in Figure 2.4.1c. Under the mapping $u + iv = k(x + iy)$, points on L_z are mapped onto the curve

$$u = kx, \qquad v = kc_1x + kc_2$$

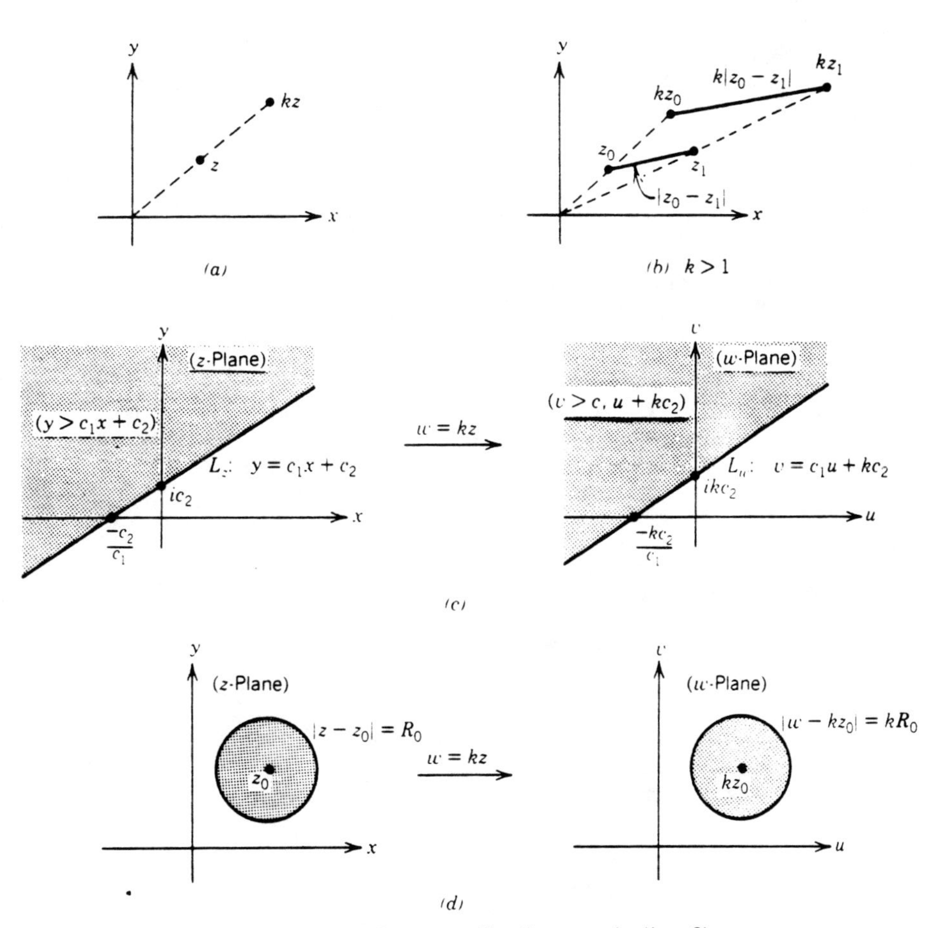

Figure 2.4.1 The magnification $w = kz$ ($k > 0$).

These points make up the line L_w shown in the figure and given by

$$L_w: \quad v = c_1 u + kc_2$$

This line is parallel to L_z, but its vertical intercept is changed from c_2 to kc_2. We also note that the half-plane

$$y > c_1 x + c_2$$

shown shaded in Figure 2.4.1c has as its image the half-plane

$$v > c_1 u + kc_2$$

Finally, consider the disk

(2.4.4) $$|z - z_0| \le R_0$$

The image of this region is

$$|w - kz_0| \le kR_0$$

This is another disk, but with its center moved radially to kz_0 and its radius stretched ($k > 1$) or contracted ($k < 1$) to kR_0. The situation is shown in Figure 2.4.1d.

■

(2.4.5) ***Example.*** Consider the special linear mapping

$$w = e^{i\theta_0} z$$

This is called a **rotation**, for if $z = |z| e^{i\theta}$, we have

(2.4.6) $$|w| = |z| \qquad \text{and} \qquad \arg(w) = \theta + \theta_0$$

Hence, points in the z-plane are simply *rotated about the origin* by an amount θ_0. Thus, any region in the z-plane will also be rotated about the origin, just as a rigid body would be. This is depicted in Figure 2.4.2.

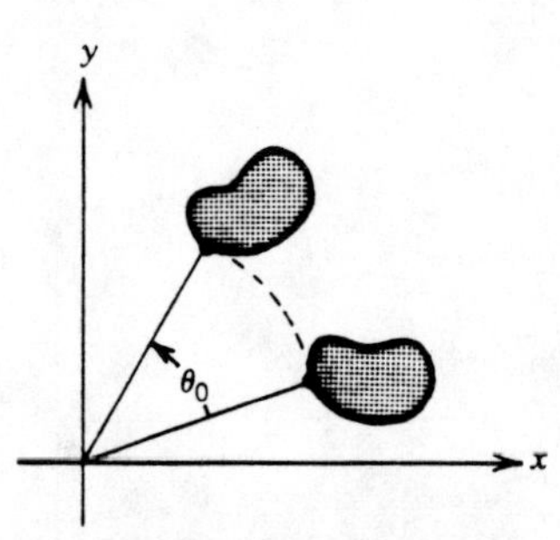

Figure 2.4.2 The rotation $w = e^{i\theta_0} z$.

■

Collecting the above results, we find that the general linear map (2.4.1) can be written as

(2.4.7)
$$w = e^{i\theta_a}|a|\left(z + \frac{b}{a}\right), \qquad \theta_a = \arg(a)$$

and is composed of

(2.4.8)

First a **translation** $w_1 = z + \dfrac{b}{a}$

Then a **magnification** $w_2 = |a|w_1$

Then a **rotation** $w = e^{i\theta_a} w_2$

(2.4.9) ***Example.*** Find the image of the unit disk $|z| < 1$ under the linear mapping $w = (1 + i)z + 2i$.

Solution. We can do this by simply solving for z and substituting into the inequality $|z| < 1$. Instead, we will follow the development of (2.4.8) above, since this is a procedure which is quite useful in mapping problems. We first factor out $1 + i$ and write the mapping in the form

(2.4.10)
$$w = e^{\pi i/4}\sqrt{2}[z + (i + 1)]$$

From this, the unit disk $|z| < 1$ is first *translated* under the mapping

$$w_1 = z + (i + 1)$$

into a disk in the w_1-plane of radius 1 and center at $w_1 = i + 1$. This disk is then *magnified* under

$$w_2 = \sqrt{2}w_1$$

into a disk in the w_2-plane of radius $\sqrt{2}$ and center at $w_2 = \sqrt{2}(1 + i)$. Finally, the rotation

$$w = e^{\pi i/4} w_2$$

produces a disk in the w-plane. This series of mappings is shown in Figure 2.4.3 (page 58).

■

In the above, we might also have found the image of the unit disk using the property that a *linear map takes a disk in the z-plane onto another disk in the w-plane, with the centers being images of each other*. Hence, we merely have to find the images under $w = (1 + i)z + 2i$, of the center $z = 0$ and any other point on the unit disk, say, $z = i$. Since these images are $w = 2i$ (which will be the new center) and $w = 3i - 1$, the image disk will be

$$|w - 2i| = |3i - 1 - 2i| = \sqrt{2}$$

Can you think of another way to do this? [*Hint:* Three points determine a circle.]

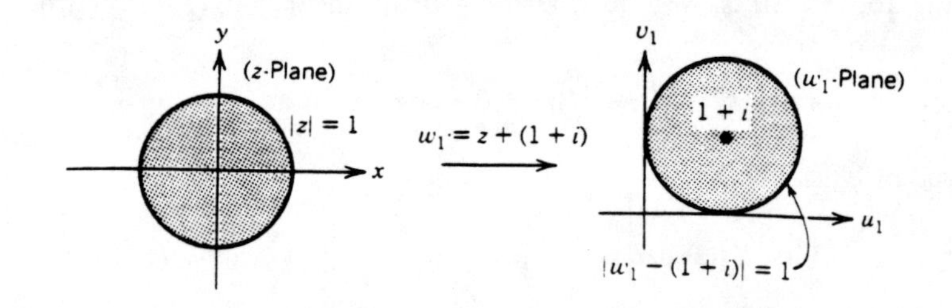

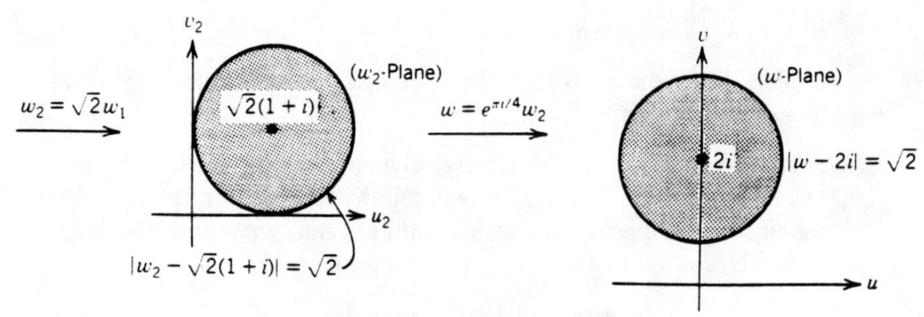

Figure 2.4.3 Example (2.4.9).

The next example shows how we can combine our knowledge of some of the mapping properties of exponential and linear functions in order to deduce the properties of somewhat more complicated mappings.

(2.4.11) ***Example.*** Under the mapping $w = -2(e^{z-1} + i)$, find the image of the strip shown in Figure 2.4.4.

Solution. We first rewrite w as $w = 2e^{\pi i}(e^{z-1} + i)$ and then note that it is

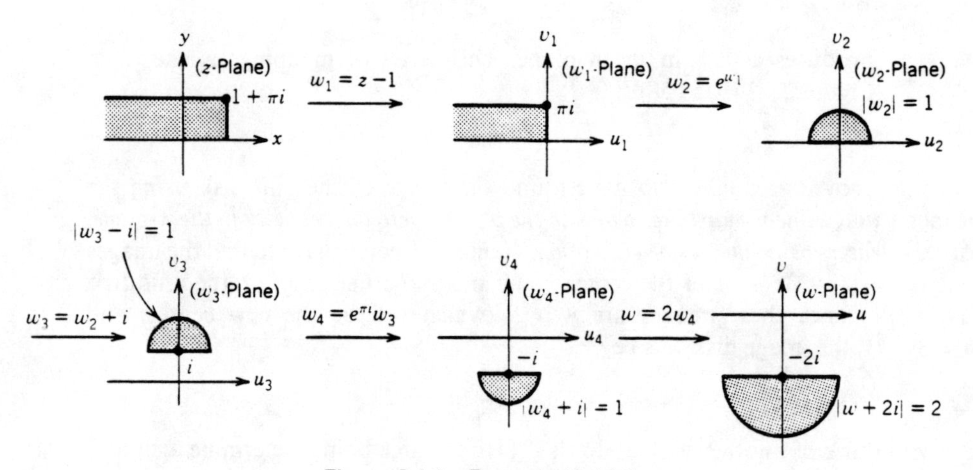

Figure 2.4.4 Example (2.4.11).

composed of the

Translation	$w_1 = z - 1$
Exponentiation	$w_2 = e^{w_1}$
Translation	$w_3 = w_2 + i$
Rotation	$w_4 = e^{\pi i} w_3$
Magnification	$w = 2w_4$

The properties of each one of these has already been discussed, and the sequence of mappings is shown in the figure, with the final result being the half-disk in the w-plane.

■

Positive Integer Powers. Here, we will consider the mapping properties of the entire functions

(2.4.12) $$w = z^n, \qquad n = 2, 3, 4, \ldots$$

(For $n = 1$, we have a linear function which has already been discussed.)

(2.4.13) ***Example.*** Under the mapping $w = z^2$, find the image of the wedge shown in the z-plane in Figure 2.4.5*a*.

Solution. First describe the wedge in polar coordinates by

$$0 < |z| < \infty \quad \text{and} \quad 0 < \theta < \theta_0 \qquad [\theta = \arg(z)]$$

Then, since $w = z^2 = |z|^2 e^{2i\theta}$, we have

$$|w| = |z|^2, \qquad \arg(w) = 2\theta$$

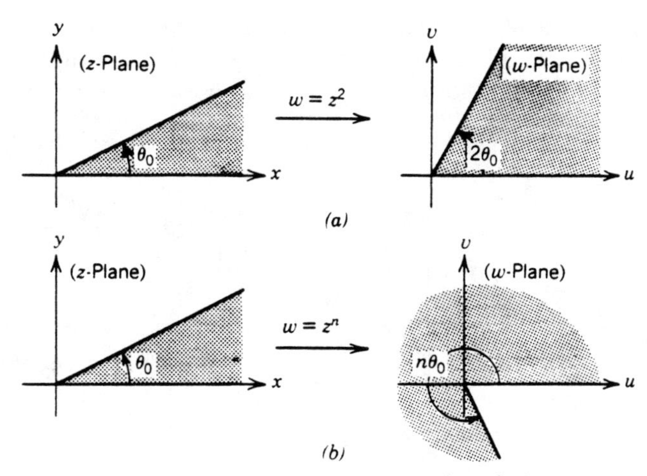

Figure 2.4.5 Mapping properties of z^n.

Thus, from the inequalities on z, we get

$$0 < |w| < \infty \qquad \text{and} \qquad 0 < \arg(w) < 2\theta_0$$

Hence, the image is another wedge in the w-plane, but with double the angle, as shown in Figure 2.4.5*a*. Note that if $\theta_0 > \pi$, the image is the whole w-plane without the origin, a piece of which is covered twice.

■

The reader should observe that the general power map $w = z^n$ will transform a wedge with vertex at the origin into another wedge with vertex at the origin but with its angle multiplied n times (see Figure 2.4.5*b*). In Problem 8, the reader is asked to show that this result need not be true if the vertex of the wedge is not at the origin.

PROBLEMS (The answers are given on page 423.)

1. Under the magnification $w = kz$, $k > 0$, what is the image of
   (a) The ellipse $x^2 + (y^2/4) = 1$?
   (b) The rectangular region $x_0 \le x \le x_1$, $y_0 \le y \le y_1$?
   (c) The region $\text{Re}(z) > 1$, $\text{Im}(z) > 1$, $|z| < 2$?
2. Find the image of the prescribed regions under the given mappings. [*Hint:* View each as a composite map.]
   (a) The strip $1 < \text{Re}(z) < 1 + (\pi/2)$ under $w = e^{2i(z-1)}$.
   (b) The strip $\text{Re}(z) < 0$, $0 < \text{Im}(z) < \pi$ under $w = e^{-z} + 1$.
   (c) The disk $|z - i| < 2$ under $w = 2i(z + 1)$.
   (d) The half-plane $x - y < 1$ under $w = [(1 + i)/\sqrt{2}(z + i)] + i$.
   (e) The strip shown in Figure 2.4.6*a* under $w = \exp\left[\dfrac{(1 + i)(z - 1 - i)}{\sqrt{2}}\right]$.
3. Describe geometrically the effect of the transformation $w = e^{i\theta_0}(z - \alpha) + \alpha$.
4. (a) Find the most general linear map $w = az + b$ mapping the circle C_z: $|z - z_0| = r_0$ onto the circle C_w: $|w - w_0| = \rho_0$.

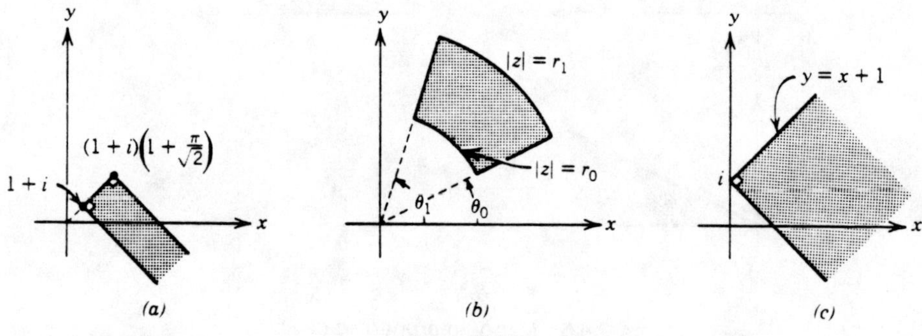

Figure 2.4.6

(b) Repeat part (a), but now require in addition that the image of the point $z = z_1$ be $w = w_1$, where z_1 is on C_z and w_1 is on C_w.

(c) The same as in (b), but now also require that the image of z_2 on C_z be w_2 on C_w.

5. Find a transformation mapping the half-plane $x + y > 1$ onto the half-plane $u - v < 1$. [*Hint:* Map the boundaries onto each other.]
6. When does there exist a linear function mapping a given rectangle in the z-plane onto another given rectangle in the w-plane?
7. Under $w = z^2$, find the image of
   (a) The quarter disk $|z| < 4$, $\text{Re}(z) > 0$, $\text{Im}(z) > 0$.
   (b) The annular sector shown in Figure 2.4.6*b*.
8. What is the image of the quarter plane $\text{Re}(z) > 1$, $\text{Im}(z) > 0$ under $w = z^2$? This result shows that a wedge whose vertex is not at the origin will in general not transform onto another wedge. [*Hint:* Solve for x and y in terms of u and v.]
9. Use the result of Problem 8 to find the image of the strip $0 < \text{Re}(z) < 1$, $\text{Im}(z) > 0$ under $w = z^2$. [*Hint:* The strip is the "difference" of two wedges.]
10. Find the image of the wedge shown in Figure 2.4.6*c* under each of the following (view as composite maps):
    (a) $w = i(z - i)^2$ (b) $w = i(\bar{z} + i)^2$ (c) $w = -i(z - i)^3$
11. (a) Show that the general quadratic map $w = az^2 + bz + c$ $(a \neq 0)$ can be written as a composition of a power map and linear maps. [*Hint:* Complete the squares.]
    (b) Find the image of the quarter plane $\text{Re}(z) > 1$, $\text{Im}(z) > 0$ under $w = z^2 - 2z + i$.
12. (a) When can the general cubic $w = z^3 + az^2 + bz + c$ be written as a composition of a power map and linear maps?
    (b) Can you generalize the result in (a) to the polynomial $w = z^n + a_{n-1}z^{n-1} + \cdots + a_1 z + a_0$?

2.5 INVERSIONS AND BILINEAR FUNCTIONS (OPTIONAL AT THIS TIME[1])

The mapping

$$w = \frac{1}{z} \tag{2.5.1}$$

is called an **inversion**. Since

$$|w| = \frac{1}{|z|} \quad \text{and} \quad \arg(w) = -\arg(z) \tag{2.5.2}$$

[1] The reader can delay the reading of this section until after Section 4.4.

we see that the exterior of the unit disk $|z| > 1$ is mapped onto the unit disk $|w| < 1$, while the unit disk $|z| < 1$ is mapped onto $|w| > 1$. The unit circle $|z| = 1$ itself remains invariant. The point $z^* = 1/\bar{z}$ is called the **inverse point of z with respect to the unit circle**. Inverse points with respect to general circles will be discussed in Problems 1, 2, and 3.

(2.5.3) ***Example.*** Show that under the inversion $w = 1/z$, the image of the disk D_z with boundary C_z, described by

(2.5.4)
$$D_z:\ |z - z_0| < R_0 \qquad \text{and} \qquad C_z:\ |z - z_0| = R_0$$

is

(a) A disk if $z = 0$ lies outside of D_z.
(b) A half-plane if $z = 0$ lies on the boundary C_z.
(c) Or the exterior of a disk if $z = 0$ lies inside of D_z.

In addition, show that the image of C_z is the boundary of the image of D_z and hence is another circle if $z = 0$ does not lie on C_z, or a line if $z = 0$ lies on C_z.

Solution. Setting $z = 1/w$ in (2.5.4), we find that the image of the disk is described by

$$|1 - z_0 w| \le R_0 |w|$$

with equality giving the image of the circle C_z. With $w = u + iv$ and $z_0 = x_0 + iy_0$, the above yields (after squaring).

(2.5.5)
$$(R_0^2 - |z_0|^2)(u^2 + v^2) + 2(x_0 u - y_0 v) \ge 1$$

Note from Figure 2.5.1 that $R_0^2 - |z_0|^2$, the coefficient of $u^2 + v^2$, is

(2.5.6)
(i) Negative if and only if $z = 0$ lies outside of D_z.
(ii) Zero if and only if $z = 0$ lies on C_z.
(iii) Positive if and only if $z = 0$ lies inside of D_z.

For $R_0^2 < |z_0|^2$, we first divide both sides of (2.5.5) by the negative quantity $R_0^2 - |z_0|^2$, hence changing the sense of the inequality, and then complete the squares in u and v to get

(2.5.7)
$$\left| w - \frac{\bar{z}_0}{|z_0|^2 - R_0^2} \right| \le \frac{R_0}{|z_0|^2 - R_0^2} \qquad \text{for } R_0^2 < |z_0|^2$$

Thus, the circle C_z is mapped onto a circle C_w given by the equality above, while the disk D_z is mapped onto a disk D_w described by the inequality. This, together with (2.5.6i), yields the result (2.5.4*a*). Also, as $R_0^2 - |z_0|^2$ gets smaller and smaller through negative values, the center of the disk in (2.5.7) moves toward $w = \infty$ while its radius gets larger and larger. (For $R_0^2 = |z_0|^2$ the image of C_z will be a line, as we will see below.) This

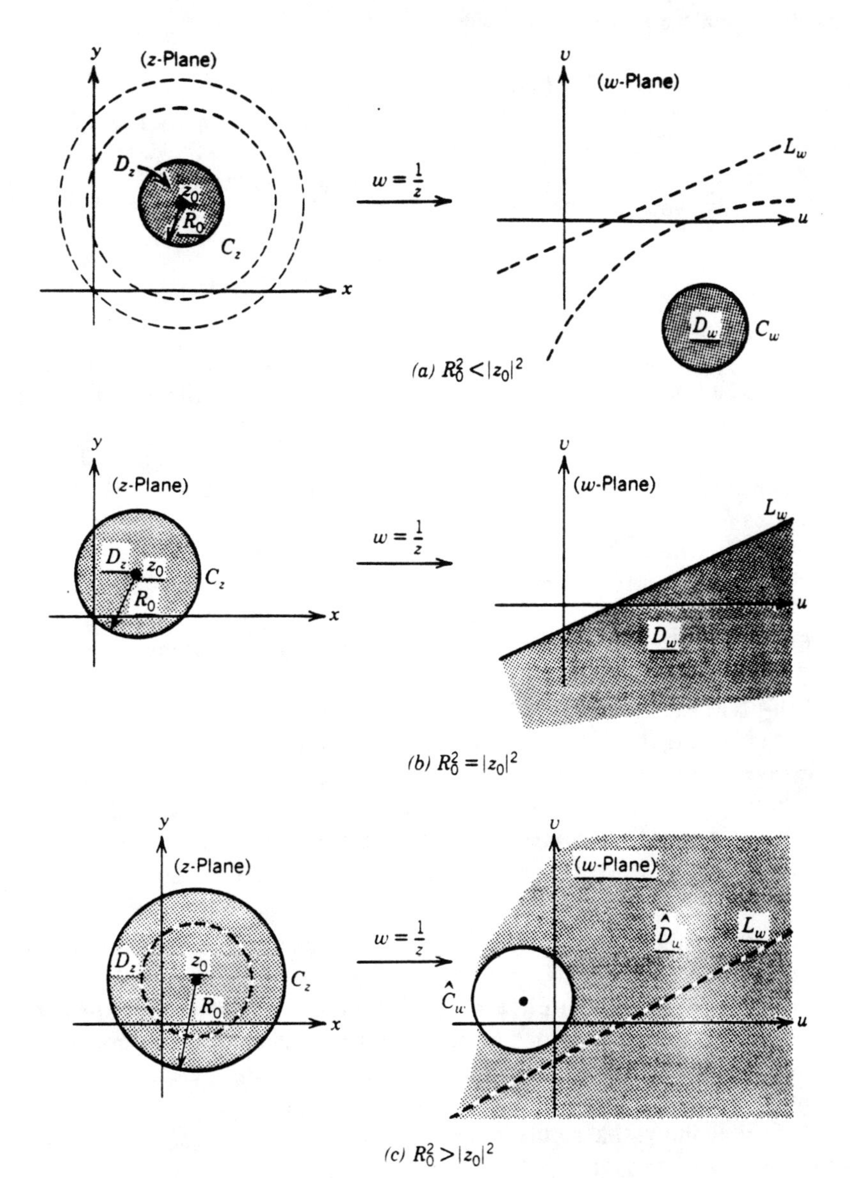

Figure 2.5.1 Mapping properties of the inversion $w = 1/z$.

situation is depicted in Figure 2.5.1*a* while z_0 is chosen to lie in the first quadrant. Note here that the dashed circle closest to C_z in the z-plane corresponds to a small value of $R_0^2 - |z_0|^2$ and has as its image the (large) dashed circle in the w-plane.

For $R_0^2 = |z_0|^2$, (2.5.5) becomes $x_0 u - y_0 v \geq 1/2$, which describes the half-plane D_w shown shaded in Figure 2.5.1*b*. This, together with (2.5.6ii), verifies part (b) of 2.5.4. Also, the circle C_z has as its image the line L_w given by equality above. (The reader should verify that this line lies above all the circles C_w in Figure 2.5.1*a*. In this figure, L_w is shown as the dashed image of the dashed circle passing through the origin in the *z*-plane.)

Finally, when $R_0^2 > |z_0|^2$, (2.5.5) yields

(2.5.8)
$$\left| w + \frac{\bar{z}_0}{R_0^2 - |z_0|^2} \right| \geq \frac{R_0}{R_0^2 - |z_0|^2} \qquad \text{for } R_0^2 > |z_0|^2$$

Hence, the circle C_z is again mapped onto a circle $\hat{C}_w$ (given by equality above), but now the disk D_z is mapped onto $\hat{D}_w$, which is the outside of the disk bounded by $\hat{C}_w$, thus verifying part (c) of (2.5.4). This is depicted in Figure 2.5.1*c* where the line L_w is also shown as the image of the dashed circle through the origin in the *z*-plane. (Here, L_w lies *below* all of the circles $\hat{C}_w$.)

Geometrically, what is happening is very interesting. As can be seen from the sequence of diagrams in Figure 2.5.1, as the circles C_z in the *z*-plane get closer to the origin, the image circles C_w get larger in radius until finally they merge into the line L_w. The disks D_z, bounded by C_z, have as their images larger and larger disks in the *w*-plane which evolve into the half-plane shown in Figure 2.5.1*b*. As the circles C_z pass through the origin and enclose it, the line L_w wraps itself around the origin in the *w*-plane and evolves into the circles $\hat{C}_w$ shown in Figure 2.5.1*c*. This "wrapping around" has the effect of taking the half-plane of Figure 2.5.1*b* and converting it into the exterior of the disk in Figure 2.5.1*c*.

■

Let us make one further remark concerning the above example. As we showed, the image of the disk depended largely upon the location of the origin $z = 0$. Note also that the inversion $w = 1/z$ is infinite at $z = 0$. From this last fact we would expect the image of the disk D_z to be a bounded set if $z = 0$ is not inside or on its boundary, while it should be an unbounded set if the opposite is true. This is clearly indicated by our results.

The next example shows that similar results hold when one considers the mapping of a half-plane by an inversion.

(2.5.9) ***Example.*** Show that under the inversion $w = 1/z$, the image of the half-plane D_z given by

(2.5.10)
$$D_z\colon\; ax + by \geq c$$

is

(a) A disk, if $z = 0$ lies outside of D_z.
(b) A half-plane if $z = 0$ lies on the boundary of D_z.
(c) Or the exterior of a disk if $z = 0$ lies inside of D_z.

Solution. From $z = 1/w$, we have

$$x = \frac{u}{u^2 + v^2} \quad \text{and} \quad y = \frac{-v}{u^2 + v^2}$$

Hence, the image of D_z is described by

$$c(u^2 + v^2) - au + bv \leq 0 \tag{2.5.11}$$

First, note from (2.5.10) that $z = 0$

(i) Lies outside of D_z if and only if $c > 0$.
(ii) Lies on the boundary of D_z if and only if $c = 0$.
(iii) Lies inside of D_z if and only if $c < 0$.

If $c > 0$, we can divide (2.5.11) by c and not change the sense of the inequality. If we then complete the squares (as in the previous example), we find that this inequality describes a disk. Hence, with (i) above, we get the result (a). If $c = 0$, then (2.5.11) describes a half-plane, thus verifying (b). We leave it to the reader to prove (c).

Finally, note that equality in (2.5.10) yields equality in (2.5.11), thus showing that the boundary of D_z will map onto the boundary of its image.

■

(2.5.12) ***Example.*** What is the image of the disk D_z: $|z - i| < 2$ bounded by the circle C_z: $|z - i| = 2$ under the mapping

$$w = \frac{1 + i}{z + i} \tag{2.5.13}$$

Solution. We first write (2.5.13) as the composition of the

Translation	$w_1 = z + i$
Inversion	$w_2 = 1/w_1$
Rotation and magnification	$w = (1 + i)w_2$

The translation will produce a disk D_{w_1} in the w_1-plane. The inversion, by Example (2.5.3), will map this new disk onto either the interior of a disk, exterior of a disk, or half-plane, depending upon whether $w_1 = 0$ lies outside, inside, or on the boundary of the disk D_{w_1} respectively. Finally, the rotation and magnification will produce a similar figure in the w-plane. Noting from the above that $w_1 = 0$ when $z = -i$, we have that the image of D_z is

(i) A disk if $z = -i$ is outside of D_z.
(ii) The exterior of a disk if $z = -i$ is in D_z.
(iii) A half-plane if $z = -i$ is on C_z.

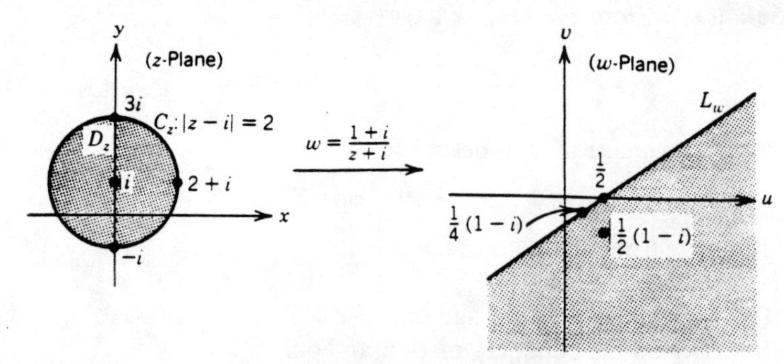

Figure 2.5.2 Example (2.5.12).

Here, $z = -i$ is on the circle C_z (see Figure 2.5.2), and hence the image of D_z is a half-plane whose boundary line L_w must be the image of the circle C_z. To find the line, we need only find the image, under the mapping (2.5.13), of any two points on C_z. Choosing $z = 3i$ and $z = 2 + i$, we find that L_w is the line joining the points $w = \frac{1}{4}(1 - i)$ and $\frac{1}{2}$. This is shown in Figure 2.5.2. To determine which of the half-planes bounded by L_w is the image of D_z, we need only determine the image of a single point inside of D_z. Hence, since the image of $z = i$, the center of D_z, is $w = \frac{1}{2}(1 - i)$, the image of D_z is the half-plane shown shaded in the figure.

■

Bilinear Functions. The previous examples showed that the inversion $w = 1/z$ maps lines and circles into lines or circles and the corresponding half-planes and disks into either half-planes, disks, or exteriors of disks. The situation which prevails depends upon whether the value of z which makes w infinite, in this case $z = 0$, lies inside or outside the boundary of the region to be mapped. The next example shows how we can use this information to construct a function which performs a pre-assigned mapping task, and will lead us into bilinear mappings.

(2.5.14) ***Example.*** Find a function $w = f(z)$ which maps the upper half-plane $\text{Im}(z) \geq 0$ onto the unit disk $|w| \leq 1$.

Solution. Our procedure will be to

(a) Map the half-plane onto a disk using an appropriate inversion map.
(b) Expand or contract the disk so that its radius is 1.
(c) Translate the new disk so that its center is at the origin.

First of all, we now know that an inversion is a likely candidate to map the half-plane onto the disk. However, since $z = 0$ lies on the boundary of the half-plane, the inversion $w = 1/z$ will just produce another half-plane. Thus, guided by the results of Example (2.5.9), we will first translate the upper half-plane so that the origin does not lie inside of the translated half-plane. One such translation (there are obviously many, and hence *the mapping we are looking for is not unique*) is

(2.5.15) $$w_1 = z + i$$

This produces the half-plane shown in the w_1-plane of Figure 2.5.3a. Now, the inversion

(2.5.16) $$w_2 = \frac{1}{w_1}$$

maps this half-plane onto a disk. (Why?) The circle bounding this disk is the image of the line $\text{Im}(w_1) = 1$ under (2.5.16). Since three points determine a circle, and since $z = i, i + 1, \infty$ lie on this line, the circle we are looking for is that one passing through $w_2 = -i, \frac{1}{2}(1 - i)$, and 0 and is shown in Figure 2.5.3a. Since the radius of this disk is $\frac{1}{2}$, the stretching

(2.5.17) $$w_3 = 2w_2$$

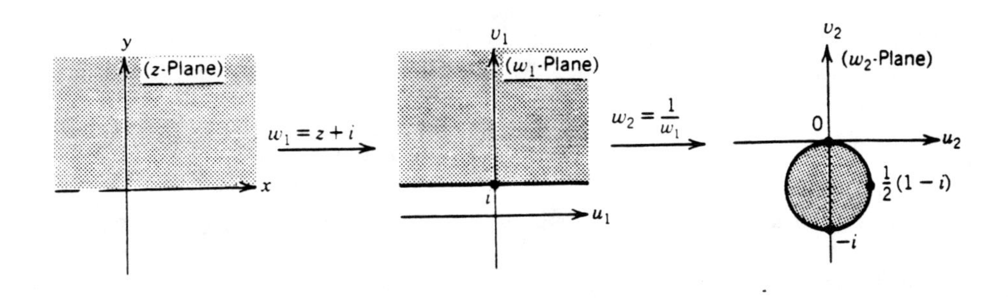

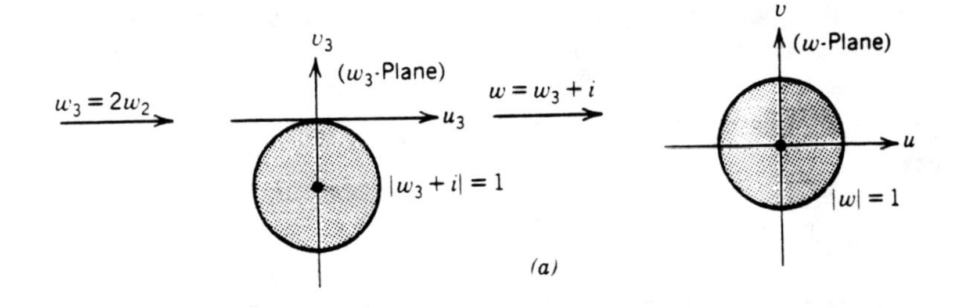

(a)

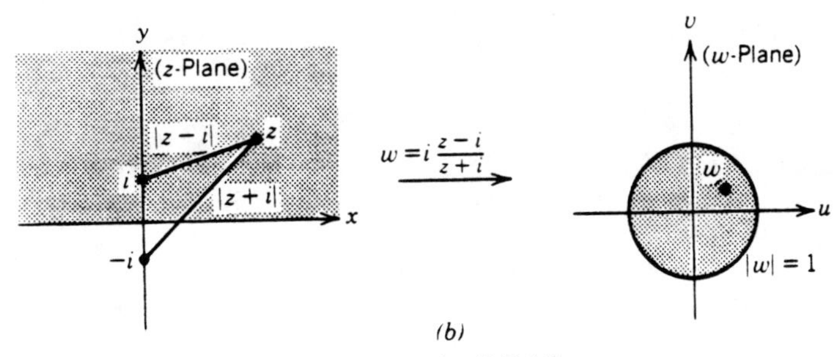

(b)

Figure 2.5.3 Example (2.5.14).

will produce a disk in the w_3-plane of radius 1 but with center at $-i$. Finally, the translation

(2.5.18)
$$w = w_3 + i$$

produces the unit disk in the w-plane. This sequence of mappings is shown in Figure 2.5.3a. Combining (2.5.15) through (2.5.18), we find that

(2.5.19)
$$w = i\left(\frac{z - i}{z + i}\right)$$

maps the upper half-plane $\text{Im}(z) \geq 0$ onto the unit disk $|w| \leq 1$. This mapping is a special case of the more general bilinear transformation to be discussed below.

Note from Figure 2.5.3b that for $\text{Im}(z) > 0, |z - i| < |z + i|$, and hence $|w| < 1$ from (2.5.19). Also, for z on the boundary of the half-plane, that is, on the real axis, $|z - i| = |z + i|$, and thus $|w| = 1$.

■

The general **bilinear transformation**, or **linear fractional transformation**, is defined by

(2.5.20)
$$w = \frac{az + b}{cz + d}, \qquad ad - bc \neq 0$$

where a, b, c, and d are complex constants. (Note that this is a ratio of linear functions, and hence the name bilinear.) Let us show that this transformation is composed of inversions and linear maps, and in the process show why the condition $ad - bc \neq 0$ is needed.

First, note that both c and d cannot be zero, since then w is not defined (this is excluded by the condition $ad - bc \neq 0$). Also note that if either $a = c = 0$, $b = d = 0$, or $a = b = 0$ (all of which are excluded by $ad - bc \neq 0$), then w is constant—a not very interesting mapping. Now, if $c = 0$ and $d \neq 0$, (2.5.20) becomes

$$w = \left(\frac{a}{d}\right)z + \frac{b}{d}$$

which is just our old friend the linear map. Also, if $a = 0, b \neq 0$, and $c \neq 0$, we have

$$w = \left(\frac{b}{c}\right)\left(\frac{1}{z + \frac{d}{c}}\right)$$

which is composed of a translation $w_1 = z + (d/c)$, an inversion $w_2 = 1/w_1$, and a rotation and magnification $w = (b/c)w_2$. Finally, if both $a \neq 0$ and $c \neq 0$, we can write (2.5.20) as

(2.5.21)
$$w = \left(\frac{a}{c}\right)\left[1 + \left(\frac{bc - ad}{ac}\right)\left(\frac{1}{z + \frac{d}{c}}\right)\right]$$

which consists of an inversion and linear maps. Also note in the above that if $ad = bc$, w again reduces to the uninteresting constant. Hence, the condition $ad \neq bc$ in (2.5.20).

(2.5.22) ***Theorem.*** The bilinear transformation (2.5.20), with $c \neq 0$, has the following properties:

(a) It is analytic everywhere except at $z = -d/c$, where $w = \infty$.

(b) It is invertible (that is, it can be solved for z) everywhere except in a neighborhood of $z = -d/c$. The inverse transformation is bilinear in w and given by $z = (-dw + b)/(cw - a)$

(c) The composition of two bilinear functions is again a bilinear function.

(d) A circle C_z in the z-plane will transform under (2.5.20) into either a line L_w or circle C_w in the w-plane, depending respectively upon whether the point $z = -d/c$ (where $w = \infty$) lies on C_z or not. Also, the image of the disk D_z bounded by C_z will be

(i) A disk bounded by C_w if $z = -d/c$ lies outside of D_z (Figure 2.5.4*a*).

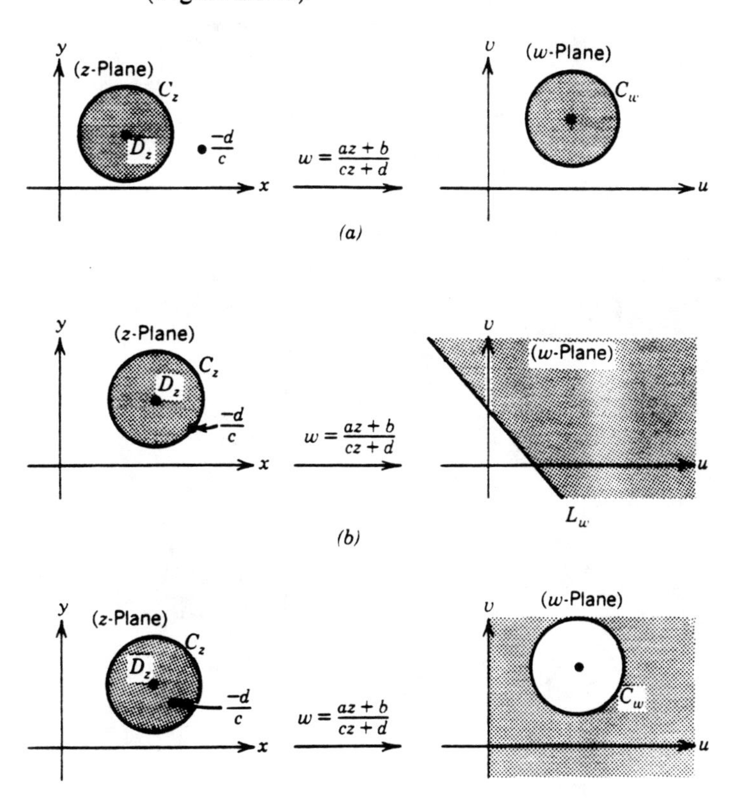

Figure 2.5.4 Mapping of a disk and circle under a bilinear map.

(2.5.23)

(ii) A half-plane bounded by L_w if $z = -d/c$ lies on C_z (Figure 2.5.4*b*).

(iii) The exterior of the disk bounded by C_w if $z = -d/c$ lies inside of C_z (Figure 2.5.4*c*).

(e) A line L_z in the z-plane will transform into either a line L_w or circle C_w depending respectively upon whether $z = -d/c$ lies on L_z or not. In addition, the image of a half-plane D_z bounded by L_z will be

(i) A disk bounded by C_w if $z = -d/c$ lies outside of D_z (Figure 2.5.5*a*).

(2.5.24)

(ii) A half-plane bounded by L_w if $z = -d/c$ lies on L_z (Figure 2.5.5*b*).

(iii) The exterior of the disk bounded by C_w if $z = -d/c$ lies inside of D_z (Figure 2.5.5*c*).

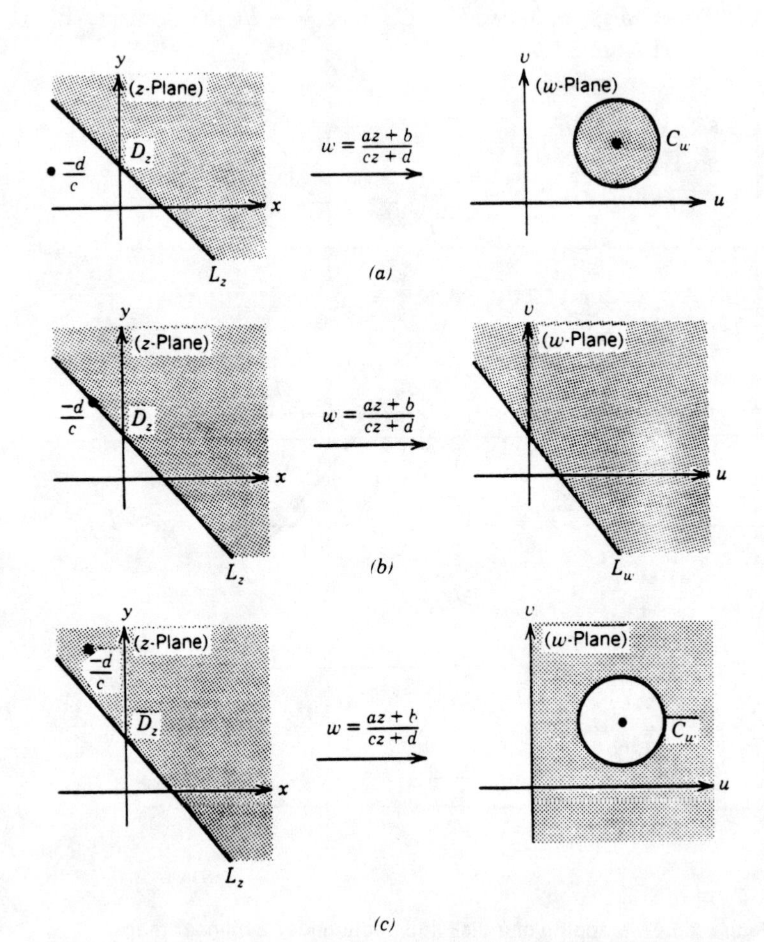

Figure 2.5.5 Mapping of a half-plane and line under a bilinear map.

Proof. The proofs of these results are rather immediate. Statement (a) follows directly from (2.5.20), while (b) follows by simply solving for z in terms of w in this same equation. To produce the result in (c), simply substitute the bilinear transformation

$$z = \frac{\alpha s + \beta}{\gamma s + \delta}, \qquad \alpha\delta \neq \beta\gamma$$

into (2.5.20) and verify that the outcome is again a bilinear transformation in s. Finally, (d) and (e) follow from the fact that the general bilinear function is a composition of an inversion and linear functions, as was exhibited above. The linear mappings do not change the shapes of lines, circles, disks, and half-planes, while the inversion has the properties noted in (d) and (e).

■

(2.5.25) ***Example.*** What is the image of the half-plane $x + y > 0$ under the bilinear mapping $w = (z - 1)/(z + i)$?

Solution. Since $z = -i$ is neither in, nor on, the boundary of the half-plane (see Figure 2.5.6), its image must be a disk whose circular boundary is the image of the line $x + y = 0$. Now, $z = 0, 1 - i, \infty$ all lie on the line $x + y = 0$ and have images $w = i, -i$, and 1, respectively. Hence, the boundary of the disk is therefore the unit circle (why?), and thus the image of the half-plane is the unit disk.

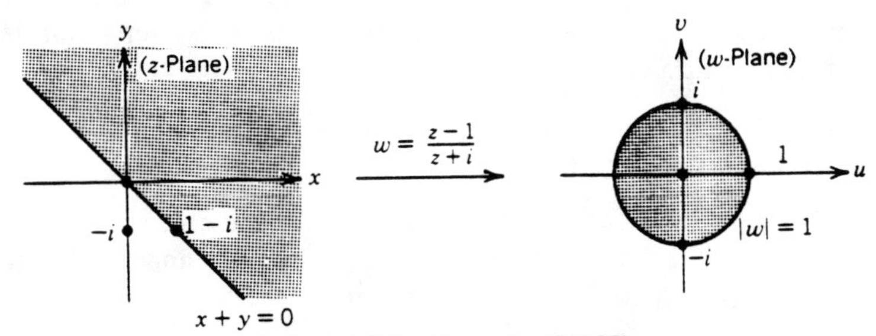

Figure 2.5.6 Example (2.5.25).

■

(2.5.26) ***Example.*** Find the image of the half-disk shown in Figure 2.5.7 under the bilinear mapping $w = z/(z + 1)$.

Solution. This half-disk is the intersection of the unit disk $|z| < 1$ with the upper half-plane $\text{Im}(z) > 0$, and hence its image will be the intersection of the images of these two regions. Since $w = \infty$ when $z = -1$, and $z = -1$ lies on the boundary of both the unit disk and the upper half-plane, the images of both these regions will be half-planes. According to

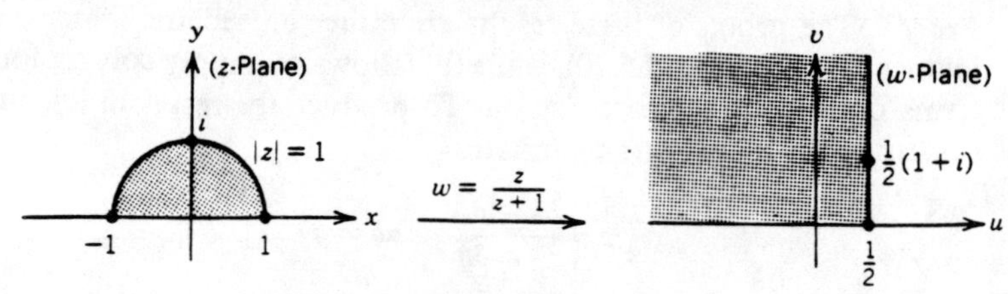

Figure 2.5.7 Example (2.5.26).

this, to compute the image of the unit disk, for example, we have only to find the images of two points on the unit circle and a point inside the disk. Now, the images of $z = 1$ and $z = i$ are $w = 1/2$ and $w = 1/2(1 + i)$, respectively. Thus, the unit disk is mapped onto a half-plane bounded by the line joining these two w-points. To determine this half-plane, we need only compute the image of any point inside the unit disk. Since $z = 0$ implies that $w = 0$, the image of the unit disk is the half-plane $\mathrm{Re}(w) < \frac{1}{2}$.

A similar analysis shows that the upper half of the z-plane is mapped onto the upper half of the w-plane. Hence, the image of the half-disk is the intersection of these two half-planes and is the quarter plane shown in Figure 2.5.7.

■

Since, in general, circles are mapped onto circles under bilinear maps (a line being considered a circle with infinite radius), and since three points determine a circle, we should expect to be able to find a bilinear map mapping three given points in the z-plane into three given points in the w-plane.

(2.5.27) ***Example.*** Find a bilinear map mapping the points $z = 1, i, 2 + i$ into the points $w = i, 1, \infty$ in this order.

Solution. The objective is to find values of a, b, c, and d in (2.5.20) so that when $z = 1$, i, and $2 + i$, w is respectively i, 1, and ∞. These three conditions then yield the three equations

$$\frac{a+b}{c+d} = i, \qquad \frac{ia+b}{ic+d} = 1, \qquad (2+i)c + d = 0$$

which has solutions

$$a = (2+i)c, \qquad b = -(2i+1)c, \qquad d = -(2+i)c$$

When these are substituted back into (2.5.20) the common factor of c will cancel, and the bilinear map we are looking for is

(2.5.28)
$$w = \frac{(2+i)z - (2i+1)}{z - (2+i)}$$

■

The reader is asked to show that the *unique* bilinear transformation which takes the points z_1, z_2, z_3 into w_1, w_2, w_3, respectively, is given by solving for w in

(2.5.29)
$$\frac{(w - w_1)(w_2 - w_3)}{(w - w_3)(w_2 - w_1)} = \frac{(z - z_1)(z_2 - z_3)}{(z - z_3)(z_2 - z_1)}$$

If any of the z_k or w_k $(k = 1, 2, 3)$ are infinite, the above formula still holds in the appropriate limit. For example, in the previous example we had

$$z_1 = 1, \qquad z_2 = i, \qquad z_3 = 2 + i,$$
$$w_1 = i, \qquad w_2 = 1, \qquad w_3 = \infty$$

Hence, (2.5.29) becomes, after taking the limit of the left side as $w_3 \to \infty$,

$$\frac{w - i}{1 - i} = \frac{(z - 1)(-2)}{[z - (2 + i)](i - 1)}$$

which, after solving for w, yields (2.5.28).

(2.5.30) ***Example.*** Find a bilinear (but *not* linear) map taking the upper half of the z-plane onto the right half of the w-plane.

Solution. Clearly, a simple rotation $w = -iz$ will do the job, but this is a linear map. To find what we want, we note that any bilinear map doing the job must also map the boundaries of the half-plane onto each other. Hence, we first seek a bilinear map

(2.5.31)
$$w_1 = \frac{az + b}{cz + d}$$

mapping the real z-axis onto the imaginary w_1-axis. To find a, b, c, and d, we choose three real values of z, and then three imaginary values of w_1 as their images under the above map. For example, if we choose

$$z = 0 \to w_1 = 0$$
$$z = 1 \to w_1 = i$$
$$z = \infty \to w_1 = 2i$$

we find

(2.5.32)
$$w_1 = \frac{2iz}{z + 1}$$

We leave it to the reader to verify that this maps the whole real z-axis onto the whole imaginary axis. Hence, the image of the upper half of the z-plane under this mapping is either the right or the left half-plane. To see which, note that $z = i$ (which is in the upper half-plane) maps into $w_1 = -1 + i$ (which is in the left half-plane). Thus, (2.5.32) produces the left, and not the right, half-plane. A final rotation $w = e^{i\pi}w_1 = -w_1$ will do the job, and we get

$$w = \frac{-2iz}{z+1}$$

■

PROBLEMS (The answers are given on page 424.)

1. The following construction leads to a definition of an **inverse point with respect to a general circle C_0: $|z - z_0| = R_0$.**
   (a) Show that the circle C_0 is transformed into the unit circle under the mapping $w = (1/R_0)(z - z_0)$.
   (b) Let w be the image of a point z under the mapping in (a). In terms of z, what is the inverse point w^* of w with respect to the unit circle in the w-plane?
   (c) Invert the mapping in (a) and solve for z. Define z^*, the inverse point of z with respect to C_0, to be the image of w^* under this inverse mapping. What is z^* in terms of z (see Figure 2.5.8a)?
2. Here, we will examine some interesting properties of the inverse point defined as in Problem 1. The circle C_0 refers to $|z - z_0| = R_0$.
   (a) Show that the product of the distances of z and z^* to the center of C_0 is R_0^2.
   (b) Show that z^* lies along the ray joining z to z_0. [*Hint:* Compute $\arg(z^* - z_0)$.]
   (c) What is the inverse point of the center z_0 of C_0 with respect to C_0?
   (d) Show that every circle passing through both z and z^* is orthogonal to C_0.
3. The points z and z^* in Problem 1 are also said to be **symmetric with respect to the circle $|z - z_0| = R_0$.** A reason for this terminology will now be outlined. Consider the particular circle C: $|z + iy_0| = y_0, y_0 > 0$, which passes through the origin and has center on the negative imaginary axis.

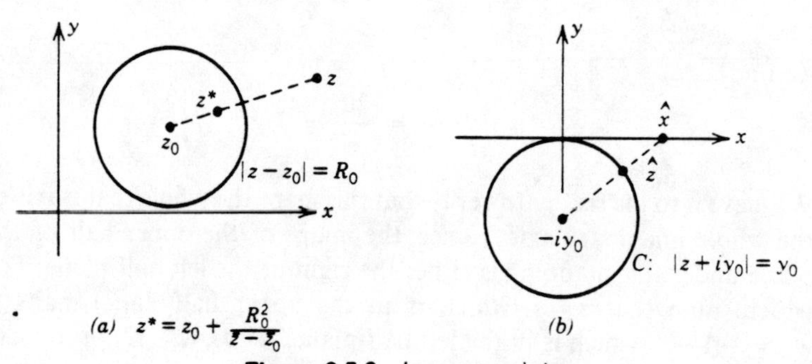

Figure 2.5.8 Inverse points.

(a) Let $\hat{x}$ be any point on the positive real axis, and denote by $\hat{z}$ the point of intersection of C and the line segment joining $\hat{x}$ and the center $-iy_0$ (see Figure 2.5.8*b*). Show that $\hat{z} = -iy_0 + [y_0(\hat{x} + iy_0)]/|\hat{x} + iy_0|$. (A similar equation will result if $\hat{x} < 0$).

(b) Show that $\lim_{y_0 \to \infty} (\hat{z}) = \hat{x}$. It is in this sense that the circles C approach the real axis as their centers recede to infinity. [*Hint:* Expand $1/|\hat{x} + iy_0| = y_0^{-1}(\hat{x}^2 y_0^{-2} + 1)^{-1/2}$ in a series of powers of y_0^{-1}.]

(c) Let z^* denote the point symmetric to z with respect to C. What is $\lim_{y_0 \to \infty} z^*$?

(d) Relative to the real axis, what is the relationship between z and the limit in (c)?

4. Find the images of the following regions under the prescribed mappings. It might help to view some of them as composite maps.
   (a) $|z| > 1$, $\text{Im}(z) < 0$ under $w = i/(z - i)$.
   (b) The wedge $\text{Re}(z) > 1$, $\text{Im}(z) > 0$ under $w = 1/(\bar{z})$.
   (c) The wedge $0 < \arg(z) < \pi/4$ under $w = 1 + (1/z^2)$.
   (d) The strip $0 < \text{Im}(z) < \pi$ under $w = (e^z + i)^{-1}$.
   (e) The wedge $x - y < 2$, $x + y > 2$ under $w = 1/z$. [*Hint:* the wedge is the intersection of two half-planes.]
5. Find the images of the following regions under the prescribed mappings. It might help to view some of them as composite maps.
   (a) The half-plane $x + 2y > 1$ under $w = (z + 2i)/(z - 1)$.
   (b) The wedge $y < 1$, $x + y > 2$ under $w = z/(z - 1 - i)$. [*Hint:* Treat it as the intersection of two half-planes.]
   (c) The disk $|z - i| \leq 1$ under $w = 2z/[2(1 + i) - z]$.
   (d) The disk $|z - 1| \leq 2$ under $w = (z - 1)/(z + 1)$.
   (e) The unit circle under $w = [(z - i)/(z + i)]^2$.
   (f) The first quadrant under $w = (z^2 + i)/(z^2 - i)$.
6. Find the images of the following regions under the prescribed mappings.
   (a) The strip $\text{Re}(z) > 0$, $0 < \text{Im}(z) < \pi$ under $w = e^z/(1 + e^z)$.
   (b) The quarter-plane $\text{Re}(z) > -1$, $\text{Im}(z) > 0$ under

$$w = \frac{1 + \exp\left[-\dfrac{\pi i}{4}(z + 1)^2\right]}{2^{-1/2}(1 - i) + \exp\left[-\dfrac{\pi i}{4}(z + 1)^2\right]}$$

7. Let $w = f(z)$ be a bilinear mapping and D the region in the z-plane between two (not necessarily concentric) circles. What are the possibilities for the images of D?
8. Find the bilinear transformation mapping, in the order given, the three z-points into the three w-points:
   (a) $z = i, 0, -1$ into $w = -1, i, -i$.
   (b) $z = 1, i, -i$ into $w = \infty, 0, -1$.
   (c) $z = i, \infty, 2$ into $w = 1, 1 + i, 2$.

9. (a) Find a bilinear map mapping $z = 0, 1 + i, 2i$ into $w = 1, i, \infty$, respectively.
(b) Show that the mapping in (a) takes the disk $|z - i| < 1$ onto the half-plane $u + v < 1$.
(c) Use the result in (b) to find an analytic mapping of the disk $|z - i| < 1$ onto the half-plane $u + v > 1$. [*Hint:* Use translation and rotations on the half-plane in (b).]
(d) Redo (a) and map $z = 0, 1 + i, 2i$ into $w = i, 1, \infty$, respectively. Show that the result accomplishes the mapping in (c). Can you draw any conclusions?

10. Find a bilinear mapping taking
(a) The disk $|z - i| < 2$ onto the half-plane $u + 2v > 2$.
(b) The half-plane $\mathrm{Im}(z) < 1$ onto the disk $|w - (1 + i)| < 1$.

11. Let D_z be the disk $|z - i| < 1$ and D_w the disk $|w - 1| < 2$.
(a) Find a *linear mapping* of D_z onto D_w.
(b) Find a bilinear (but not linear) mapping of D_z onto D_w (this shows that there are many mappings doing a specific job).

12. Find a mapping $w = f(z)$ which takes
(a) The first quadrant onto the unit disk.
(b) The semidisk $|z| < 1, \mathrm{Im}(z) > 0$ onto the upper half-plane [*Hint:* Pick any bilinear map taking $z = 1$ into $w = \infty$. (Why do you think this choice was made?) Then, translations, rotations, and integer powers will finish the job.]
(c) The strip $\pi < \mathrm{Re}(z) < 2\pi$ onto the unit disk.

13. Do you think it possible to find a bilinear mapping taking an annulus onto the upper half-plane?

14. (a) Use the following series of steps to show that the **most general bilinear map of the upper half-plane onto the unit disk** is given by

$$w = e^{i\theta_0}\left(\frac{z - z_0}{z - \bar{z}_0}\right) \qquad \mathrm{Im}(z_0) > 0$$

(i) If $w = (az + b)/(cz + d)$ $(ad \neq bc)$, find the most general values of a, b, c, and d so that the real axis maps onto the unit circle. [*Hint:* Let $z = 0, 1, \infty$ map onto $w = w_1, w_2, w_3$, respectively, where $|w_k| = 1$ $(k = 1, 2, 3)$. Then, write

$$w = \left(\frac{a}{c}\right)\left(\frac{z + b/a}{z + d/c}\right)$$

where a/c, b/a, d/c are given in terms of w_1, w_2, w_3.]
(ii) Show that $|a/c| = 1$ and $b/a = \overline{d/c}$. (Use the fact that $\bar{w}_k = 1/w_k$.)
(iii) Finally, since $w = 0$ when $z = -b/a$, show that $\mathrm{Im}(-b/a) > 0$ in order that the upper half-plane map onto the disk.

(b) Invert the result in (a) to find the most general bilinear transformation mapping the unit disk onto the upper half-plane.
(c) Use the results of (a) and (b) to derive the most general bilinear

transformation taking the unit disk onto itself. [*Hint:* Go from the unit disk to the upper half-plane and then back to the unit disk.]

15. (a) Show that a bilinear map $w = (az + b)/(cz + d)$ $(ad - bc \neq 0)$ maps the real axis onto the real axis if and only if a, b, c, and d are either all real or all real multiples of the same complex number.

(b) Show that if $w = (az + b)/(cz + d)$, where a, b, c, and d are real and $ad - bc \neq 0$, then the upper half-plane is mapped onto the upper half-plane if and only if $ad - bc > 0$. [*Hint:* What is the image of $z = i$?]

2.6 TRIGONOMETRIC AND HYPERBOLIC FUNCTIONS

From equation (2.3.14) of Section 2.3, we know that

$$\cos y = \frac{1}{2}(e^{iy} + e^{-iy}) \quad \text{and} \quad \sin y = \frac{1}{2i}(e^{iy} - e^{-iy})$$

With this as motivation, we are led to the following definitions of the **complex trigonometric functions** in terms of complex exponentials:

(2.6.1)

$$\text{(a)} \quad \sin z = \frac{1}{2i}(e^{iz} - e^{-iz})$$

$$\text{(b)} \quad \cos z = \frac{1}{2}(e^{iz} + e^{-iz})$$

The other trigonometric functions are defined in terms of $\sin z$ and $\cos z$ exactly as in calculus.

(2.6.2) ***Properties of sin z and cos z***

(a) In terms of the real trigonometric functions and the hyperbolic functions $\cosh y = \frac{1}{2}(e^y + e^{-y})$, $\sinh y = \frac{1}{2}(e^y - e^{-y})$, we have

(2.6.3)

$$\sin z = \sin x \cosh y + i \cos x \sinh y$$

$$\cos z = \cos x \cosh y - i \sin x \sinh y$$

(b) $\sin z$ and $\cos z$ are entire, and

$$\frac{d}{dz}(\sin z) = \cos z, \qquad \frac{d}{dz}(\cos z) = -\sin z$$

(c) $|\sin z|^2 = \sin^2 x + \sinh^2 y$,
$|\cos z|^2 = \cos^2 x + \sinh^2 y$.

(d) $\sin z$ and $\cos z$ are both periodic with the same period as their real counterparts, that is,

$$\cos(z + 2\pi n) = \cos z, \qquad n = 0, \pm 1, \pm 2, \ldots$$
$$\sin(z + 2\pi n) = \sin z,$$

(e) $\sin^2 z + \cos^2 z = 1$, for all z.
(f) $\sin(-z) = -\sin z$ and $\cos(-z) = \cos z$.
(g) $\sin(z_1 + z_2) = \sin z_1 \cos z_2 + \cos z_1 \sin z_2$,
$\cos(z_1 + z_2) = \cos z_1 \cos z_2 - \sin z_1 \sin z_2$.
(h) The only zeros of $\sin z$ and $\cos z$ are their real zeros, that is,

$$\sin z = 0 \Rightarrow z = n\pi, \qquad n = 0, \pm 1, \ldots$$

$$\cos z = 0 \Rightarrow z = \left(n + \frac{1}{2}\right)\pi, \qquad n = 0, \pm 1, \ldots$$

(i) $\sin(iy) = i \sinh y$ and $\cos(iy) = \cosh y$.

■

These results are easily derived from (2.6.1). For example, to derive (2.6.3) for $\sin z$, we write

$$\begin{aligned}\sin z &= \frac{1}{2i}[e^{ix}e^{-y} - e^{-ix}e^{y}] \\ &= \frac{1}{2i}[(e^{-y} - e^{y})\cos x + i(e^{-y} + e^{y})\sin x] \\ &= \frac{1}{2}(e^{y} + e^{-y})\sin x + \frac{i}{2}(e^{y} - e^{-y})\cos x \\ &= \cosh y \sin x + i \sinh y \cos x\end{aligned}$$

Let us make note of an important property of these functions. From calculus, $\sin x$ and $\cos x$ are bounded in magnitude by 1. From (e) above, it might appear that the same is true for the complex functions. That this is not true follows from (c), where we see that $|\sin z|$ and $|\cos z|$ can become unbounded if $y = \text{Im}(z)$ becomes large.

(2.6.4) ***Example.*** Find all solutions of

$$\sin z = 4 \tag{2.6.5}$$

Solution. First note that in calculus the equation $\sin x = 4$ would have no solution. However, there are complex solutions which we will now exhibit. First, using (2.6.3) and equating real and imaginary parts we get

(i) $\sin x \cosh y = 4$
(ii) $\cos x \sinh y = 0$

From the second equation, we find

$$y = 0 \quad \text{or} \quad x = \left(n + \frac{1}{2}\right)\pi, \qquad n = 0, \pm 1, \ldots$$

Substituting $y = 0$ into (i) above, we get $\sin x = 4$, which has no solutions (remember, x is real!). Hence, we must have $x = (n + \frac{1}{2})\pi$, and (i) now yields

$$\cosh y = 4(-1)^n$$

where we have used $\sin(n + \frac{1}{2})\pi = (-1)^n$. Since $\cosh y \geq 0$, n must be even and the solution of the above is

$$y = \cosh^{-1} 4 \qquad \text{and} \qquad n = 2k \qquad k = 0, \pm 1, \pm 2, \ldots$$

where either of the two choices of the inverse function $\cosh^{-1} 4$ is permissible. Hence, the solutions of (2.6.5) are

(2.6.6)
$$z = \frac{\pi}{2} + i \cosh^{-1} 4 + 2\pi k \qquad k = 0, \pm 1, \ldots$$

Another approach to solving (2.6.5) is to substitute $\sin z = (1/2i)(e^{iz} - e^{-iz})$ into (2.6.5), derive a quadratic equation for e^{iz}, and then solve for z using the definition of the exponential. We urge the reader to carry this out.

■

The next example illustrates some of the mapping properties of $\sin z$.

(2.6.7) ***Example.*** What is the image of the line $x = x_0$ under the mapping $w = \sin z$? Use these results to find the image of the vertical strip shown in Figure 2.6.1b.

Solution. From (2.6.3), with $w = u + iv$, we have

(2.6.8)
$$u = \sin x \cosh y \qquad \text{and} \qquad v = \cos x \sinh y$$

Hence, the line $x = x_0$ is mapped onto the curve

(2.6.9)
$$\begin{aligned} u &= \sin x_0 \cosh y \\ v &= \cos x_0 \sinh y \end{aligned} \qquad -\infty < y < \infty$$

which has y as its parameter and will depend upon the value of x_0. For example, if $x_0 = 0$, then

(2.6.10)
$$u = 0, \qquad v = \sinh y \qquad -\infty < y < \infty$$

Hence, the image of the vertical line $x = 0$ is the imaginary axis in the w-plane, with the positive imaginary axis in the z-plane (designated by ① in Figure 2.6.1a), mapping onto the positive imaginary axis in the w-plane (also designated by ①). This is because $y \geq 0$ implies that $\sinh y \geq 0$. Similarly, the half-line designated by ② in the z-plane is mapped onto the half-line designated by ② in the w-plane.

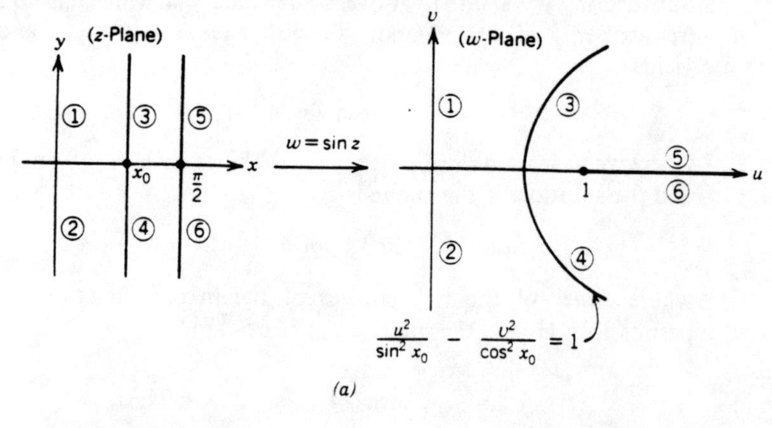

(a)

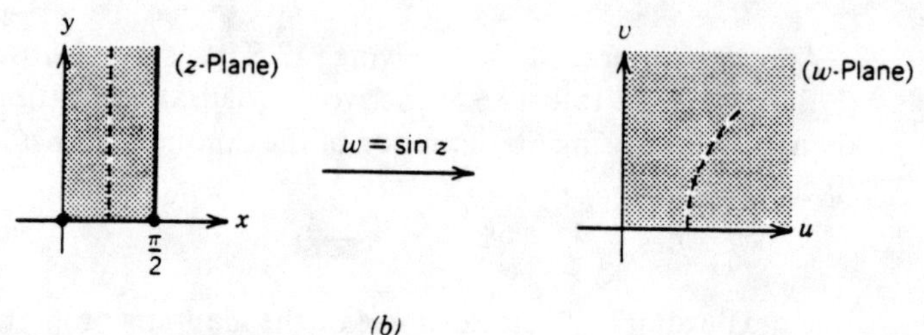

(b)

Figure 2.6.1 Mapping property of $w = \sin z$.

If $0 < x_0 < \pi/2$, then neither $\sin x_0$ nor $\cos x_0$ vanish; and from (2.6.9), we can eliminate y to get

(2.6.11)
$$\frac{u^2}{\sin^2 x_0} - \frac{v^2}{\cos^2 x_0} = 1 \qquad \text{for } 0 < x_0 < \pi/2$$

Hence, the image of the line $x = x_0$ for this range of x_0 is all or part of the two branches of the hyperbola given above. To determine which part, first note that for this range of x_0, both $\sin x_0$ and $\cos x_0$ are positive. Hence. from (2.6.9), we must have $u > 0$, while v is positive (negative) when y is positive (negative). Thus, only the right half-branch of the hyperbola is taken on. The situation is also shown in Figure 2.6.1a with the numbers ③ and ④ designating corresponding images.

Finally, if $x_0 = \pi/2$, then from (2.6.9),

(2.6.12)
$$v = 0, \qquad u = \cosh y \qquad -\infty < y < \infty$$

Hence, the image of this vertical line is now that piece of the real axis lying to the right of $w = 1$ (since $\cosh y \geq 1$) and covered twice (because $\cosh y$ is even in y). This is also shown in Figure 2.6.1a with ⑤ and ⑥ designating corresponding images.

Thus, we see that as the vertical line $x = x_0$ moves from $x_0 = 0$ to $x_0 = \pi/2$, its image begins as the vertical line $\mathrm{Re}(w) = 0$, bends around to become the hyperbolas of (2.6.11), and finally (when $x_0 = \pi/2$) is bent completely onto the part of the real axis described by (2.6.12).

To determine the image of the strip in Figure 2.6.1*b*, we observe that it can be generated by moving the half-vertical line (shown dotted) from left to right, and hence its image will be generated by the images of these half-vertical lines (shown dotted on the *w*-plane). This yields the whole first quadrant in the *w*-plane. In Problem 11, the reader is asked to map the same vertical strip in the figure using horizontal lines.

■

The reader should note from (2.6.2g) that, since

(2.6.13)
$$\sin\left(z + \frac{\pi}{2}\right) = \cos z$$

the mapping properties for the cosine function can be derived from those for the sine function by first using the translation $z + \pi/2$ (see Problem 12b).

Hyperbolic Functions. As in elementary calculus, we define the **complex hyperbolic functions** by

(2.6.14)
$$\sinh z = \frac{1}{2}(e^z - e^{-z}) \quad \text{and} \quad \cosh z = \frac{1}{2}(e^z + e^{-z})$$

From this and the definition of $\sin z$ and $\cos z$, we have

(2.6.15)
$$\sinh z = -i\sin(iz), \qquad \cosh z = \cos(iz)$$

Hence, properties and results concerning the hyperbolic functions can be derived from those for the trigonometric functions (see Problem 13).

PROBLEMS (The answers are given on page 425.)

1. Derive the following identities:

(a) $\tan(z_1 \pm z_2) = \dfrac{\tan z_1 \pm \tan z_2}{1 \mp \tan z_1 \tan z_2}$

(b) $\sin z_1 \pm \sin z_2 = 2\sin[\frac{1}{2}(z_1 \pm z_2)]\cos[\frac{1}{2}(z_1 \mp z_2)]$

(c) $\sin^2 z = \frac{1}{2}(1 - \cos(2z))$ and $\cos^2 z = \frac{1}{2}[1 + \cos(2z)]$

(d) $\sin\left(z + \dfrac{n\pi}{2}\right) = \begin{cases} (-1)^{(1/2)(n-1)}\cos z & n \text{ odd} \\ (-1)^{n/2}\sin z, & n \text{ even} \end{cases}$

(e) $\cos\left(z + \dfrac{n\pi}{2}\right) = \begin{cases} (-1)^{(1/2)(n+1)}\sin z, & n \text{ odd} \\ (-1)^{n/2}\cos z, & n \text{ even} \end{cases}$

2. Express the following in Cartesian form:

(a) $\sin\left(\frac{\pi}{2} + i\right)$ (b) $\exp\left[\sin\left(\frac{\pi i}{2}\right)\right]$ (c) $\cos(e^i)$

3. Show that $\sin z_1 = \sin z_2$ if and only if $z_1 = z_2 + 2\pi n$ or $z_1 = -z_2 + (2n + 1)\pi$, $n = 0, \pm 1, \ldots$.
4. Show that $\sin z$ and $\cos z$ can be expressed as compositions of linear maps, the exponential, and the mappings $z \pm (1/z)$.
5. Find all z satisfying the following equations:
(a) $\cos(iz) = 0$ (b) $\sin z = -2i$
(c) $\sin(e^z) = 0$ (d) $e^{\sin z} = i$
(e) $\sin z = 2\cos z$ (f) $\sin(i/z) = i$
6. Consider the equation $\tan z = \alpha$.
(a) For which α are there solutions of this equation? [*Hint:* Use Definition (2.6.1).]
(b) Show that the solutions are real if and only if α is real.
7. If $f(z)$ stands for either $\sin z$ or $\cos z$, show that $f(\bar{z}) = \overline{f(z)}$.
8. Where is $f(z) = \cos \bar{z}$ differentiable? Analytic? Compute $f'(z)$ wherever it exists.
9. Without using the Cauchy–Riemann equations, state why $f(z) = \cos(x^2 - y^2) \cdot \sinh(2xy) - i \sin(x^2 - y^2) \cosh(2xy)$ is entire. [*Hint:* Can you write f as the composition of two entire functions?]
10. Prove that $|1 - 2\sin^2 z| \geq 1 - 2\sin^2 x$. [*Hint:* $1 - 2\sin^2 z = \cos^2 z - \sin^2 z$.]
11. (a) What is the image of the horizontal line segment $y = y_0$, $0 < x < \pi/2$ under $w = \sin z$?
(b) Use the result in (a) to find the image of the strip in the z-plane of Figure 2.6.1b under $w = \sin z$.
12. (a) Under $w = \sin z$, find the images of the strips
(i) $-\pi/2 < \mathrm{Re}(z) < 0$, $\mathrm{Im}(z) > 0$.
(ii) $\pi/2 < \mathrm{Re}(z) < \pi$, $\mathrm{Im}(z) > 0$.
[*Hint:* Use the result in (i) together with linear maps.]
(b) Under $w = \cos z$, find the images of the strips
(i) $-\pi/2 < \mathrm{Re}(z) < 0$, $\mathrm{Im}(z) > 0$.
(ii) $0 < \mathrm{Re}(z) < \pi/2$, $\mathrm{Im}(z) > 0$.
[*Hint:* Use $\cos z = \sin(z + \pi/2)$.]
13. Verify the following results for the hyperbolic functions defined by (2.6.14). [*Hint:* Use (2.6.15).]
(a) $\sinh z = \sinh x \cos y + i \cosh x \sin y$
$\cosh z = \cosh x \cos y + i \sinh x \sin y$
(b) $\cosh^2 z - \sinh^2 z = 1$
(c) $|\cosh z|^2 = \cosh^2 x - \sin^2 y$
$|\sinh z|^2 = \cosh^2 x - \cos^2 y$
(d) $\cosh(-z) = \cosh z$, $\sinh(-z) = -\sinh z$
(e) $|\cosh z| \leq \cosh x$

14. Are $\cosh z$ and $\sinh z$ periodic? If so, what are their periods?

15. Find all values of z satisfying

(a) $\cosh z = 0$ (b) $\sinh z = 0$ (c) $\cosh z = i$

(d) $\cosh z + \sinh z = \alpha$ (α is any nonzero complex number)

2.7 LOGARITHMS AND COMPLEX POWERS

Using calculus as our guide, we might expect to define the complex logarithm $w = \log z$ as that function which is inverse to the exponential. That is, for each z, $\log z = u + iv$ is defined by the equation $z = e^{u+iv}$. With this, we find that $e^u = |z|$ and $v = \arg(z)$. Hence, we are led to the following definition.

(2.7.1) ***Definition.*** The **complex logarithm log z** is defined for $z \neq 0$ by

(2.7.2)
$$\log z = \ln|z| + i\arg(z) \qquad z \neq 0$$

where $\ln|z|$ denotes the usual *real natural logarithm* of the positive quantity $|z|$.

■

Note that *since arg(z) is multiple valued, so is log z.*

(2.7.3) ***Example.*** The possible values of log 1 are

$$\log 1 = 2\pi i n \qquad n = 0, \pm 1, \pm 2, \ldots$$

since $\ln 1 = 0$ and $\arg(1) = 2\pi n$. Also, since $1 + i = \sqrt{2} \cdot e^{\pi i/4}$, we have

$$\log(1 + i) = \frac{1}{2}\ln 2 + i\left(\frac{\pi}{4} + 2\pi n\right) \qquad n = 0, \pm 1, \ldots$$

■

The multivalued logarithm defined by (2.7.2) can be made single-valued if we restrict $\arg(z)$ to lie in some definite 2π range. In so doing, we will say that we have chosen a **branch of the logarithm**. In particular, if the principal argument is chosen, we have the following definition.

(2.7.4) ***Definition.*** The **principal branch of the logarithm**, or the **principal logarithm**, denoted by $\operatorname{Log} z$, is defined by

$$\operatorname{Log} z = \ln|z| + i\operatorname{Arg}(z) \qquad z \neq 0$$

where $\operatorname{Arg}(z)$ is the principal argument of z and lies in the range

(2.7.5)
$$-\pi < \operatorname{Arg}(z) \leq \pi$$

■

Note in the above that we use *capital L* to denote the principal logarithm.

(2.7.6) ***Example.*** For real and positive z, $\operatorname{Arg}(z) = 0$ and hence,

$$\operatorname{Log} x = \ln x \qquad x > 0$$

Thus, in this case the principal logarithm reduces to the real natural logarithm from calculus, and $\text{Log}\, z$ indeed can be viewed as a *continuation of* $\ln x$ *into the complex plane and onto the negative real axis.* In fact, for x real and negative, $\text{Arg}(x) = \pi$ and

$$\text{Log}\, x = \ln(-x) + \pi i \qquad x < 0$$

Finally, we have

$$\text{Log}(-i) = -\frac{\pi i}{2}$$

■

(2.7.7) ***Example.*** If we choose a range for $\arg(z)$ different from that in (2.7.5), then we will get a **different branch of the logarithm**. For example, if we choose the usual polar coordinate range

$$0 < \arg(z) \le 2\pi$$

we find from (2.7.2) that for x real and positive,

(2.7.8)
$$\log x = \ln x + 2\pi i \qquad x > 0$$

Note that this branch of the logarithm *does not reduce* to the usual natural logarithm for positive x.

■

Let us show that *$\text{Log}\, z$ is not continuous on the negative real axis.* If $x_0 < 0$, then from (2.7.5) and Figure 2.7.1a, we have

$$\lim_{z \to x_0} \text{Arg}(z) = \begin{cases} \pi & \text{if } \text{Im}(z) > 0 \\ -\pi & \text{if } \text{Im}(z) < 0 \end{cases}$$

Hence, the imaginary part of $\text{Log}\, z$ has different limits as you approach the negative real axis, thus proving our contention. The wiggly line in Figure 2.7.1*a* is called a **branch cut** for $\text{Log}\, z$.

If instead of the principal logarithm we choose the **general branch**

(2.7.9)
$$\log z = \ln|z| + i\arg(z) \qquad \theta_0 < \arg(z) \le \theta_0 + 2\pi$$

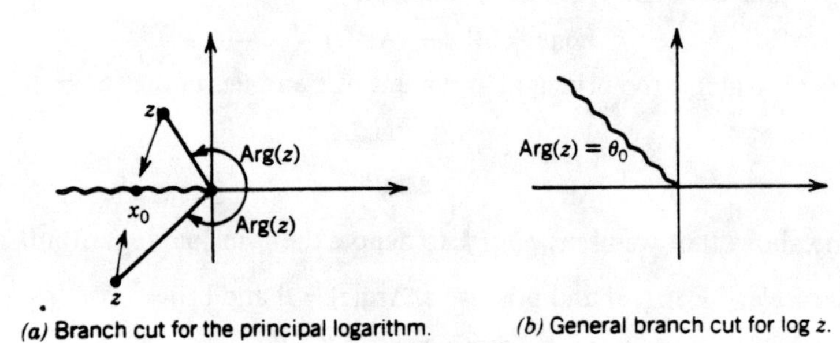

(a) Branch cut for the principal logarithm. *(b)* General branch cut for log *z*.

Figure 2.7.1

for some θ_0, then this $\log z$ will be discontinuous along the ray $\arg(z) = \theta_0$. This ray is now the branch cut for *this* branch of $\log z$ (see Figure 2.7.1*b*). Note that no matter what the value of θ_0, $z = 0$ lies on this ray. We say that $z = 0$ is a **branch point** for $\log z$. (We will have more to say about branch points and branch cuts in Section 3.2.) Finally, the general branch of $\log z$ in (2.7.9) is continuous for all z not on the ray $\arg(z) = \theta_0$.

(2.7.10) ***Example.*** Where is the function $f(z) = \text{Log}(1 - z^2)$ discontinuous? These curves are the "branch cuts for $f(z)$."

Solution. Since we are dealing with the principal logarithm, we know that $\text{Log}(1 - z^2)$ is discontinuous only when $1 - z^2$ is real and negative. Now, with $z = x + iy$,

$$1 - z^2 \text{ real} \Rightarrow xy = 0$$

$$1 - z^2 \text{ negative} \Rightarrow 1 - x^2 + y^2 < 0$$

Hence, from the first equation, $x = 0$ or $y = 0$. If $x = 0$, then the second equation yields $1 + y^2 < 0$, which is impossible. Thus, we must have that $y = 0$ and hence $x^2 > 1$. Therefore the "branch cuts" for this more complicated logarithm function is that piece of the real axis lying outside of the interval $-1 < x < 1$, as shown in Figure 2.7.2.

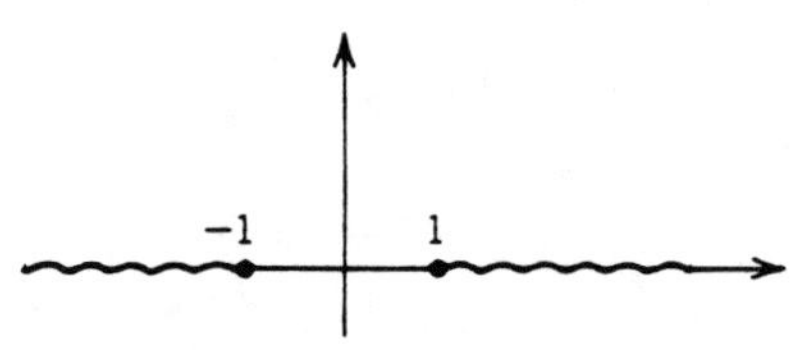

Figure 2.7.2 "Branch cuts" for $\text{Log}(1 - z^2)$.

(2.7.11) ***Theorem.*** The multiple-valued logarithm has the following properties:

(a) $e^{\log z} = z$, for any branch.

(b) $\log(e^z) = z + 2\pi i n$, $n = 0, \pm 1, \pm 2, \ldots$. Hence, in general, $\log(e^z) \neq z$.

(c) Branches of $\log z_1$, $\log z_2$, and $\log(z_1 z_2)$ can be chosen so that

$$\log(z_1 z_2) = \log z_1 + \log z_2$$

If the same branch is chosen for all three, then this result is not true in general.

(d) The general branch (2.7.9) is analytic for $\arg(z) \neq \theta_0$, and here,

$$\frac{d}{dz}[\log z] = \frac{1}{z} \qquad z \neq 0, \quad \theta_0 < \arg(z) < \theta_0 + 2\pi \tag{2.7.12}$$

Proof. Let us prove (a) and (d) [the proof of (c) follows directly from the corresponding result in Theorem (1.2.16)]. For any branch, we have

$$\log z = \ln|z| + i[\arg(z) + 2\pi n]$$

where $\arg(z)$ is some specific value of the argument. Hence, to prove (a), we have

$$\begin{aligned} e^{\log z} &= e^{\ln|z|} \cdot e^{i\arg(z)} \cdot e^{2\pi i n} \\ &= |z|e^{i\arg(z)} \qquad (\text{since } e^{2\pi i n} = 1 \text{ and } e^{\ln|z|} = |z|) \\ &= z \end{aligned}$$

Perhaps the simplest way to prove (d) is to verify that the Cauchy–Riemann equations are satisfied in polar coordinates. First of all, from (2.7.9), we have $\log z = \hat{u}(r,\theta) + i\hat{v}(r,\theta)$, where $\hat{u}(r,\theta) = \ln r$ and $\hat{v}(r,\theta) = \theta = \arg(z)$ are continuously differentiable for $r > 0$. Moreover, they satisfy

$$\hat{u}_r = \frac{1}{r}\hat{v}_\theta \quad \text{and} \quad \hat{u}_\theta = -r\hat{v}_r$$

Thus, the Cauchy–Riemann equations (2.2.18) are satisfied and $\log z$ is therefore analytic. Also [see (2.2.19)],

$$\begin{aligned} \frac{d}{dz}[\log z] &= (\hat{u}_r + i\hat{v}_r)(\cos\theta - i\sin\theta) \\ &= \frac{1}{r}(\cos\theta - i\sin\theta) \\ &= \frac{1}{r(\cos\theta + i\sin\theta)} \\ &= \frac{1}{z} \end{aligned}$$

■

(2.7.13) ***Example.*** Refer to part (c) of the above theorem and show that, in general, *even* for the principal logarithm,

$$\operatorname{Log}(z_1 z_2) \neq \operatorname{Log} z_1 + \log z_2 \tag{2.7.14}$$

For which z_1 and z_2 would you be assured of equality above?

Solution. Let $z_1 = z_2 = -1 + i$. Then, $z_1 z_2 = -2i$ and $\operatorname{Log} z_1 = \operatorname{Log} z_2 = (1/2)\ln 2 + (3\pi i/4)$, while $\operatorname{Log}(z_1 z_2) = \ln 2 - (\pi i/2)$. Thus, (2.7.14) is verified for this special choice of z_1 and z_2. We note that the lack of equality in (2.7.14) is due to the fact that when the principal logarithms $\operatorname{Log} z_1$ and $\operatorname{Log} z_2$ are added, the sum of their imaginary parts is $3\pi i/2$, which exceeds the principal argument range and thus must be

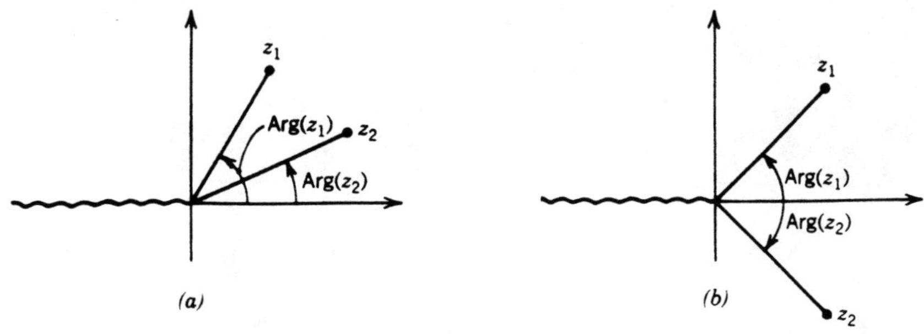

Figure 2.7.3 Values of z_1 and z_2 for which $\text{Log}(z_1z_2) = \text{Log}\, z_1 + \text{Log}\, z_2$.

brought back into this range by subtracting $2\pi i$. To avoid this possibility, we note from Figure 2.7.3 that if z_1 and z_2 are both in the right half of the z-plane, then

$$\text{Arg}(z_1 z_2) = \text{Arg}(z_1) + \text{Arg}(z_2)$$

and hence equality in (2.7.14) will hold. Of course, equality can also hold for other possibilities, such as when $z_1 = -1 + i$, $z_2 = e^{\pi i/8}$.

■

(2.7.15) ***Example.*** Under the mapping $w = \text{Log}(z)$, what is the image in the w-plane of the wedge and pie-shaped regions shown respectively in Figures 2.7.4*a* and *b*?

Solution. The wedge can be described by $0 < |z| < \infty$, $0 < \text{Arg}(z) < \theta_0$. Hence, with $u = \ln|z|$ and $v = \text{Arg}(z)$, the image in the $w = u + iv$ plane is described by

$$-\infty < u < \infty, \qquad 0 < v < \theta_0 \tag{2.7.16}$$

and is the strip shown in Figure 2.7.4*a*.

Now the region in the z-plane of Figure 2.7.4*b* can be described by

$$0 < |z| < R_0, \qquad 0 < \text{Arg}(z) < \theta_0,$$

and hence,

$$-\infty < u < \ln R_0 \qquad 0 < v < \theta_0$$

Thus, the image is the half-strip shown in the w-plane.

Note that these *mapping properties hold only for wedges whose vertices are at the origin.*

■

Complex Powers. If the real quantities a and x are positive, then from calculus we know that $x^a = e^{a \ln x}$. Motivated by this, we have the following definition.

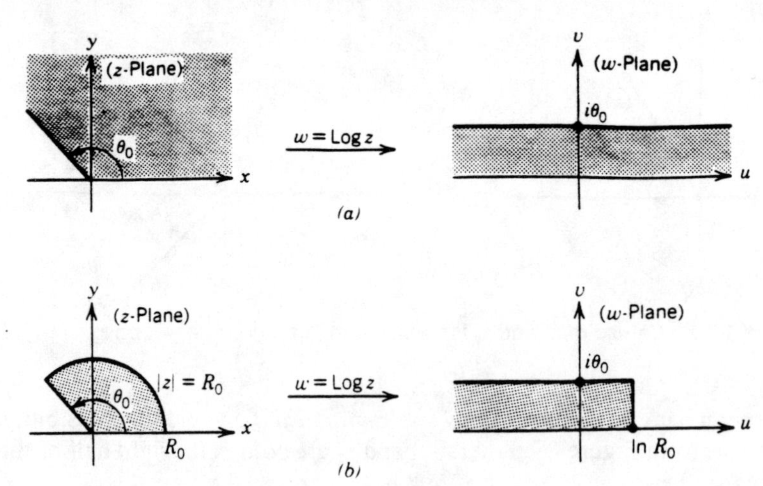

Figure 2.7.4 Example (2.7.15).

(2.7.17) ***Definition.*** If α is a complex constant, we define the **complex power function z^α** by

$$z^\alpha = e^{\alpha \log z} \qquad z \neq 0$$

Note that z^α is multiple-valued since $\log z$ is. If the principal branch $\text{Log}\, z$ is used above, then we have the **principal branch of z^α**.

(2.7.18) ***Example.*** If $k = 0, 1, 2, \ldots$, then z^k, defined as above, is not really multiple-valued but is just the integer power function of Section 2.4. To see this, let $\arg_1(z)$ and $\arg_2(z)$ be different choices of $\arg(z)$. Then, since $\arg_1(z) = \arg_2(z) + 2\pi n$ for some integer n,

$$\begin{aligned} z^k &= e^{k[\ln|z| + i\arg_1(z)]} \\ &= e^{k[\ln|z| + i\arg_2(z)]} \cdot e^{2\pi i n k} \\ &= e^{k[\ln|z| + i\arg_2(z)]} \end{aligned}$$

Hence, z^k is single-valued since it does not depend on the choice of $\arg(z)$. This result is also true for the negative integers $k = -1, -2, \ldots$.

Now, if $\alpha = 1/k$, $k = 2, 3, \ldots$, then $z^{1/k}$ defined as in (2.7.17) has precisely k values which are just the k roots discussed in Section 1.2. In fact, we have that

(2.7.19)
$$w = z^{1/k} \qquad \text{if and only if } z = w^k$$

We leave the verification of these results to the reader.

(2.7.20) ***Example.*** To compute the "strange" number i^i, we have

$$\begin{aligned} i^i &= e^{i \log i} \\ &= \exp\left\{i\left[\ln 1 + i\left(\frac{\pi}{2} + 2\pi n\right)\right]\right\} \\ &= \exp\left[-\left(\frac{\pi}{2} + 2\pi n\right)\right] \qquad n = 0, \pm 1, \pm 2, \ldots \end{aligned}$$

Hence, there are an infinite number of values of i^i, *but they are all real.*

■

(2.7.21) ***Example.*** Show that if the principal branch is used,

$$(z^2)^{1/2} = \begin{cases} z & \text{for } \mathrm{Re}(z) > 0, \text{ or } \mathrm{Re}(z) = 0 \text{ and } \mathrm{Im}(z) > 0 \\ -z & \text{elsewhere} \end{cases} \tag{2.7.22}$$

Solution. Since $\log z^2 = \ln|z|^2 + i \arg(z^2)$, we have for the principal branch

$$(z^2)^{1/2} = |z| e^{(1/2)\mathrm{Arg}(z^2)} \qquad \text{since } e^{(1/2)\ln|z|^2} = |z|$$

Using the range of the principal argument, we can show that

$$\mathrm{Arg}(z^2) = \begin{cases} 2\,\mathrm{Arg}(z) & \text{for } \mathrm{Re}(z) > 0, \text{ or } \mathrm{Re}(z) = 0 \text{ and } \mathrm{Im}(z) > 0 \\ 2\,\mathrm{Arg}(z) - 2\pi & \text{for } \mathrm{Re}(z) < 0 \text{ and } \mathrm{Im}(z) \geq 0 \\ 2\,\mathrm{Arg}(z) + 2\pi & \text{for } \mathrm{Re}(z) \leq 0 \text{ and } \mathrm{Im}(z) < 0 \end{cases} \tag{2.7.23}$$

Hence, the result (2.7.22) follows. Note from this that if z is real, then we get the usual equality

$$\sqrt{x^2} = |x|$$

■

Using the chain rule, we can easily prove the following.

(2.7.24) ***Theorem.*** The general branch of the power function

$$w = z^\alpha \qquad \theta_0 < \arg(z) \leq \theta_0 + 2\pi$$

is single-valued and analytic for $\arg(z) \neq \theta_0$, and

$$\frac{d}{dz}(z^\alpha) = \alpha z^{\alpha - 1} \qquad \theta_0 < \arg(z) < \theta_0 + 2\pi$$

where the same branch of $z^{\alpha-1}$ *is used as for* z^α. Note that the ray $\arg(z) = \theta_0$ is a branch cut for this function.

■

(2.7.25) ***Example.*** For the principal branch of the square root, where is the function $f(z) = [(z-1)/(z-2)]^{1/2}$ analytic?

Solution. From (2.7.17), the only region in the z-plane where $f(z)$ might not be analytic is where $\text{Log}\,[(z-1)/(z-2)]$ is not continuous, that is, where

$$\text{Im}\left(\frac{z-1}{z-2}\right) = 0 \qquad \text{and} \qquad \text{Re}\left(\frac{z-1}{z-2}\right) < 0 \tag{2.7.26}$$

Hence, with $z = x + iy$, a simple calculation shows that this region is described by

$$y = 0 \qquad \text{and} \qquad x^2 - 3x + 2 < 0$$

and is the line segment joining $z = 1$ and $z = 2$.

■

(2.7.27) ***Example.*** For the principal branch of

$$w = z^{1/n} \qquad n = 2, 3, \ldots$$

what is the image in the w-plane of the wedges shown in the z-planes of Figures 2.7.5*a* and *b*?

Solution. Since $w = e^{(1/n)\ln|z|} \cdot e^{(1/n)\text{Arg}(z)}$, we have

$$|w| = |z|^{1/n} \qquad \text{and} \qquad \arg(w) = \frac{1}{n}\text{Arg}(z)$$

Now, the wedge of Figure 2.7.6*a* is described by

$$0 < |z| < \infty \qquad 0 < \text{Arg}(z) < \theta_0$$

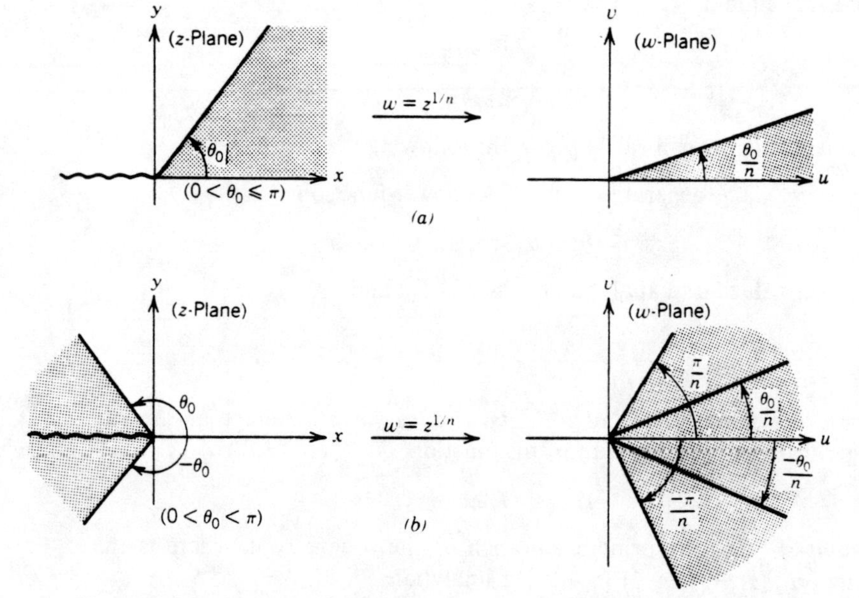

Figure 2.7.5 Mapping properties of the principal branch of $w = z^{1/n}$.

and hence its image is

$$0 < |w| < \infty \qquad 0 < \arg(w) < \frac{1}{n}\theta_0$$

and is shown in the figure. Note that the wedge angle is reduced by a factor of $1/n$.

The wedge in the z-plane of Figure 2.7.5b must be described by *two* sets of inequalities, since we are using the principal branch, namely,

$$0 < |z| < \infty \qquad \theta_0 < \operatorname{Arg}(z) \leq \pi$$

and

$$0 < |z| < \infty \qquad -\pi < \operatorname{Arg}(z) < -\theta_0$$

Hence, its image in the w-plane consists of the two wedges

$$0 < |w| < \infty \qquad \frac{1}{n}\theta_0 < \arg(w) \leq \frac{\pi}{n}$$

and

$$0 < |w| < \infty \qquad -\frac{\pi}{n} < \arg(w) < -\frac{\theta_0}{n}$$

The reason for this strange behavior, as contrasted with the previous mapping, is that the negative real axis in the z-plane, which is a branch cut for the principal branch of $z^{1/n}$, is contained in the interior of the wedge of Figure 2.7.5b.

■

From the above example, we see that the **principal branch of the function $z^{1/n}$ maps wedges with their vertices at the origin onto other wedges whose angles are the fraction $1/n$ of the original**—*so long as the negative real axis is not inside of the wedge to be mapped.*

PROBLEMS (The answers are given on page 425.)

1. Show that the only function $f(z) = u + iv$ which is analytic in $0 < |z| < \infty$, $\theta_0 < \arg z < \theta_0 + 2\pi$ and which satisfies $f'(z) = 1/z$ is a branch of the logarithm. [*Hint:* Express the differential equation and the Cauchy–Riemann equations in polar coordinates. Also see Example (2.3.2).]
2. What are all the possible values of the following?

   (a) $\log(5e^{-7\pi i/4})$ (b) $\log(ie^{1+2i})$ (c) $\operatorname{Log}(e^{3\pi i/2})$

   (d) $\operatorname{Log}(e^{5\pi i/2})$ (e) $\operatorname{Log}(\operatorname{Log} e^{i})$ (f) $\operatorname{Log}\left[\frac{(1+i)e}{\sqrt{2}}\right]$

(g) $\log\left(\dfrac{1+i}{1+i\sqrt{3}}\right)$ (h) $\log[\sin(e^{1+\pi i/2})]$

(i) $\sin\left\{1-\operatorname{Log}\left[\dfrac{e(1+i)}{\sqrt{2}}\right]\right\}$

3. Find $f(0)$ and $f(\pi i)$, where $f(z) = \operatorname{Log}\left(\dfrac{z}{\pi} + e^{-z}\right)$.

4. Find all z satisfying

(a) $\operatorname{Log} z = \dfrac{\pi i}{3}$ (b) $\operatorname{Log} z = 3\pi i$

(c) $\operatorname{Log} z^3 = 3\operatorname{Log} z$ (d) $\operatorname{Log}(e^z) = z$ (e) $\operatorname{Log}\bar{z} = -\operatorname{Log} z$

(f) $\log\bar{z} = -\log z$, where the branch of $\log z$ is such that $-3\pi/2 < \arg(z) \le \pi/2$

5. For which z does $\operatorname{Log}\left(\dfrac{z}{\bar{z}}\right) = \operatorname{Log} z - \operatorname{Log}\bar{z}$?

6. Compute the following limits:

(a) $\displaystyle\lim_{z\to\operatorname{Log}(\pi i)} (x^2 + y^2 + 2ixy)$ (b) $\displaystyle\lim_{z\to 0} (z\log z)$

(c) $\displaystyle\lim_{z\to\infty}\left[\frac{\log z}{z}\right]$ (d) $\displaystyle\lim_{z\to\infty}\left[\operatorname{Log}\left(\frac{z-1}{z}\right)\right]$

(e) $\displaystyle\lim_{z\to\infty}\left[\operatorname{Log}\left(\frac{1-z}{z}\right)\right]$

7. What is the most general branch of $\log z$ for which the equation $\log z = 0$ has a solution? What are the solutions?

8. Where are the following not analytic?

(a) $\operatorname{Log}(e^z + 1)$

(b) $\log(1 - z^2)$, where the branch of $\log z$ is chosen so that $-3\pi/2 < \arg(z) \le \pi/2$

(c) $\operatorname{Log}[z(1+z^2)]$ (d) $\operatorname{Log}\bar{z}$

9. Find the image of the given regions under the given mappings:

(a) The punctured disk $0 < |z| < 1$ under $w = \operatorname{Log} z$. Note the image of the unit circle. Is it the boundary of the image of the punctured disk?

(b) The first quadrant under $\operatorname{Log}(1 + z^2)$. [*Hint:* View as a composite map.]

(c) The quarter disk $|z| < 1$, $\operatorname{Re}(z) > 0$, $\operatorname{Im}(z) > 0$, under $\operatorname{Log}(-iz^2)$.

(d) The quarter disk in (c) under $\sin(e^{-\pi i/2}\operatorname{Log} z)$.

(e) The upper half-plane under the branch of $\log z$ where $-3\pi/2 < \arg(z) \le \pi/2$.

(f) The semidisk $|z| < 1$, $\operatorname{Im}(z) > 0$ under $\operatorname{Log}[(1+z)/(1-z)]$. [*Hint:* $(1+z)/(1-z)$ is a bilinear map.]

10. What are all the possible values of
(a) $(1+i)^i$ (b) $\log(i^i)$ (c) $(\log i)^{1/2}$
(d) $(-2)^{\mathrm{Log}\, i}$ (e) $(e^{\pi i})^i$ (f) i^{e^i} (g) $\mathrm{Log}\sqrt{i}$?

11. With the principal branch of the square root, find all solutions of
(a) $\sqrt{z} = 1$ (b) $\sqrt{z} = -1$
(c) $z + \sqrt{1 - z^2} = 0$ (d) $z + \sqrt{1 + z^2} = 0$

12. With the choice of the principal branch, for which z is it true that $(z^n)^{1/n} = z$, where n is a positive integer?

13. Consider the function $f(z) = (z^2 - 1)^{1/2}$, where the principal branch is used. If "$\sqrt{\ }$" denotes the usual real positive square root, and x is real, show that

$$f(x) = \begin{cases} \sqrt{x^2 - 1} & |x| \ge 1 \\ i\sqrt{1 - x^2} & |x| < 1 \end{cases}$$

14. With the principal branch of the square root, for which z does $\sin z = (1 - \cos^2 z)^{1/2}$? [*Hint:* Use $\cos^2 z + \sin^2 z = 1$ and the result of Example (2.7.21).]

15. Consider the complex-valued functions $[x^2 + (\epsilon^2 \pm i)^2]^{1/2}$, where x and ϵ are real, and the principal branch is assumed. Show that

$$\lim_{\epsilon \to 0} [x^2 + (\epsilon^2 \pm i)^2]^{1/2} = \begin{cases} \sqrt{x^2 - 1} & x^2 > 1 \\ \pm i\sqrt{1 - x^2} & x^2 < 1 \end{cases}$$

where "$\sqrt{\ }$" denotes the usual real, positive square root. [*Hint:* If ϵ is small, in what quadrant does $x^2 + (\epsilon^2 + i)^2$ lie? Now take the limit $\epsilon \to 0$.]

16. Find the images of the following regions under the given mappings. All branches are taken to be principal branches.
(a) The semidisk $|z| < 1$, $\mathrm{Im}(z) > 0$ under $z^{1/2}$.
(b) The semidisk $|z| < 1$, $\mathrm{Re}(z) < 0$ under $z^{1/2}$.
(c) The quarter-plane $\mathrm{Re}(z) > 1$, $\mathrm{Im}(z) > 0$ under $\mathrm{Log}(\sqrt{z - 1})$.

17. Find a function $w = f(z)$ analytic in the region D_z and mapping D_z onto D_w, where D_z and D_w are as follows:
(a) D_z: The half-plane $\mathrm{Im}(z) > 1$; D_w: the first quadrant.
(b) D_z: The half-plane $y + x > 1$; D_w: the first quadrant.
(c) D_z: The half-plane $x + y > 1$; D_w: the strip $0 < y < 1$.

2.8 THE INVERSE SINE FUNCTION

Here, we will discuss only the inverse sine function. Other inverse functions can be similarly defined and will be discussed in the problems.

(2.8.1) ***Definition.*** The **inverse sine function** $\sin^{-1} z$ is defined by the property

$$w = \sin^{-1} z \quad \text{if and only if} \quad z = \sin w$$

■

(2.8.2) ***Theorem.*** The inverse sine function $w = \sin^{-1} z$ is defined for all z, is multiple-valued, and is given in terms of the logarithm by

$$\sin^{-1} z = -i \log(iz + \sqrt{1 - z^2}) \tag{2.8.3}$$

where any of the branches of the logarithm and square root may be taken.

Proof. Since $w = \sin^{-1} z$, we have

$$z = \sin w = \frac{1}{2i}(e^{iw} - e^{-iw})$$

Multiplying both sides of the above by $2ie^{iw}$, we get

$$e^{2iw} - (2iz)e^{iw} - 1 = 0$$

Solving this quadratic equation for e^{iw} and then taking a general logarithm of both sides yields (2.8.3). Clearly, then, $\sin^{-1} z$ is multiple-valued. The only values of z for which it is not defined are those for which $iz + \sqrt{1 - z^2} = 0$. That there are no solutions of this equation (no matter which branch of the square root is chosen) is easily seen by bringing iz over to the right-hand side and squaring both sides.

■

(2.8.4) ***Example.*** Compute all of the possible values of $\sin^{-1} 2$.

Solution. For $z = 2$, the possible values of $\sqrt{1 - z^2}$ are $\pm i(3^{1/2})$. Hence, from (2.8.3) (since $2 \pm 3^{1/2}$ are both positive)

$$\begin{aligned}\sin^{-1} 2 &= -i \log[i(2 \pm 3^{1/2})] \\ &= -i\left[\ln(2 \pm 3^{1/2}) + i\left(\frac{\pi}{2} + 2\pi n\right)\right] \qquad n = 0, \pm 1, \ldots\end{aligned}$$

Finally, since $\ln(2 - 3^{1/2}) = -\ln(2 + 3^{1/2})$, we get

$$\sin^{-1} 2 = \frac{\pi}{2} + 2\pi n \pm i \ln(2 + 3^{1/2}) \qquad n = 0, \pm 1, \ldots$$

Note that in elementary calculus $\sin^{-1}(2)$ is not defined (as a real number).

■

(2.8.5) ***Definition.*** The **principal inverse sine, $\operatorname{Sin}^{-1} z$,** is defined by (2.8.3) when the principal branches of both the logarithm and square root are chosen.

■

(2.8.6) ***Example.*** If x is real, compute $\operatorname{Sin}^{-1} x$ and show that this reduces to the usual principal inverse sine function from elementary calculus when $-1 \le x \le 1$.

Solution. If x is real, we have

$$(\text{prin.})\sqrt{1 - x^2} = \begin{cases} (1 - x^2)^{1/2} & |x| \le 1 \\ i(x^2 - 1)^{1/2} & |x| > 1 \end{cases}$$

where $(1 - x^2)^{1/2}$ and $(x^2 - 1)^{1/2}$ denote the usual real positive square roots. Hence,

$$|ix + \sqrt{1 - x^2}| = \begin{cases} 1 & |x| \le 1 \\ |x + (x^2 - 1)^{1/2}| & |x| > 1 \end{cases}$$

and the principal argument is

$$\text{Arg}(ix + \sqrt{1 - x^2}) = \begin{cases} \text{Arg}[ix + (1 - x^2)^{1/2}] & |x| \le 1 \\ \dfrac{\pi}{2} & x > 1 \\ -\dfrac{\pi}{2} & x < -1 \end{cases}$$

Thus, for $|x| > 1$, from the above and (2.8.3), we have

$$\text{Sin}^{-1} x = \frac{-i}{2} \ln|x + (x^2 - 1)^{1/2}| + \begin{cases} \dfrac{\pi}{2} & x > 1 \\ -\dfrac{\pi}{2} & x < -1 \end{cases}$$

while for $|x| \le 1$, $\text{Sin}^{-1} x$ is real and given by

(2.8.7)
$$\text{Sin}^{-1} x = \text{Arg}[ix + (1 - x^2)^{1/2}] \qquad |x| \le 1$$

To show that this is equivalent to the real principal inverse sine from calculus, first note that

$$\frac{d}{dx}\{\text{Log}[ix + (1 - x^2)^{1/2}]\} = \frac{i}{(1 - x^2)^{1/2}}$$

and thus, from (2.8.3), we have

$$\frac{d}{dx}(\text{Sin}^{-1} x) = \frac{1}{(1 - x^2)^{1/2}}$$

which is the same equation satisfied by the real principal inverse sine function. Also, note from Figure 2.8.1 that $\theta = \text{Arg}[ix + (1 - x^2)^{1/2}]$ increases monotonically as x increases from $x = -1$ to $x = 1$ and has the property that

$$\begin{cases} \theta = -\dfrac{\pi}{2} & \text{for } x = -1 \\ -\dfrac{\pi}{2} < \theta < \dfrac{\pi}{2} & \text{for } -1 < x < 1 \\ \theta = \dfrac{\pi}{2} & \text{for } x = 1 \end{cases}$$

This is precisely the description of the real principal inverse sine function.

■

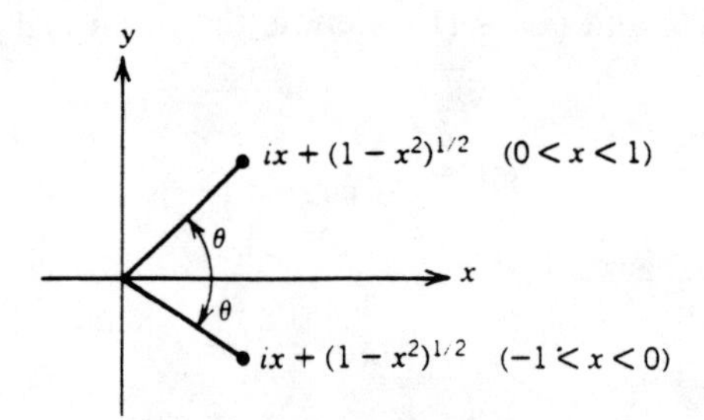

Figure 2.8.1 Example (2.8.6).

(2.8.8) ***Theorem.*** The principal inverse sine function $\mathrm{Sin}^{-1} z$ is analytic in the z-plane cut as shown in Figure 2.8.2, and here

$$\frac{d}{dz}(\mathrm{Sin}^{-1} z) = \frac{1}{\sqrt{1 - z^2}} \tag{2.8.9}$$

where the principal branch of the square root is to be used.

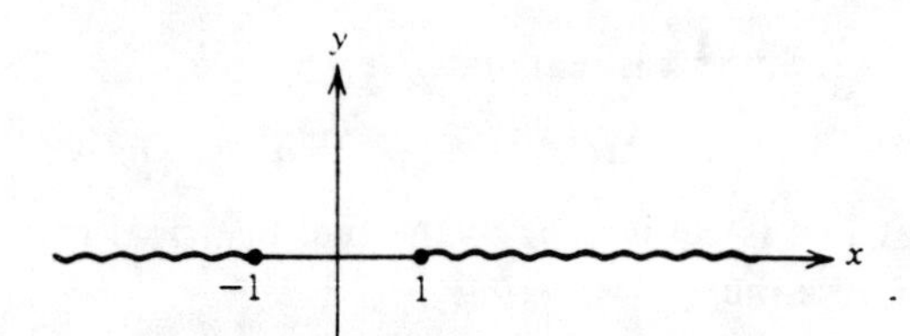

Figure 2.8.2 Branch cuts for $\mathrm{Sin}^{-1} z$.

Proof. First of all, (prin.) $\sqrt{1 - z^2}$ is analytic except when both $\mathrm{Im}(1 - z^2) = 0$ and $\mathrm{Re}(1 - z^2) \leq 0$. These are just the cuts in the figure. Also, $\mathrm{Log}(iz + \sqrt{1 - z^2})$ is analytic except when both

$$\mathrm{Im}(iz + \sqrt{1 + z^2}) = 0 \qquad \text{and} \qquad \mathrm{Re}(iz + \sqrt{1 - z^2}) \leq 0 \tag{2.8.10}$$

Since the quantity λ, defined by $\lambda = iz + \sqrt{1 - z^2}$, is not zero for any z, we have

$$\lambda = iz + \sqrt{1 - z^2} \Rightarrow z = \frac{\lambda^2 - 1}{2i\lambda}$$

Now the first of conditions (2.8.10) yields $\mathrm{Im}(\lambda) = 0$. Hence, with λ real, we see from the second equality above that z must be imaginary. Thus, we have shown that only imaginary z could possibly make $\mathrm{Im}(iz + \sqrt{1 - z^2}) = 0$. Now, with $z = iy$, (prin.) $\sqrt{1 - z^2} = (1 + y^2)^{1/2}$, where this is the real, positive square root; and hence, again with $z = iy$,

$$\mathrm{Re}(iz + \sqrt{1 - z^2}) = -y + (1 + y^2)^{1/2} > 0 \qquad \text{for all } y$$

Hence, the second of the conditions in (2.8.10) can never be satisfied, and thus $\mathrm{Sin}^{-1} z$ is analytic in the cut plane of the figure. The result (2.8.9) then follows by differentiating both sides of (2.8.3).

■

(2.8.11) ***Example.*** What is the image of the upper half-plane under the principal inverse sine map?

Solution. From the mapping properties of $w = \sin z$, we know that a half-strip described either by

$$\mathrm{Im}(z) > 0 \qquad \text{and} \qquad 2\pi n - \frac{\pi}{2} < \mathrm{Re}(z) < 2\pi n + \frac{\pi}{2}$$

or by

$$\mathrm{Im}(z) < 0 \qquad \text{and} \qquad 2\pi n + \frac{\pi}{2} < \mathrm{Re}(z) < 2\pi n + \frac{3\pi}{2}$$

for $n = 0, \pm 1, \ldots$ is mapped onto the upper half-plane (this is depicted in Figure 2.8.3*a*). Hence, we expect the inverse function to map the upper half-plane onto one of these half-strips. Which one it is will be determined by which branch of $\sin^{-1} z$ is chosen. For the principal branch, we have from (2.8.3)

(2.8.12)
$$u = \mathrm{Arg}(iz + \sqrt{1 - z^2}) \qquad \text{and} \qquad v = -\ln|iz + \sqrt{1 - z^2}|$$

Also, with $w = u + iv = \mathrm{Sin}^{-1} z$, we have $x + iy = \sin(u + iv)$, which leads to

(2.8.13)
$$x = \sin u \cosh v \qquad \text{and} \qquad y = \cos u \sinh v$$

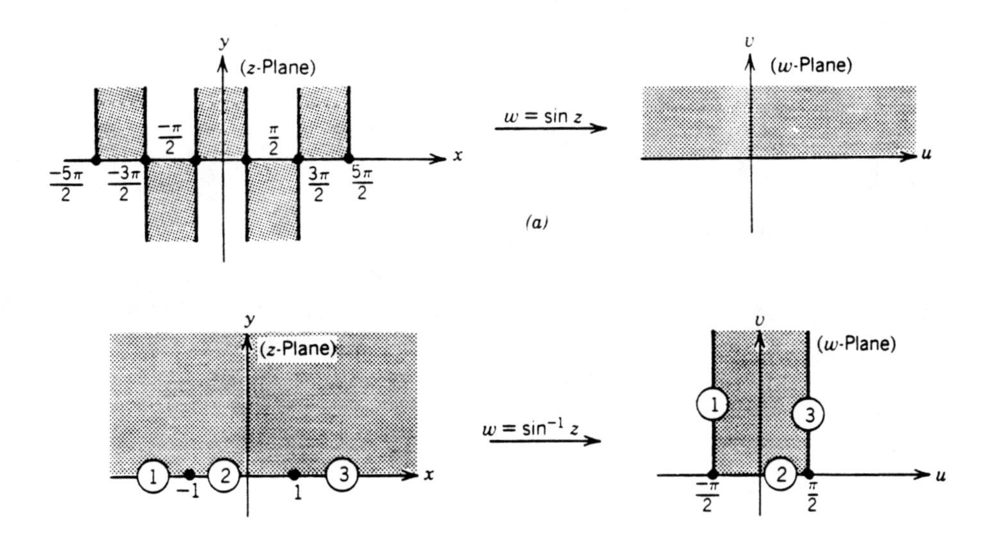

Figure 2.8.3

The following steps show that $v(x, y) > 0$ for z in the upper half-plane.

(i) Since $\mathrm{Sin}^{-1} z$ is analytic for $y > 0$, $v(x, y)$ is certainly continuous here.

(ii) From (2.8.13), $v = 0$ only when $y = 0$. Hence, with (i), $v(x, y)$ must be of one sign for $y > 0$. To see which sign, pick any point in the upper half-plane.

(iii) From (2.8.12), with $z = i$, we find $v > 0$.

Now, from (2.8.13), the image of the upper half-plane $y > 0$ is described by $\cos u \sinh v > 0$. Since $v > 0$ here, we must have $\cos u > 0$. But from (2.8.12), the range of the principal argument yields $-\pi < u \leq \pi$. The only values of u in this range for which $\cos u > 0$ is $-\pi/2 < u < \pi/2$. Hence, the mapping is as depicted in Figure 2.8.3*b*. The circled numbers there indicate how the pieces of the boundaries are mapped onto each other.

■

PROBLEMS (The answers are given on page 427.)

1. Use the procedure in the proof of Theorem (2.8.2) to derive the following formulas:

(a) $\cos^{-1} z = -i \log(z + \sqrt{z^2 - 1})$

(b) $\tan^{-1} z = \dfrac{-i}{2} \log\left(\dfrac{1 + iz}{1 - iz}\right) \qquad z \neq \pm i$

(c) $\sec^{-1} z = -i \log\left(\dfrac{1 + \sqrt{1 - z^2}}{z}\right) \qquad z \neq 0$

(d) $\cot^{-1} z = \dfrac{-i}{2} \log\left(\dfrac{z + i}{z - i}\right) \qquad z \neq \pm i$

(e) $\sinh^{-1} z = \log(z + \sqrt{1 + z^2})$

(f) $\tanh^{-1} z = \dfrac{1}{2} \log\left(\dfrac{1 + z}{1 - z}\right) \qquad z \neq \pm 1$

2. Compute all possible values of the following inverse functions, together with their principal values. (The principal values are defined by taking principal values of all logarithms and square roots which appear in Problem 1.)

(a) $\sin^{-1} 1$ (b) $\sin^{-1} 0$ (c) $\cos^{-1} x$ (d) $\sin^{-1}(iy)$

(e) $\cos^{-1}(iy)$ (f) $\tan^{-1}(e^{i\theta_0}) \qquad \theta_0 \neq \left(n + \dfrac{1}{2}\right)\pi$

3. (a) Show that (2.8.9) holds for any branch of $\sin^{-1} z$.

(b) Compute $(d/dz)(\cos^{-1} z)$ and $(d/dz)(\tan^{-1} z)$; [see Part (a).] Also, show that $(d/dz)(\tan^{-1} z)$ is in fact a single-valued function.

4. (a) If $f_1(z)$ and $f_2(z)$ are any two branches of $\sin^{-1} z$, show that either $f_1(z) = f_2(z) + 2\pi n$ or $f_1(z) = (2k+1)\pi - f_2(z)$ for $n, k = 0, \pm 1, \pm 2, \ldots$.
(b) Prove that $\sin(\sin^{-1} z) = z$, independent of the branch of $\sin^{-1} z$. [*Hint:* Write sine in terms of exponentials.]
(c) Is it true that $\text{Sin}^{-1}(\sin z) = z$?

5. (a) If $g_1(z)$ and $g_2(z)$ are two different branches of $\cos^{-1} z$, what are the possible relationships between them?
(b) Is it true that $\cos(\cos^{-1} z) = z$?

6. (a) What is the relationship between any two branches $h_1(z)$ and $h_2(z)$ of $\tan^{-1} z$?
(b) What are the branch cuts for the principal inverse tangent

$$\text{Tan}^{-1} z = \frac{-i}{2} \text{Log}\left(\frac{1+iz}{1-iz}\right)?$$

(c) Prove that for z real, $\text{Tan}^{-1} z$ reduces to the usual real principal inverse tangent from calculus. [*Hint:* First show that with $z = x$ $\text{Tan}^{-1} z = \frac{1}{2}\text{Arg}(1 - x^2 + 2ix)$. Now show that $\frac{1}{2}\text{Arg}(1 - x^2 + 2ix)$ must be *some branch* of the real function $\tan^{-1} x$ by showing that their derivatives are equal. Finally, show that the range of $\frac{1}{2}\text{Arg}(1 - x^2 + 2ix)$ is the same as the real principal inverse tangent.]

7. In terms of z, compute the most general values of
(a) $\sin(\cos^{-1} z)$ (b) $\cos(\sin^{-1} z)$

8. With $\text{Tan}^{-1} z$ designating the principal inverse tangent, verify the mapping shown in Figure 2.8.4. [*Hint:* See Problem 1b and view the mapping as a composite of a bilinear mapping followed by a principal logarithm.]

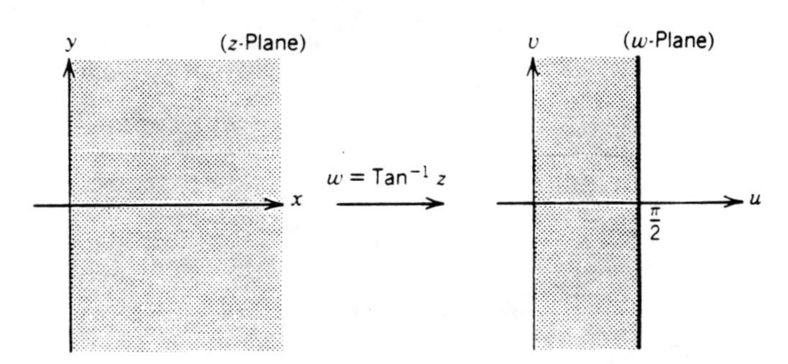

Figure 2.8.4

3

ANALYTIC FUNCTIONS AND SOME APPLICATIONS

3.1 ISOLATED SINGULARITIES; PARTIAL FRACTION DECOMPOSITION

To recapitulate, a function is analytic at z_0 if it is differentiable at z_0 *and* at each point in some neighborhood of z_0.

(3.1.1) ***Definition.*** The point z_0 is a **singular point of $f(z)$** if $f(z)$ is not analytic at z_0 but is analytic somewhere in every neighborhood of z_0. We also say that z_0 is a **singularity of $f(z)$**.

■

It would take us too far afield and defeat the purpose of this text, if we were to try to indicate why the above is an appropriate definition of a singular point. To this end, we would refer the interested reader to more advanced treatises on the subject. **For our purposes, a singularity will be any point where f is not analytic.**

(3.1.2) ***Example.*** The function $f(z) = 1/(z - 1)$ has $z = 1$ as its only singular point. The principal logarithm Log z has singular points at the origin and along the whole negative real axis.

■

In this text, we will be interested in only two types of singular points—isolated singular points and branch points, with the latter discussed in the next section.

(3.1.3) ***Definition.*** A point z_0 is an **isolated singular point** of $f(z)$ if f is analytic at every point in some neighborhood of z_0, except at z_0 itself.

■

(3.1.4) ***Example.*** The function

$$f(z) = \frac{z^2(z-1)}{(z^2+1)^2(z^2+z)}$$

has isolated singular points at $z = -1, \pm i$, and 0. Indeed, any rational function $f(z) = p(z)/q(z)$, where p and q are polynomials in z, will have isolated singular points at the zeros of the denominator $q(z)$.

■

(3.1.5) ***Example.*** The principal logarithm $\operatorname{Log} z$ has a singular point at $z = 0$. However, it is *not isolated*, since every neighborhood of $z = 0$ contains points on the negative real axis, where f is also not analytic. In the next section, we will show that $z = 0$ is a branch point of $\operatorname{Log} z$.

■

(3.1.6) ***Example.*** The function $f(z) = 1/\sin z$ has singular points at $z = n\pi$, $n = 0, \pm 1, \pm 2, \ldots$, and they are all isolated. The function $g(z) = 1/\sin(1/z)$ has singular points at $z = 0$ and at $z = 1/n\pi$, $n = \pm 1, \pm 2, \ldots$. However, $z = 0$ is not isolated because every neighborhood of it must contain points of the form $1/n\pi$ for some integer n.

■

Partial Fraction Decomposition[1] A full discussion of the behavior of functions in a neighborhood of isolated singularities will have to be delayed until Section 6.5, where we will have infinite series techniques at our disposal. However, here we will review the **partial fraction decomposition** procedure from calculus and show how it is used to separate the effect of the isolated singular points of a rational function.

(3.1.7) ***Example.*** What is the partial fraction decomposition for

$$f(z) = \frac{1}{(z-i)(z+1)}?$$

Solution. Note that f has isolated singularities at $z = i$ and $z = -1$. The form of its partial fraction decomposition is

$$\frac{1}{(z-i)(z+1)} = \frac{A}{z-i} + \frac{B}{z+1} \tag{3.1.8}$$

[1] This material should be familiar to the reader and can be skipped over. It will however be used in Chapters 5 and 6.

where the constants A and B can be computed in two ways. First, we can combine the fractions on the right-hand side into $[(A + B)z + (A - iB)]/[(z - i)(z + 1)]$, equate the numerators, and then equate coefficients of like powers of z. Doing this, we get the two equations $A + B = 0$ and $A - iB = 1$, which yield

$$A = \frac{1 - i}{2} \quad \text{and} \quad B = \frac{-(1 - i)}{2}$$

Another approach is to observe that if we multiply both sides of (3.1.8) by $z - i$ and then take the limit $z \to i$, we immediately get $A = 1/(1 + i) = \frac{1}{2}(1 - i)$. The result for B follows in the same way.

Note that what this partial fraction decomposition accomplishes is to *separate the effects of the singularities*. Thus, the term $A/(z - i)$ in (3.1.8) has a singularity only at $z = i$, while the term $B/(z + 1)$ takes care of the singularity at $z = -1$.

■

(3.1.9) ***Example.*** Find a partial fraction decomposition for

$$f(z) = \frac{1}{(z + 2i)^2(z - 1)}$$

Solution. Following the approach of the previous example, we might attempt an expansion of the form

$$f(z) = \frac{A}{z - 1} + \frac{B}{(z + 2i)^2} \tag{3.1.10}$$

However, this will not work as the reader is asked to verify. What goes wrong is related to the fact that the quadratic $(z + 2i)^2$ in the denominator of $f(z)$ dictates that we must have a first-order polynomial in the numerator of this term in (3.1.10). Thus, we next try an expansion of the form

$$\frac{1}{(z + 2i)^2(z - 1)} = \frac{A}{z - 1} + \frac{Bz + C}{(z + 2i)^2} \tag{3.1.11}$$

The numbers A, B, and C can be evaluated by combining the two fractions and equating coefficients of like powers of z. However, we will proceed differently. First, if we multiply both sides of the above by $z - 1$ and then set $z = 1$, we get $A = 1/(1 + 2i)^2$. To evaluate B and C, we cannot just multiply by $z + 2i$ and set $z = -2i$, since this limit won't exist. Instead, we first multiply both sides of (3.1.11) by $(z + 2i)^2$ to get

$$\frac{1}{z - 1} = \frac{A(z + 2i)^2}{z - 1} + Bz + C \tag{3.1.12}$$

Now setting $z = -2i$ yields

$$-2iB + C = -(2i + 1)^{-1}$$

If we now differentiate both sides of (3.1.12) and evaluate the result at $z = -2i$, we get

$$B = -(2i + 1)^{-2}$$

From these last two results we can evaluate the remaining constants B and C in (3.1.11).

■

Let us note in (3.1.11) that if we write

$$Bz + C = B(z + 2i) + (C - 2iB)$$

then

(3.1.13)
$$\frac{1}{(z + 2i)^2(z - 1)} = \frac{A}{z - 1} + \frac{B}{z + 2i} + \frac{\hat{C}}{(z + 2i)^2}$$

where $\hat{C} = C - 2iB$. In Problem 3, the reader is asked to show that if $q(z)$ is the polynomial

(3.1.14)
$$q(z) = (z - z_1)^{n_1}(z - z_2)^{n_2}\ldots(z - z_N)^{n_N}$$

where the $z_k\,(k = 1, 2, \ldots, N)$ are the zeros of q, and if $p(z)$ is another polynomial with $p(z_k) \neq 0\ (k = 1, 2, \ldots, N)$ and such that

$$\text{degree}\,(p) < \text{degree}\,(q)$$

then we have the partial fraction decomposition

(3.1.15)
$$\frac{p(z)}{q(z)} = \sum_{k=1}^{n_1} \frac{A_k^{(1)}}{(z - z_1)^k} + \sum_{k=1}^{n_2} \frac{A_k^{(2)}}{(z - z_2)^k} + \cdots + \sum_{k=1}^{n_N} \frac{A_k^{(N)}}{(z - z_1)^k}$$

The complex constants $\{A_k^{(j)}\}$ can be computed as in the examples. If the degree of p is not strictly less than that of q, then a long division is first needed.

PROBLEMS (The answers are given on page 428.)

1. What are the singular points of the following functions? Which are isolated?

(a) $\dfrac{z^2 - 4}{z + 2}$ (b) $\dfrac{z^2 + 5z + 6}{z^3 + 8}$ (c) $\dfrac{e^z}{1 + z^2}$

(d) $\dfrac{1}{z + \sqrt{1 + z^2}}$ (principal branch)

(e) $\dfrac{z + 4}{z^3 - 6z^2 - 27z}$ (f) $\dfrac{z^4 + 2}{z^4 + 3z^2}$ (g) $\dfrac{\text{Log}(1 + z)}{z}$

(h) $\dfrac{\sqrt{z}}{z + 1}$ (principal branch) (i) $\dfrac{\sqrt{z^2}}{z}$ (principal branch)

2. Write a partial fraction expansion for the following rational functions:

(a) $\dfrac{z}{(z-1)(z+2i)^2}$ (b) $\dfrac{z^2+z+1}{z^3-iz^2+4z-4i}$ (c) $\dfrac{5z+2}{z^3+3z^2+3z+1}$

3. Let $q(z) = (z-z_1)^{n_1}(z-z_2)^{n_2}\ldots(z-z_N)^{n_N}$, where $n_1, \ldots, n_N$ are positive integers and no two of the numbers z_k, $k = 1, 2, \ldots, N$, are equal. Let $p(z)$ be a polynomial with degree strictly less than that of q and not vanishing at any of the z_k. Show that

(a) degree $(q) = \sum_{k=1}^{N} n_k$

(b) The partial fraction decomposition of p/q is as given in (3.1.15), with the numbers $A_k^{(j)}$ given by

$$A_k^{(j)} = \frac{1}{(n_j-k)!}\left(\frac{d}{dz}\right)^{n_j-k}\left[\frac{(z-z_j)^{n_j}p(z)}{q(z)}\right]\Bigg|_{z=z_j}$$

4. If $p(z)$ and $q(z)$ are polynomials with degree $(p) \geq$ degree (q), show that p/q can be decomposed into the sum of a polynomial and a partial fraction decomposition of the form (3.1.15). [*Hint:* First do a long division.]

5. Use the result of Problem 4 to decompose the following:

(a) $\dfrac{z^3+3z+1}{z^2-5z+4}$ (b) $\dfrac{z^4+2}{(z-2)^2z^2}$ (c) $\dfrac{z^3+1}{(z+1)(z-i)}$

3.2 BRANCH POINTS AND BRANCH CUTS

In Chapter 6, we will discuss in detail the behavior of analytic functions in the neighborhood of an isolated singular point. Here, we will consider a special, but important, type of nonisolated singularity—a branch point.

(3.2.1) ***Definition.*** The point $z = z_0$ is a **branch point** of $f(z)$ if there is some small enough circle surrounding z_0 such that when z continuously traverses the circle once, then $f(z)$

(a) Changes continuously with z

(b) But does not return to its original value.

■

(3.2.2) ***Example.*** Show that $z = 0$ is the only branch point of $\log z$.

Solution. First, as in Figure 3.2.1*a*, pick any $z_0 \neq 0$ and consider the circle shown there. For any starting point z_1 on this circle, choose a particular value of $\theta_1 = \arg(z_1)$. Hence, $\log z_1$ is now uniquely defined. From the figure, for z on the circle we have

$$\log z = \ln|z| + i\theta \tag{3.2.3}$$

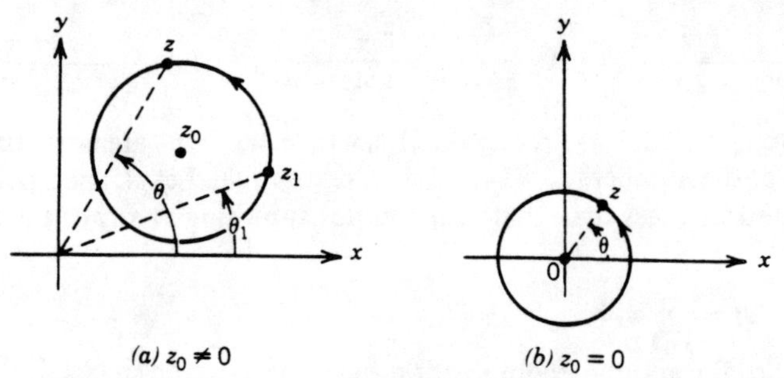

Figure 3.2.1

Now, as z continuously traverses the circle (say, counterclockwise), θ and $\ln|z|$, starting at θ_1 and $\ln|z_1|$, respectively, will change continuously and return to their original values. Thus, $\log z$ undergoes no net change, and $z_0 \neq 0$ is not a branch point.

For $z_0 = 0$, the same analysis as above, and Figure 3.2.1*b*, reveal that θ now changes by 2π in one traversal, and hence $\log z$ suffers a 2π change in its imaginary part. Thus, $z_0 = 0$ is a branch point.

∎

Note from the above example that if z continues to circle the origin, the imaginary part of $\log z$ will continue to increase and the function will *never* return to any of its previous values. Such a point is called a **logarithmic branch point**.

(3.2.4) ***Example.*** Discuss the branch points of the power function $f(z) = z^\alpha$. In particular, consider the case when α is an integer.

Solution. From its definition,

$$f(z) = e^{\alpha \log z} = e^{\alpha[\ln|z| + i\theta]} \tag{3.2.5}$$

Thus, for exactly the same reasons as in the previous example, the only possible branch point is $z = 0$. To treat this case, consider a circle surrounding the origin. Now, start at some point $z = z_1$ on this circle and define $f(z_1) = \exp[\alpha \ln|z_1| + i\theta_1]$ where θ_1 is some choice of $\arg(z_1)$. As z traverses the circle continuously from z_1, $f(z)$ in (3.2.5) will change continuously from its value $f(z_1)$. As the circuit is completed, θ will have increased by 2π, and $f(z)$ will have arrived at the value

$$\exp\{\alpha[\ln|z_1| + i(\theta_1 + 2\pi)]\} = e^{2\pi i\alpha} f(z_1)$$

Hence, if we denote by $\Delta f(z_1)$ the difference between the value of f after one traversal and its original value $f(z_1)$, we have

$$\Delta f(z_1) = (e^{2\pi i\alpha} - 1) f(z_1) \tag{3.2.6}$$

Note that if α is not an integer, then $e^{2\pi i\alpha} \neq 1$, and $\Delta f(z_1) \neq 0$. Thus, the origin *is* a branch point of z^α. Conversely (as should be expected), if α is an integer, the origin is not a branch point.

■

Note in the above example that if we encircle the origin N times counterclockwise and denote by $\Delta_N f(z_1)$ the change in f, then (3.2.6) becomes

(3.2.7)
$$\Delta_N f(z_1) = (e^{2\pi i N\alpha} - 1)f(z_1)$$

Now, if α is the rational number m_0/n_0, where m_0 and n_0 have no common divisors, then f will return to its original value $f(z_1)$ after n_0 circuits (take $N = n_0$ above). In fact, for these α's, f has exactly n_0 distinct values at z_1 (see Sections 1.2 and 2.7). In this case, we say that $z = 0$ is a **branch point of finite order**.

(3.2.8) ***Example.*** Discuss the branch points of $f(z) = \sqrt{z(z-1)}$.

Solution. All of the possible values of $f(z)$ are given by

(3.2.9)
$$f(z) = \exp\left\{\frac{1}{2}\left[\ln(|z||z-1|) + i\arg(z) + i\arg(z-1)\right]\right\}$$

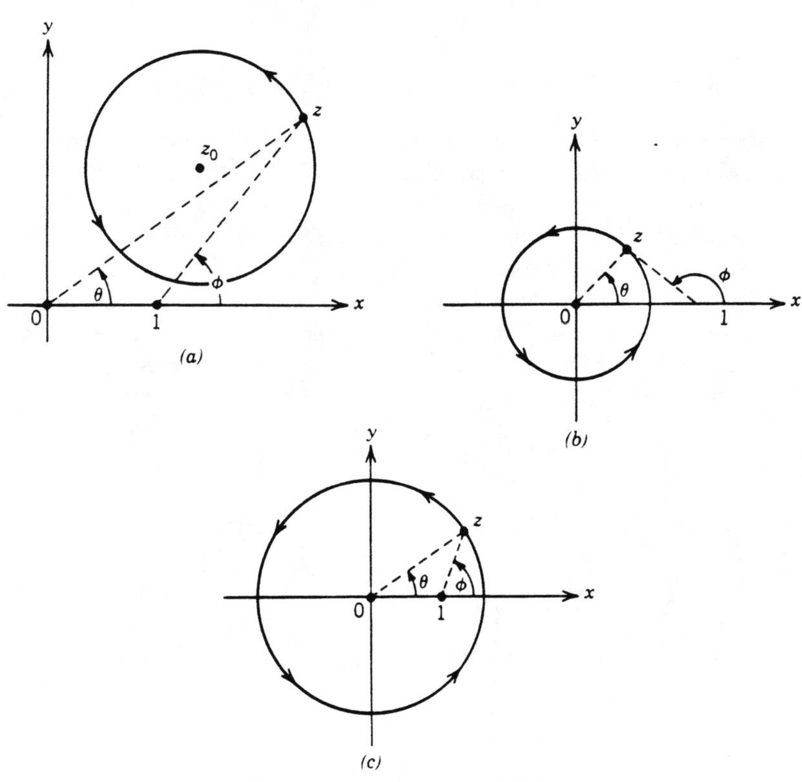

Figure 3.2.2 Example (3.2.8).

First, let us traverse a circle surrounding $z_0 \neq 0$ and small enough so that $z = 0$ and $z = 1$ are not included within it. Then, as in Example (3.2.2), $\theta = \arg(z)$ and $\phi = \arg(z - 1)$ will return to their previous values (see Figure 3.2.2*a*). Hence, the only possible branch points are $z = 0$ and $z = 1$. Now, let us complete a circuit about the origin once along a small enough circle which does not include $z = 1$ within it. Then, starting at some point, $\arg(z)$ will increase by 2π while $\arg(z - 1)$ will return it its original value (see Figure 3.2.2*b*). Hence, by (3.2.9), the value of $f(z)$ after one complete circuit will be the negative of its original value. Exactly the same result will be true as $z = 1$ is encircled once. Hence, both $z = 0$ and $z = 1$ are branch points.

■

In the above example, let us ask what will happen if we encircle both points together. Thus, consider the circle of Figure 3.2.2*c*. In one counterclockwise traversal, both θ and ϕ will increase by 2π, and hence so will $\frac{1}{2}[\arg(z) + \arg(z - 1)]$. Thus, by (3.2.9), there will be no change in $f(z)$.

(3.2.10) ***Example.*** Show that the only possible branch points of

(3.2.11)
$$f(z) = \sqrt{1 - e^z}$$

are at the zeros of $1 - e^z$, that is at $z = 2\pi i n$, $n = 0, \pm 1, \pm 2, \ldots$.

Solution. Let $z_0 \neq 2\pi i n$ for any $n = 0, \pm 1, \ldots$, and consider a domain D_z, as shown in Figure 3.2.3, which does not include within it any of the points $z = 2\pi i n$. Hence, its image D_w under the mapping $w = 1 - e^z$ cannot contain the origin $w = 0$. Now consider a circle C_z which surrounds z_0. Its image must be a curve C_w such as shown in the figure. Hence, because $w = 1 - e^z$ is a continuous, single-valued function of z, as z continuously traverses the circle C_z, w will continuously traverse the curve C_w. Also, since $w = 0$ is not in D_w, it cannot lie inside of C_w. Hence, $\phi = \arg(w)$ (see the figure) will undergo no net change in one circuit. Thus,

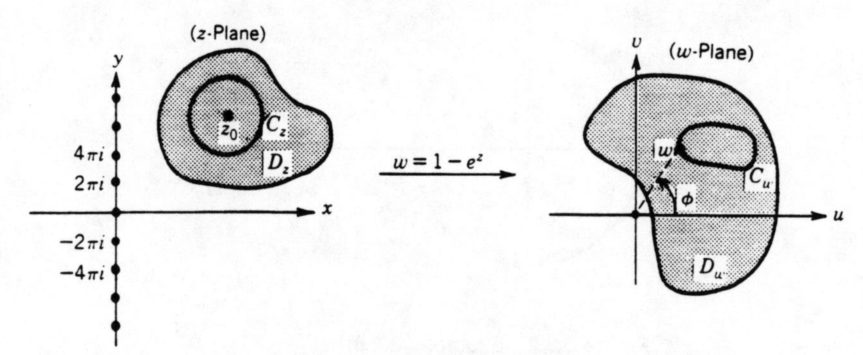

Figure 3.2.3 Example (3.2.10).

$\sqrt{w} = \sqrt{1 - e^z}$ will return to its original value, and z_0 cannot be a branch point.

■

The point at infinity can also be a branch point with the following definition.

(3.2.12) ***Definition.*** The **point at infinity is a branch point** of $f(z)$ if $z = 0$ is a branch point of $f(1/z)$.

■

(3.2.13) ***Example.*** Show that $z = \infty$ is a branch point of $f(z) = \sqrt{1 - z}$ but not a branch point of $g(z) = \sqrt{z(z - 1)}$.

Solution. All of the possible values of $f(1/z)$ are given by

$$f\left(\frac{1}{z}\right) = \exp\left\{\frac{1}{2}\left[\ln\left|\frac{z-1}{z}\right| + i\arg(z-1) - i\arg(z)\right]\right\} \tag{3.2.14}$$

In the above (since we are concerned with $z = 0$), we let z traverse a small circle around $z = 0$. Then, $\arg(z)$ will increase by 2π while $\arg(z - 1)$ undergoes no change. Thus, $-\frac{1}{2}\arg(z)$ in (3.2.14) suffers a change of $-\pi$ and $f(1/z)$ does not return to its original value. Hence, $z = \infty$ is a branch point for $f(z)$.

For $g(z)$, we have

$$g\left(\frac{1}{z}\right) = \exp\left\{\frac{1}{2}\left[\ln\left|\frac{1-z}{z^2}\right| + i\arg(1-z) - 2i\arg(z)\right]\right\} \tag{3.2.15}$$

Here, as z encircles the origin, $\frac{1}{2}[-2\arg(z)]$ suffers a change of -2π, and $g(1/z)$ *does* return to its original value. Hence, $z = \infty$ is *not* a branch point of $g(z)$.

■

Branch Cuts. In Section 2.7, we discussed how a single-valued analytic branch can be chosen for the multiple-valued function $\log z$. For example, the principal branch $\operatorname{Log} z$ was defined by choosing the argument range $-\pi < \arg(z) \leq \pi$. The effect of this is to "cut" the z-plane along the negative real axis, with the resulting single-valued function being analytic in the **cut plane** $-\pi < \arg(z) < \pi$. By cutting the plane in this manner, we prohibit closed circuits around the branch point $z = 0$, thus rendering $\operatorname{Log} z$ single-valued; see Example (3.2.2). Note that we could have cut the z-plane along any other line joining $z = 0$ to $z = \infty$. This, together with a choice of a 2π-range of argument, would then define another branch of $\log z$. Hence, by choosing a single-valued branch of $\log z$, we are performing two separate tasks.

- **(i)** We are choosing a **branch cut** to prevent complete circuits about the branch point, thus rendering the function single-valued.
- **(ii)** We are choosing a range of argument to unambiguously define the function at each point in the cut plane.

Often, the choice of branch cut and range of argument (or arguments) are determined in order to satisfy some additional conditions, as the following examples illustrate.

(3.2.16) ***Example.*** Find a single-valued branch for the function $f(z) = \sqrt{z}$ which is analytic and negative on the positive real axis. For this branch, what is $\sqrt{z}$ when z is real and negative?

Solution. Here, since $z = 0$ is the only branch point, any branch cut for $f(z)$ must extend from $z = 0$ to $z = \infty$ to preclude a closed circuit about $z = 0$. Also, the cut is not to intersect the positive real axis, since f is to be analytic there. Hence, consider the cut of Figure 3.2.4*a*. With $z = |z| e^{i\theta}$,

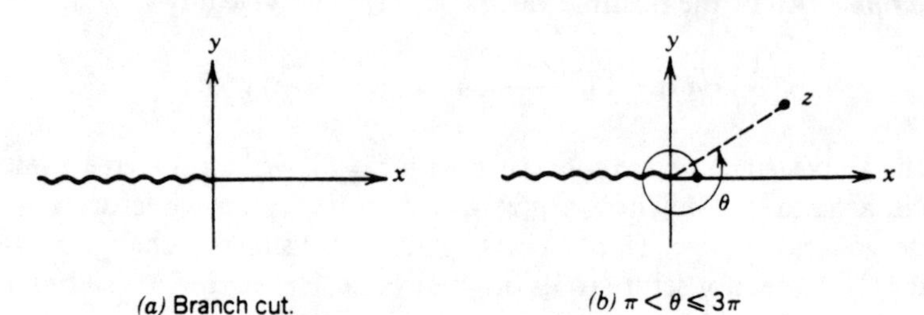

(a) Branch cut. *(b)* $\pi < \theta \leqslant 3\pi$

Figure 3.2.4 Example (3.2.16).

we have

$$\sqrt{z} = |z|^{1/2} e^{i\theta/2}$$
$$-\pi + 2\pi k < \theta \leq \pi + 2\pi k \qquad k = 0, \pm 1, \pm 2, \ldots \tag{3.2.17}$$

where $|z|^{1/2}$ denotes the positive square root of z and the range of θ above is a very general one consistent with the cut. For any value of the integer k in (3.2.17), this branch of $\sqrt{z}$ is analytic on the positive real axis. We must now choose a value of k which makes it negative here. To this end, note from the range of θ that

$$z \text{ is positive when } \theta = 2\pi k. \tag{3.2.18}$$

Hence, from this and (3.2.17), for positive z we have $\sqrt{z} = |z|^{1/2} e^{\pi i k}$. The requirement that this be negative leads to $e^{\pi i k} = -1$. One choice is $k = 1$, and this leads to the branch

$$\sqrt{z} = |z|^{1/2} e^{i\theta/2} \qquad \pi < \theta \leq 3\pi \tag{3.2.19}$$

as depicted in Figure 3.2.4*b*. For this branch, when z is negative we have $\theta = 3\pi$, and hence,

$$\sqrt{z} = -i|z|^{1/2} \qquad \text{(for } z \text{ negative)}$$

■

(3.2.20) ***Example.*** Is it possible to find a branch of $f(z) = \sqrt{z}$ which satisfies all of the following?

(i) It is analytic on the positive real axis.
(ii) It is positive on the positive real axis.
(iii) It is negative imaginary on the negative real axis.

Solution. Note that this example is very similar to the previous one except that now there are more requirements. Hence, we might not be able to *start with a given cut*, as we did before, but may have to choose in the process both the cut and the argument range in order to accomplish our goals. Now the most general cut of the z-plane from $z = 0$ to $z = \infty$ is along the ray $\theta = \theta_0$, $0 \le \theta_0 < 2\pi$, as shown in Figure 3.2.5*a*. Thus, with $z = |z|e^{i\theta}$, the most general branch of $\sqrt{z}$ is

$$\sqrt{z} = |z|^{1/2}e^{i\theta/2}$$
$$\theta_0 + 2\pi k < \theta \le \theta_0 + 2\pi(k+1) \qquad k = 0, \pm 1, \pm 2, \ldots \tag{3.2.21}$$

[The result (3.2.19) of the previous example corresponds in the above to $\theta_0 = \pi$ and $k = 0$.] Now from Figure 3.2.5*a*, this general branch is analytic on the positive real axis only if

$$\theta_0 \neq 0 \tag{3.2.22}$$

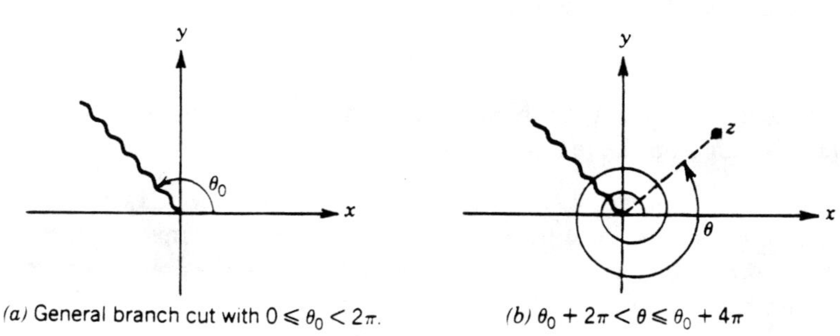

(*a*) General branch cut with $0 \leq \theta_0 < 2\pi$.   (*b*) $\theta_0 + 2\pi < \theta \leq \theta_0 + 4\pi$

Figure 3.2.5 Example (3.2.20).

Also, from (3.2.21) and the figure (and after a few seconds of thought), we have

(a) z positive $\Rightarrow \theta = 2\pi(k+1)$

(b) z negative $\Rightarrow \theta = \begin{cases} \pi + 2\pi k & \text{If } 0 < \theta_0 < \pi \\ \pi + 2\pi(k+1) & \text{If } \pi \le \theta_0 \end{cases}$

Using this in (3.2.21) together with requirement **(ii)** yields

$$\sqrt{z} \text{ positive for } z \text{ positive} \Rightarrow e^{\pi i(k+1)} = 1 \tag{3.2.23}$$

while the demand **(iii)** that $\sqrt{z}$ be negative-imaginary when z is negative leads to

$$e^{(\pi i/2)(2k+1)} = -i \qquad \text{for } 0 < \theta_0 < \pi$$

and

$$e^{(\pi i/2)(2k+3)} = -i \qquad \text{for } \theta_0 \geq \pi$$

This in turn yields

(3.2.24) $$e^{\pi i k} = \begin{cases} -1 & \text{for } 0 < \theta_0 < \pi \\ +1 & \text{for } \pi \leq \theta_0 \end{cases}$$

Now, combining the above with (3.2.23) leads to the following restrictions on θ_0 (the position of the cut) and k (the argument range):

$$e^{\pi i k} = -1 \qquad \text{and} \qquad 0 < \theta_0 < \pi$$

For example, if $k = 1$, then the branch

$$\sqrt{z} = |z|^{1/2} e^{i\theta/2}$$

$$\theta_0 + 2\pi < \theta \leq \theta_0 + 4\pi \qquad 0 < \theta_0 < \pi$$

satisfies all of the desired conditions. The situation is shown in Figure 3.2.5*b*.

Finally, note that since $\theta_0 < \pi$, the branch cut of the previous example would not have sufficed.

■

(3.2.25) ***Example.*** Find a branch of $f(z) = \sqrt{z(z-1)}$ which is analytic and positive for z on the real intervals $x < 0$ and $x > 1$.

Solution. Here, we have two branch points $z = 0, 1$ [see Example (3.2.8)] and, in general, we will need two branch cuts, one emanating from each point (see Figure 3.2.6*a*). From the development in the previous example, if we write

$$z = |z| e^{i\theta} \qquad \text{and} \qquad z - 1 = |z-1| e^{i\phi}$$

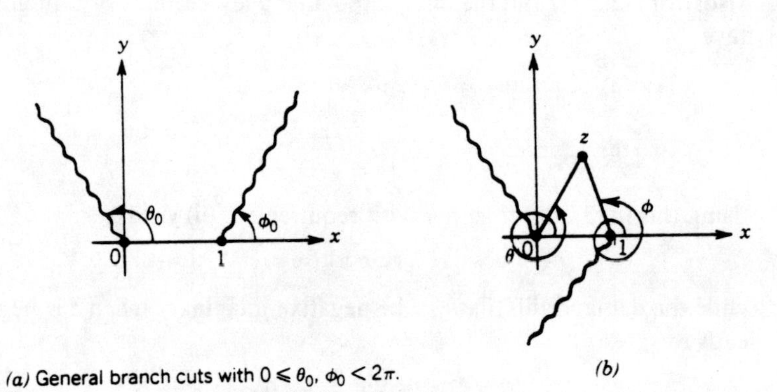

(a) General branch cuts with $0 \leq \theta_0, \phi_0 < 2\pi$.

Figure 3.2.6 Example (3.2.25).

then

$$f(z) = |z|^{1/2}|z - 1|^{1/2} e^{(i/2)(\theta + \phi)} \tag{3.2.26}$$

where the most general ranges of the arguments θ and ϕ are

$$\begin{aligned} &\theta_0 + 2\pi k < \theta \le \theta_0 + 2\pi(k + 1) \qquad 0 \le \theta_0 < 2\pi \\ &\phi_0 + 2\pi l < \phi \le \phi_0 + 2\pi(l + 1) \qquad 0 \le \phi_0 < 2\pi \\ &k, l = 0, \pm 1, \pm 2, \ldots \end{aligned} \tag{3.2.27}$$

Now we seek values of θ_0, ϕ_0, k, and l so that $f(z)$ satisfies the conditions of our problem.

First, since f is to be analytic for $x < 0$ and $x > 1$, we must have (see the figure)

$$\theta_0 \ne 0, \pi \qquad \text{and} \qquad \phi_0 \ne 0, \pi \tag{3.2.28}$$

Now note (as in the pevious example) that, *on the interval* $x > 1$,

$$\theta = 2\pi(k + 1) \qquad \text{and} \qquad \phi = 2\pi(l + 1) \tag{3.2.29}$$

while *for* $x < 0$,

$$\theta = \begin{cases} \pi + 2\pi k & \text{for } 0 < \theta_0 < \pi \\ \pi + 2\pi(k + 1) & \text{for } \theta_0 > \pi \end{cases} \qquad \text{and}$$
$$\phi = \begin{cases} \pi + 2\pi l & \text{for } 0 < \phi_0 < \pi \\ \pi + 2\pi(l + 1) & \text{for } \phi_0 > \pi \end{cases} \tag{3.2.30}$$

Hence, using (3.2.26) and (3.2.29), the condition that $f(z)$ be positive for z in the real interval $x > 1$ yields

$$e^{\pi i(k + l)} = 1 \tag{3.2.31}$$

Now, from (3.2.30), we find that for z *in the interval* $x < 0$,

$$\frac{1}{2}(\theta + \phi) = \pi(k + l) + \begin{cases} \pi & \theta_0 \text{ and } \phi_0 < \pi \\ 2\pi & \theta_0 < \pi \text{ and } \phi_0 > \pi, \text{ or} \\ & \theta_0 > \pi \text{ and } \phi_0 < \pi \\ 3\pi & \theta_0 \text{ and } \phi_0 > \pi \end{cases} \tag{3.2.32}$$

With this, and (3.2.31), $f(z)$ in (3.2.26) becomes, for $x < 0$,

$$f(z) = |z|^{1/2}|z - 1|^{1/2} \cdot \begin{cases} -1 & \theta_0, \phi_0 < \pi \text{ or } \theta_0, \phi_0 > \pi \\ +1 & \theta_0 < \pi \text{ and } \phi_0 > \pi, \text{ or} \\ & \theta_0 > \pi \text{ and } \phi_0 < \pi \end{cases}$$

Thus, if $f(z)$ is to be positive on the interval $x < 0$, we must choose the plus sign above. This, coupled with the specific choice $k = l = 0$ in (3.2.31), leads to the following solution to our problem:

$$\sqrt{z(z-1)} = |z|^{1/2}|z-1|^{1/2}e^{(i/2)(\theta+\phi)}$$

(3.2.33)
$$\theta_0 < \theta \leq \theta_0 + 2\pi \qquad 0 < \theta_0 < \pi$$
$$\phi_0 < \phi \leq \phi_0 + 2\pi \qquad \pi < \phi_0 < 2\pi$$

This situation is depicted in Figure 3.2.6*b*. Note that this branch has the property that for z on the interval $0 < x < 1$, $\theta = 2\pi$ and $\phi = 3\pi$. Thus,

$$\sqrt{x(x-1)} = i[x(1-x)]^{1/2} \qquad \text{for } 0 < x < 1$$

■

(3.2.34) ***Example.*** In the previous example, what would be the situation if we relaxed the conditions on $f(z) = \sqrt{z(z-1)}$ and looked for a branch which was *only to be positive on the interval* x > 1?

Solution. Here, we proceed exactly as before. However, now we first of all replace (3.2.28) by

(3.2.35)
$$\theta_0 \neq 0 \qquad \text{and} \qquad \phi_0 \neq 0$$

since $f(z)$ need not be analytic for $x < 0$. In addition, since $f(z)$ is not required to be positive for z in the interval $x < 0$, we need only impose condition (3.2.31)

(3.2.36)
$$e^{\pi i(k+l)} = 1$$

If we choose $k = l = -1$ above, and $\theta_0 = \phi_0 = \pi$ in (3.2.27), then (3.2.26) and (3.2.27) yield the branch

(3.2.37)
$$\sqrt{z(z-1)} = |z|^{1/2}|z-1|^{1/2}e^{(i/2)(\theta+\phi)} \qquad -\pi < \theta, \phi \leq \pi$$

This branch is depicted in Figure 3.2.7*a*, where we see part of the branch cut which starts at $z = 1$ superimposed on the one starting at $z = 0$. Thus, there appears to be a branch cut on the negative real axis on which $f(z)$ is not analytic. However, note from Figure 3.2.7*a* that for $x_0 < 0$,

$$\lim_{z\to x_0} \theta = \lim_{z\to x_0} \phi = \begin{cases} \pi & \text{Im}(z) > 0 \\ -\pi & \text{Im}(z) < 0 \end{cases}$$

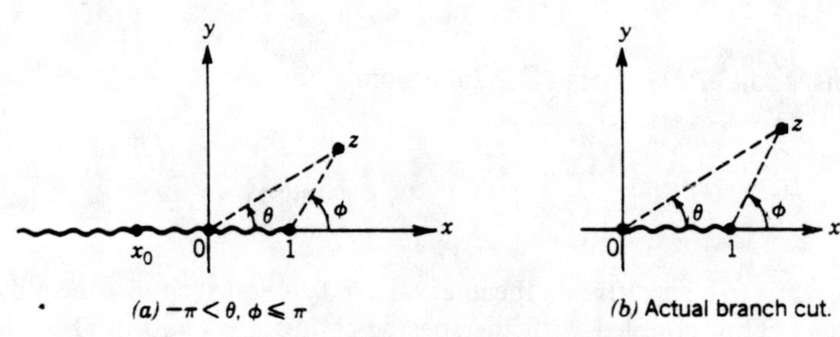

Figure 3.2.7 Example (3.2.34).

Hence, from (3.2.37) we conclude that this branch of $\sqrt{z(z-1)}$ is in fact continuous on the negative real axis. In Problem 5, the reader is asked to show that in fact $f(z)$ is analytic here and thus the branch cut consists of only that part of the real axis joining the two branch points $z = 0$ and $z = 1$, as shown in Figure 3.2.7*b*.

Finally, note from (3.2.37) that

$$\sqrt{z(z-1)} = \sqrt{z} \cdot \sqrt{z-1}$$

where the branches of $\sqrt{z}$ and $\sqrt{z-1}$ on the right-hand side are the principal branches (since $-\pi < \theta, \phi \le \pi$). However, (3.2.37) does not provide the values of the principal branch of $\sqrt{z(z-1)}$. By this we mean the following. For example, if $z_0 = -2$, we have

$$\text{(principal branch)} \sqrt{z_0} = i\,2^{1/2}$$

and

$$\text{(principal branch)} \sqrt{z_0 - 1} = i\,3^{1/2}$$

Thus, from (3.2.37),

$$\sqrt{z_0(z_0-1)} = -6^{1/2}$$

However,

$$\text{(principal branch)} \sqrt{z_0(z_0-1)} = 6^{1/2}$$

■

Let us make the following observations. In Examples (3.2.16) and (3.2.20), both $z = 0$ and $z = \infty$ are branch points of $\sqrt{z}$, and the branch cuts joined these two branch points (see Figures 3.2.4 and 3.2.5). In Example (3.2.25), only $z = 0$ and $z = 1$ are branch points of $\sqrt{z(z-1)}$, and the possible branch cuts are as shown in Figure 3.2.6. Here, the cuts start from the two branch points and extend to infinity. We can view this in some sense as a line joining the two branch points, but going through $z = \infty$. Finally, in Example (3.2.34), where again only $z = 0$ and $z = 1$ are branch points, the cut was that piece of the real axis joining the two branch points (see Figure 3.2.7*b*). Hence, these examples point out that, in general, branch cuts must start and end at branch points.

(3.2.38) ***Example.*** Consider the function

$$f(z) = [z(z-1)(z-i)]^{1/4}$$

Clearly, $z = 0, 1$, and i are branch points. In addition, $z = \infty$ is also a branch point, since $f(1/z)$ has a branch point at $z = 0$. From Figure 3.2.8*a*, we see that with

(3.2.39)
$$f(z) = |z(z-1)(z-i)|^{1/4} e^{i(\theta_1+\theta_2+\theta_3)/4}$$

no branch cut can be chosen which would permit a closed circuit around any pair of branch points, nor around all three, and still yield a single-valued function. Thus, the branch cuts must join all three branch points

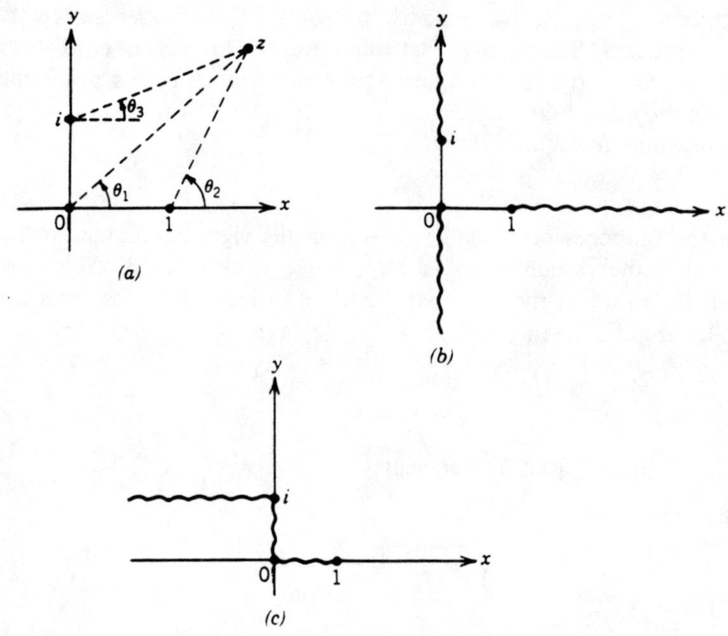

Figure 3.2.8 Example (3.2.28).

and go off to infinity. Part (*b*) of the figure shows cuts which accomplish these ends. Note that the three branch points are joined to $z = \infty$ and can be viewed as being connected to each other through infinity. Also note that the branch cuts in part (*c*) of the figure serve equally well to define a single-valued branch of $f(z)$.

■

The subject of branch points and branch cuts can be given a beautiful geometric interpretation with the introduction of **Riemann surfaces**. The reader is referred to a more advanced textbook for a discussion of this topic.

PROBLEMS (The answers are given on page 428.)

1. What are the branch points (including $z = \infty$) for the following functions?

(a) $\sqrt{1+z^2}$ (b) $[(z+1)/(z-1)]^{1/2}$
(c) $[z^2(z-1)]^{1/3}$ (d) $\log[z(1-z)]$
(e) $[(z-\alpha_1)(z-\alpha_2)\cdots(z-\alpha_n)]^{1/2}$ $\alpha_i \neq \alpha_j$
(f) $\log(1+z^2)$ (g) $\sqrt{\log z}$
(h) $[z(z-i)]^{3/2}$ (i) $\log(1+\sqrt{z})$
(j) $(z^2+\sqrt{z^4-1})^{1/3}$ (k) $\log(1+\log z)$

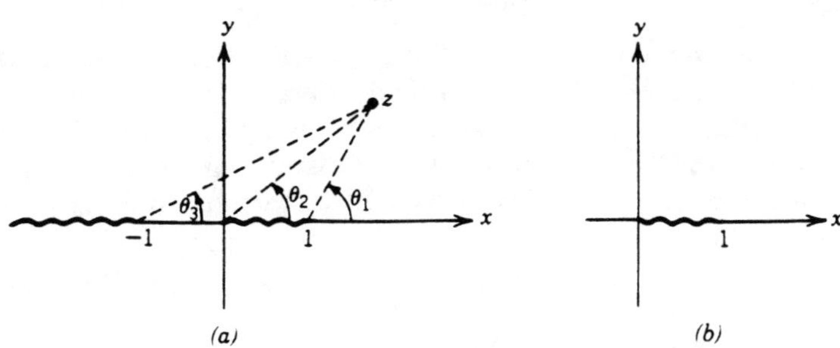

Figure 3.2.9

2. (a) Show that, for any branch, $\sqrt{z}$ is real only when z is real and positive, while $z^{1/3}$ is real only when z is real.
   (b) Generalize the results in (a) to the function $z^{1/n}$.
3. (a) Use the evenness of the cosine function to show that $\cos\sqrt{z}$ has no branch points.
   (b) Show that $z = 0$ is a branch point for $\sin\sqrt{z}$.
   (c) Show that $z = 0$ is not a branch point for $[\sin(\sqrt{z}]/\sqrt{z}$.
4. Let $f(z) = \sqrt{z^2}$.
   (a) Show that $f(z)$ has no branch points.
   (b) If the principal branch of the square root is chosen, show that $f(z)$ is not entire [see Example (2.7.21) of Section 2.7].
5. Consider that branch of $\sqrt{z(z-1)}$ defined as in equation (3.2.37) and depicted in Figure 3.2.7*a*. Show that this branch is analytic on the negative real axis.
6. Consider the function $f(z) = \sqrt{z(z^2-1)}$.
   (a) What are the branch points of f?
   (b) Show that the cuts in Figure 3.2.9*a* are permissible branch cuts.
   (c) Choose ranges of θ_1, θ_2, and θ_3 consistent with the cuts in the figure and such that $f(x) > 0$ for $x > 1$. For these ranges, what are the values of $f(x)$ for $0 < x < 1$? For $-1 < x < 0$? For $x < -1$?
7. Consider $g(z) = -i\log(z + \sqrt{z^2-1})$.
   (a) What are the branch points of g?
   (b) Choose a branch of g so that $0 \leq g(x) \leq \pi$ when $-1 \leq x \leq 1$.
8. Consider the function $h(z) = \log(z + \sqrt{z})$.
   (a) In addition to $z = 0$, when is $z = 1$ also a branch point?
   (b) For the case in (a), why is the branch cut shown in Figure 3.2.9*b* inappropriate?

3.3 HARMONIC FUNCTIONS AND HARMONIC CONJUGATES

Here, we will investigate some additional consequences of the Cauchy–Riemann equations

(3.3.1) $$u_x = v_y \quad \text{and} \quad u_y = -v_x$$

which were discussed in Section 2.2. We will show later (Section 5.3) that an analytic function $f(z) = u + iv$ possesses derivatives of all orders, and thus its real part $u(x, y)$ and imaginary part $v(x, y)$ will possess partial derivatives of all orders. For the time being, however, let us assume that both u and v possess continuous second partial derivatives. Thus, from advanced calculus, their mixed second partial derivatives are equal; and from (3.3.1), we find

(3.3.2)
$$u_{xx} = v_{xy} = -u_{yy} \Rightarrow u_{xx} + u_{yy} = 0$$

with a similar result for v.

(3.3.3) ***Definition.*** A real-valued function $\phi(x, y)$ is said to be **harmonic in a domain** D if in D it satisfies **Laplace's equation** $\nabla^2\phi \equiv \phi_{xx} + \phi_{yy} = 0$, where the symbol $\nabla^2\phi$, the **Laplacian** of ϕ, is defined by the first equality above.[1]

■

With this definition and (3.3.2), we have the following result.

(3.3.4) ***Theorem.*** If the real and imaginary parts $u(x, y)$ and $v(x, y)$ of an analytic function $f(z) = u + iv$ possess continuous second partial derivatives in a domain D, then they are both harmonic in D.

■

(3.3.5) ***Example.*** Show that the function $\phi(x, y) = e^{\alpha x} \sin(\alpha y)$, where α is real, is harmonic everywhere, while $\psi(x, y) = \ln(x^2 + y^2)^{1/2}$ is harmonic in every domain not containing the origin.

Solution. For the function ϕ, we have $\phi_{xx} = \alpha^2 e^{\alpha x} \sin(\alpha y)$ and $\phi_{yy} = -\alpha^2 e^{\alpha x} \sin(\alpha y)$. Thus, $\phi_{xx} + \phi_{yy} = 0$ everywhere. For ψ, we can proceed in the same way, but let us verify the result using the above theorem. We first note that the principal logarithm, $\operatorname{Log} z$, is analytic except when $z = 0$ or when z lies on the negative real axis. Thus, since $\psi = \ln(x^2 + y^2)^{1/2}$ is the real part of $\operatorname{Log} z$, by the above Theorem ψ is harmonic except possibly in a neighborhood of the negative real axis and the origin. Let us now show that ψ is also the real part of another function which *is* analytic on the negative real axis. Such a function is $g(z) = \log z$, where we now choose that branch of the logarithm so that the argument of z lies in the range $0 < \arg(z) \leq 2\pi$.

■

One consequence of Theorem (3.3.4) is that *not every sufficiently differentiable, real-valued function* u(x, y) *can be the real (or imaginary) part of an analytic function.* Now, by necessity, it must be harmonic.

[1] Oftentimes, $\Delta\phi$ is used as a symbol for the Laplacian of ϕ.

(3.3.6) ***Example.*** The function $u_1(x, y) = x \sin y$ cannot possibly be the real part of an analytic function since it is *not* harmonic. However, the function $u_2(x, y) = xy$, since it is harmonic, *might* be the real (or imaginary) part of an analytic function.

■

(3.3.7) ***Definition.*** Let $u(x, y)$ be real-valued and harmonic in a domain D. Then, $v(x, y)$ is said to be a **harmonic conjugate** of u in D if the complex function $f(z) = u + iv$ is analytic in D.

■

(3.3.8) ***Example.*** Find a harmonic conjugate $v(x, y)$ for the harmonic function $u(x, y) = x + y$.

Solution. Since u and v are to be the real and imaginary parts, respectively, of an analytic function, the Cauchy–Riemann equations (3.3.1) must hold. These lead to the following two equations for v:

$$v_x = -1 \quad \text{and} \quad v_y = 1 \tag{3.3.9}$$

We integrate the first equation with respect to x to find

$$v(x, y) = -x + g(y)$$

where $g(y)$ is an as yet undetermined function of the remaining variable y. Using this result in the second equation, we get

$$g'(y) = 1 \Rightarrow g(y) = y + k \qquad k = \text{constant}$$

Thus, a harmonic conjugate is $v(x, y) = -x + y + k$ (*note that* v *is not unique*). Finally, as is easily shown, u and v are the real and imaginary parts of the analytic function

$$f(z) = (1 - i)z + ik \tag{3.3.10}$$

■

In Problem 18 of Section 5.2, we will show, in much the same way as in the previous example but using contour integrals, that single-valued **harmonic conjugates can always be found in simply connected domains, and are unique up to an additive constant.** (A simply connected domain is, loosely speaking, a domain with no "holes" in it.)

(3.3.11) ***Example.*** Consider again the real-valued function $u(x, y) = x \sin y$ of Example (3.3.6). From there, we know that u cannot be the real part of an analytic function since it is not harmonic. It is informative to see what goes wrong if we attempt to find a harmonic conjugate $v(x, y)$ using the procedure of the last example. Here, the Cauchy–Riemann equation $v_x = -u_y$ yields

$$v_x = -x \cos y \Rightarrow v = -\frac{1}{2}x^2 \cos y + g(y)$$

and the second Cauchy–Riemann equation $v_y = u_x$ yields

$$\frac{1}{2}x^2 \sin y + g'(y) = \sin y$$

Thus, $g'(y) = \sin y - \frac{1}{2}x^2 \sin y$. Since g is a function *only* of y, and x and y are independent, this equation cannot possibly be satisfied in any domain, and thus to complex conjugate $v(x, y)$ can exist.

■

The following two interesting results often prove useful in applications.

(3.3.12) ***Theorem.*** Let $u(x, y)$ be harmonic and possess continuous first partial derivatives in some neighborhood of the point (x_0, y_0). Let $v(x, y)$ be a harmonic conjugate of u, and consider the **level curves**

$$\begin{aligned} C_1&: \quad u(x, y) = u(x_0, y_0) \\ C_2&: \quad v(x, y) = v(x_0, y_0) \end{aligned} \tag{3.3.13}$$

which intersect at (x_0, y_0). Then these curves are orthogonal at (x_0, y_0) if *not both* $u_x(x_0, y_0)$ and $v_x(x_0, y_0)$ vanish (see Figure 3.3.1).

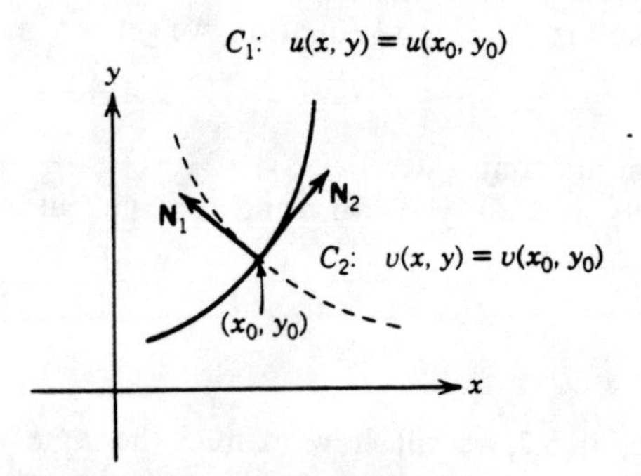

Figure 3.3.1 Theorem (3.3.12).

Proof. First consider the function $u_x^2(x, y) + v_x^2(x, y)$. Since this is assumed continuous and nonvanishing at (x_0, y_0), there is a whole neighborhood of (x_0, y_0) where it does not vanish. Also, since $v_x = -u_y$ and $u_x = v_y$, we have $u_x^2 + v_x^2 = u_x^2 + u_y^2 = v_x^2 + v_y^2$. Hence, there is a neighborhood of (x_0, y_0) in which $u_x^2 + u_y^2$ and $v_x^2 + v_y^2$ do not vanish. Thus, from calculus, C_1 and C_2 possess normal vectors $\mathbf{N}_1$ and $\mathbf{N}_2$, respectively, at (x_0, y_0) given by

$$\begin{aligned} \mathbf{N}_1 &= u_x(x_0, y_0)\mathbf{i} + u_y(x_0, y_0)\mathbf{j} \\ \mathbf{N}_2 &= v_x(x_0, y_0)\mathbf{i} + v_y(x_0, y_0)\mathbf{j} \end{aligned} \tag{3.3.14}$$

Taking the dot product, we have $\mathbf{N}_1 \cdot \mathbf{N}_2 = u_x v_x + u_y \cdot v_y = 0$, where the second equality follows from the Cauchy–Riemann equations. Hence, $\mathbf{N}_1$ and $\mathbf{N}_2$, and thus C_1 and C_2, are orthogonal at (x_0, y_0).

■

This result can be interpreted in terms of analytic functions as follows:

(3.3.15) ***Theorem.*** Let $f(z) = u(x, y) + iv(x, y)$ be analytic and twice continuously differentiable in a neighborhood of $z_0 = x_0 + iy_0$, with $f'(z_0) \neq 0$.[2] Then, the curves $u(x, y) = u(x_0, y_0)$ and $v(x, y) = v(x_0, y_0)$ are orthogonal at (x_0, y_0).

Proof. Since f is analytic and f' continuous, $u(x, y)$ and $v(x, y)$ are harmonic conjugates with continuous first partial derivatives. Also, since $f'(z_0) = u_x(x_0, y_0) + iv_x(x_0, y_0) \neq 0$, not both u_x and v_x vanish at (x_0, y_0). Hence, the result follows from the previous theorem.

■

(3.3.16) ***Example.*** Consider $f(z) = z^2 = x^2 - y^2 + 2ixy$. At the point $z_0 = 1 + i$, the curves of the above theorem are

$$x^2 - y^2 = 0 \quad \text{and} \quad xy = 1$$

In Figure 3.3.2, we sketch these curves in a neighborhood of $z_0 = 1 + i$. Clearly, they are orthogonal at z_0. Note that $f'(1 + i) = 2(1 + i) \neq 0$.

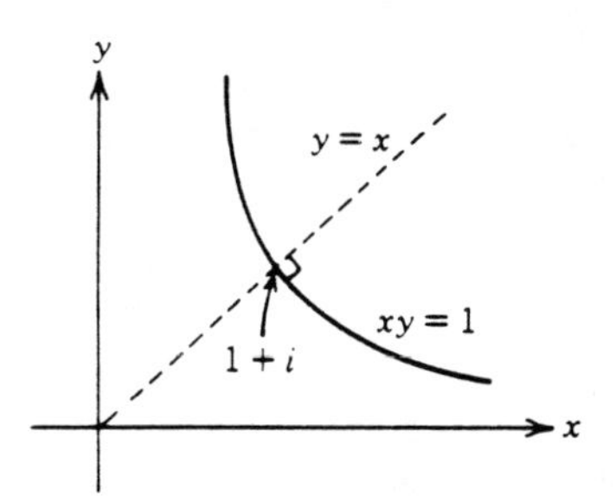

Figure 3.3.2 Example (3.3.16).

■

As we have shown, the real and imaginary parts of a twice continuously differentiable analytic function must be harmonic. Hence, we might expect there to be a profound relationship between solutions of Laplace's equation and analytic functions. In the applications which follow, we shall show that indeed there is, and in Chapter 4 we show how we can exploit this and solve some boundary value problems for Laplace's equation.

[2] Such functions are called **conformal**, and in Chapter 4 we shall discuss some additional consequences of this.

PROBLEMS (The answers are given on page 428.)

(Problems with an asterisk are more theoretical.)

1. Compute the Laplacian, where it exists, of the following functions. Indicate which are harmonic.
   (a) $ax + by$ (b) $x^2 + y^2$ (c) $x^2 - y^2$
   (b) $\dfrac{x^2 - y^2}{x^2 + y^2}$ (e) $\ln(xy)$ (f) $e^{ax}\sin(by)$

2. (a) Let $\phi(x, y)$ possess continuous second partial derivatives in some domain D. Under the polar coordinate change of variables $x = r\cos\theta$ and $y = r\sin\theta$, we can define the composite function $\hat{\phi}(r,\theta) = \phi(r\cos\theta, r\sin\theta)$. Use the chain rule to show that

$$\phi_{xx} + \phi_{yy} = \frac{1}{r}\frac{\partial}{\partial r}\left(\frac{r\partial\hat{\phi}}{\partial r}\right) + \frac{1}{r^2}\frac{\partial^2\hat{\phi}}{\partial\theta^2}$$

   (b) Use the result in (a) to show that

$$\phi = \frac{y(1 + x^2 + y^2)}{1 - 2(x^2 - y^2) + (x^2 + y^2)^2}$$

   is harmonic.

3. (a) Use two methods to verify that $e^{x^2-y^2}\cdot\sin(2xy)$ is harmonic. [*Hint:* Consider e^{z^2}.]
   (b) If $f(z)$ is entire with f'' continuous, why is $\ln|f(z)|$ harmonic in some neighborhood of every point where $f \neq 0$?

4. (a) If $\phi(x, y)$ is harmonic, show that

$$u(x, y) = \phi\left(\frac{y}{x^2 + y^2}, \frac{x}{x^2 + y^2}\right)$$

   is harmonic. [*Hint:* Use polar coordinates and the result of Problem 2a, together with the chain rule.]
   (b) Let $\phi(\alpha)$ be a function of one argument and have a continuous derivative. For which ϕ's are the following composite functions harmonic? [*Hint:* Derive a differential equation for $\phi(\alpha)$.]
   (i) $u(x, y) = \phi(xy)$
   (ii) $u(x, y) = \phi(x^2 + y^2)$

5. Since $x = \frac{1}{2}(z + \bar{z})$ and $y = (1/2i)(z - \bar{z})$, we can view any real-valued function $\phi(x, y)$ as a function $\psi(z, \bar{z})$ of the two "independent" variables z and $\bar{z}$ defined by $\psi(z, \bar{z}) = \phi(\frac{1}{2}(z + \bar{z}), (1/2i)(z - \bar{z}))$ (also see Problem 16 of Section 2.2). In the following, assume ϕ is sufficiently differentiable.
   (a) Use the chain rule to show that

$$\frac{\partial^2\psi}{\partial z\,\partial\bar{z}}(z, \bar{z}) = \frac{1}{4}\nabla^2\phi$$

(b) If ϕ is harmonic, integrate the equation for $\psi(z, \bar{z})$ in (a) and show that $\phi(x, y)$ can be written as the sum $f(z) + g(\bar{z})$ of a function of z and a function of $\bar{z}$.

(c) Verify the results in (b) for the following harmonic functions ϕ, and show that $g(\bar{z}) = \overline{f(z)}$.

(i) $ax + by$ (ii) xy (iii) $x^2 - y^2$
(iv) $e^x \cos y$ (v) $e^y \cos x$

(d) Use the result of part (a) to find a particular solution to the following inhomogeneous Laplace equations. [*Hint:* Express the right-hand sides in terms of z and $\bar{z}$ and then integrate the equation for $\partial^2\psi/(\partial z\, \partial\bar{z})$.]

(i) $\nabla^2\phi = 1$ (ii) $\nabla^2\phi = ax + by$ (iii) $\nabla^2\phi = xy$

6. A function $\phi(x, y)$, which possesses continuous partial derivatives through fourth order, is **biharmonic** if $\nabla^4\phi = \phi_{xxxx} + 2\phi_{xxyy} + \phi_{yyyy} = 0$, where the symbol ∇^4 is the **biharmonic operator**.

(a) Show that $\nabla^4\phi = \nabla^2(\nabla^2\phi)$

(b) If ϕ_1, ϕ_2, and ϕ_3 are all harmonic, show that $x\phi_1 + y\phi_2 + \phi_3$ is biharmonic.

(c) If $\phi(x, y)$ is biharmonic and we define $\psi(z, \bar{z})$ as in Problem 5, show that $\partial^4\psi/(\partial z^2\, \partial\bar{z}^2) = 0$.

7. Find harmonic conjugates $v(x, y)$ for the following functions.

(a) $u = x^2y - (1/3)y^3 + xy$ (b) $u = e^{2y}\sin(2x)$
(c) $u = x^2 - (y - 1)^2$ (d) $u = e^{ax}\cos(by)$
(e) $u = x^2 + axy + y^2$ (f) $u = y + 3xy^2 - x^3$
(g) $u = (x\cos y - y\sin y)e^x$ (h) $u = \sin(x)\sinh(y)$
(i) $u = y/(x^2 + y^2)$ (j) $u = (x^2 - y^2)/(x^2 + y^2)$

[*Hint:* In (i) and (j), use the Cauchy–Riemann equations in polar coordinates to find the harmonic conjugates.]

8.* Show that if v is a harmonic conjugate of u, then $-u$ is a harmonic conjugate of v.

9.* Prove that if a harmonic conjugate of $u(x, y)$ exists in a domain D, then it is unique up to an additive constant. [*Hint:* If v_1 and v_2 are two harmonic conjugates, show that $v_1 - v_2 =$ constant. Use the fact that a domain is connected.]

10. Let $\phi(\alpha)$ be a real-valued function of one argument which possesses a continuous second derivative with respect to that argument. Let $u(x, y)$ be a harmonic function which vanishes nowhere and such that $u_x^2 + u_y^2 \neq 0$. Let $v(x, y)$ be a harmonic conjugate of u in a domain D. Consider the function $\psi(x, y) = \phi(u(x, y) \cdot v(x, y))$.

(a) Show that $\nabla^2\psi = [(uv_x + vu_x)^2 + (uv_y + vu_y)^2]\phi''$.

(b) Conclude that ψ is harmonic if and only if $\phi(\alpha) = a + b\alpha$. [*Hint:* Show that the term in brackets in (a) can never vanish.]

11.* Here, we prove that harmonic conjugates always exist in a disk. Assume that $u(x, y)$ is harmonic with continuous second partial derivatives inside a disk, and let (x_0, y_0) be any point in the disk.

(a) Integrate the Cauchy–Riemann equation $v_x = -u_y$ to get $v(x, y) = g(y) - \int_{x_0}^{x} u_y(\xi, y)\,d\xi$, where g is an arbitrary function of y. [We can do this since the line segment joining (x_0, y) and (x, y) lies in the disk.]

(b) Use Leibnitz's rule from calculus to conclude that $v_y(x, y) = g'(y) - \int_{x_0}^{x} u_{yy}(\xi, y)\,d\xi$.

(c) Use the fact that u is harmonic to conclude from (b) that $v_y = g'(y) + u_x(x, y) - u_x(x_0, y)$.

(d) Finally, use (c) and the Cauchy–Riemann equation $v_y = u_x$ to conclude that $v(x, y) = \int_{y_0}^{y} u_x(x_0, \eta)\,d\eta - \int_{x_0}^{x} u_y(\xi, y)\,d\xi$ + constant.

12. Here, we will show how to "formally" construct an analytic function from its real part $u(x, y)$. The reader is urged to read Problem 5 first.

(a) If $f(z) = u(x, y) + iv(x, y)$ is analytic, use the Cauchy–Riemann equations to show that $\overline{f(\bar{z})}$ is also analytic in z. [*Hint:* $\overline{f(\bar{z})} = u(x, -y) - iv(x, -y)$.]

(b) Let $u(x, y)$ be the real part of an analytic function $f(z)$. Then, $u(x, y) = \frac{1}{2}[f(z) + \overline{f(z)}]$. Use part (a) to show that $u(x, y) = \frac{1}{2}[f(z) + g(\bar{z})]$, where g is an analytic function of its argument.

(c) Replace x and y in (b) in terms of z and $\bar{z}$ (as in Problem 5) and set $\bar{z} = 0$ to show that $f(z) = 2u(z/2, z/2i)$ + constant.

(d) Verify the result in (c) for the harmonic functions $u(x, y)$ given by
(i) $x + y$ (ii) $x^2 - y^2$ (iii) $e^x \cos y$

13. (a) Write the following functions in the form $u + iv$ and verify that the level curves $u(x, y) = u(0, 0)$ and $v(x, y) = v(0, 0)$ are orthogonal at $z = 0$.
(i) z (ii) $z + z^2$ (iii) $\sin z$ (iv) e^z

(b) Consider $f(z) = z^2$. At what angle do the level curves $u(x, y) = u(0, 0)$ and $v(x, y) = v(0, 0)$ intersect? Why doesn't the result of Theorem (3.3.15) hold?

3.4 STEADY, IDEAL, IRROTATIONAL, INCOMPRESSIBLE FLUID FLOWS

In Appendix C, we derive the basic equations and boundary conditions governing the fluid velocity $\mathbf{v}(x, y)$ of a fluid flowing in a source-free region past a rigid stationary obstacle. These equations are

(3.4.1)

(a) $\mathbf{v}(x, y) = v_1(x, y)\mathbf{i} + v_2(x, y)\mathbf{j}$

(b) $\dfrac{\partial v_1}{\partial x} + \dfrac{\partial v_2}{\partial y} = 0$ (in a source-free region)

(c) $\dfrac{\partial v_1}{\partial y} - \dfrac{\partial v_2}{\partial x} = 0$ (irrotational fluid)

(d) $\mathbf{v} \cdot \mathbf{n} = 0$ (on the boundary of a rigid body)

In (d), $\mathbf{n}$ is the unit outer normal field to a curve C which serves as the boundary of a stationary, rigid, two-dimensional body (see Figure 3.4.1). This equation expresses the fact that no fluid can penetrate a rigid body.

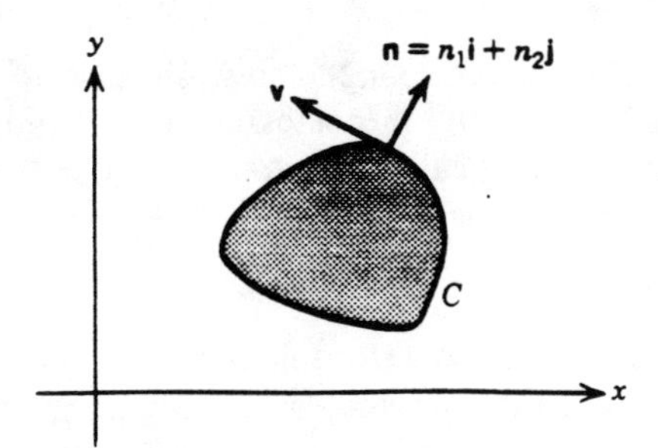

Figure 3.4.1 Velocity vector on the boundary of a rigid body.

One of the difficulties in solving any fluid flow problem is that equations (3.4.1b and c) are coupled partial differential equations for the two scalar functions $v_1(x, y)$ and $v_2(x, y)$. However, we note that *if there were to exist* a sufficiently differentiable scalar function $\phi(x, y)$ such that

(3.4.2) $$v_1 = \phi_x \quad \text{and} \quad v_2 = \phi_y$$

then (3.4.1c) would automatically be satisfied (such a function is called a **velocity potential function**[1]). With the above, the remaining differential equation (3.4.1b) becomes Laplace's equation

$$\phi_{xx} + \phi_{yy} = 0 \qquad \text{(in a source-free region)}$$

Thus, we see that *the potential function* $\phi(x, y)$ *is harmonic in a source-free region.* Finally, the boundary condition (3.4.1d) becomes, in terms of ϕ,

(3.4.3) $$\frac{\partial \phi}{\partial n} \equiv n_1\phi_x + n_2\phi_y = 0 \qquad \text{(on the boundary of a stationary, rigid body)}$$

where $\partial\phi/\partial n$ denotes the **normal derivative** of ϕ, that is, the rate of change of ϕ in the direction of a normal to this boundary curve. In Chapter 4, we shall use conformal mappings to solve **boundary value problems** such as above. However, here we shall be less ambitious.

The Stream Function and Complex Potential. What we have accomplished by our manipulations so far is, first of all, to have changed our problem from one of finding *two* scalar functions v_1 and v_2 satisfying (3.4.1b and c) to one of finding a single harmonic function ϕ, from which we can compute v_1 and v_2 by (3.4.2). Second, since ϕ is harmonic, we can use the results of Section 3.3 concerning the relationship between real harmonic functions and complex analytic functions.

To begin with, we represent the real two-dimensional fluid velocity vector $\mathbf{v} = v_1\mathbf{i} + v_2\mathbf{j}$ as a **complex fluid velocity function** $V(z)$ defined by

(3.4.4) $$\begin{aligned} V(z) &= v_1(x, y) + i\,v_2(x, y) \\ &= \phi_x + i\phi_y \end{aligned}$$

[1] In advanced calculus we learn that such a function always exists in certain regions.

where the second equality follows from (3.4.2). Now, since ϕ is harmonic, by Section 3.3 [see paragraph after (3.3.10)] it possesses, in a suitable domain, a harmonic conjugate $\psi(x, y)$ which here is called the **stream function**. Together, ϕ and ψ satisfy the Cauchy–Riemann equations

$$\phi_x = \psi_y \quad \text{and} \quad \phi_y = -\psi_x$$

Hence, we can combine the two real-valued harmonic conjugates $\phi(x, y)$ and $\psi(x, y)$ into a single analytic function $\Phi(z)$, called the **complex potential**, and which is defined by

(3.4.5) $$\Phi(z) = \phi(x, y) + i\psi(x, y)$$

The derivative of $\Phi(z)$ is given by $\Phi'(z) = \phi_x + i\psi_x = \phi_x - i\phi_y$, where the second equality follows from the Cauchy–Riemann equations. This and (3.4.4) lead to the following relationship between the complex velocity and complex potential:

(3.4.6) $$V(z) = \overline{\Phi'(z)}$$

Let us collect our results.

(3.4.7) ***Complex Function Formulation***

- **(a)** $\mathbf{v}(x, y) = v_1(x, y)\mathbf{i} + v_2(x, y)\mathbf{j}$ (fluid velocity field)
  $\mathbf{v} \cdot \mathbf{n} = 0$ (on a rigid body)
- **(b)** $v_1 = \phi_x, \quad v_2 = \phi_y$ (velocity potential ϕ)
  $\phi_{xx} + \phi_{yy} = 0$ (in a source-free region)
- **(c)** $\phi_x = \psi_y, \quad \phi_y = -\psi_x$ (stream function ψ)
- **(d)** $V(z) = v_1(x, y) + i v_2(x, y)$ (complex velocity)
- **(e)** $\Phi(z) = \phi(x, y) + i\psi(x, y)$ (complex potential)
- **(f)** $V(z) = \overline{\Phi'(z)}$
- **(g)** $\dfrac{\partial \phi}{\partial n} = 0$ and $\psi = \text{constant}$ (on the boundary of a rigid body)

■

The second part of (g) above (which will be proved below) relates the stream function $\psi(x, y)$ in a very interesting way to the boundary of a rigid stationary body.

(3.4.8) ***Theorem.*** The *stream function* ψ(x, y) *is constant on the boundary of a rigid stationary body*. Conversely, the level curves $\psi(x, y) = $ constant can be considered to be the boundary of a rigid body immersed in a fluid flow.

Proof. We will present the proof in a step-by-step format.

(a) From (3.4.1d), $\mathbf{v}$ must, at each point, be tangent to the boundary of a rigid body. Hence, by (3.4.2), the vector $\phi_x\mathbf{i} + \phi_y\mathbf{j}$ is tangent to this boundary curve.

(b) From the Cauchy–Riemann equations (3.4.7c), we have $\phi_x\psi_x + \phi_y\psi_y = 0$. Thus, the vector $\psi_x\mathbf{i} + \psi_y\mathbf{j}$ is orthogonal to

$\phi_x\mathbf{i} + \phi_y\mathbf{j}$. This, together with **(a)**, shows that $\psi_x\mathbf{i} + \psi_y\mathbf{j}$ is orthogonal to the boundary curve.

(c) From advanced calculus, $\psi_x\mathbf{i} + \psi_y\mathbf{j}$ is also normal to the level curves $\psi(x, y) = \text{constant}$.

(d) Finally, with **(b)** and **(c)** we conclude that ψ is constant on the boundary of a rigid body.

■

From this, we can see why ψ is called a stream function—nonviscous fluid flows (*streams*) along the curves $\psi(x, y) = \text{constant}$. These curves are called **streamlines**.

Let us now state the basic mathematical problem which we will pose for these types of fluid flows.

(3.4.9) ***Basic Problem***

(a) A given flow field $V_0(z)$ is set up in the z-plane (such as a uniform flow to be discussed later).

(b) Associated with the *given* flow $V_0(z)$ is a *given* complex potential $\Phi_0(z)$ satisfying $V_0 = \overline{\Phi_0'(z)}$.

(c) A rigid body with bounding curve C is immersed in the fluid. This will distort the given flow and produce a new velocity field $V(z)$.

(d) Corresponding to this (as yet unknown) velocity field $V(z)$ is a (as yet unknown) complex potential $\Phi(z)$ which we will write in the form

$$\Phi(z) = \Phi_0(z) + \Phi_d(z)$$

where $\Phi_d(z)$ denotes the **disturbed potential** due to the rigid body.

(e) The objective is to find $\Phi_d(z)$ so that $\text{Im}[\Phi(z)] = \text{constant}$ on C (since the stream function $\psi = \text{Im}(\Phi)$ is constant on the boundary of a rigid body). Hence, from **(d)**, we wish to find $\Phi_d(z)$ so that

$$\text{Im}[\Phi_0(z) + \Phi_d(z)] = \text{constant} \qquad (\text{on } C)$$

(f) Having now determined $\Phi_d(z)$, and hence $\Phi(z)$, the final velocity field $V(z)$ is given by

$$V(z) = \overline{\Phi'(z)}$$

■

A typical problem of this sort is depicted in Figure 3.4.2. There, a uniform flow field $V_0(z)$ [see Example (3.4.11) below] is set up in the z-plane (Figure 3.4.2*a*), and a rigid body D is immersed in this flow. In Figure 3.4.2*b*, we show the streamlines which can be expected for the final flow field.

The "Basic Problem" described above is what we would *like* to solve, namely, we wish to construct the disturbed potential $\Phi_d(z)$. This, however, is often a formidable task. Instead, what we will do here is to reverse this procedure. Namely, we first note that any given function $\Phi(z) = \phi(x, y) + i\psi(x, y)$ will give rise to a fluid flow $V(z) = \overline{\Phi'(z)}$ in any region of analyticity of Φ. We can then interpret this as a flow

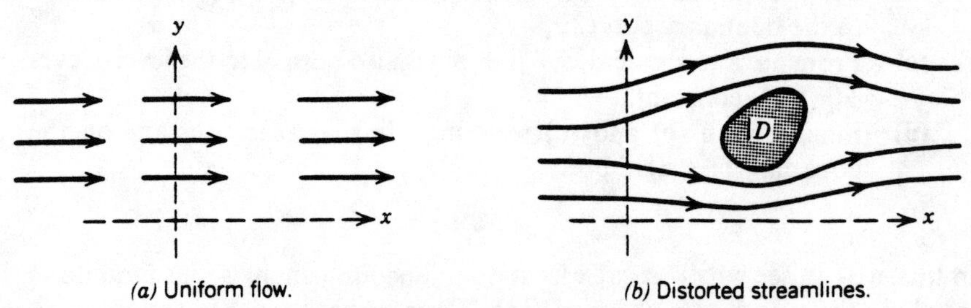

(a) Uniform flow. (b) Distorted streamlines.

Figure 3.4.2 Uniform flow past on obstacle.

about any stationary rigid body whose boundary is described by an equation of the form $\psi(x, y) = \text{constant}$. In this manner, namely, *by considering specific functions* $\Phi(z)$, we will compile a list of simple flows which might be able to be combined later to form more complicated ones. Also, as we will see in the examples and problems, the singularities of $\Phi(z)$ are often physically significant.

Finally, we state (without proof) the following results concerning the pressure and forces on a body immersed in a fluid flow. These follow from the full equations of fluid dynamics and can be found in any textbook on the subject.

(3.4.10) ***Theorem.*** Let ρ be the density of a fluid and $p(z)$ the **pressure** at $z = x + iy$. If there are no outside forces (such as gravity), then we have

(a) (Bernoulli's Law)

$$\frac{p}{\rho} + \frac{1}{2}\left|V(z)\right|^2 = \text{constant} \qquad (\textit{along a streamline})$$

The constant may be different for different streamlines. The quantity $|V(z)|$ is the **speed of the fluid.**

(b) The **net force F** due to the fluid acting on a rigid body bounded by a curve C is given by the line integral

$$\mathbf{F} = -\oint_C p\mathbf{n}\, ds$$

where $\mathbf{n}$ is the unit outer normal to C and thus *points into the fluid.*

■

(3.4.11) ***Example.*** Discuss and sketch the fluid flow associated with the complex potential $\Phi(z) = a_0 e^{-i\alpha} z$, where $a_0 > 0$.

Solution. With $\Phi(z) = \phi + i\psi$, we have

$$\phi(x, y) = a_0(x \cos \alpha + y \sin \alpha)$$

and

$$\psi(x, y) = a_0(y \cos \alpha - x \sin \alpha)$$

while the complex velocity field $V = \overline{\Phi'(z)}$ is given by

$$V = a_0 e^{i\alpha} \tag{3.4.12}$$

Hence, the velocity field is constant, and thus this complex potential gives rise to a **uniform fluid flow**. The streamlines ψ = constant are the straight lines

(3.4.13) $$y \cos \alpha - x \sin \alpha = \text{constant} \qquad (\textit{streamlines})$$

From (3.4.12), the direction of the flow makes an angle α with the positive real axis. These results are sketched in Figure 3.4.3. Also, the direction of the flow is parallel to the streamlines in (3.4.13), as is to be expected. If we

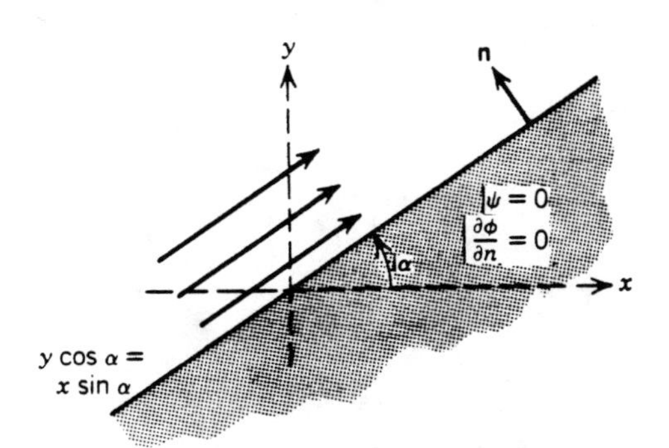

Figure 3.4.3 Uniform flow over flat plate.

choose the constant on the right-hand side of (3.4.13) to be zero, then this flow can be interpreted as that of a fluid flowing over a rigid plate in the form of a half-plane bounded by the line $y \cos \alpha = x \sin \alpha$ (see the figure). Also, since the vector $\mathbf{n} = -(\sin \alpha)\mathbf{i} + (\cos \alpha)\mathbf{j}$ is normal to the line, we have, from the expression for ϕ, that on the boundary of the rigid plate

$$\frac{\partial \phi}{\partial n} = \mathbf{n} \cdot (\phi_x \mathbf{i} + \phi_y \mathbf{j}) = 0$$

Thus, we have verified the boundary condition in (3.4.7g).

■

(3.4.14) ***Example.*** Discuss and sketch the flow field associated with the complex potential $\Phi(z) = z^2$.

Solution. The potential ϕ, stream function ψ, and velocity V are, respectively,

$$\phi = x^2 - y^2, \qquad \psi = 2xy, \qquad V(z) = 2(x - iy)$$

Note that the streamlines ψ = constant are the hyperbolas

$$xy = \text{constant}$$

In the special case when the constant is zero, the streamlines are the x- and/or y-axes. Thus, we can consider this complex potential function as giving rise to fluid flow in a region whose boundaries consist of parts of the x- and y-axes. In particular, we can interpret the flow as that in a right angle wedge, as shown in Figure 3.4.4a. The direction of the flow is as shown in the figure and can be deduced from $V(z)$.

We can also interpret this as a flow toward a flat plate (the x-axis), as shown in Figure 3.4.4*b*. Note that $V(z)$ vanishes when $z = 0$. Such a point is called a **stagnation point**.

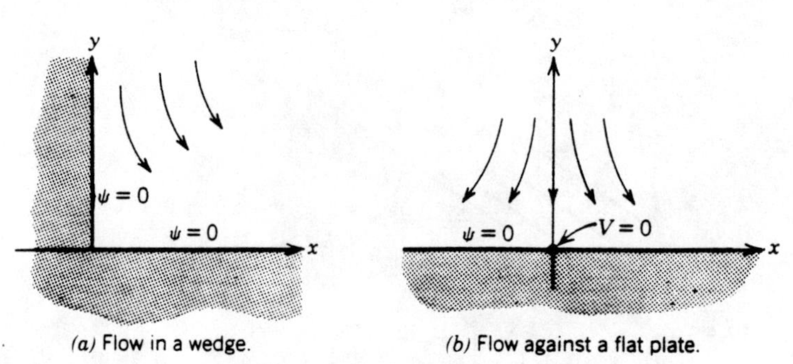

(a) Flow in a wedge. *(b)* Flow against a flat plate.

Figure 3.4.4 Example (3.4.14).

Finally, as the reader is urged to verify, $\partial\phi/\partial n = 0$ on the boundaries in both of these cases.

■

(3.4.15) ***Example.*** Discuss and sketch the fluid flow associated with the complex potential

$$\Phi(z) = a_0 \log z$$

where a_0 is a real constant.

Solution. With $\Phi(z) = \phi + i\psi$, we have in polar coordinates

$$\phi = a_0 \ln r \quad \text{and} \quad \psi = a_0(\theta + 2\pi n) \tag{3.4.16}$$

where the term $2\pi n$ comes from the multivaluedness of $\log z$. The complex velocity is given by

$$V(z) = \frac{a_0}{\bar{z}} \tag{3.4.17}$$

Note that $z = 0$ is a singular point for V (as it is also for Φ), and the velocity field is independent of which branch of the logarithm is chosen. The streamlines are the rays $\theta =$ constant, and hence the *fluid flows along radial lines*. From (3.4.17),

$$V(z) = \frac{a_0}{r} e^{i\theta} \quad \text{and} \quad \text{speed} = |V(z)| = \frac{|a_0|}{r} \tag{3.4.18}$$

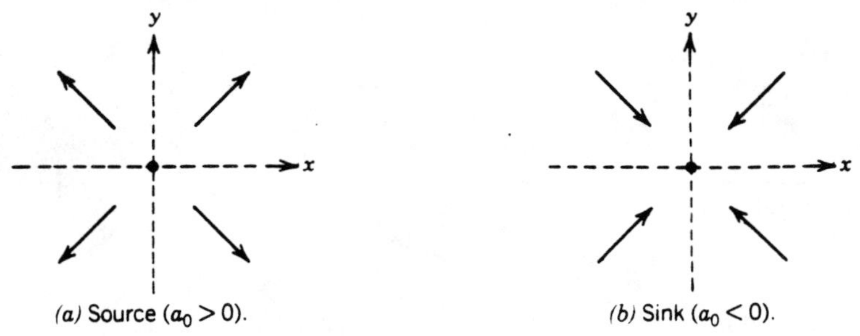

(a) Source ($a_0 > 0$). (b) Sink ($a_0 < 0$).

Figure 3.4.5 $\Phi(z) = a_0 \log z$.

The first of these equations shows that the velocity field $\mathbf{v} = (a_0/r)[(\cos\theta)\mathbf{i} + (\sin\theta)\mathbf{j}]$ and thus the direction of this radial flow is away from the origin if $a_0 > 0$ and toward the origin if $a_0 < 0$. The flow field is shown in Figure 3.4.5. In addition, the second of the above equations shows that the speed becomes unbounded as the origin is approached. For $a_0 > 0$, we say that this potential gives rise to a **source** at the origin with **source strength** $2\pi a_0$, while for $a_0 < 0$, we have a **sink** at the origin. While the choice of the nomenclature "source" and "sink" is motivated by the above description of the flow field, it can also be justified on the basis of equation (C. 11e) in Appendix C. There, we learn that the quantity $\rho \oint_C \mathbf{v} \cdot \mathbf{n}\, ds$ is the fluid produced per unit time in a region bounded by the closed curve C. As a special case, let C be a circle of radius R_0 centered at the origin. Then the unit outer normal $\mathbf{n}$ to C is directed away from the origin; and from $V(z)$ in (3.4.18), we have that on C, $\mathbf{v} \cdot \mathbf{n} = a_0/R_0$. Thus, since $ds = R_0\, d\theta$ on C,

P.458

$$\rho \oint_C \mathbf{v} \cdot \mathbf{n}\, ds = \frac{a_0 \rho}{R_0} \int_0^{2\pi} R_0\, d\theta = 2\pi a_0 \rho$$

Hence, the amount of fluid produced inside C is constant and independent of the radius R_0, and it is positive for $a_0 > 0$ (a source) but negative for $a_0 < 0$ (a sink).

Finally, we note that if we had taken the complex potential $\Phi(z)$ to be proportional to $\log(z - z_0)$ instead of $\log z$, the description of the flow would be exactly the same save for the fact that the singularity would be at $z = z_0$ instead of the origin.

■

(3.4.19) ***Example.*** Discuss the complex potential $\Phi(z) = i\kappa_0 \log z$, where κ_0 is a real constant.

Solution. Here, $\phi = -\kappa_0(\theta + 2\pi n)$ and $\psi = \kappa_0 \ln r$. Thus, the streamlines and velocity are given in polar coordinates by

(3.4.20)
$$r = \text{constant} \quad \text{and} \quad V(z) = \frac{-i\kappa_0}{r} e^{i\theta} = \frac{\kappa_0}{r} e^{i(\theta - \pi/2)}.$$

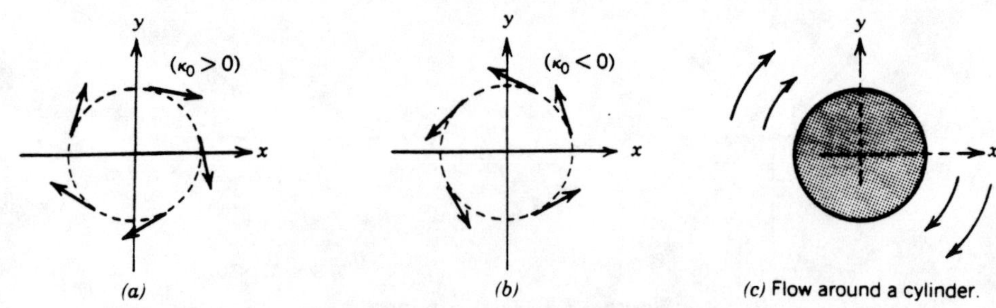

Figure 3.4.6 Vortex flows $\Phi(z) = i\kappa_0 \log z$.

From the streamlines, we see that the fluid moves in circles about the origin. The direction of flow is determined from $V(z)$ and depends upon the sign of κ_0, as indicated in Figures 3.4.6*a* and *b*. Such a flow is called a **vortex flow**, and the quantity $-2\pi\kappa_0$ is the **vortex strength**.

Note that since the circles r = constant are streamlines, we can also interpret this as a fluid flow around a rigid cylinder, as depicted in Figure 3.4.6*c*. With this interpretation, let us compute the net force on the cylinder. From (3.4.20), on the circle $|z| = R_0$, we have $|V(z)| = |\kappa_0|/R_0$. Hence, from Bernoulli's law (3.4.10a), the pressure on the cylinder is constant; and by (3.4.10b), the force on this body vanishes, since

(3.4.21)
$$\mathbf{F} = -p\oint_C \mathbf{n}\, ds = \mathbf{0}$$

We will have more to say about this in Problem 10.

■

(3.4.22) ***Example.*** What interpretation can be given to the complex potential

$$\Phi(z) = z + \frac{1}{z}?$$

Sketch the flow and compute the net force on the body.

Solution. In polar coordinates, the streamlines are

(3.4.23)
$$\psi = \left(r - \frac{1}{r}\right)\sin\theta = \text{constant}$$

We note that if the constant is chosen to be zero, the set of streamlines includes (as in the previous example) the unit circle $r = 1$. Also, for z large in magnitude, $1/z$ is small compared to z, and $\Phi(z)$ "behaves like" z, which, from Example (3.4.11) (with $\alpha = 0$), corresponds to a uniform horizontal flow to the right. Hence, we can view this as a uniform flow to the right impinging upon the unit disk. In Figure 3.4.7*a*, we have drawn some of the other streamlines, showing the paths that fluid particles take around the cylinder. Now the fluid velocity is given by

$$V(z) = \overline{\Phi'(z)} = 1 - \frac{1}{\bar{z}^2}$$

and from this we see that the stagnation points (where $V = 0$) are on the boundary of the disk at $z = \pm 1$. We also have that, on the unit circle $z = e^{i\theta}$,

$$\text{speed} = |V(z)| = [2(1 - \cos 2\theta)]^{1/2}$$

and hence, from Bernoulli's law (3.4.10a), the pressure is given by

$$p(z) = \rho(k + \cos 2\theta) \qquad \text{(on the unit circle)}$$

where k is a constant. Now the unit normal $\mathbf{n}$ at each point $e^{i\theta}$ on the circle is given by $\mathbf{n} = (\cos\theta)\mathbf{i} + (\sin\theta)\mathbf{j}$. Hence, from the above and the force law (3.4.10b), we have

$$\begin{aligned} \mathbf{F} &= -\int_0^{2\pi} \rho[k + \cos(2\theta)][(\cos\theta)\mathbf{i} + (\sin\theta)\mathbf{j}]\, d\theta \\ &= \mathbf{0} \end{aligned}$$

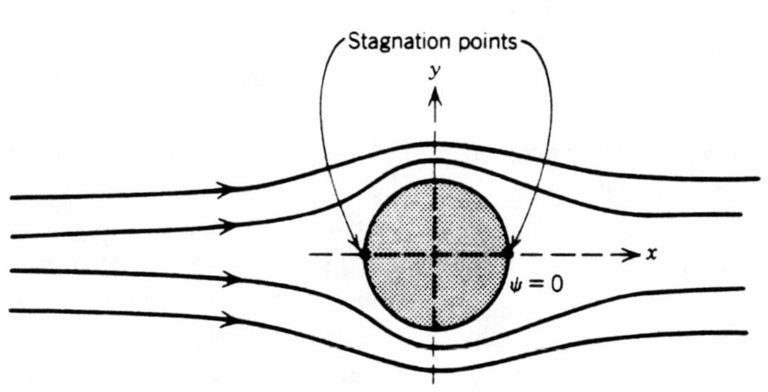

(a) Flow around a cylinder.

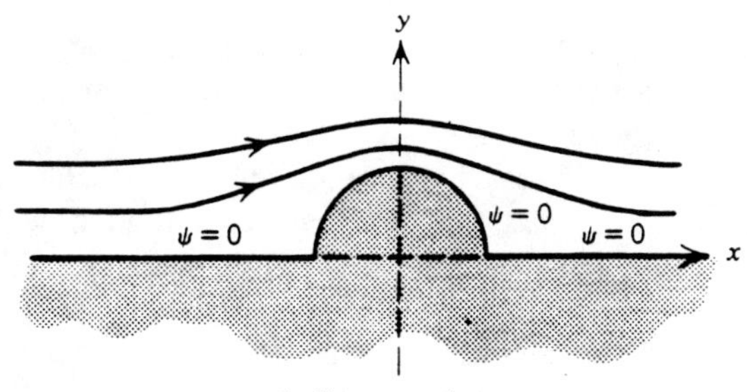

(b) Flow over a bump.

Figure 3.4.7 Flow with potential $\Phi(z) = z + (1/z)$.

Thus, as in the previous example, there is no net force on the body. In Problem 10, we will show that when this flow is combined with the circulation flow of the previous example, we do get a net force on the cylinder. This is related to the **lift** experienced by an airplane in flight.

Finally, we note from (3.4.23) that with the constant again set to zero, the rays $\theta = 0$ and $\theta = \pi$ are both streamlines. Hence, any curve consisting of pieces of both the unit circle and the x-axis can be considered a streamline. Such a **flow over a "bump" on a half-plane** is shown in Figure 3.4.7b.

■

PROBLEMS (The answers are given on page 429.)

(Problems with an asterisk are more theoretical.)

1.* In this problem, we show that the streamlines $\psi(x, y) =$ constant can indeed be interpreted as paths along which fluid particles travel. The **partical paths** are those curves $C: x = x(t)$, $y = y(t)$, parametrized by time t, along which fluid particles flow.
   (a) From the definition of v_1 and v_2 in (3.4.7a), show that the particle paths are characterized by the solutions of the differential equations $dx/dt = v_1(x, y)$ and $dy/dt = v_2(x, y)$.
   (b) Use the potential representation (3.4.7b) to show that the slope at each point of C is given by $dy/dx = \phi_y/\phi_x$.
   (c) Use the Cauchy–Riemann equations (3.4.7c), and differentials, to show that the curves $\psi(x, y) =$ constant are solutions of the equation in (b).
2. (a) Use equations (3.4.1b and c) to show that the real functions $v_1(x, y)$ and $-v_2(x, y)$ satisfy the Cauchy–Riemann equations.
   (b) Use part (a) to conclude that if v_1 and v_2 are continuously differentiable, then $\bar{V}(z)$, the conjugate of the complex velocity, is analytic. Why could we have expected this from (3.4.7f)?
3. From Bernoulli's law (3.4.10a), what is the relationship between pressure and speed?
4. Discuss the complex potential $\Phi(z) = (a_0 + i\kappa_0)\log z$, where a_0 and κ_0 are real. This gives rise to a **swirling flow** and is a combination of the two flows in Examples (3.4.15) and (3.4.19).
5. (a) Let $\Phi(z) = z^{1/2}$, where that branch of $z^{1/2}$ is chosen so that $0 < \arg(z) \leq 2\pi$. Show that this gives rise to a fluid flow around a semi-infinite flat plate. Sketch some of the streamlines (see Figure 3.4.8a) and verify that on the plate, $\mathbf{v} \cdot \mathbf{n} = 0$. [*Hint:* Use polar coordinates.]
   (b) Let $a > \frac{1}{2}$ and consider that branch of $\Phi(z) = z^a$ such that $\theta_0 - 2\pi < \theta \leq \theta_0$, where θ_0 is any angle in the range $\pi/a < \theta_0 < 2\pi$.
      (i) Show that this potential gives rise to a fluid flow in the wedge-shaped region $0 < \arg(z) < \pi/a$.

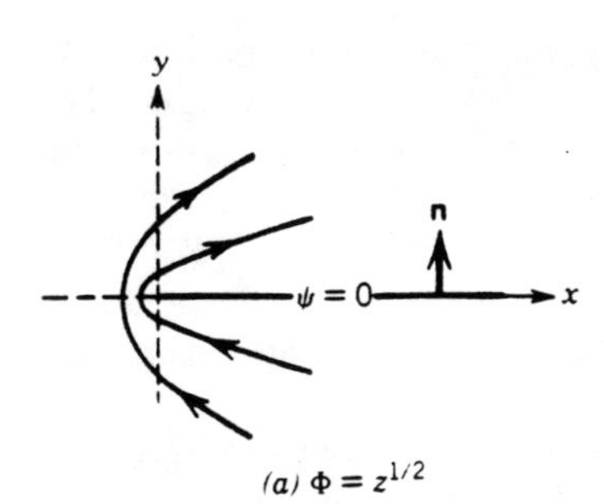

(a) $\Phi = z^{1/2}$

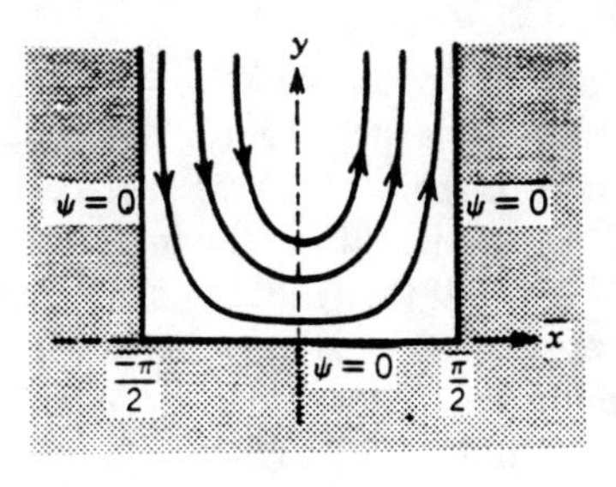

(b) $\Phi = \sin z$

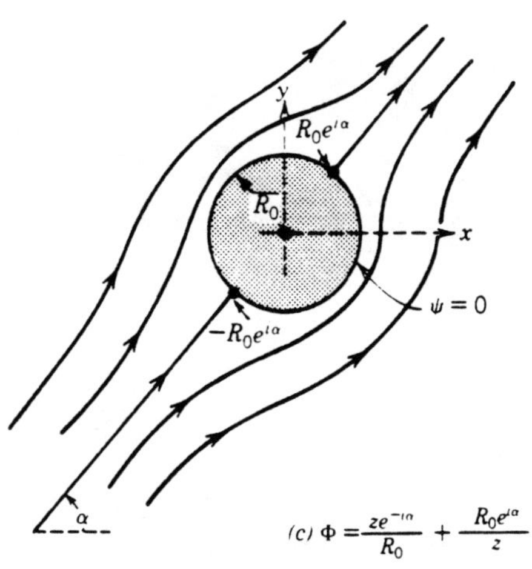

(c) $\Phi = \frac{ze^{-i\alpha}}{R_0} + \frac{R_0e^{i\alpha}}{z}$

Figure 3.4.8

(ii) Sketch some streamlines and verify that $\mathbf{v} \cdot \mathbf{n} = 0$ on the boundaries.

(iii) Note that, for $\frac{1}{2} < a < 1$, the boundary has a *sharp edge* protruding into the fluid flow. Compare the velocity fields near the vertex of the wedge for the two cases $\frac{1}{2} < a < 1$ and $a \geq 1$.

6. Verify that the complex potential $\Phi(z) = \sin z$ gives rise to a fluid flow as shown in Figure 3.4.8*b*, with the streamlines having the general shape shown there. [*Hint:* If $y = f(x)$ is a streamline, where is $y' > 0$? Where <0? Where 0?]

7. Consider the complex potential $\Phi(z) = ze^{-i\alpha}/R_0 + R_0e^{i\alpha}/z$, where $R_0 > 0$.

(a) Show that, for $z \to \infty$, this approaches the potential for a uniform flow from the left inclined at angle α to the positive real axis [see Example (3.4.11)].

(b) Show that this can be interpreted as a uniform flow impinging at angle α upon a circular cylinder of radius R_0, as shown in Figure 3.4.8*c*.

(c) Verify that ψ is constant on the boundary of the cylinder.

(d) Where are the stagnation points?

8. Consider the potential $\Phi(z) = \log(z - z_0) + \log(z - \bar{z}_0)$, where $\mathrm{Im}(z_0) > 0$. Note that this is the potential due to sources at $z = z_0$ and at $z = \bar{z}_0$, one being the image of the other in the real axis. What we describe below is the solution of a flow problem using the **method of images**.
   (a) Show that the real axis is a streamline.
   (b) Interpret Φ as giving rise to a fluid flow in the upper half-plane due to a point source located at z_0. Where are the stagnation points? Sketch some streamlines (see Figure 3.4.9).
   (c) Assume that on the wall $y = 0$, the pressure vanishes as $x \to \pm\infty$. Show that the constant in Bernoulli's law (3.4.10a) is zero, and compute the net force on the wall.

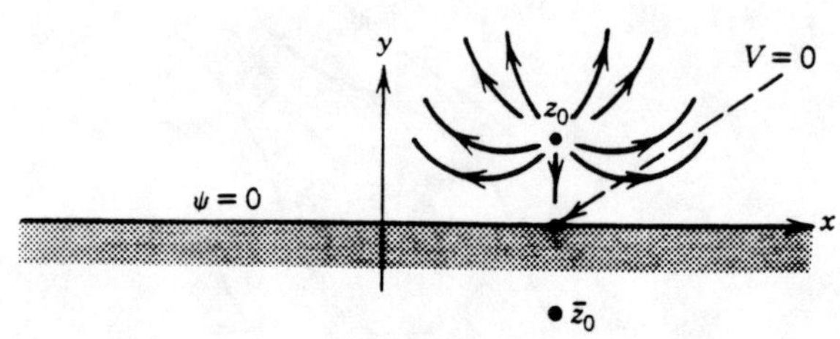

Figure 3.4.9

9. A source at z_1 superimposed with a sink at z_2 gives rise to the potential $\Phi(z) = \log(z - z_1) - \log(z - z_2)$.
   (a) Show that the streamlines are either circles or the line passing through z_1 and z_2 (see Figure 3.4.10*a*). [*Hint:* If $z_1 = x_1 + iy_1$, then $\arg(z - z_1) = \tan^{-1}[(y - y_1)/(x - x_1)]$, with a similar expression for $\arg(z - z_2)$. Now take the tangent of both sides of the equation $\psi = \text{constant}$.]
   (c) Let $z_2 = z_1 + \epsilon e^{i\theta_0}$ and consider the potential $\Phi_\epsilon(z) = (1/\epsilon)\Phi(z)$. Show that $\lim_{\epsilon \to 0} \Phi_\epsilon(z) = e^{i\theta_0}/(z - z_1)$. This is called a **dipole potential at z_1 in the direction $\theta = \theta_0$**. Show that the streamlines are either circles passing through z_1 and tangent to the line through z_1 of slope $\tan\theta_0$ (see Figure 3.4.10*b*) or this line itself.

10. The vortex flow potential of Example (3.4.19) and the potential of Example (3.4.22) both gave rise to flow past a cylinder with no net force on the cylinder. Consider the combination of these two potentials

$$\Phi(z) = a_0\left(z + \frac{1}{z}\right) + \frac{i\Gamma}{2\pi}\log z$$

where a_0 and Γ are positive numbers.
   (a) Show that this can be interpreted as giving rise to a uniform horizontal fluid flow from the left impinging on the unit disk. [*Hint:* What is the behavior of Φ for $z \to \infty$, and what are the streamlines?]
   (b) What is the velocity field?

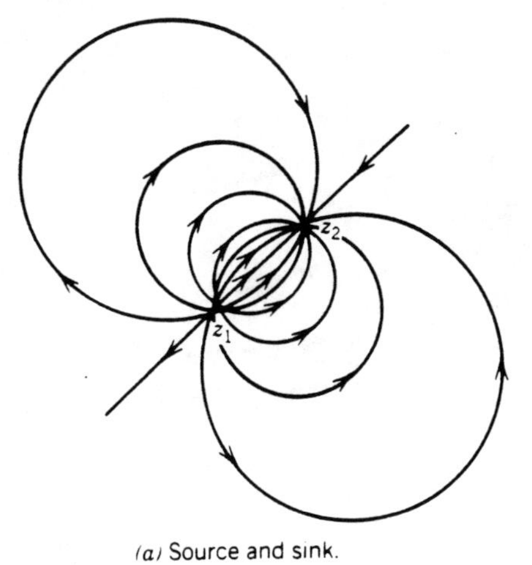

(a) Source and sink.

(b) Dipole.

Figure 3.4.10

(c) What is the speed on the boundary of the disk? The pressure [see (3.4.10a)]?

(d) From (c), show that the pressure is not symmetric with respect to the real axis. What can be concluded about the net force?

(e) On the unit circle $z = e^{i\theta}$, the outer normal is $\mathbf{n} = (\cos\theta)\mathbf{i} + (\sin\theta)\mathbf{j}$. Use (3.4.10b) to show that the net force on the disk is upward. What is it? Hence, there is a **lift** but no **drag**.

(f) Where are the stagnation points? Show that the situation changes for different ranges of the vortex parameter Γ.

(g) Show that for the various cases in part (f), the flow picture is as in Figure 3.4.11.

11. With $\zeta = |\zeta|e^{i\theta}$, consider that branch of $\sqrt{\zeta}$ defined by $\sqrt{\zeta} = |\zeta|^{1/2}e^{i\theta/2}$, $0 < \theta \le 2\pi$.

(a) Show that for $\mathrm{Im}(\zeta) > 0$, $\sqrt{\zeta^2} = \zeta$.

(b) With the above choice of branch, show that the function $\Phi(z) = \sqrt{1 + z^2}$ is analytic everywhere in the z-plane cut as indicated by the wavy lines of Figure 3.4.12*a*.

(c) For $z \to \infty$ in the upper half-plane, show that Φ in part (b) approaches the potential of a uniform horizontal flow to the right.

(d) Show that $\Phi(z)$ is real on the cuts in part (a), and hence verify that the flow due to Φ can be interpreted as flow around the plate $x = 0, 0 < y < 1$, in the upper half-plane (see Figure 3.4.12*b*). Where is the velocity singular?

(e) Show that $\psi = \mathrm{Im}(\sqrt{1 + z^2})$ vanishes only on the cut in the figure. Then use the continuity of ψ and the result of part (c) to show that $\psi > 0$ in the upper half-plane cut along the line $\mathrm{Re}(z) = 0, 0 < \mathrm{Im}(z) \le 1$.

(f) Show that the streamlines $\psi(x, y) = k > 0$ have the general shape shown in Figure 3.4.12*b*. [*Hint:* From the equation $(\phi + i\psi)^2 = 1 + (x + iy)^2$,

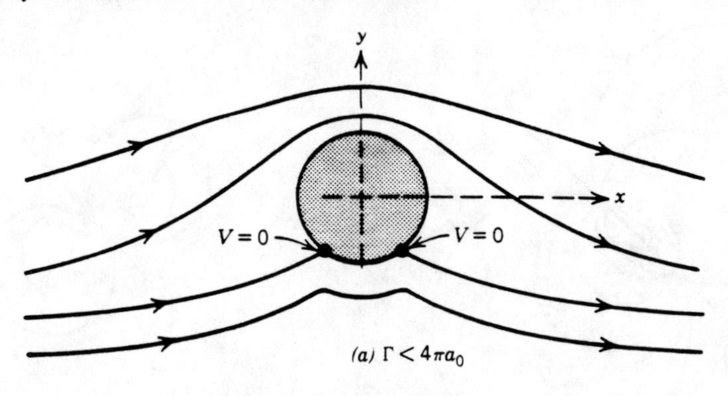

(a) $\Gamma < 4\pi a_0$

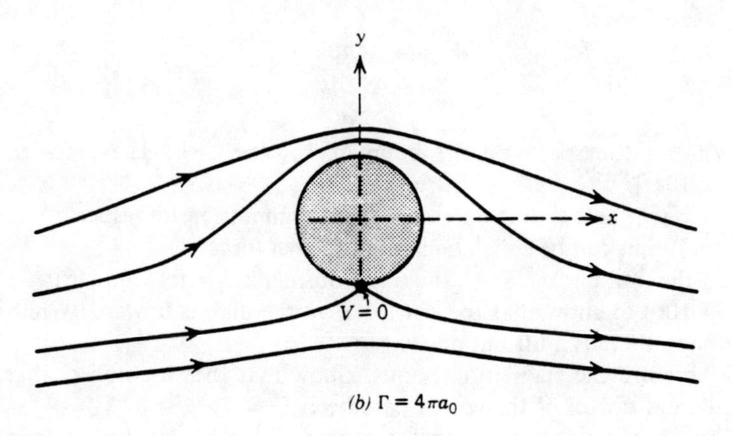

(b) $\Gamma = 4\pi a_0$

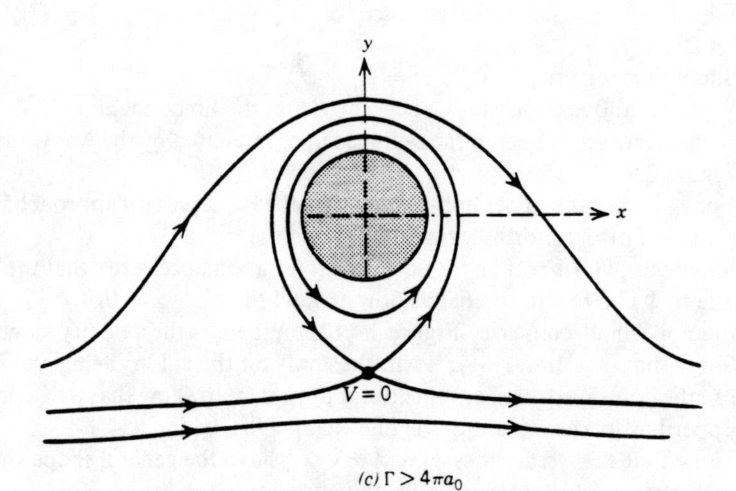

(c) $\Gamma > 4\pi a_0$

Figure 3.4.11 Flow with lift past a cylinder.

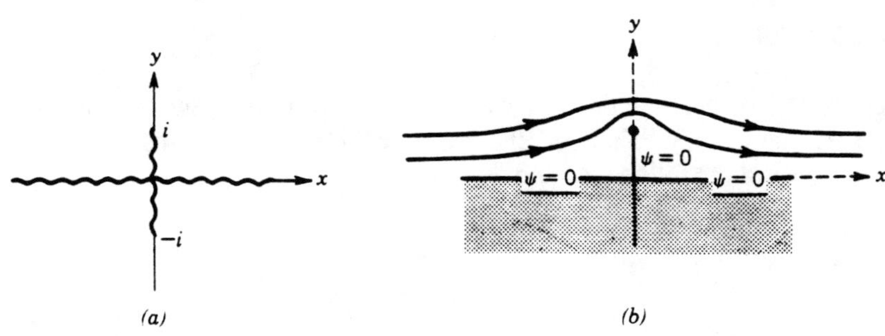

Figure 3.4.12

show that on each streamline of the form $y = y(x)$, y' is positive if $x < 0$, is negative if $x > 0$, and vanishes if $x = 0$.]

12. Consider the complex potential $\Phi(z) = z\cos\alpha - i(z^2 - 1)^{1/2} \cdot \sin\alpha$, where the branch of $\sqrt{z^2 - 1}$ is chosen so that (see Figure 3.4.13*a*)

$$\sqrt{z^2 - 1} = |z^2 - 1|^{1/2} e^{(i/2)(\theta_1 + \theta_2)} \qquad -\pi < \theta_1, \theta_2 \leq \pi$$

(a) Show that this branch of $\sqrt{z^2 - 1}$ is analytic except on the cut $-1 \leq x \leq 1$ shown in the figure.

(b) Show that $\sqrt{z^2 - 1}$ is positive imaginary for z on the cut.

(c) Show that for $z \to \infty$, $\sqrt{z^2 - 1} \approx z$. [*Hint:* Show that, in the limit, both θ_1 and θ_2 are approximately equal to $\arg(z) + 2\pi n$, for some n.]

(d) Interpret Φ as giving rise to a uniform flow at infinity impinging at angle α on a finite plate (see Figure 3.4.13*b*). What happens in the special case $\alpha = 0$? [*Hint:* Show that Φ is real on the plate and hence conclude that it is a streamline.]

(e) Show that there is a stagnation point on the plate located at $z = \cos\alpha$, if $0 < \alpha < \pi$, and at $z = -\cos\alpha$, if $\pi < \alpha < 2\pi$ (see Figure 3.4.13*b*).

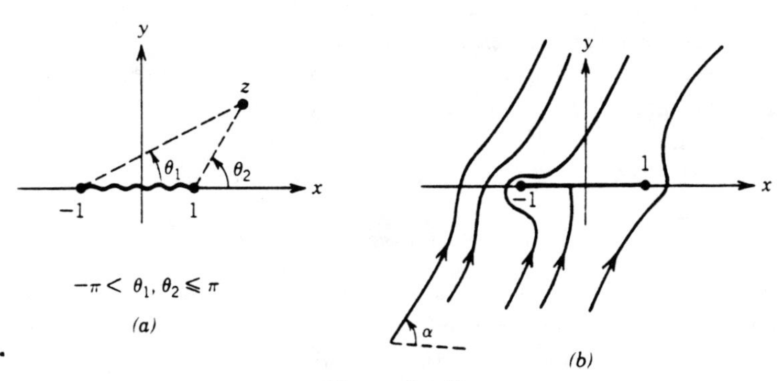

Figure 3.4.13

3.5 ELECTROSTATICS

In Appendix D, we derived the basic equations governing a two-dimensional, steady, electric field $\mathbf{E}(x, y)$. There we showed that $\mathbf{E}$ is related to a scalar potential function $\phi(x, y)$ and the following hold.

(3.5.1) ***Basic Equations***

- **(a)** $\phi_{xx} + \phi_{yy} = 0$ (in a charge-free region)
- **(b)** $\phi = \text{constant}$ (in, or on, a **conductor**)
- **(c)** $\mathbf{E} = -(\phi_x\mathbf{i} + \phi_y\mathbf{j})$ and $\mathbf{F} = q_0\mathbf{E}$, where $\mathbf{F}$ is the force exerted on the charge q_0. Note that **E is orthogonal to the equipotential curves** $\boldsymbol{\phi(x, y) = \text{constant}}$ (see Problem 1*a*).
- **(d)** $\oint_C (\partial\phi/\partial n)\, ds = -4\pi q$, where q is the total charge inside and on a curve C [see (D.14c) and (D.15c) in Appendix D]. The unit normal $\mathbf{n}$ points away from the interior of C (see Figure 3.5.1) and the normal derivative $\partial\phi/\partial n$ is given by $\partial\phi/\partial n = \mathbf{n} \cdot \nabla\phi$.

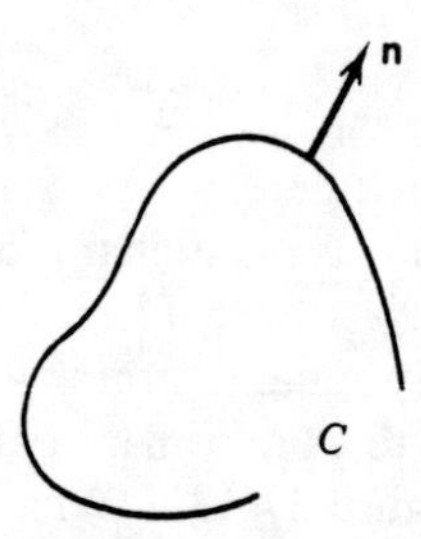

Figure 3.5.1

In Chapter 4, we will show how conformal mappings can be used to solve **boundary value problems** such as (a) and (b) above. However, here we will be less ambitious and simply show how we can apply our knowledge of complex function theory to determine some specific electrostatic fields.

To begin with, we know from Section 3.3 that in appropriate regions of the xy-plane we can construct, from a *given* harmonic function $\phi(x, y)$, an analytic function $\Phi(z)$ having ϕ as its real part. If we reverse this process, that is, *start* with a *given* analytic function, then we are led to the following observation.

(3.5.2) ***Observation***

If $\Phi(z) = \phi(x, y) + i\psi(x, y)$ is a *given* analytic function, then we can *interpret* its (harmonic) real part $\phi(x, y)$ as an electrostatic potential giving rise to an electric field $\mathbf{E} = -(\phi_x\mathbf{i} + \phi_y\mathbf{j})$.

We call the function $\Phi(z)$ the **complex electrostatic potential.** This is related to the electric field as follows. First, note that the electric field vector $\mathbf{E} = -(\phi_x \mathbf{i} + \phi_y \mathbf{j})$ determines an equivalent complex function, the **complex electrostatic field,** given by

(3.5.3) $$E(z) = -(\phi_x + i\phi_y) = -(\phi_x - i\psi_x)$$

where the second equality follows since ϕ and ψ in (3.5.2) satisfy the Cauchy–Riemann equations. Now since $\Phi(z)$ is analytic, we have $\Phi'(z) = \phi_x + i\psi_x$, and we find from the above that $E(z) = -\overline{\Phi'(z)}$. Hence, we proceed as follows to generate electrostatic fields.

(3.5.4) ***Procedure***

(a) Start with a *given* analytic function $\Phi(z) = \phi(x, y) + i\psi(x, y)$.
(b) From $\Phi(z)$, generate an electrostatic field $E = -\overline{\Phi'(z)}$.
(c) Then *interpret* the equipotential curves $\phi(x, y) = \text{constant}$ as conductors or the boundaries of conductors.

■

With this type of procedure, we are able to compile a list of electrostatic fields having specific properties. The hope then is that these might be able to be combined so as to produce more complicated fields. Before we proceed with examples illustrating just how this is carried out, we have the following definition.

(3.5.5) ***Definition.***
(a) A **condensor** is made up of two conductors carrying equal, but opposite, charge (see Figure 3.5.2).

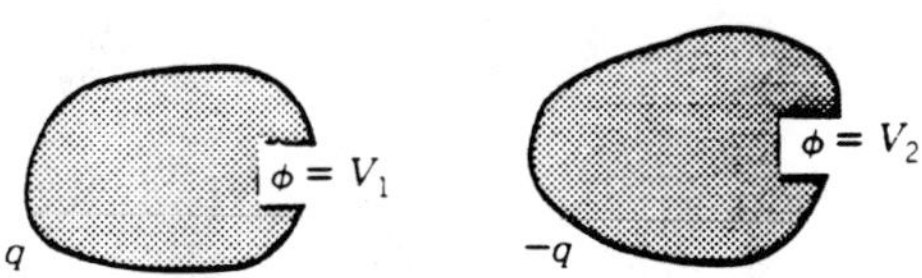

Figure 3.5.2 Condensor.

(b) The **capacity** of a condensor which consists of a conductor with charge q and at potential V_1 and of a conductor with charge $-q$ at potential V_2 is

$$C_{ap} = \frac{q}{V_1 - V_2}$$

■

(3.5.6) ***Example.*** Discuss the complex potential

$$\Phi(z) = -E_0 z e^{-i\alpha} \qquad E_0 > 0$$

Solution. By (3.5.4b), this gives rise to the electric field $E = -\overline{\Phi'}$ and potential $\phi = \text{Re}(\Phi)$ given by

$$\text{(3.5.7)} \qquad \begin{aligned} &\textbf{(a)} \quad E(z) = E_0 e^{i\alpha} \\ &\textbf{(b)} \quad \phi(x, y) = -E_0(x\cos\alpha + y\sin\alpha) \end{aligned}$$

This can be interpreted in several different ways. First, note from (a) that E is constant. Hence, Φ can be described as giving rise to a **uniform electric field** making angle α with the positive real axis [since $\arg(E) = \alpha$ by (a)]. This is depicted in Figure 3.5.3*a*. Second, note from (b) that the equipotentials $\phi(x, y) = k$ are the lines

$$\text{(3.5.8)} \qquad x\cos\alpha + y\sin\alpha = -\frac{k}{E_0} \qquad \text{(equipotentials)}$$

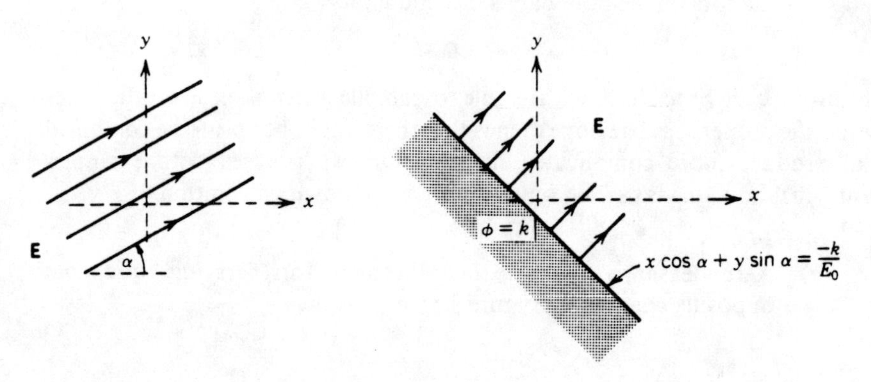

(a) Uniform field $E(z) = E_0 e^{i\alpha}$

(b) Uniform field above a conducting half-plane.

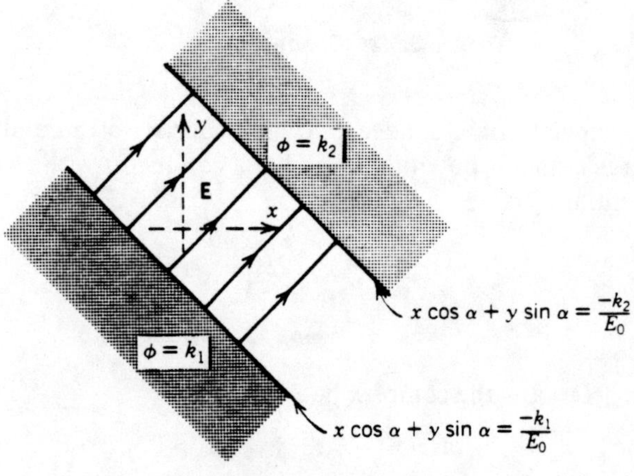

(c) Uniform field between conducting half-planes.

Figure 3.5.3 $\Phi(z) = -E_0 e^{-i\alpha z}$.

$\Phi(z) = -E_0 z e^{-i\alpha}$ with $E_0 > 0$, $\alpha \in \mathbb{R}$

Hence, each such line can be interpreted as the boundary of a conductor. Thus, as shown in Figure 3.5.3*b*, this complex potential might describe a uniform electric field in the region above a conducting half-plane, or a uniform electric field between two conducting half-planes, as shown in Figure 3.5.3*c*. Note that in both of these interpretations, the electric field is orthogonal to the boundaries of the conductors, as it must be by (3.5.1c).

■

(3.5.9) ***Example.*** Discuss the electric field generated by the logarithm $\Phi(z) = -2q_0 \log z$, where q_0 is real.

Solution. This gives rise to the following potential $\phi(x, y)$ and electric field $E(z)$, both of which are singular at the origin:

(3.5.10)

$$\textbf{(a)} \quad \phi(x, y) = -2q_0 \ln r$$

$$\textbf{(b)} \quad E(z) = \frac{2q_0}{\bar{z}} = \frac{2q_0}{x^2 + y^2}(x + iy)$$

Note that if $q_0 > 0$, the field points away from the origin (see Figure 3.5.4*a*) while if $q_0 < 0$, the field points toward the origin. Let us now show that this is in fact the potential of a point charge q_0 at the origin (indeed, the above discussion is nothing more than a description of **Coulomb's law**). To do this, consider a circle C of arbitrary radius r_0 centered at the origin (see Figure 3.5.4*b*). The normal $\mathbf{n}$ to C is in the radial direction; and hence from (3.5.10a), the normal derivative $\partial\phi/\partial n$ is given by

$$\frac{\partial \phi}{\partial n} = \frac{\partial \phi}{\partial r} = -\frac{2q_0}{r_0} \qquad \text{(on } C\text{)}$$

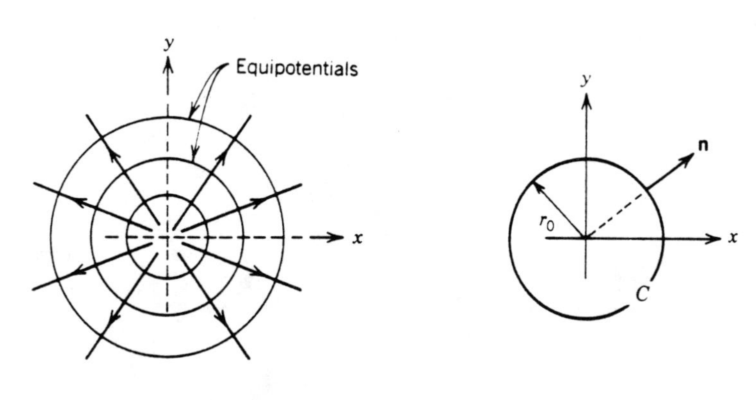

Figure 3.5.4 Point charge.

Thus,

$$\oint_C \frac{\partial \phi}{\partial n} ds = -\frac{2q_0}{r_0} \cdot (\text{length of } C) = -4\pi q_0$$

If we equate this to the result in (3.5.1d), we have

$$q_0 = \text{total charge inside and on } C$$

Since this result is independent of the radius r_0 of the circle and thus is the same no matter how small the circle, we have our point charge interpretation. Also, note from (3.5.10a) that the equipotentials are circles centered at the charge q_0. These are shown in Figure 3.5.4*a*.

If, instead of the present situation, we had the complex potential

$$\Phi(z) = -2q_0 \log(z - z_0)$$

the above interpretation would be the same save for the fact that the source would now be located at $z = z_0$.

■

(3.5.11) ***Example.*** A *positive charge* of strength q_0 is located above a conducting half-plane. Show that the resulting electrostatic field can be derived by the **method of images**.

Solution. As in Figure 3.5.5, take the conductor to be the lower half-plane $\text{Im}(z) \le 0$, with the charge located at $z = z_0$ in the upper half-plane. The **method of images** tells us to replace the whole half-plane by an equal and opposite charge $-q_0$ located at the point $\bar{z}_0$, which is the image of z_0 in the conductor. Now, the potential $\Phi(z)$ due to these two charges is just the sum of the potentials due to each one separately, and each of these is given by the results of the previous example. Hence,

$$\Phi(z) = -2q_0 \log(z - z_0) + 2q_0 \log(z - \bar{z}_0) \tag{3.5.12}$$

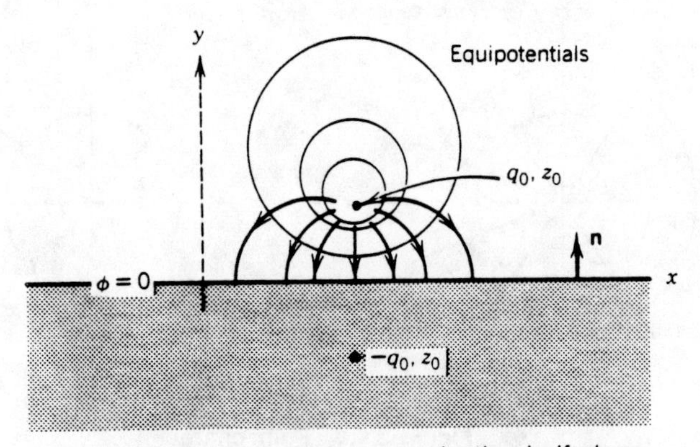

Figure 3.5.5 Charge over a conducting half-plane.

from which the potential $\phi(x, y)$ and field $E(z)$ are

$$\phi(x, y) = -2q_0 \ln\left|\frac{z - z_0}{z - \bar{z}_0}\right| \quad \text{and} \quad E(z) = \frac{2q_0(\bar{z}_0 - z_0)}{(\bar{z} - z_0)(\bar{z} - \bar{z}_0)} \tag{3.5.13}$$

Note from the figure that for z on the real axis,

$$\left|\frac{z - z_0}{z - \bar{z}_0}\right| = 1$$

and hence $\phi = 0$ here. Thus, the real axis, which is the boundary of the conductor, is an equipotential, and the complex potential in (3.5.12) is the solution of our problem in the upper half-plane. Note also that $\Phi(z)$ is singular at the image charge $-q_0$ which is located at $z = \bar{z}_0$ but that *this is not in the region in which we wish to solve our problem*. In fact, since by (3.5.1b) the potential inside a conductor is constant (which here, by continuity, we take to be its boundary value 0), the solution in both the upper and lower half-planes is given by

$$\phi(x, y) = \begin{cases} -2q_0 \ln\left|\dfrac{z - z_0}{z - \bar{z}_0}\right| & \text{Im}(z) > 0 \\ 0 & \text{Im}(z) \le 0 \end{cases} \tag{3.5.14}$$

Now, from the above, the equipotentials in the upper half-plane are described by

$$\left|\frac{z - z_0}{z - \bar{z}_0}\right| = k \qquad 0 < k < 1$$

(Note that $k < 1$ since z and z_0 both lie in the upper half-plane.) A simple computation shows that these curves are the circles

$$|z - \hat{z}| = \frac{2ky_0}{1 - k^2} \tag{3.5.15}$$

where

$$\hat{z} = x_0 + i\left(\frac{1 + k^2}{1 - k^2}\right) y_0$$

Note that the centers of these circles lie directly above the charge at z_0. In the previous example, the equipotentials due to a single charge were also circles but they were *centered* at the charge. What we are seeing here is a shift of the equipotentials due to the presence of the conducting half-plane. Some of these circles are shown in Figure 3.5.5. Also plotted are the directions of the electric field (the **field lines**) which at each point must be orthogonal to an equipotential. That their directions are as indicated follows either from (3.5.13) or from the fact that since $q_0 > 0$, $\mathbf{E}$ points away from q_0 (see the previous example). Also note that $\mathbf{E}$ is orthogonal to the conducting plane.

Finally, let us use (3.5.1d) to compute the total charge induced by q_0 on the surface of the conductor (can you guess as to what it will turn out to be?). Now, on the real axis, $\mathbf{n} = \mathbf{j}$, and hence, by (3.5.13),

$$\left.\frac{\partial \phi}{\partial n}\right|_{y=0} = \left.\frac{\partial \phi}{\partial y}\right|_{y=0}$$

$$= -2q_0\left[\frac{y - y_0}{(x - x_0)^2 + (y - y_0)^2} - \frac{y + y_0}{(x - x_0)^2 + (y + y_0)^2}\right]_{y=0}$$

$$= \frac{4q_0 y_0}{(x - x_0)^2 + y_0^2}$$

Thus, by (3.5.1d),

$$\text{Charge on surface} = \frac{4q_0 y_0}{-4\pi}\int_{-\infty}^{\infty}\frac{dx}{(x - x_0)^2 + y_0^2}$$

$$= -q_0$$

This is just the amount of the "fictitious" image charge.

■

In the problems, the reader is asked to solve other problems involving point charges and wedges using images. In Section 4.6, we will show how conformal mappings can be used to extend the class of problems to which the method of images can be applied.

PROBLEMS (The answers are given on page 429.)

(Problems with an asterisk are more theoretical. Problems 1–3 deal with the material in Appendix D.)

1.* (a) From (D.8a), why can you conclude that $\mathbf{E}$ is orthogonal to the equipotentials $\phi(x, y) = \text{constant}$?

(b) Prove Theorem (D.11). Note that the work done in moving a charge through a field is the negative of the work done by the field. [*Hint:* If C is a curve joining P_1 to P_2, with unit tangent vector field $\mathbf{T}$, then the work is the line integral $-\oint_{P_1}^{P_2} \mathbf{F}\cdot\mathbf{T}\,ds$. Now use (D.8a) and the result from advanced calculus that $\mathbf{T}\cdot\nabla\phi = \partial\phi/\partial s$.]

2. Use equations (D.8a) and (D.10) to show that

$$\nabla\cdot\mathbf{E} = \begin{cases} 0 & \text{in a charge-free region} \\ 4\pi\rho & \text{in a region with charge density } \rho \end{cases}$$

3.* Prove Theorem (D.14). [*Hint:* In part (c), surround S with a larger surface and then apply the divergence theorem to (D.10). Now let the new surface approach S.]

4.* Here, we will discuss the physical relevance of the imaginary part $\psi(x, y)$ of the complex potential $\Phi = \phi + i\psi$. We first note that ϕ and ψ satisfy the Cauchy–Riemann equations $\phi_x = \psi_y$ and $\phi_y = -\psi_x$. Also, note from Figure 3.5.6*a* the relationship between the normal vector **n** and tangent vector **T** to a curve C.

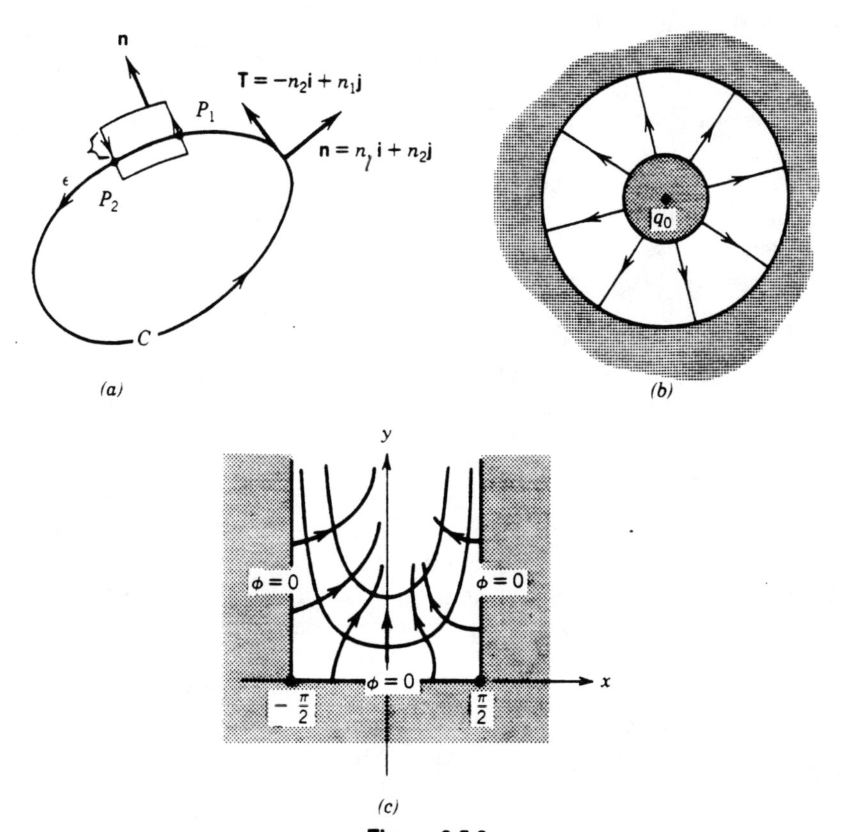

Figure 3.5.6

(a) Prove that **E** is tangent to the curves $\psi(x, y) =$ constant. These curves are called **field lines**, or **lines of force**. [*Hint:* Use the Cauchy–Riemann equations and the fact that **E** is orthogonal to the equipotentials. Also see Theorem (3.3.15).] P. 121

(b) Prove that on any curve C, $\partial\phi/\partial n = \partial\psi/\partial s$, where $\partial\phi/\partial n = \mathbf{n} \cdot \nabla\phi$, and $\partial\psi/\partial s$, the rate of change of ψ with respect to arc length, is defined by $\mathbf{T} \cdot \nabla\psi$ [*Hint:* Use Figure 3.5.6*a* and the Cauchy–Riemann equations.]

(c) In Figure 3.5.6*a*, assume that the region inside C is occupied by a conductor, and consider two points P_1 and P_2 on C. If $q_{1,2}$ is the surface charge on C between these two points, show that

$$q_{1,2} = \frac{\psi(P_1) - \psi(P_2)}{4\pi}$$

[*Hint:* Consider the "rectangular type" curve shown in the figure, where the top piece is the same shape as that part of C between P_1 and P_2. Now use (i) (3.5.1d), applied to this "rectangle"; (ii) the fact that ϕ = constant in a conductor; and (iii) the result of part (b). Then take the limit $\epsilon \to 0$.]

5. Use the result of Problem 4c to show, as in Example (3.5.11), that the total charge induced on a half-plane due to a charge q_0 located above it is $-q_0$. [*Hint:* From (3.5.12), $\psi = -2q_0[\arg(z - z_0) - \arg(z - \bar{z}_0)]$. Note that this difference is independent of the choice of the branch of the logarithm. Evaluate these arguments for z on the real axis in the limits $x \to \pm\infty$.]
6. For $q_0 > 0$, interpret the potential of Example (3.5.9) as that which gives rise to an electric field between two cylindrical conductors, with field lines as shown in Figure 3.5.6*b*.
7. Consider the complex potential $\Phi(z) = i \sin z$.
   (a) Show that the lines $x = \pm\pi/2$ and $y = 0$ are equipotentials corresponding to $\phi(x, y) = 0$.
   (b) Interpret this complex potential as giving rise to an electric field in a semiinfinite strip bounded by conductors as shown in Figure 3.5.6*c*.
   (c) Show that the shape of the equipotentials are as shown in the figure and the electric field lines are as drawn with the given directions. [*Hint:* In the equation $\phi(x, y)$ = constant, show that the slope dy/dx of this curve is positive for $x > 0$, zero for $x = 0$, negative for $x < 0$. Also, use the fact that **E** is orthogonal to the equipotentials.]
8. Show that the complex potential $\Phi(z) = iz^n$, $n = 2, 3, \ldots$, gives rise to an electric field inside a wedge of angle π/n whose boundaries are **grounded** (potential = 0) conductors.
9. In this problem, we will show how to generate the field due to a **dipole**. Consider two charges, one of strength q_0 at $z = z_0$ and the other of strength $-q_0$ at $z = z_1$ (see Figure 3.5.7*a*).

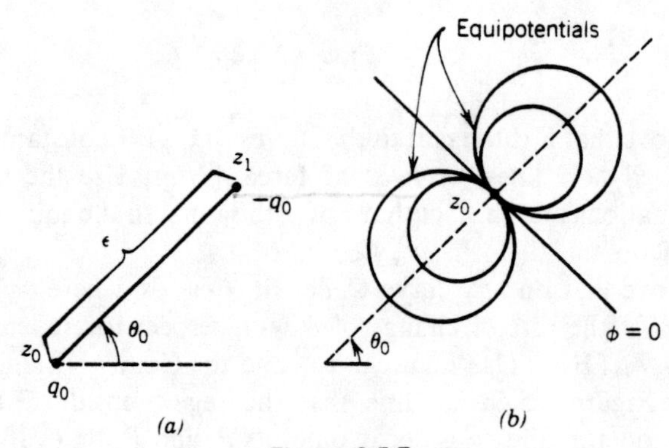

Figure 3.5.7

(a) What is the complex potential due to these two charges?
(b) In part (a), let $z_1 = z_0 + \epsilon e^{i\theta_0}$ and $q_0 = m_0/\epsilon$. Now take the limit $\epsilon \to 0$ and show that the limiting potential is $\Phi_d(z) = -(2m_0 e^{i\theta_0})/(z - z_0)$. This is the complex potential for a **dipole of strength m_0 at $z = z_0$ with direction θ_0**.
(c) Show that the equipotentials for the dipole are as shown in Figure 3.5.7*b* and, with the exception of the solid line shown, are circles through z_0 and tangent to the line.

10. Consider the complex potential

$$\Phi(z) = -2q_0[\log(e^z - e^{z_0}) - \log(e^z - e^{\bar{z}_0})]$$

where $0 < \mathrm{Im}(z_0) < \pi$.
(a) Show that the lines $\mathrm{Im}(z) = 0, \pi$ are equipotentials.
(b) Find the complex field $E(z)$.
(c) Show that for "z near z_0", $E(z)$ is "close to" the function $2q_0/(\bar{z} - \bar{z}_0)$. [*Hint:* Why is it true that $e^{\bar{z}} - e^{\bar{z}_0} = (\bar{z} - \bar{z}_0)\ [(e^{\bar{z}} - e^{\bar{z}_0})/(\bar{z} - \bar{z}_0)] \approx (\bar{z} - \bar{z}_0)e^{\bar{z}_0}$ for z "close to z_0".]
(d) Why can you interpret this potential as giving rise to a field due to a point charge q_0 located between two conducting half-planes?

11. We can use the result of Example (3.5.11) to construct the electric field generated by a conducting cylinder raised to potential V_0 and lying above a grounded, conducting half-plane as shown in Figure 3.5.8. Let the center of the cylinder be at $z = ih_0$ ($h_0 > 0$) and have radius r_0 ($r_0 < h_0$).

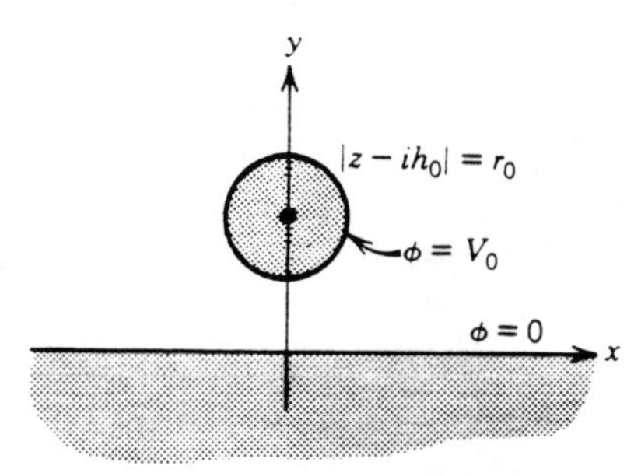

Figure 3.5.8

(a) In equation (3.5.15), each equipotential circle can be considered to be the boundary of a conducting cylinder. Use the center and radius of the cylinder given in the figure to find three equations between x_0, y_0, and k.
(b) On each circle in (3.5.15), the potential $\phi(x, y)$ is the constant $-2q_0 \ln k$. This now leads to another equation between q_0 and k. What is it?
(c) Solve the equations in (a) and (b) for $z_0 = x_0 + iy_0$ and for q_0 in terms of V_0 and h_0. Then, (3.5.12) should provide the complex potential for our problem. Check your answer. [*Hint:* Use the fact that k in (3.5.15) lies in the range $0 < k < 1$.]

12. Consider the complex potential $\Phi(z) = -E_0(z - r_0^2/z)$, $E_0 > 0$.
(a) Show that for large $|z|$, the field is a uniform field.

(b) Interpret this potential as giving rise to an electric field produced when a grounded conducting cylinder of radius r_0 is placed in the uniform electric field due to the potential $-E_0 z$.

cf pp. 209–210

13. (a) The method of images can be applied to cases other than the half-plane of Example (3.5.11). Thus, consider the situation depicted in Figure 3.5.9*a*. Use the image system shown there to compute the complex potential due to a charge q_0 surrounded by the grounded conductors shown. Verify that the real potential $\phi(x, y)$ vanishes on the conductors.

(b) Consider the wedge-shaped conductor shown in Figure 3.5.9*b*, where n is some positive integer. What is the image system for the charge q_0? Construct the complex potential. [*Hint:* Keep reflecting in the boundaries of the wedge and show that the image system eventually repeats.]

(c) Why won't the method of images work in part (b) if n is not a positive integer? In Section 4.6, we will show how conformal mapping techniques can circumvent this difficulty.

14. The method of images can also be used when dealing with a point charge in the presence of a conducting cylinder. Thus, consider Figure 3.5.9*c*. Place an image charge of strength $-q_0$ inside the cylinder at z_1. By symmetry, we might expect z_1 to lie along the line joining the center of the cylinder with z_0. Find z_1 so that the resulting (real) potential is zero on the cylinder surface. [*Hint:* The real potential ϕ is of the form $-2q_0[\ln|z - z_0| - \ln|z - z_1|] + k$. Find the constant k.]

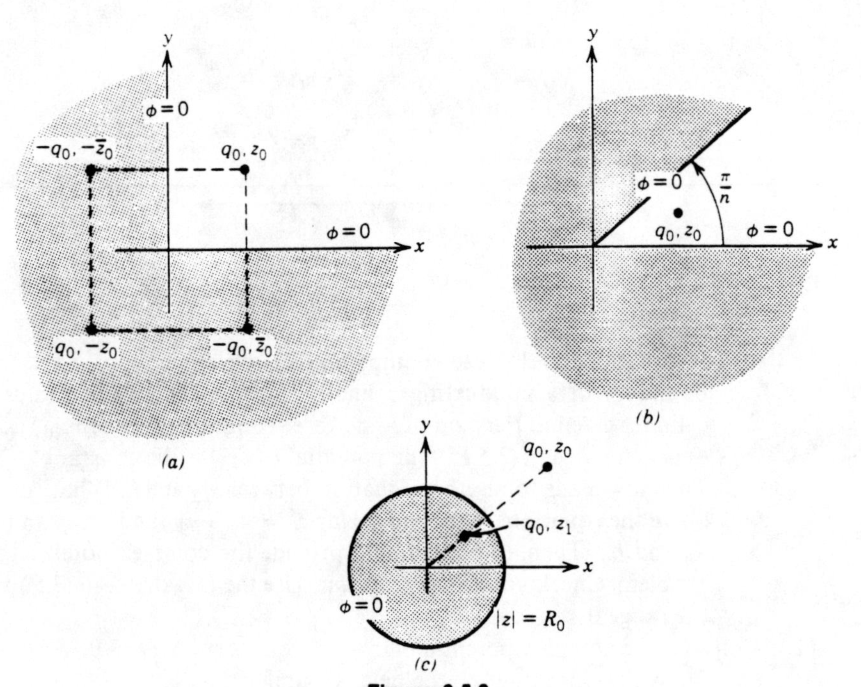

Figure 3.5.9

4

CONFORMAL MAPPING

4.1 CURVES AND THEIR PARAMETRIZATION; EQUIVALENT CURVES[1]

A **curve** C in the xy-plane is defined by two real functions $x(t)$ and $y(t)$ and a real interval $t_0 \le t \le t_1$, and is denoted by

(4.1.1)
$$C: \begin{array}{l} x = x(t) \\ y = y(t) \end{array} \qquad t_0 \le t \le t_1 \qquad (t_1 > t_0)$$

In complex form, we can write C in terms of a complex-valued function $z(t)$ as

(4.1.2)
$$C: \quad z = z(t) = x(t) + iy(t) \qquad t_0 \le t \le t_1 \qquad (t_1 > t_0)$$

The real variable t is called the **parameter of the curve**, and the above describes the **parametric representation** of C. The **initial point** of C is

$$z(t_0) = x(t_0) + iy(t_0)$$

while the **final point** is

$$z(t_1) = x(t_1) + iy(t_1)$$

[1] This section will also be needed in Chapter 5. For our present purposes, it can be discussed briefly, with an emphasis placed on the terminology and definitions.

A curve is **closed** if its final and initial points are the same, that is, if

$$z(t_0) = z(t_1) \qquad \text{(closed curved)}$$

The **trace of C** is the set of points in the z-plane satisfying (4.1.2). Also, as the examples will show, inherent in prescribing a curve is a direction from the initial to the final point. This is called the **direction induced by the parametrization.**

(4.1.3) ***Example.*** Consider the curve

$$C: \quad z = it \qquad -1 \leq t \leq 1$$

Its trace is that part of the imaginary axis $x = 0$, $-1 \leq y \leq 1$, and its direction is from the initial point $t = -i$ *upward* to the final point $z = i$.

■

(4.1.4) ***Example.*** Consider the curve

$$C_1: \quad z = z_1(t) = e^{it} \qquad 0 \leq t \leq 2\pi$$

It is closed (why?); and since $|z_1(t)| = 1$ for all t, its trace lies on the unit circle. In fact, the whole unit circle is covered exactly once (except for the final and initial points), since for each z on the unit circle we can find a unique value of t in the given interval satisfying $z = e^{it}$ (the reader is asked to verify this). Finally, as is shown in Figure 4.1.1a, the direction induced on this trace is counterclockwise. To see this, we write the curve as

$$x = \cos t \qquad y = \sin t \qquad 0 \leq t \leq 2\pi$$

Then, $dx/dt = -\sin t$, $dy/dt = \cos t$, and we have

$$\frac{dx}{dt} \text{ is } \begin{cases} \text{negative if} & 0 < t < \pi \\ \text{positive if} & \pi < t < 2\pi \end{cases}$$

(4.1.5) and

$$\frac{dy}{dt} \text{ is } \begin{cases} \text{positive if} & 0 < t < \pi/2 \\ \text{negative if} & \pi/2 < t < 3\pi/2 \\ \text{positive if} & 3\pi/2 < t < 2\pi \end{cases}$$

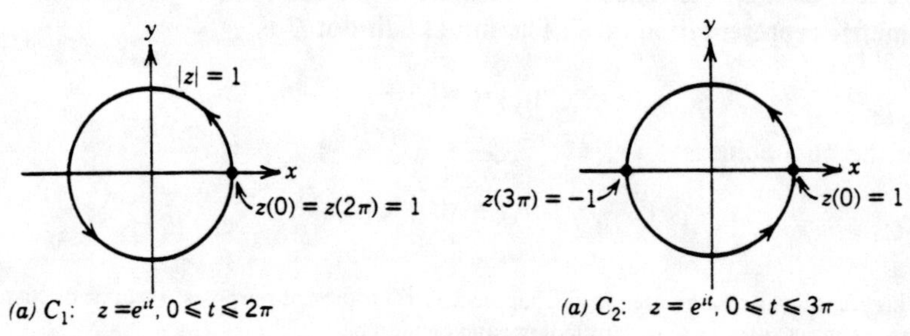

Figure 4.1.1 Example (4.1.4).

Thus, starting from the initial point $z(0) = 1$ and increasing t from $t = 0$ to $t = \pi/2$, we see that x must decrease while y increases. Performing this type of analysis in each of the ranges $\pi/2 < t < \pi$, $\pi < t < 3\pi/2$, and $3\pi/2 < t < 2\pi$, we can easily verify the counterclockwise direction in Figure 4.1.1*a*.

If instead of C_1 we consider the curve C_2 defined by

(4.1.6)
$$C_2: \quad z = e^{it} \qquad 0 \le t \le 3\pi$$

then the trace of C_2 is again the whole unit circle, but now the upper half is covered twice. However, C_2 is *not* closed (why?) even though its trace seems to be. This is shown in Figure 4.1.1*b*.

■

We now state some results concerning complex-valued functions $z(t)$ of a single real variable. With the concepts of limit, continuity and derivative defined as in Sections 2.1 and 2.2, we have the following theorem.

(4.1.7) ***Theorem.*** Let $z(t)$ be given by $x(t) + i\,y(t)$, $t_0 \le t \le t_1$. Then,

(a) $z(t)$ has a limit at $\hat{t}$ if and only if *both* $x(t)$ and $y(t)$ have limits, and then

(4.1.8)
$$\lim_{t\to\hat{t}} [z(t)] = \lim_{t\to\hat{t}} [x(t)] + \lim_{t\to\hat{t}} [y(t)]$$

At the end points of the interval $[t_0, t_1]$, the limit is to be taken as a **one-sided limit**.

(b) $z(t)$ is continuous if and only if *both* $x(t)$ and $y(t)$ are.

(c) $z(t)$ is differentiable if and only if *both* $x(t)$ and $y(t)$ are, and then

(4.1.9)
$$z'(t) = x'(t) + i\,y'(t) \qquad \left(' = \frac{d}{dt}\right)$$

(d) Let $f(z) = u(x, y) + i\,v(x, y)$ be a complex-valued function defined in a domain D which includes the trace of the curve $z = z(t)$.

(i) If $f(z)$ and $z(t)$ are continuous, so is the **composite function**

(4.1.10)
$$f(z(t)) = u(x(t), y(t)) + i\,v(x(t), y(t))$$

(ii) If $f(z)$ is analytic in D and $z(t)$ is differentiable for $t_0 \le t \le t_1$, then the composite function $f(z(t))$ is differentiable for $t_0 \le t \le t_1$, and we have the **chain rule**

(4.1.11)
$$\frac{d}{dt}[f(z(t))] = f'(z(t)) \cdot z'(t)$$

where $f'(z(t))$ means $f'(z)|_{z=z(t)}$. At the end points $t_{0,1}$ of the interval, the derivatives are defined by one-sided limits.

■

The curve (4.1.2) is said to be **continuous** if the complex function $z(t)$ is a continuous function of t. It is **piecewise continuous** if it consists of a finite number of continuous pieces, that is, if there are a finite number of subintervals $\hat{t}_{k-1} < t < \hat{t}_k$, $k = 1, 2, \ldots, N$, with $\hat{t}_0 = t_0$ and $\hat{t}_N = t_1$, on each of which $z(t)$ is continuous.

(4.1.12) ***Example.*** The curve

$$C: \quad z = z(t) = \begin{cases} (1+i)t & 0 \le t \le 1 \\ (3-i) + (i-1)t & 1 < t \le 2 \end{cases} \tag{4.1.13}$$

is piecewise continuous since $z(t)$ is continuous on both intervals $0 < t < 1$ and $1 < t < 2$, but is discontinuous at $t = 1$ since $\lim_{t \uparrow 1} [z(t)] = 1 + i$ while $\lim_{t \downarrow 1} [z(t)] = 2$. The trace of C is shown in Figure 4.1.2 together with its direction. Note from the figure that even though the curve is discontinuous at $t = 1$, its trace is a connected set of points.

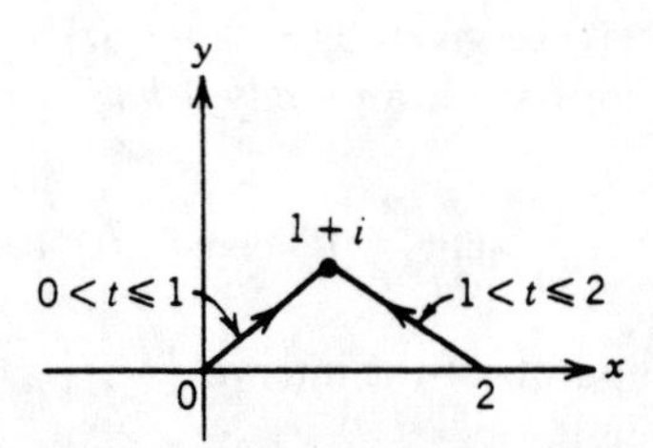

Figure 4.1.2 Example (4.1.12).

■

(4.1.14) ***Definition.*** Consider the curve $z = z(t)$, $t_0 \le t \le t_1$.

(a) It is **differentiable** if $z'(t)$ exists for $t_0 \le t \le t_1$. If $z'(t)$ is continuous, it is said to be **continuously differentiable**.

(b) It is **smooth** if it is continuously differentiable and if, in addition,

$$z'(t) \ne 0 \qquad t_0 \le t \le t_1 \tag{4.1.15}$$

(In Section 4.3, we will see that this last condition guarantees that $z'(t)$ can be represented by a vector which is tangent to the trace.)

(c) It is called a **contour** if it is **piecewise smooth**, that is, consists of a finite number of smooth pieces. (*We will interchangeably use the words "contour" and "piecewise smooth"*.)

■

(4.1.16) ***Example.*** The curve C: $z = i|t|$, $-1 \le t \le 1$, is continuous, but not differentiable, at $t = 0$ since

$$\frac{dz(0)}{dt} = \lim_{t \to 0} \left[\frac{z(t) - z(0)}{t} \right] = i \lim_{t \to 0} \left(\frac{|t|}{t} \right) \tag{4.1.17}$$

and this last limit does not exist (why?). However, C is a contour, that is, piecewise smooth, since $z' = -i$ for $-1 < t < 0$ while $z' = i$ for $0 < t < 1$. The trace of this curve is shown together with its direction in Figure 4.1.3.

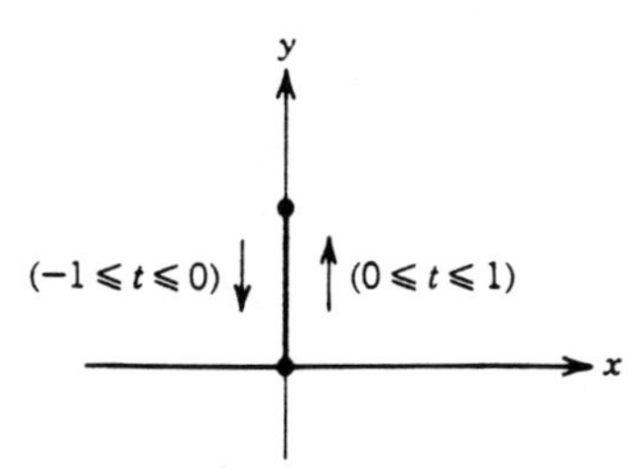

Figure 4.1.3 Example (4.1.16).

Note that C is differentiable at the end points since the left- and right-hand derivatives exist at $t = -1$ and $t = 1$, respectively, and we have

$$z'(-1) = \lim_{t \downarrow -1} \left[\frac{i|t| - i}{t + 1} \right] = -i$$

while

$$z'(1) = \lim_{t \uparrow 1} \left[\frac{i|t| - i}{t - 1} \right] = i$$

■

A curve is called **simple** if, geometrically, it doesn't intersect itself. Analytically, it is simple if the function $z(t)$ in (4.1.2) is **one to one**, that is,

(4.1.18) $$z(t) = z(\tau) \quad \text{if and only if} \quad t = \tau$$

A curve is a **simple closed curve** if it is closed and if it satisfies the above for all t and τ except for the initial and final values, respectively. Finally, we define the curve $-C$ by

(4.1.19) $$-C\colon \quad z = z(-t) \qquad -t_1 \le t \le -t_0$$

In Problem 4, the reader is asked to verify that **C and $-C$ have the same trace but opposite direction.**

(4.1.20) ***Example.*** Discuss the curve

$$C\colon \quad z = z(t) = \begin{cases} (1 + i)t & 0 \le t \le 2 \\ 4i + (1 - i)t & 2 < t \le 3 \\ 6 - t + i & 3 < t \le 6 \end{cases}$$

Solution. That C is continuous is easily verified. Also, since

$$z'(t) = \begin{cases} 1 + i & 0 < t < 2 \\ 1 - i & 2 < t < 3 \\ -1 & 3 < t < 6 \end{cases}$$

we see that C is piecewise smooth and thus a contour. Its trace and direction are shown in Figure 4.1.4, where we see that C is not simple [since $z(1) = z(5) = 1 + i$]. Finally, the curve $-C$ is given by

(4.1.21)
$$-C: \quad z = \begin{cases} 6 + t + i & -6 \le t < -3 \\ 4i - (1 - i)t & -3 \le t < -2 \\ -(1 + i)t & -2 \le t < 0 \end{cases}$$

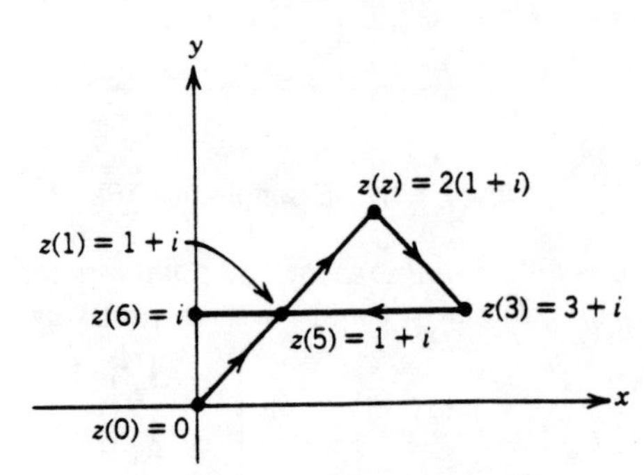

Figure 4.1.4 Example (4.1.20).

■

Often we must *construct* a curve having given properties.

(4.1.22) ***Example.*** Find a simple, smooth curve whose trace is the line segment $y = 2x$ with initial point $1 + 2i$ and final point $2 + 4i$.

Solution. We can choose $x = t$ as the parameter and write one such curve as

$$C: \quad z = t + 2it \qquad 1 \le t \le 2$$

If instead we wanted a simple curve having $2 + 4i$ as its initial point and $1 + 2i$ as its final point, then the curve $-C$, that is, $z = -(1 + 2i)t$, $-2 \le t \le -1$, would suffice.

■

(4.1.23) ***Example.*** Find a simple, closed, smooth curve whose trace is the ellipse $x^2/4 + y^2 = 1$ traversed *clockwise*.

Solution. Note that the left-hand side of the above can be written as a sum of squares $(x^2/2) + y^2 = 1$, an equation "reminiscent of a circle". Thus, we might try the parametrization

(4.1.24)
$$\hat{C}: \quad \frac{x}{2} = \cos t, \qquad y = \sin t \qquad 0 \le t \le 2\pi$$

where geometrically t is the polar angle. This curve is simple and closed, and its trace is the given ellipse [since $\cos^2(t) + \sin^2(t) = 1$]. However, the induced direction is counterclockwise and not clockwise, as the reader is urged to verify. Thus, a candidate for the curve we want is $-\hat{C}$ given by

$$C: \quad z = 2\cos t - i\sin t \qquad -2\pi \le t \le 0$$

■

As the next example illustrates, curves may be defined for an infinite parameter interval. To these cases, one can *extend all of the concepts introduced above if we demand that the various conditions are met on every bounded* t *interval.*

(4.1.25) ***Example.*** Consider the curve

$$C_1: \quad z = (1 - e^{-t}) + i(1 - 2e^{-t}) \qquad 0 \le t < \infty$$

Since $1 - 2e^{-t} = 2(1 - e^{-t}) - 1$, the trace of C_1 must be part of the line $y = 2x - 1$, and in fact is that part of this line lying between $z_0 = -i$ and $z_1 = 1 + i$. The point $1 + i$ *does not lie on the curve*, hence there is no final point, but it is approached as $t \to \infty$, while z_0 serves as the initial point of the curve. Also, C_1 is simple and smooth, and it is depicted in Figure 4.1.5*a*.

The curve

(4.1.26)
$$C_2: \quad z = (1 - e^{-t}) + i(1 - 2e^{-t}) \qquad -\infty < t < \infty$$

is also simple and smooth, and its trace is shown in Figure 4.1.5*b*. Note that there is neither an initial nor a final point. However, we might say that C_2 starts out at $z = \infty$ and approaches $z = 1 + i$ from the left along the line $y = 2x - 1$.

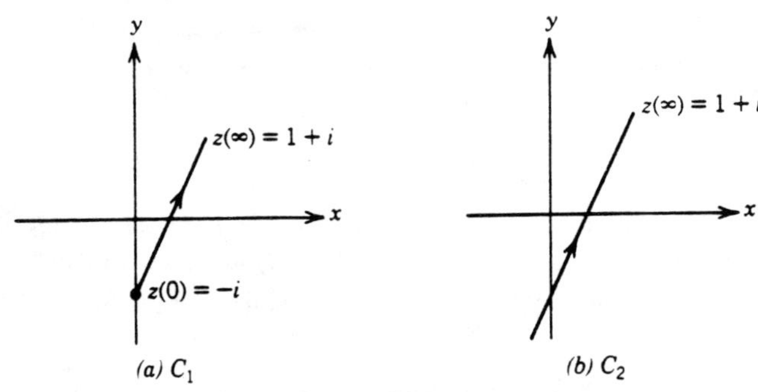

Figure 4.1.5 Example (4.1.25).

■

We now state, without proof, a rather obvious, but actually a very deep, result.

(4.1.27) ***Theorem*** **(Jordan Curve Theorem).** Let C be a simple, closed contour. Then C separates the plane into two separate and distinct regions, the **inside and outside of** C, one of which is bounded, the other unbounded.

■

Finally, we have the following definitions, the first of which is repeated from Section 1.1.

(4.1.28) ***Definition***

(a) A set is **connected** if every pair of points in it can be joined by a broken line lying in the set.

(b) A connected set D is **simply connected** if the inside of any simple closed contour, whose trace is in D, also lies in D. Otherwise, D is said to be **multiply connected**.

■

Loosely speaking, **a simply connected set has no holes in it**, as the next example shows.

(4.1.29) ***Example.*** The inside of a simple, closed contour is simply connected, as shown in Figure 4.1.6*a*. The domain shown in Figure 4.1.6*b* is multiply connected. To see this, consider the inside of that simple, closed contour having the dashed set of points in the figure as its trace. This set does not lie in the domain (the hole is included in it).

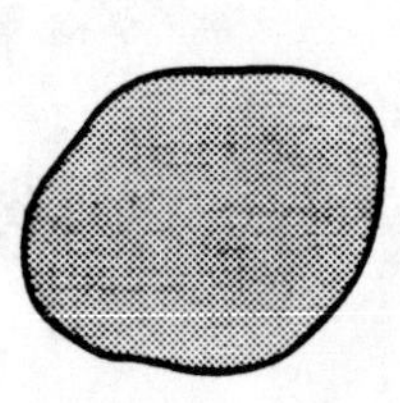

(a) Simply connected domain.

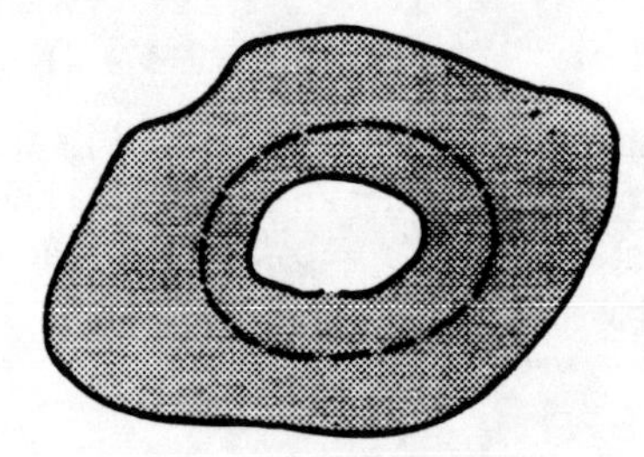

(b) Multiply connected domain.

Figure 4.1.6

■

Equivalent Curves. It seems intuitively clear that two curves having the same initial and final points, the same trace and the same direction, are virtually indistinguishable. We will now show that this is a consequence of the following definitions.

(4.1.30) ***Definition.*** Two piecewise smooth curves

$$C_1: \quad z = z_1(t) \qquad t_0 \le t \le t_1$$
$$C_2: \quad z = z_2(\tau) \qquad \tau_0 \le \tau \le \tau_1$$

are **equivalent** if there exists a real function $\phi(\tau)$ possessing the following properties:

(a) For $\tau_0 < \tau < \tau_1$, $\phi(\tau)$ is continuously differentiable and $\phi'(\tau) > 0$.

(b) $\phi(\tau_0) = t_0$ and $\phi(\tau_1) = t_1$.

(c) $z_1(\phi(\tau)) = z_2(\tau)$ for $\tau_0 \le \tau \le \tau_1$.

The equation $t = \phi(\tau)$ is called a **change of parameter** for the curve C_1.

■

(4.1.31) *Theorem*

(a) If C_1 and C_2 are equivalent, piecewise smooth curves, then they have the same initial and final points, trace, and direction.

(b) If C_1 and C_2 are simple, smooth, *nonclosed* curves having the same trace and initial and final points, then they are equivalent.

(c) If C_1 and C_2 are simple, smooth, *closed* curves with the same trace *and direction*, they are equivalent.

■

Proof. We will only prove (a) and leave the proof of (b) to Problem 10. First, it follows from (4.1.30b and c) that the initial and final points and the traces of C_1 and C_2 are the same. That the traces are the same follows from the fact that the equation $t = \phi(\tau)$ can be inverted since $\phi' \neq 0$. Hence for each t there is a unique τ, and vice versa.

Now, if we represent C_1 and C_2 by

$$C_1:\quad z = z_1(t) = x_1(t) + iy_1(t) \qquad t_0 \le t \le t_1$$

$$C_2:\quad z = z_2(\tau) = x_2(\tau) + iy_2(\tau) \qquad \tau_0 \le \tau \le \tau_1$$

then, from (4.1.30c),

$$x_2(\tau) = x_1(\phi(\tau)) \qquad \text{and} \qquad y_2(\tau) = y_1(\phi(\tau))$$

Since ϕ is continuously differentiable, by the chain rule

$$\frac{dx_2}{d\tau} = \phi' \frac{dx_1}{dt} \qquad \text{and} \qquad \frac{dy_2}{d\tau} = \phi' \frac{dy_1}{dt}$$

By (4.1.30a), $\phi' > 0$, and hence the signs of x_2' and x_1' (and y_2' and y_1') are the same, and thus C_1 and C_2 have the same direction at corresponding points of their traces. (If ϕ' in (4.1.30a) were not required to be of one sign, then we could not draw this conclusion.)

■

(4.1.32) *Example.* Consider the curves

(4.1.33)

$$C_1:\quad z = z_1(t) = -t + i\sqrt{1 - t^2} \qquad -1 \le t \le 1,$$

$$C_2:\quad z = z_2(\tau) = e^{i\tau} \qquad 0 \le \tau \le \pi$$

These are simple and smooth and have the same trace and initial and final points. Hence, by (b) above, they must be equivalent. To see this, set

$$\phi(\tau) = -\cos\tau \qquad 0 \le \tau \le \pi$$

The reader is asked to verify that this function meets all of the requirements of Definition (4.1.30).

■

In Section 5.1, we will see that contour integrals of analytic functions over equivalent curves are equal.

PROBLEMS (The answers are given on page 430.)

(Problems with an asterisk are more theoretical.)

1. Determine the initial and final points, z_0 and z_1, of the following curves, and describe each curve using the words closed, simple, smooth, and piecewise smooth.
   (a) $z = e^{2it} + 1 \quad 0 \le t \le \pi$
   (b) $z = (1 + i)(1 - |t|) \quad -1 \le t \le 1$
   (c) $z = \begin{cases} t & -1 \le t \le 1 \\ e^{i(t-1)} & 1 < t \le \pi + 1 \end{cases}$
   (d) $z = e^{i/t} \quad 0 < t \le 1/\pi$
   (e) $z = (1 + i)t^4 \quad -1 \le t \le 1$
   (f) $z = \begin{cases} (1 + i)t & 0 \le t \le 1 \\ 3 + i(t - 3) & 1 < t \le 3 \end{cases}$
   (g) $z = \dfrac{t}{1 + t^2} + i \quad -\infty < t < \infty$
   (h) $z = \begin{cases} t + i(1 + t^2) & -2 \le t \le 1 \\ 2 - t + i[3 - (t - 2)^2] & 1 < t \le 4 \end{cases}$
   (i) $z = (1 - \cos t)e^{it} \quad 0 \le t \le 2\pi$
   (j) $z = \sin(2t)e^{it} \quad 0 \le t \le 2\pi$
2. Sketch the trace and direction of the curves in Problem 1.
3. If C denotes the curves in Problem 1, what is $-C$ for those in (a), (b), (c), and (i)?
4.* Let C be the piecewise smooth curve $z = z(t)$, $t_0 \le t \le t_1$.
   (a) Show that the curve $-C$ defined by (4.1.19) has the same trace as C but *opposite* direction.
   (b) Let $\hat{C}$ be the curve $z = z(t_0 + t_1 - t)$, $t_0 \le t \le t_1$. Show that $\hat{C}$ and $-C$ have the same trace *and* direction.
5. If α and β are complex constants, with $\alpha \ne \beta$, show that the curve $z = \alpha t + \beta(1 - t)$, $0 \le t \le 1$, is simple and has as its trace the **line segment** joining $z = \beta$ to $z = \alpha$.
6. Show that the curve given in Example (4.1.20) crosses itself *only* when $t = 1$ and $t = 5$.
7. Compute the Cartesian form of the composite functions $f(z(t))$ and verify the **chain rule** where it is applicable.
   (a) $f = z^2, \quad z(t) = \cos t + i \sin t$
   (b) $f = e^{\bar{z}}, \quad z(t) = te^{it}$
   (c) $f = \sin z, \quad z(t) = \begin{cases} t & -1 \le t \le 1 \\ 3 + it & 1 < t \le 3 \end{cases}$
8. Find a simple, piecewise smooth curve whose trace and direction is described below:

(a) The circle $|z - i| = 2$ traversed clockwise.
(b) The triangle with vertices at $z = 0, 1 + i$, traversed counterclockwise.
(c) The upper half of the ellipse

$$\frac{(x - x_0)^2}{a^2} + \frac{(y - y_0)^2}{b^2} = 1$$

traversed from left to right.
(d) The right half of the hyperbola

$$x^2 - \frac{y^2}{4} = 1$$

traversed upwards.
(e) That branch of the hyperbola $xy = 1$ lying in the first quadrant and traversed from right to left.
(f) The boundary of the half-disk $|z| < 1$, $\text{Im}(z) > 0$, traversed counterclockwise.
(g) The square with vertices at $z = \pm 1, \pm i$ traversed counterclockwise.

9. Show that the curve $z = e^{(a+i)t}$, $0 \leq t < \infty$, describes a spiral for a real. [*Hint:* Use polar coordinates.]

10.* The following series of steps can be used to prove part (b) of Theorem (4.1.3). Let C_1: $z = z_1(t)$, $t_0 \leq t \leq t_1$, and C_2: $z = z_2(\tau)$, $\tau_0 \leq \tau \leq \tau_1$, be simple and smooth and possess the same trace and initial and final points.
(a) Pick any $\hat{\tau}$ in the interval (τ_0, τ_1). Then, $\hat{z} = z_2(\hat{\tau})$ lies on the trace of C_2, and hence on the trace of C_1. Show that there must be a unique $\hat{t}$ in the interval (t_0, t_1) corresponding to $\hat{\tau}$.
(b) The correspondence in (a) defines a function $t = \phi(\tau)$ on $\tau_0 < \tau < \tau_1$. If we define $\phi(\tau_0) = t_0$ and $\phi(\tau_1) = t_1$, show that this function is invertible and continuous on $\tau_0 \leq \tau \leq \tau_1$.
(c) Let τ and $\hat{\tau}$ be in the interval (τ_0, τ_1). Use the mean value theorem from calculus and the result in (b) to show that

$$z_1(\phi(\tau)) - z_1(\phi(\hat{\tau})) = [\phi(\tau) - \phi(\hat{\tau})] \cdot \begin{bmatrix} x_1'(\phi(\tau^*)) \\ +iy_1'(\phi(\tilde{\tau})) \end{bmatrix}$$

where x_1 and x_2 are defined by $z_1(t) = x_1(t) + iy_1(t)$ and τ^* and $\tilde{\tau}$ are in the interval $(\tau, \hat{\tau})$.
(d) Use the result in (c) and the smoothness of the curves to show that the function $\phi(\tau)$ in (b) is differentiable. [*Hint:* $z_2(\tau) = z_1(\phi(\tau))$ is differentiable in τ.]
(e) Finally, prove that $\phi(\tau)$ is continuously differentiable and $\phi'(\tau) > 0$.

11. Show that each of the following pairs of curves are equivalent.
(a) C_1: $z = e^{it}$ $\quad 0 \leq t \leq \pi$
C_2: $z = e^{2i\tau}$ $\quad 0 \leq \tau \leq \pi/2$

(b) C_1: $z = \dfrac{1}{2}(1 + i)(e^t - ie^{-t})$ $\quad 0 \leq t < \infty$

C_2: $z = \tau + i\sqrt{\tau^2 - 1}$ $\quad 1 \leq \tau < \infty$

12. Let C_0: $z = z_0(t)$ and C_1: $z = z_1(t)$, $t_0 \le t \le t_1$, be two contours in a domain D. We say that **C_0 can be continuously deformed into C_1 in D** if, *for each value* of s in the interval $0 \le s \le 1$, there is a contour C_s: $z = z(t; s)$, $t_0 \le t \le t_1$, in D so that **(i)** $z(t; s)$ is continuous in s, and **(ii)** $z(t; 0) = z_0(t)$ and $z(t; 1) = z_1(t)$. Exhibit a continuous deformation of the following curves into one another. Unless otherwise defined, D is the whole z-plane. It might help to draw the traces.
    (a) C_0: $z = t + i$, C_1: $z = t$ $0 \le t \le 1$
    (b) C_0: $z = t + i$, C_1: $z = 2t$ $0 \le t \le 1$
    (c) C_0: $z = t + i$, C_1: $z = i(1 - t)$ $0 \le t \le 1$
    (d) C_0: $z = e^{it}$, C_1: $z = 2e^{it}$ $0 \le t \le 2\pi$
    D is the annular domain between the two circles.
    (e) C_0: $z = \cos t + 3i \sin t$, C_1: $z = \cos t + 2i \sin t$ $0 \le t \le 2\pi$
    D is the domain between the two ellipses.
    (f) C_0: $z = \cos t + 3i \sin t$, C_1: $z = \frac{1}{2}e^{it}$ $0 \le t \le 2\pi$
    D is the domain between the ellipse and the circle.
13. (a) Construct a continuous deformation of a simple, closed contour $z = z(t)$, $t_0 \le t \le t_1$, $z(t_0) = z(t_1)$ into a point A_0 inside of C.
    (b)* Prove that a domain D is simply connected if and only if every simple closed contour in D can be continuously deformed in D to a point.

4.2 INVERSE FUNCTIONS[1]

As we shall see in the next few sections, we are often confronted with the need to *find a mapping* w = f(z) *which takes a specified region in the z-plane into a specified region in the* w*-plane*. In doing so, we will naturally be concerned with the inverse relation between z and w.

(4.2.1) ***Definition.*** Let $w = f(z)$ be a single-valued function mapping a region D_z in the z-plane into a region D_w in the w-plane (see Figure 4.2.1). Then, we say that

(a) f is **onto** if for *each* w_0 in D_w there is a z_0 in D_z so that $f(z_0) = w_0$.

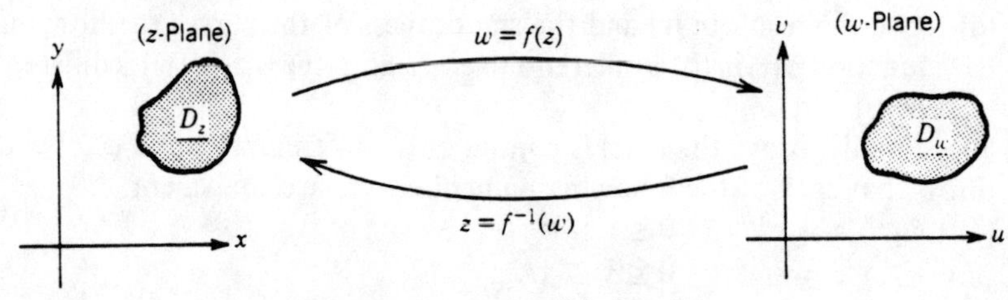

Figure 4.2.1 Invertible function.

[1] The material in this section is theoretical in nature.

(b) ***f* is one to one (1-1) on D_z** if

$$f(z_1) = f(z_2) \text{ if and only if } z_1 = z_2$$

We also say that f is **simple**.

(c) ***f* is invertible on D_z** if there is a single-valued inverse function $f^{-1}(w)$ mapping D_w onto D_z such that *both* of the following hold:

(4.2.2)

(i) $f(f^{-1}(w)) = w$ for all w in D_w

(ii) $f^{-1}(f(z)) = z$ for all z in D_z

(d) f is **locally invertible at a point** z_0 if there is some neighborhood of z_0 on which f is invertible.

■

Note that if $w = f(z) = u(x, y) + i\,v(x, y)$ is invertible, then the real transformation from (x, y) to (u, v) defined by

(4.2.3)
$$u = u(x, y) \qquad \text{and} \qquad v = v(x, y)$$

is invertible and defines two real functions $x(u, v)$ and $y(u, v)$ so that

$$f^{-1}(w) = x(u, v) + i\,y(u, v)$$

In Problem 1, the reader is asked to show that the following relationships hold between the four real functions $u(x, y)$, $v(x, y)$, $x(u, v)$, and $y(u, v)$:

(4.2.4)

(a) $u\big(x(u, v), y(u, v)\big) = u \qquad v\big(x(u, v), y(u, v)\big) = v$

(b) $x\big(u(x, y), v(x, y)\big) = x \qquad y\big(u(x, y), v(x, y)\big) = y$

(4.2.5) ***Example.*** Consider the function

$$w = f(z) = \frac{z - 1}{z + 1} \qquad z \neq -1$$

If D_z denotes the z-plane excepting the point $z = -1$, then f is invertible on D_z with the inverse function given by

$$z = f^{-1}(w) = \frac{1 + w}{1 - w}$$

An easy computation shows that

$$u(x, y) = \frac{x^2 + y^2 - 1}{(x + 1)^2 + y^2} \qquad \text{and} \qquad v(x, y) = \frac{2y}{(x + 1)^2 + y^2}$$

while

$$x(u, v) = \frac{1 - u^2 - v^2}{(u - 1)^2 + v^2} \qquad \text{and} \qquad y(u, v) = \frac{2v}{(u - 1)^2 + v^2}$$

The reader is asked to verify (4.2.4).

■

The next result shows that the concepts defined in (4.2.1) are not independent.

(4.2.6) ***Theorem.*** A function $f(z)$ is invertible on a domain D_z if and only if it is one to one and onto.

Proof. First, assume that f is invertible on D_z. Then, by (4.2.1c), it is onto, and there is a function f^{-1} so that, for all z in D_z, $f^{-1}(f(z)) = z$. Now,

$$f(z_1) = f(z_2) \Rightarrow f^{-1}(f(z_1)) = f^{-1}(f(z_2))$$
$$\Rightarrow z_1 = z_2$$

and hence f is one to one. The reader is asked to prove the converse of this in Problem 4.

■

It may be that a function is invertible on some regions but not others, as the next example shows.

(4.2.7) ***Example.*** The function $f(z) = z^2$ maps both the first and third quadrants *onto* the upper half-plane as shown in Figure 4.2.2*a*. It is not 1–1 on these two quarter-planes taken together since if z is in this region, so is $-z$, and we have $f(z) = f(-z)$. Hence, by the above theorem, f is not invertible here. In fact, f is not invertible on any region which contains pairs of points which are images in the origin of each other.

However, f *is* invertible on the first quadrant alone, with one such single-valued inverse given by (see Figure 4.2.2*b*)

$$z = f^{-1}(w) = (\text{prin.})\ w^{1/2}$$

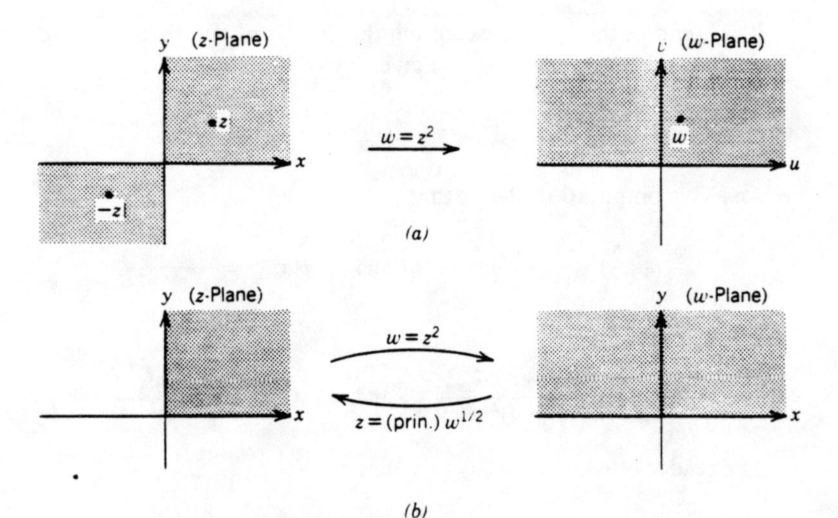

Figure 4.2.2 Example (4.2.7).

The reader is asked to construct a function inverse to $f(z)$ on the third quadrant (the principal branch of $w^{1/2}$ will not work).

■

The next result provides us with a simple criterion for determining when a function is locally invertible.

(4.2.8) ***Theorem.*** Let $w = f(z)$ be analytic with a continuous derivative at z_0. Then,

(a) f is **locally invertible** at z_0 if $f'(z_0) \neq 0$.

(b) The inverse function $f^{-1}(w)$ is analytic at $w_0 = f(z_0)$, and

$$\frac{d}{dw}\left[f^{-1}(w_0)\right] = \frac{1}{f'(z_0)}$$

Proof. Since $f(z) = u(x, y) + i\,v(x, y)$ is continuously differentiable at z_0, so are $u(x, y)$ and $v(x, y)$. Also,

$$\begin{aligned} |f'|^2 &= |u_x + i\,v_x|^2 = u_x^2 + v_x^2 \\ &= u_x v_y - v_x u_y \end{aligned}$$

where the last equation follows from the Cauchy–Riemann equations. The term on the right of this equation is just the **Jacobian of the transformation** (4.2.3) and does not vanish since we are assuming $f'(z_0) \neq 0$. Thus, from advanced calculus the transformation (4.2.3) is invertible in some neighborhood of z_0, and the inverse functions $x = x(u, v)$, $y = y(u, v)$ are also continuously differentiable in some neighborhood of $w_0 = f(z_0)$. To show that $f^{-1}(w)$ is analytic, we must now show that $x(u, v)$ and $y(u, v)$ satisfy the Cauchy–Riemann equations $x_u = y_v$ and $x_v = -y_u$ at w_0. This follows from (4.2.4a) by differentiating both equations there first with respect to u and then with respect to v, using the chain rule. Doing this, we get

(4.2.9)
$$\begin{aligned} u_x x_u + u_y y_u &= 1, \qquad & u_x x_v + u_y y_v &= 0 \\ v_x x_u + v_y y_u &= 0, \qquad & v_x x_v + v_y y_v &= 1 \end{aligned}$$

If we solve these for x_u, y_u, x_v, and y_v, and then use the Cauchy–Riemann equations for $u(x, y)$ and $v(x, y)$, we can conclude that $f^{-1}(w)$ is analytic in some neighborhood of w_0. Thus, the composite function $f(f^{-1}(w))$ is analytic, and we can differentiate both sides of the equation

$$f(f^{-1}(w)) = w$$

with respect to w, using the chain rule, to get

(4.2.10)
$$f'(f^{-1}(w)) \cdot \frac{d}{dw}(f^{-1}(w)) = 1$$

from which (4.2.8b) follows.

■

The next example illustrates what might happen in a neighborhood of a point where $f' = 0$.

(4.2.11) ***Example.*** Consider

$$w = f(z) = z^3$$

By the above result, f is locally invertible in a neighborhood of any $z_0 \neq 0$ (since $f' \neq 0$ here). That f is *not invertible* in any neighborhood of $z = 0$ is easily seen, since the mapping $w = z^3$ maps any disk D_z: $|z| < r_0$ onto the disk D_w: $|w| < r_0^3$, and D_w is covered three times. This means that for each $\hat{w}$ in D_w, there are three values of $\hat{z}$ (the three roots $w^{1/3}$) in D_z so that $\hat{w} = \hat{z}^3$. Hence, f is not one to one and thus not invertible. Note that the *real function* $y = x^3$ *is invertible* on any interval containing $x = 0$, the inverse being the real-valued function $y^{1/3}$.

■

The next result shows that certain properties of curves remain invariant under analytic maps.

(4.2.12) ***Theorem.*** Let C_z be the curve

$$C_z: \quad z = z(t) \qquad t_0 \le t \le t_1$$

and let $f(z)$

(i) Be analytic in some domain containing the trace of C_z.
(ii) Be invertible, and hence one to one on C_z.
(iii) Satisfy $f' \neq 0$ on C_z.

Let C_w denote the image in the w-plane of C_z under the mapping $w = f(z)$, that is,

$$C_w: \quad w = w(t) = f(z(t)) \qquad t_0 \le t \le t_1 \tag{4.2.13}$$

then

(a) If C_z is simple, so is C_w.
(b) If C_z is smooth, so is C_w.
(c) If C_z is piecewise smooth, so is C_w.

Proof. By (4.1.18), to prove (a), we must show that $w(t) = w(\tau)$ if and only if $t = \tau$. Now,

$$\begin{aligned} w(t) = w(\tau) &\Leftrightarrow f(z(t)) = f(z(\tau)) && \text{[from (4.2.13)]} \\ &\Leftrightarrow z(t) = z(\tau) && \text{(since } f \text{ is one to one)} \\ &\Leftrightarrow t = \tau && \text{(since } C_z \text{ is simple)} \end{aligned}$$

For (b), by Definition (4.1.14b) of smoothness, we assume $z'(t)$ is continuous and $|z'(t)| \neq 0$. Hence, the composite function $w(t) = f(z(t))$ is also continuously differentiable and, by the chain rule,

$$w'(t) = z'(t) \cdot f'(z(t)) \neq 0 \qquad (\text{since } f' \neq 0)$$

Thus, C_w is smooth. Part (c) is proved in much the same way.

■

PROBLEMS (The answers are given on page 431.)

(Problems with an asterisk are more theoretical.)

1.* Use the fact that $f(f^{-1}(w)) = w$ and $f^{-1}(f(z)) = z$ to verify (4.2.4).

2. Where in the z-plane are the following functions one to one?
   (a) $\sin z$ (b) e^z (c) (prin.) $\sqrt{z}$ (d) $\operatorname{Log} z$
   (e) z^4 (f) $\sinh z$ (g) $\dfrac{1+z^2}{1-z^2}\ (z \neq \pm 1)$.

3. Where in the z-plane can you be assured that the functions in Problem 2 are locally invertible?

4.* Here, we will complete the proof of Theorem (4.2.6). Thus, we assume that $w = f(z)$ is one to one on D_z and maps D_z onto D_w. We wish to show that f is invertible on D_z.
   (a) Why is it true that for each w in D_w there is a z in D_z so that $w = f(z)$?
   (b) If we designate the correspondence between w and z in part (a) by $z = g(w)$, show that $g(w)$ is a single-valued function having the property that $f(g(w)) = w$ for w in D_w.
   (c) Finally, show that $g(w)$ is the inverse function we are looking for.

5. Consider the function $f(z) = e^z$.
   (a) What is the image D_w of the strip D_z: $\pi/2 < \operatorname{Im}(z) < 3\pi/2$?
   (b) Is e^z invertible on D_z? If so, what is a single-valued inverse $f^{-1}(w)$ which maps D_w onto D_z?
   (c) Show that e^z is *locally invertible* everywhere.
   (d) Show that e^z is *not* invertible everywhere. Hence, **local invertibility does not imply global invertibility**.

6. Let D_w be the image of a region D_z under the principal logarithm $w = \operatorname{Log} z$.
   (a) Use Theorem (4.2.6) to show that $\operatorname{Log} z$ is invertible on D_z.
   (b) What is the inverse function?

7. Discuss the invertibility of $w = f(z) = x + i(x + y)$.

8. Discuss the invertibility of
   (a) $w = x + iy^2$ (b) $w = x + iy^3$

9.* Let $w_1 = g(z)$ map a region D_z onto a region D_{w_1}, and $w = f(w_1)$ map D_{w_1} onto a region D_w. Assume that g is invertible on D_z and f is invertible on D_{w_1}, and consider the composite function $h(z) = f(g(z))$.
   (a)* Why does h map D_z onto D_w?
   (b) Prove that h is invertible on D_z.

4.3 CONFORMAL MAPPINGS

From Theorem (4.2.8) of the previous section, we learn that an analytic function with a nonvanishing derivative is locally invertible. Here, we will develop some further geometric consequences of the condition $f' \neq 0$; but first a result having to do with the **tangent vector to a smooth curve**.

(4.3.1) ***Theorem.*** Let C be the smooth curve

$$C: \quad z = z(t) \qquad t_0 \leq t \leq t_1$$

(We remind the reader that, for a smooth curve, $z(t)$ is continuously differentiable with $z'(t) \neq 0$.) For each value of t, let $\mathbf{T}(t)$ be the vector in the plane represented by $z'(t)$, that is, the vector from $z = 0$ to $z = z'(t)$ (see Figure 4.3.1*a*). Then,

(a) $\mathbf{T}(t)$ is tangent to the trace of C at the point $z(t)$ and points in the direction of C (see Figure 4.3.1*b*).

(b) If $\theta(t)$ is the **local polar angle** for $\mathbf{T}(t)$, then there is a value of $\arg[z'(t)]$ for which

(4.3.2)
$$\theta(t) = \arg[z'(t)] \qquad 0 \leq \theta(t) < 2\pi$$

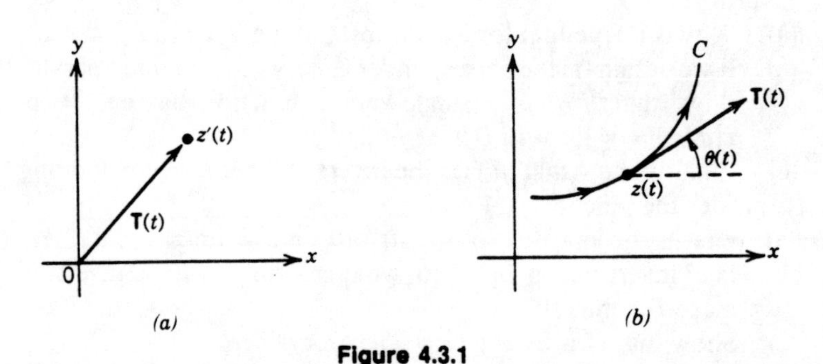

Figure 4.3.1

Proof. Since C is smooth, $(x')^2 + (y')^2 \neq 0$; and from calculus, the vector $\mathbf{T}(t) = x'(t)\mathbf{i} + y'(t)\mathbf{j}$ is defined to be tangent to the curve C. This is just the vector represented by $z'(t) = x'(t) + i\,y'(t)$. That $\mathbf{T}(t)$ points in the direction of C at $z = z(t)$ follows from the definition of derivative and is shown in Problem 1. Also, the angle $\theta(t)$ between $\mathbf{T}(t)$ and the positive x-axis is given by $\theta(t) = \tan^{-1}(y'/x')$. Since this is also the expression for the most general argument of $z'(t)$, we can certainly find a value of $\arg(z')$ in the range $0 \leq \theta < 2\pi$ so that (4.3.2) is satisfied.

■

In Problem 2, we show how tangents can sometimes be defined at points where $z'(t) = 0$.

(4.3.3) ***Example.*** Consider the curve

$$C: \quad z = z(t) = \begin{cases} t + i(1 + t^2) & -2 \le t \le 1 \\ 2 - t + i[3 - (t - 2)^2] & 1 < t \le 4 \end{cases}$$

Find tangent vectors to C when $t = -1$ and $t = 3$ and discuss what happens at $t = 1$.

Solution. First, note that the trace of C consists of pieces of the two parabolas $y = 1 + x^2$ and $y = 3 - x^2$. This, together with the direction of C, are shown in Figure 4.3.2*a*. Also, since $z'(t)$ exists (except when $t = 1$) and is continuous and nonvanishing, C is piecewise smooth. In addition, C is *not simple* since $z(-1) = z(3) = -1 + 2i$. Now,

$$z'(-1) = 1 - 2i \qquad \text{and} \qquad z'(3) = -1 - 2i$$

Hence, as shown in part (b) of the figure, at the nonsimple point $z = -1 + 2i$, there are two vectors, $\mathbf{T}(-1)$ and $\mathbf{T}(3)$.

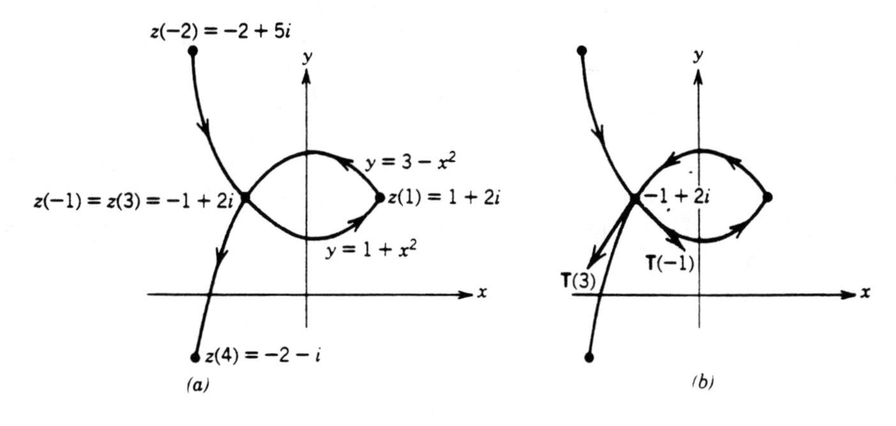

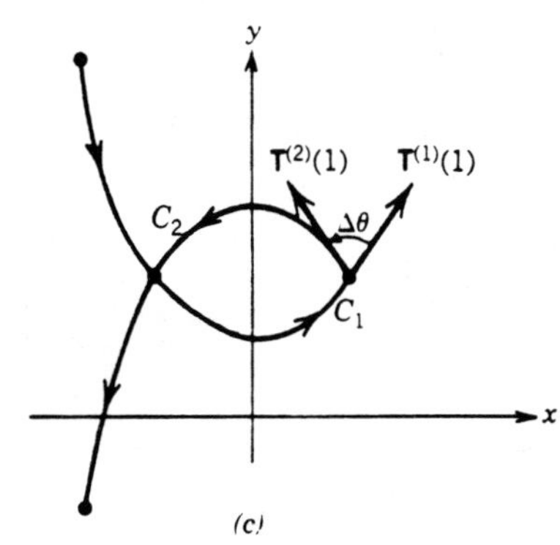

Figure 4.3.2 Example (4.3.3).

At $t = 1$, where $z = 1 + 2i$, C is not smooth but consists of two nearby smooth pieces, C_1 and C_2, defined by

(4.3.4)
$$C_1: \quad z = z_1(t) = t + i(1 + t^2). \qquad -2 \le t \le 1,$$
$$C_2: \quad z = z_2(t) = 2 - t + i[3 - (t - 2)^2] \qquad 1 \le t \le 4$$

Each of the functions $z_1(t)$ and $z_2(t)$ has a one-sided derivative at $t = 1$ given by

(4.3.5)
$$z_1'(1) = \lim_{t \uparrow 1} [z_1'(t)] = 1 + 2i \qquad z_2'(1) = \lim_{t \downarrow 1} [z_2'(t)] = -1 + 2i$$

The complex numbers $z_1'(1)$ and $z_2'(1)$ can be represented by vectors $\mathbf{T}^{(1)}(1)$ and $\mathbf{T}^{(2)}(1)$, respectively. These in turn are the tangent vectors to C_1 and C_2, respectively, at $t = 1$ and are shown in Figure 4.3.2*c*.

■

In the above example, one might deem it strange that two *different* tangent vectors, $\mathbf{T}(-1)$ and $\mathbf{T}(3)$, exist at the same smooth point on the trace. However, if we view a curve as describing the position at time t of a particle traveling in a plane, then the tangent vector $\mathbf{T}(t)$ points in the direction of motion at time t; and if the particle happens to pass through the same point at different times, it might very well be traveling in different directions at these two times.

In this same example, we also had two tangents, $\mathbf{T}^{(1)}(1)$ and $\mathbf{T}^{(2)}(1)$, at the point $z = 1 + 2i$, where C was *not smooth*. These were the tangents to the curves C_1 and C_2 in (4.3.4). Hence, we might say that C takes an abrupt turn here, with the angle of rotation $\Delta\theta$ (shown in Figure 4.3.2*c*) defined as that given by rotating $\mathbf{T}^{(1)}(1)$ into $\mathbf{T}^{(2)}(1)$. Motivated by this, let us now make more precise the idea of the angle between two curves.

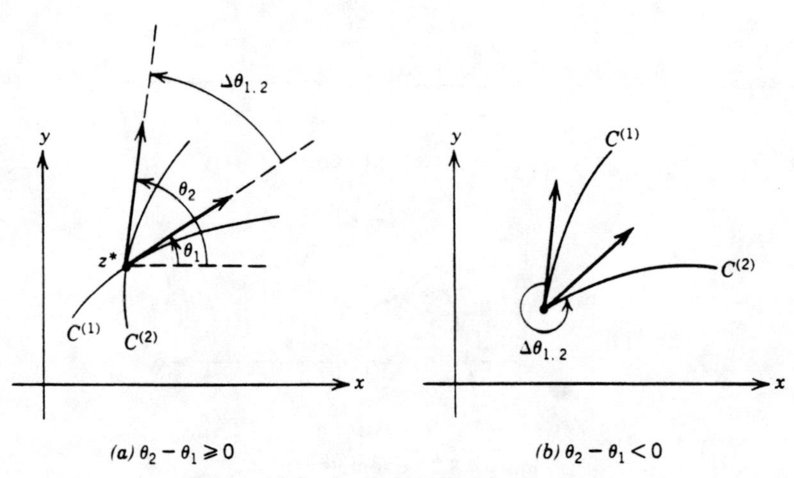

Figure 4.3.3 Angle between two curves.

(4.3.6) ***Definition.*** Let $C^{(1)}$ and $C^{(2)}$ be two simple curves which

(i) Intersect at $z = z^*$.

(ii) Are smooth in a neighborhood of z^*.

(iii) Have tangent vectors $\mathbf{T}^{(1)}$ and $\mathbf{T}^{(2)}$, respectively, at z^* (see Figure (4.3.3).

If θ_1 and θ_2 are the local polar angles for $\mathbf{T}^{(1)}$ and $\mathbf{T}^{(2)}$, respectively, we define $\Delta\theta_{1,2}|_{z=z^*}$, **the angle from $C^{(1)}$ to $C^{(2)}$**, to be

(4.3.7)
$$\Delta\theta_{1,2}|_{z=z^*} = \begin{cases} \theta_2 - \theta_1 & \text{if } \theta_2 - \theta_1 \geq 0 \\ 2\pi + (\theta_2 - \theta_1) & \text{if } \theta_2 - \theta_1 < 0 \end{cases}$$

where $0 \leq \theta_1, \theta_2 < 2\pi$.

■

The reader will observe from Figure 4.3.3 that **$\Delta\theta_{1,2}$ is the positive angle through which $C^{(1)}$ must be rotated in order to locally line up with $C^{(2)}$**. Also, note from (4.3.7) that, since both θ_1 and θ_2 are to be in the range $0 \leq \theta_1, \theta_2 < 2\pi$, we must have

(4.3.8)
$$0 \leq \Delta\theta_{1,2}|_{z=z^*} < 2\pi$$

(4.3.9) ***Example.*** For the three simple, smooth curves

(4.3.10)
$$\begin{aligned} &C^{(1)}: \quad z = z_1(t) = t && 0 \leq t \leq 1 \\ &C^{(2)}: \quad z = z_2(\tau) = 1 + i\tau && 0 \leq \tau \leq 2 \\ &C^{(3)}: \quad z = z_3(\sigma) = \frac{1}{4}\sigma^2 - i\sigma && -2 \leq \sigma \leq 0 \end{aligned}$$

compute $\Delta\theta_{1,2}, \Delta\theta_{2,3}$, and $\Delta\theta_{3,1}$ at the respective points of intersection.

Solution. The curves are shown in Figure 4.3.4. Now, $C^{(1)}$ and $C^{(2)}$ intersect at $z = 1$ which, from (4.3.10), corresponds to $t = 1$ for $C^{(1)}$ and $\tau = 0$ for $C^{(2)}$. Hence, if θ_1 and θ_2 represent the respective polar angles of

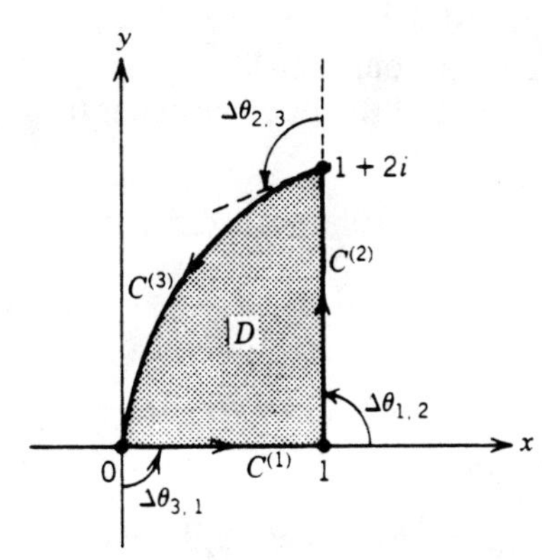

Figure 4.3.4 Example (4.3.9).

the tangents to $C^{(1)}$ and $C^{(2)}$, we have

$$\theta_1(1) = \arg[z_1'(1)] = 0$$

and

$$\theta_2(0) = \arg[z_2'(0)] = \pi/2$$

Thus, from (4.3.7), since $\theta_2 - \theta_1 > 0$,

$$\Delta\theta_{1,2}|_{z=1} = \theta_2 - \theta_1 = \pi/2$$

At $z = 1 + 2i$, the point of intersection of $C^{(2)}$ and $C^{(3)}$, we have $z_2'(2) = i$ and $z_3'(-2) = -1 - i$. Hence, $\theta_2(2) = \pi/2$ and $\theta_3(-2) = 5\pi/4$, and thus

$$\Delta\theta_{2,3}|_{z=1+2i} = \theta_3 - \theta_2 = \frac{3\pi}{4}$$

We leave it to the reader to show that

(4.3.11)
$$\Delta\theta_{3,1}|_{z=0} = \frac{\pi}{2}$$

These angle changes are shown in the figure.

Now consider the region D, shown shaded in Figure 4.3.4 and bounded by the lines $y = 0$ and $x = 1$, and by the parabola $x = \frac{1}{4}y^2$. The boundary of D consists of the traces of the curves $C^{(1)}$, $C^{(2)}$, and $C^{(3)}$ in (4.3.10). It is clear from the figure that as we traverse this boundary counterclockwise starting at $z = 0$, we will make a 90° turn at $z = 1$ ($\Delta\theta_{1,2} = \pi/2$), a 135° turn at $z = 1 + 2i$ ($\Delta\theta_{2,3} = 3\pi/4$), and a 90° turn at $z = 0$ ($\Delta\theta_{3,1} = \pi/2$).

■

We are now in a position to prove the main result of this section, namely, that under certain analytic maps (called conformal), angles between curves remain invariant.

(4.3.12) ***Theorem.*** Let $C_z^{(1)}$ and $C_z^{(2)}$ be two simple curves which intersect at z^* and which are smooth in some neighborhood of z^*. Let $C_w^{(1)}$ and $C_w^{(2)}$ be the images in the w-plane of these curves under the mapping $w = f(z)$ (see Figure 4.3.5). Assume that

(i) $f(z)$ is analytic and continuously differentiable at z^*.
(ii) $f'(z^*) \neq 0$

Then,

(4.3.13)
$$\Delta\theta_{1,2}|_{z=z^*} = \Delta\phi_{1,2}|_{w=w^*}$$

where (see the figure)

$$\Delta\theta_{1,2}|_{z=z^*} = \text{angle between } C_z^{(1)} \text{ and } C_z^{(2)} \text{ at } z^*$$

$$\Delta\phi_{1,2}|_{w=w^*} = \text{angle between } C_w^{(1)} \text{ and } C_w^{(2)} \text{ at } w^* = f(z^*)$$

Such a mapping is said to be **conformal at z^***.

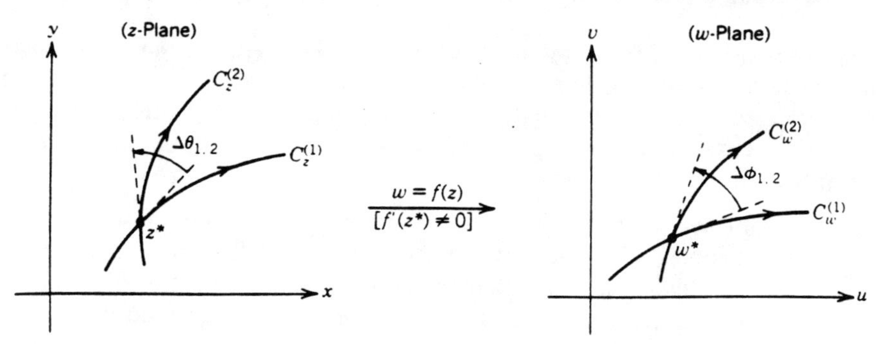

Figure 4.3.5 Conformality: $\Delta\theta_{1,2} = \Delta\phi_{1,2}$.

Proof. With the above properties of $f(z)$, it is a simple matter to show that the image curves $C_w^{(1)}$ and $C_w^{(2)}$ are also simple and smooth in some neighborhood of w^*; see Theorem (4.2.12). Thus, the local polar angles ϕ_1 and ϕ_2 for the tangent vectors to $C_w^{(1)}$ and $C_w^{(2)}$ are well defined, as is also $\Delta\phi_{1,2}$. If $C_z^{(1)}$ and $C_z^{(2)}$ are parametrized by

$$C_z^{(1)}: \quad z = z_1(t) \qquad \text{and} \qquad C_z^{(2)}: \quad z = z_2(t) \qquad t_0 \le t \le t_1$$

then $C_w^{(1)}$ and $C_w^{(2)}$ are parametrized by

$$C_w^{(1)}: \quad w = f(z_1(t)) \qquad \text{and} \qquad C_w^{(2)}: \quad w = f(z_2(t)) \qquad t_0 \le t \le t_1$$

From (4.3.2), we find that at z^* the local polar angles θ_1 and θ_2 for $C_z^{(1)}$ and $C_z^{(2)}$ are given by

$$\theta_1 = \arg(z_1') \qquad \text{and} \qquad \theta_2 = \arg(z_2') \tag{4.3.14}$$

while the local polar angles ϕ_1 and ϕ_2 for $C_w^{(1)}$ and $C_w^{(2)}$ are given by

$$\phi_1 = \arg[z_1' \cdot f'(z^*)] \qquad \text{and} \qquad \phi_2 = \arg[z_2' \cdot f'(z^*)] \tag{4.3.15}$$

The above results follow from the chain rule $(d/dt)[f(z(t))] = z' \cdot f(z)$. From these and (4.3.14), we also find that

$$\phi_1 = \theta_1 + \arg[f'(z^*)] + 2\pi n_1 \qquad \text{and} \qquad \phi_2 = \theta_2 + \arg[f'(z^*)] + 2\pi n_2$$

where n_1 and n_2 are integers. Hence,

$$\phi_2 - \phi_1 = \theta_2 - \theta_1 + 2\pi(n_2 - n_1)$$

This, combined with the definition of $\Delta\theta_{1,2}$ and $\Delta\phi_{1,2}$ in (4.3.7), leads to

$$\Delta\phi_{1,2} = \Delta\theta_{1,2} + 2\pi k \tag{4.3.16}$$

where k is an integer. Now, both $\Delta\phi_{1,2}$ and $\Delta\theta_{1,2}$ lie in the range $0 \le \Delta\phi_{1,2}, \Delta\theta_{1,2} < 2\pi$. Hence, from the above, if $k \ge 1$, then $\Delta\phi_{1,2}$ will be greater than or equal to 2π, while if $k \le -1$, $\Delta\phi_{1,2}$ will be negative. Thus, the only possible choice for k is $k = 0$, therefore verifying (4.3.13).

■

This theorem tells us that **local angle changes remain invariant in both magnitude and sense under a continuously differentiable, analytic map with $f' \neq 0$.** Hence, if a curve is rotated through some angle to be locally aligned with another curve, the images of the curves will be locally aligned by a similar rotation through exactly the same angle.

(4.3.17) ***Example.*** Consider the rectangle in the z-plane of Figure 4.3.6. Under the mapping $w = e^z$, its image is the boundary of the semiannular region shown in the w-plane. Pick a simple, smooth parametrization of the pieces of the boundary of the rectangle which has the indicated directions and then verify the directions shown on the curves in the w-plane. Note that e^z satisfies all of the conditions of the above theorem.

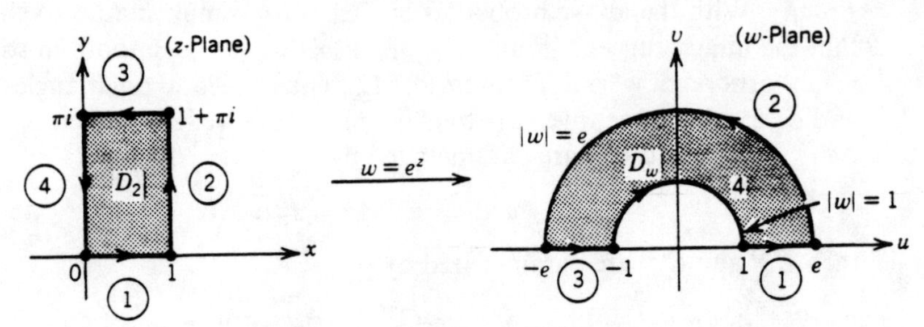

Figure 4.3.6 Example (4.3.17).

Solution. The pieces of the rectangle can be parametrized by

① $C_z^{(1)}$: $z = t$ $\quad 0 \leq t \leq 1$, ② $C_z^{(2)}$: $z = 1 + it$ $\quad 0 \leq t \leq \pi$

③ $C_z^{(3)}$: $z = -t + i\pi$ $\quad -1 \leq t \leq 0$, ④ $C_z^{(4)}$: $z = -it$ $\quad -\pi \leq t \leq 0$

Under the mapping $w = e^z$, the images of these curves are

(4.3.18)

① $C_w^{(1)}$: $w = e^t$ $\quad 0 \leq t \leq 1$, ② $C_w^{(2)}$: $w = e \cdot e^{it}$ $\quad 0 \leq t \leq \pi$

③ $C_w^{(3)}$: $w = -e^{-t}$ $\quad -1 \leq t \leq 0$, ④ $C_w^{(4)}$: $w = e^{-it}$ $\quad -\pi \leq t \leq 0$

The reader can now easily verify the directions shown on the curves in the w-plane of the figure.

Note that at $z = 1$, where $C_z^{(1)}$ meets $C_z^{(2)}$, you must make an abrupt angle change of $\pi/2$ to the left in order to continue along the rectangle in the prescribed direction. This is also true in the w-plane at $w = e$, where $C_w^{(1)}$ meets $C_w^{(2)}$. This is nothing more than an illustration of the above theorem.

■

(4.3.19) ***Example.*** Show that local angle changes at $z = 0$ are *not preserved* under the mapping $f(z) = z^2$. Note here that $f'(0) = 0$.

Solution. Consider the two simple, smooth curves

$$C_z^{(1)}:\quad z_1(t) = t \qquad 0 \le t \le 1$$

$$C_z^{(2)}:\quad z_2(t) = it \qquad 0 \le t \le 1$$

They intersect at $z = 0$ and their directions are shown in Figure 4.3.7. Their images under $w = z^2$ are

(4.3.20)
$$C_w^{(1)}:\quad w_1(t) = t^2 \qquad 0 \le t \le 1$$
$$C_w^{(2)}:\quad w_2(t) = -t^2 \qquad 0 \le t \le 1$$

These curves intersect at $w = 0$; they are simple, but are not smooth at $w = 0$ $[w_1'(0) = w_2'(0) = 0]$. Hence, here we cannot define tangent vectors as in Theorem (4.3.1). However, tangents, and therefore angles $\phi_1(t)$ and $\phi_2(t)$, *can* be defined at points for which $t \neq 0$. Then an appropriate definition of the local angles $\phi_1(0)$ and $\phi_2(0)$ to $C_w^{(1)}$ and $C_w^{(2)}$ at $w = 0$ might be the limits (see Problem 2)

$$\phi_1(0) = \lim_{t \downarrow 0} [\phi_1(t)] \qquad \text{and} \qquad \phi_2(0) = \lim_{t \downarrow 0} [\phi_2(t)]$$

if these limits exist. Now, for $t \neq 0$, from (4.3.20) we have $\phi_1(t) = 0$ and $\phi_2(t) = \pi$. Hence, from the above definition we get $\phi_1(0) = 0$ and $\phi_2(0) = \pi$, which by (4.3.7) yield

$$\Delta\phi_{1,2}|_{w=0} = \pi \qquad \text{(see Figure 4.3.7)}$$

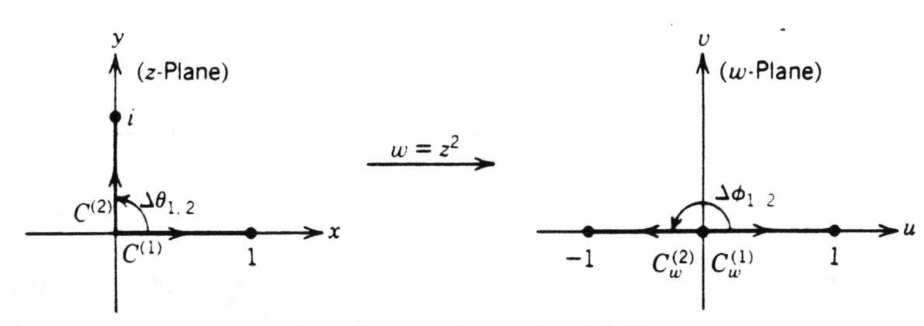

Figure 4.3.7 Example (4.3.19).

If we compute the local angles at $z = 0$ to $C_z^{(1)}$ and $C_z^{(2)}$, we find that

$$\Delta\theta_{1,2}|_{z=0} = \frac{\pi}{2}$$

Thus,

$$\Delta\phi_{1,2}|_{w=0} = 2\Delta\theta_{1,2}|_{z=0}$$

and hence the local angle is doubled. This of course was to be expected from the mapping properties of $w = z^2$.

■

Sometimes, in our search for a mapping function, the properties of conformal functions permit us to eliminate certain candidates, as the next example shows.

(4.3.21) ***Example.*** Suppose we seek a function $w = f(z)$ which is analytic and invertible in the half-disk and on its boundary, as shown in Figure 4.3.8, and which respectively maps the half-disk and its boundary onto the upper half of the w-plane and *its* boundary, the real axis. Then, by necessity this function must have a vanishing derivative somewhere on the boundary of the half-disk. For if not, f would be conformal everywhere on this boundary, and the 90° angles shown in the figure between the two pieces of the boundary would map into pieces of the real w-axis also having relative angles of 90°. Clearly this cannot be. Hence, it would be silly to search for, say, a bilinear map to accomplish the mapping since it does not have a vanishing derivative. In fact, one function mapping the half-disk onto the half-plane (the reader might wish to verify this) is

(4.3.22)
$$f(z) = \sin\left[e^{-\pi i/2}\,\mathrm{Log}\,z - \frac{\pi}{2}\right]$$

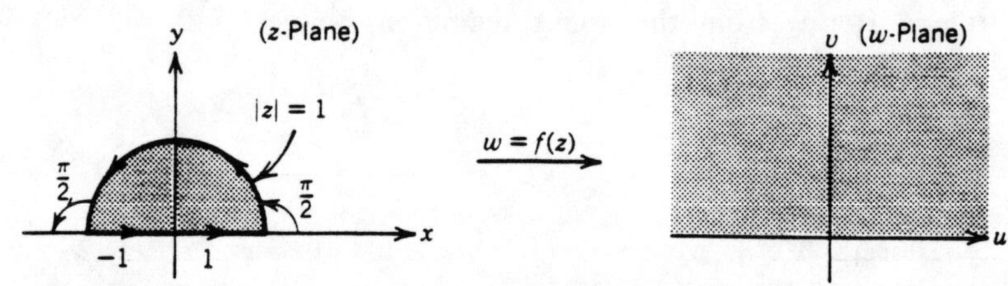

Figure 4.3.8 Example (4.3.21).

Note from Example (4.3.17) and Figure 4.3.6 that the region D_z bounded by the rectangle is mapped by $w = e^z$ onto the semiannular region D_w bounded by the image of the rectangle. It is a common misconception on the part of beginning students in this subject that such a situation is always true. The next example shows that even when the mapping function seems to be "super nice" (such as e^z), **it is not always true that the image of the boundary of a region is the boundary of the image of that region.**

(4.3.23) ***Example.*** Consider the rectangle shown in the z-plane of Figure 4.3.9, together with the mapping $w = e^z$, which is analytic and conformal everywhere. The reader is asked to verify that the images (and directions) of the pieces of the rectangle parametrized by

① $z = t \quad 0 \le t \le 1,$ ② $z = 1 + it \quad 0 \le t \le 3\pi$

③ $z = 3\pi i - t \quad -1 \le t \le 0,$ ④ $z = -it \quad -3\pi \le t \le 0$

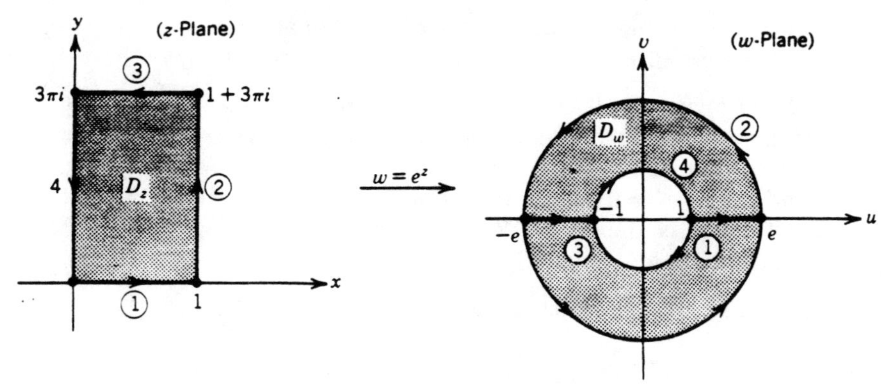

Figure 4.3.9 Example (4.3.23).

are as given in the figure, with the circles in the w-plane covered one and one-half times. Note that the image of the region D_z bounded by the rectangle is the whole annular region D_w whose boundary consists of just the two circles shown and not these circles plus the two pieces ① and ③ of the real axis, as shown in the figure.

■

The reason why the image of the boundary was not the boundary of the image in the above example was that the mapping function e^z was not one to one, and hence did not possess a single-valued inverse (note, for example, that $e^{\pi i/2} = e^{5\pi i/2}$). In general, we have the following result, which will be stated without proof.

(4.3.24) ***Theorem.*** Let D be a domain bounded by a simple, closed, piecewise smooth curve C. Let $f(z)$ be analytic on C and inside D, and be *invertible* on C. Then,

(a) $f(z)$ is conformal and invertible on C and in D.
(b) The image of D under $w = f(z)$ is the interior of the image of C.
(c) If C is traversed counterclockwise, so is its image.

■

From this we learn, that, if $f(z)$ **takes no value more than once on the boundary of a region, the boundary of the image is the image of the boundary.**

PROBLEMS (The answers are given on page 431.)

(Problems with an asterisk are more theoretical.)

1.* Let a curve C *be* described in vector notation by $\mathbf{r}(t) = x(t)\mathbf{i} + y(t)\mathbf{j}$, $t_0 \le t \le t_1$ where $\mathbf{r}$ denotes the **position vector** from the origin to the point $x\mathbf{i} + y\mathbf{j}$. Assume that $x(t)$ and $y(t)$ are continuously differentiable functions with $(x')^2 + (y')^2 \neq 0$.

Show that the tangent vector $\mathbf{T}(t) = x'(t)\mathbf{i} + y'(t)\mathbf{j}$ at the point $x(t)\mathbf{i} + y(t)\mathbf{j}$ points in the same direction as C. [*Hint:* $\mathbf{T}(t) = \mathbf{r}'(t) = \lim_{h \to 0} 1/h \cdot [\mathbf{r}(t+h) - \mathbf{r}(t)]$. Refer to Figure 4.3.10 and compute the limits for $h > 0$ and $h < 0$.]

2. Here, we show how we can define a tangent to a curve at a point where $z' = 0$. Consider the continuously differentiable curve C: $z = z(t)$, $t_0 \le t \le t_1$. Assume that $z'(t^*) = 0$ and that $z'(t)$ does not vanish at any other point in some small interval containing t^*. We define the tangent vector $\mathbf{T}(t^*)$ at $z(t^*)$ to be that vector having local polar angle

$$\theta(t^*) = \lim_{t \to t^*} [\theta(t)] \qquad 0 \le \theta(t^*) < 2\pi$$

if the limit exists. Here, $\theta(t)$ is the local polar angle of the tangent vector $\mathbf{T}(t)$ for t "close to" but not equal to t^*. Since $z'(t) \ne 0$ for these values of t, $\theta(t)$ is well defined. If t^* is an end point, then the above is to be a one-sided limit. For the following curves, compute the tangent vectors at all points where $z' = 0$.

(a) $z = t^2$ $\quad 0 \le t \le 1$ $\qquad$ (b) $z = t^2$ $\quad -1 \le t \le 0$
(c) $z = t^2$ $\quad -1 \le t \le 1$ $\qquad$ (d) $z = \alpha t^3$ $\quad 0 \le t \le 1$
(e) $z = t^2 + it^4$ $\quad 0 \le t \le 1$ $\qquad$ (f) $z = t^4 + it^2$ $\quad 0 \le t \le 1$
(g) $z = t^2 + it^3$ $\quad 0 \le t \le 1$ $\qquad$ (h) $z = t^2 + it^3$ $\quad -1 \le t \le 0$
(i) $z = t^3 + it^2$ $\quad 0 \le t \le 1$ $\qquad$ (j) $z = t^3 + it^2$ $\quad -1 \le t \le 0$

3. What are the points of intersection for each of the following pairs of curves? At each point, compute the angle $\Delta\theta_{1,2}$ between $C_z^{(1)}$ and $C_z^{(2)}$.
(a) $C_z^{(1)}$: $z = e^{it}$ $\quad 0 \le t \le \pi$ $\quad$ and $\quad$ $C_z^{(2)}$: $z = -\tau,\ -1 \le \tau \le 2$
(b) $C_z^{(1)}$: $z = 1 + e^{-it}$ $\quad -\pi \le t \le 0$ $\quad$ and $\quad$ $C_z^{(2)}$: $z = i\tau,\ -2 \le \tau \le 2$
(c) $C_z^{(1)}$: $z = t$ $\quad 0 \le t \le 2$ $\quad$ and $\quad$ $C_z^{(2)}$: $z = -\tau + i(\tau + 1)^2$, $-2 \le \tau \le -1$

4. Under the mapping $w = e^z$, each of the curves $C_z^{(1)}$ and $C_z^{(2)}$ in Problem 3 is transformed respectively into curves $C_w^{(1)}$ and $C_w^{(2)}$ in the w-plane. Verify, for each pair of curves, that the angles between $C_w^{(1)}$ and $C_w^{(2)}$ at their respective points of intersection are the same as those between $C_z^{(1)}$ and $C_z^{(2)}$. Why must this be?

5. (a) The curve C: $z = \sqrt{t}$, $0 \le t \le 1$, is not even differentiable at $t = 0$. Nevertheless, its trace, the real interval $0 \le x \le 1$, is traversed from left to right, and it would seem possible to define a tangent vector to C at $t = 0$.

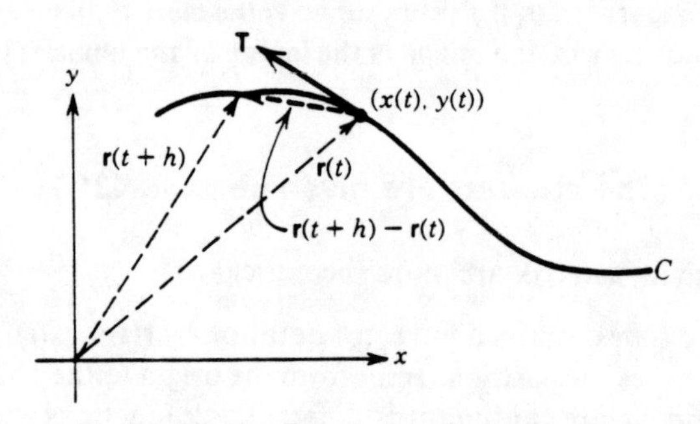

Figure 4.3.10 Tangent vector to a curve.

We do this in much the same way as in Problem 2, by defining the polar angle $\theta^*(0)$ of the tangent to be

$$\theta^*(0) = \lim_{t \downarrow 0} \{\arg[z'(t)]\} \qquad 0 \le \theta^*(0) < 2\pi$$

if the limit exists. Show that here, $\theta^*(0) = 0$ (note that $\lim_{t \downarrow 0} z'(t)$ does not exist).

(b) Consider the two curves $C_z^{(1)}$: $z = t$, $0 \le t \le 1$, and $C_z^{(2)}$: $z = i\tau$, $0 \le \tau \le 1$.

(i) What is the angle $\Delta\theta_{1,2}$ between $C_z^{(1)}$ and $C_z^{(2)}$ at their point of intersection?

(ii) Under the mapping $w = z^{1/n}$ $(n = 1, 2, \ldots)$, where the principal branch is used, the curves $C_z^{(1)}$ and $C_z^{(2)}$ are mapped into curves $C_w^{(1)}$ and $C_w^{(2)}$. Use the definition in part (a) to compute the angle $\Delta\phi_{1,2}$ between $C_w^{(1)}$ and $C_w^{(2)}$ at their point of intersection. How is this related to the angle $\Delta\theta_{1,2}$ in (i)?

(iii) Redo part (ii), but now use that branch of $z^{1/n}$ where $0 < \arg(z) \le 2\pi$.

6. Consider the shaded region D_z of Figure 4.3.11.

(a) Parametrize the pieces ①, ②, and ③ of the boundary of D_z so that the induced directions are as indicated.

(b) What is the image of D_z under the mapping $w = \operatorname{Log} z$?

(c) What are the images of ①, ②, and ③ under $w = \operatorname{Log} z$?

(d) Let D_w denote the image of D_z in (b). Show that the boundary of D_w is traversed counterclockwise, which is the same way the boundary of D_z is traversed, and with the same "sharp turns."

7. Consider the two curves $C_z^{(1)}$: $z = t$, $0 \le t \le 1$, and $C_z^{(2)}$: $z = i\tau$, $0 \le \tau \le 1$. These intersect at $z = 0$, and the angle between them is $\Delta\theta_{1,2} = \pi/2$.

(a) What are the images of these curves under the mapping $w = \bar{z}$?

(b) What is the angle $\Delta\phi_{1,2}$ between the curves in (a) at their point of intersection? Explain why it is not $\pi/2$.

8.* Let C_1 and C_2 be two simple, smooth curves which intersect at $z = z^*$. Let $\hat{C}_1$ and $\hat{C}_2$ be curves which are equivalent respectively to C_1 and C_2; see Definition (4.1.30). Show that the angle between C_1 and C_2 at z^* is the same as that between $\hat{C}_1$ and $\hat{C}_2$.

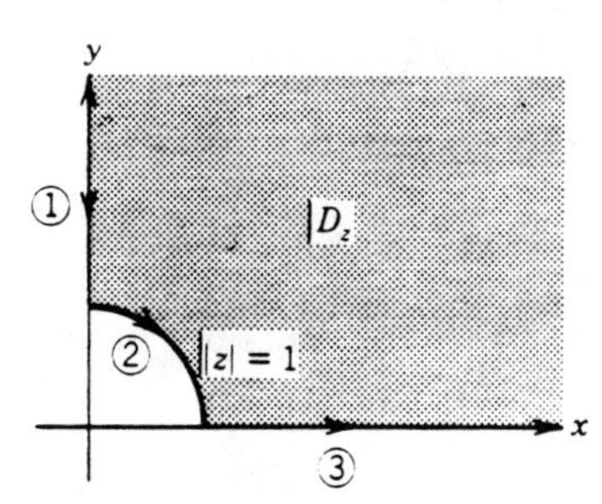

Figure 4.3.11

4.4 INVARIANCE OF LAPLACE'S EQUATION; BOUNDARY VALUE PROBLEMS FOR HARMONIC FUNCTIONS

In Appendices B and D, we show how boundary value problems for Laplace's equation arise quite naturally when considering problems in steady-state heat conduction and electrostatics [see equations (B.10) and (D.15a, b)]. In the next section, we will show how to solve some of these problems using conformal mapping techniques. However, here we will be concerned with a very basic and interesting result, which shows how Laplace's equation transforms under suitable analytic mappings.

(4.4.1) ***Theorem* (Invariance of Laplace's equation).** Let $w = f(z) = u(x, y) + i\,v(x, y)$.

(i) Be analytic, invertible, and possess a continuous second derivative on a domain D_z, and

(ii) Map D_z onto a domain D_w (see Figure 4.4.1).

We then know that $f(z)$ defines a real invertible transformation

$$u = u(x, y), \qquad v = v(x, y),$$

(4.4.2) and

$$x = x(u, v), \qquad y = y(u, v)$$

from D_z to D_w, where the functions $u(x, y)$, $v(x, y)$, $x(u, v)$, and $y(u, v)$ possess continuous second partial derivatives in their respective domains.[1] Let $U(x, y)$ be a real-valued function with continuous second partial derivatives in D_z. Then, under the invertible change of variables (4.4.2), $U(x, y)$ transforms into the **composite function** $\hat{U}(u, v)$ defined by

$$\hat{U}(u, v) = U(x(u, v), y(u, v)) \tag{4.4.3}$$

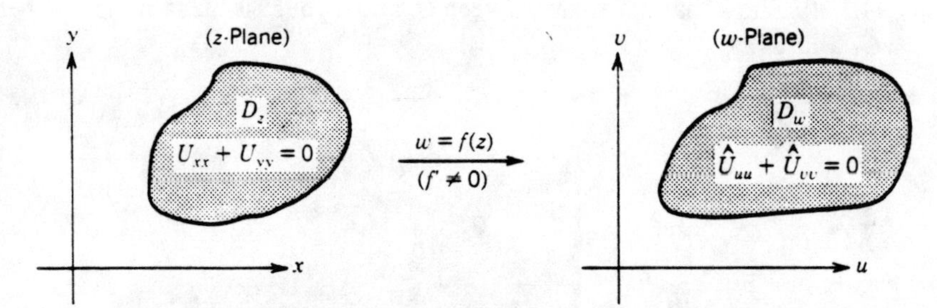

Figure 4.4.1 Invariance of Laplace's equation.

[1] These results can be found in any text on mathematical analysis or advanced calculus.

This function also possesses continuous second partial derivatives in D_w[1] and satisfies

(4.4.4) $$U(x, y) = \hat{U}(u(x, y), v(x, y))$$

Then,

(a) $U_{xx} + U_{yy} = |f'(z)|^2(\hat{U}_{uu} + \hat{U}_{vv})$

(b) Laplace's equation is invariant if $f'(z) \neq 0$ on D_z, that is,

$$\underset{(\text{in } D_z)}{U_{xx} + U_{yy} = 0} \quad \text{if and only if} \quad \underset{(\text{in } D_w)}{\hat{U}_{uu} + \hat{U}_{vv} = 0} \qquad (\text{if } f' \neq 0)$$

Proof. The chain rule applied to (4.4.4) leads to

$$U_x = \hat{U}_u u_x + \hat{U}_v v_x \qquad \text{and} \qquad U_y = \hat{U}_u u_y + \hat{U}_v v_y$$

Another application of the chain rule, which the reader is urged to carry out, yields

$$U_{xx} + U_{yy} = (u_x^2 + u_y^2)\hat{U}_{uu} + (v_x^2 + v_y^2)\hat{U}_{vv} + 2(u_x v_x + u_y v_y)\hat{U}_{uv}$$
$$+ (u_{xx} + u_{yy})\hat{U}_u + (v_{xx} + v_{yy})\hat{U}_v$$

The last two terms vanish since u and v are the real and imaginary parts of an analytic function and are thus harmonic. The third term on the right-hand side also vanishes, and u and v satisfy the Cauchy–Riemann equations. Finally, from the Cauchy–Riemann equations and the fact that $f'(z) = u_x + iv_x$, we have

$$u_x^2 + u_y^2 = v_x^2 + v_y^2 = |f'(z)|^2$$

Hence, the first two terms on the right-hand side combine to produce the result in **(a)**. The result in **(b)** follows immediately from this.

■

In Problem 1, the reader is asked to provide an alternative proof of part of these results.

(4.4.5) ***Example.*** Consider the function $U(x, y) = xy + 1$ and the transformation

$$w = u + iv = \operatorname{Log} z$$

In the half-plane $\operatorname{Im}(z) > 0$, the principal logarithm meets all of the conditions of Theorem (4.4.1), with the inverse function given by $x + iy = e^w = e^u(\cos v + i \sin v)$. Thus, $x = e^u \cos v$ and $y = e^u \sin v$, and the composite function $\hat{U}$ is

(4.4.6) $$\hat{U}(u, v) = e^{2u} \sin v \cos v + 1$$

Now the image under $w = \operatorname{Log} z$ of $\operatorname{Im}(z) > 0$ is the strip $-\infty < u < \infty$, $0 < v < \pi$. The reader is asked to verify that $U(x, y)$ is harmonic in the upper half-plane while, $\hat{U}(u, v)$ is harmonic in the strip.

■

Boundary Value Problems. The previous theorem tells us how Laplace's equation transforms under certain mappings. The problems (in the next section, for example) which we really wish to solve are **boundary value problems.** For these, we seek a real function which is harmonic in a domain and satisfies prescribed conditions on the boundary of that domain. The type of boundary conditions which are commonly of interest are

(4.4.7)

(a) **Dirichlet conditions**, where the *function* is prescribed (such as prescribing the temperature on a boundary in heat conduction, or the potential on a conductor in electrostatics).

(b) **Neumann conditions**, where the normal derivative[2] of the function is prescribed (such as is satisfied by the temperature on an insulated boundary or by the fluid potential on the boundary of a rigid body).

The next result tells us how these types of conditions transform under appropriate mappings.

(4.4.8) ***Theorem.*** Let $f(z) = u(x, y) + iv(x, y)$ be defined and invertible on a simple curve C_z in the z-plane, with the real transformation $u = u(x, y)$, $v = v(x, y)$, and its inverse represented as in (4.4.2). Let C_w be the image of C_z under the mapping $w = f(z)$.

(a) If a real-valued function $U(x, y)$ satisfies the **Dirichlet condition**

$$U(x, y) = h(x, y) \qquad \text{for } (x, y) \text{ on } C_z$$

then under the transformation, the composite function $\hat{U}(u, v)$ in (4.4.3) also satisfies a Dirichlet condition, namely,

$$U(u, v) = h(x(u, v), y(u, v)) \qquad \text{for } (u, v) \text{ on } C_w$$

(b) If $U(x, y)$ is constant on C_z, then $\hat{U}(u, v)$ is the *same constant* on C_w.

(c) Assume C_z is smooth with unit normal field $\mathbf{n}_z$ and $f(z)$ is analytic on C_z with $f' \neq 0$. Hence, C_w is also smooth and has a unit normal field $\mathbf{n}_w$ (see Figure 4.4.2). Then, **a homogeneous Neumann condition is invariant under the mapping**, that is,

$$\frac{\partial U}{\partial n_z} = 0 \text{ (on } C_z) \quad \text{if and only if} \quad \frac{\partial \hat{U}}{\partial n_w} = 0 \text{ (on } C_w) \tag{4.4.9}$$

[2] If $\mathbf{n} = n_1\mathbf{i} + n_2\mathbf{j}$ is the unit normal to a curve C, then the **normal derivative** of U is defined by

$$\frac{\partial U}{\partial n} = \mathbf{n} \cdot \nabla U = n_1 \frac{\partial U}{\partial x} + n_2 \frac{\partial U}{\partial y}$$

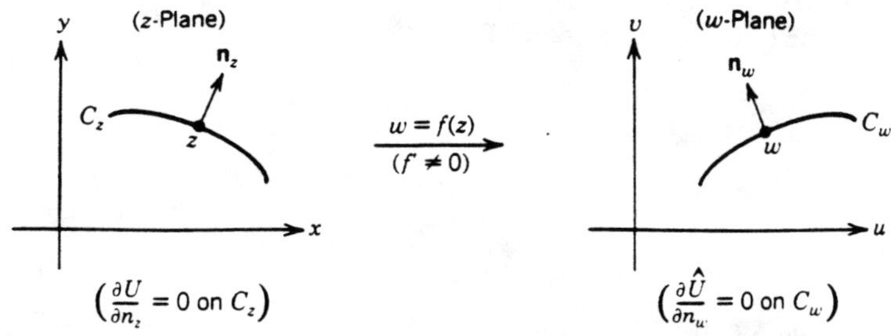

Figure 4.4.2 Invariance of homogeneous Neumann condition.

Proof. The proof of (c) is rather detailed and will be given in Problem 4. The proof of (a) is immediate; since if (x, y) is on C_z, then $(x(u, v), y(u, v))$ is on C_w, and the condition $U(x, y) = h(x, y)$ on C_z becomes, from (4.4.3),

$$\begin{aligned}\hat{U}(u, v) &= U(x(u, v), y(u, v)) \\ &= h(x(u, v), y(u, v)) \qquad (\text{on } C_w)\end{aligned}$$

Now, if $h(x, y)$ is constant, then so is $h(x(u, v), y(u, v))$, and thus (b) follows.

■

(4.4.10) ***Example.*** Consider the function

$$U(x, y) = x$$

and the mapping $w = \text{Log}\, z$. If D_z is the first quadrant of the z-plane, then its image D_w under this mapping is the strip shown in Figure 4.4.3. The circled numbers show which parts of the boundaries are mapped onto one another. Also shown are the outer normals. The mapping is invertible with a nonvanishing derivative in D_z and on its boundaries (with the exception of $z = 0$). Since $z = e^w$, the composite function is

$$\hat{U}(u, v) = e^u \cos v \tag{4.4.11}$$

Clearly, U and $\hat{U}$ are harmonic in D_z and D_w, respectively.

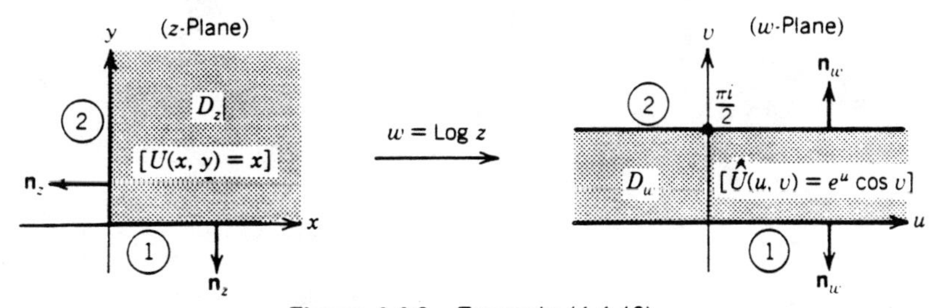

Figure 4.4.3 Example (4.4.10).

On the boundaries marked ①, we have

$$\left.\frac{\partial U}{\partial n_z}\right|_{y=0} = -\left.\frac{\partial U}{\partial y}\right|_{y=0} = 0$$

while

$$\left.\frac{\partial \hat{U}}{\partial n_w}\right|_{v=0} = -\left.\frac{\partial \hat{U}}{\partial v}\right|_{v=0} = 0 \qquad \text{from (4.4.11)}$$

Hence (as we know from the theorem), the homogeneous Neumann condition remains invariant. On the boundaries marked ②, we have

$$U|_{x=0} = 0 \qquad \text{and} \qquad \left.\frac{\partial U}{\partial n_z}\right|_{x=0} = -\left.\frac{\partial U}{\partial x}\right|_{x=0} = -1$$

while

$$\hat{U}|_{v=\pi/2} = 0 \qquad \text{and} \qquad \frac{\partial \hat{U}}{\partial n_w} = \left.\frac{\partial \hat{U}}{\partial v}\right|_{v=\pi/2} = -e^u$$

Thus (again according to the theorem), the constant Dirichlet condition $U = 0$ on the boundaries ② remains invariant. However, note that the constant Neumann condition on boundary ② is not invarient. In Problem 4d, the reader is asked to extend part (c) of the theorem to the case where the Neumann condition is not zero.

■

We are now in a position to formulate a procedure which can be used to solve the standard boundary value problems for Laplace's equation. It is based upon the previous two results which, to repeat, state that under appropriate mappings, Laplace's equation is invariant, and Dirichlet and homogeneous Neumann conditions retain their form. We outline the procedure below.

(4.4.12) ***Procedure for Dirichlet Problem***

(a) Suppose we wish to find a function $U(x, y)$ which is harmonic in a domain D_z and has prescribed (Dirichlet) boundary values $h(x, y)$ on the boundary C_z of D_z (see Figure 4.4.4).

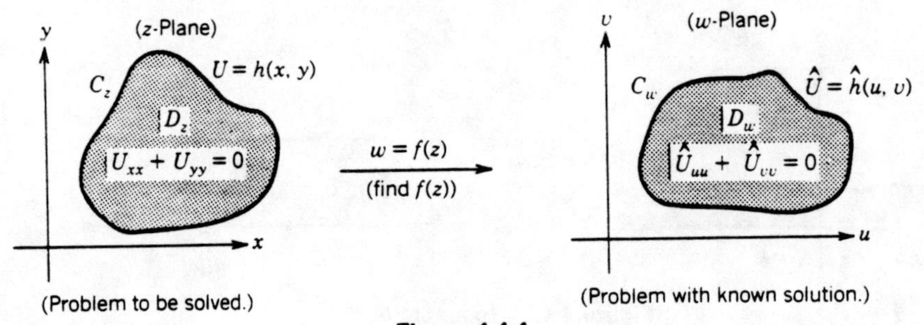

Figure 4.4.4

(b) Assume that we know how to solve a similar problem for a function $\hat{U}(u, v)$ in a domain D_w which has boundary C_w in the $w = u + iv$ plane. That is, for any prescribed boundary values on C_w, we can find a function $\hat{U}(u, v)$ which is harmonic in D_w and takes these boundary values.

(c) We now seek a function $w = f(z)$ which is analytic, conformal, and twice continuously differentiable on D_z and which maps D_z onto D_w and C_z onto C_w in an invertible manner.

(d) With the function in (c), the boundary values $h(x, y)$ on C_z transform into boundary values $\hat{h}(u, v) = h(x(u, v), y(u, v))$ on C_w.

(e) From the assumption in (b), we can find a function $\hat{U}(u, v)$ which is harmonic in D_w and takes on the boundary values $\hat{h}(u, v)$.

(f) Finally, by Theorem (4.4.1), the function

$$U(x, y) = \hat{U}(u(x, y), v(x, y))$$

must be harmonic in D_z, while by Theorem (4.4.8a) it must assume the values $h(x, y)$ on C_z.

■

Though the above procedure is stated for a Dirichlet problem, it is equally valid for a problem with a homogeneous Neumann condition or for problems where the function is prescribed on one part of the boundary, and the normal derivative vanishes on the other (such **mixed boundary value problems** have physical meaning, as we shall see in the next section). Basically, the procedure hinges upon the ability to

(4.4.13)
(a) Find a **model problem** which is solvable.
(b) Find an analytic invertible function mapping one given region onto another given region.

In Chapter 2, where we discussed the mapping properties of specific functions, the objective was to *find the image of a specified region under a given mapping*. The reader should contrast this with the much more difficult objective in (b) above. In the next section, we show how this may be accomplished in some special cases, while at the same time we solve some specific boundary value problems arising in steady-state heat conduction and electrostatics. In the meantime, let us exhibit the solutions to some very simple **model problems** which will be useful later.

(4.4.14) ***Example.*** Find a function $\hat{U}$ of u and v which is harmonic in the infinite strip $-\infty < u < \infty$, $0 < v < H$ and assumes the constant boundary-values shown in Figure 4.4.5*a*.

Solution. Note that the geometry of the strip is independent of u in the sense that every vertical cross section is the same. Also note that the boundary conditions on $v = 0$ and $v = H$ are likewise independent of u (they are constants). Hence, we are motivated to search for a function $\hat{U}$ which is also independent of u. Thus, if we assume

$$\hat{U} = \hat{U}(v)$$

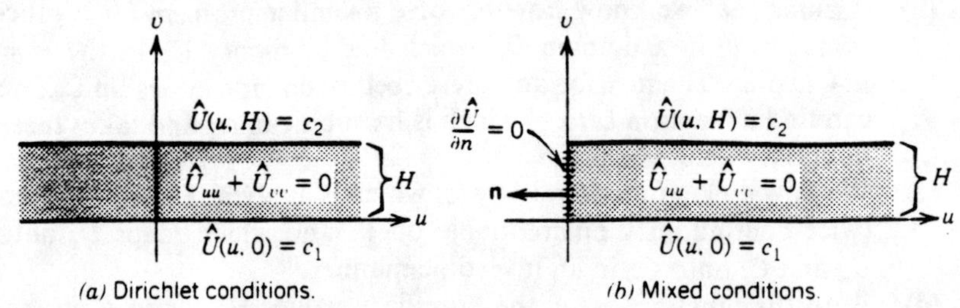

(a) Dirichlet conditions. (b) Mixed conditions.

Figure 4.4.5 Model problems for a strip.

then Laplace's equation $\hat{U}_{uu} + \hat{U}_{vv} = 0$ becomes the ordinary differential equation

(4.4.15)
$$\frac{d^2\hat{U}}{dv^2} = 0$$

for $\hat{U}(v)$, while the boundary conditions on $v = 0$ and $v = H$ become respectively

$$\hat{U}(0) = c_1 \quad \text{and} \quad \hat{U}(H) = c_2$$

If we now integrate (4.4.15) twice and impose the above boundary conditions, we find that the solution to our problem is

(4.4.16)
$$\hat{U} = c_1 + \frac{1}{H}(c_2 - c_1)v$$

Note that we can also interpret this function as the solution to the mixed boundary value problem depicted in Figure 4.4.5*b*. This is so because on the boundary $u = 0$ the unit outer normal **n** is in the direction of the negative u-axis, and hence, here $\partial\hat{U}/\partial n = -\partial\hat{U}/\partial u$. Now, since $\hat{U}$ is independent of u, $\partial\hat{U}/\partial n = 0$.

■

(4.4.17) ***Example.*** Find a function $\hat{U}$ which is harmonic in the annular region shown in Figure 4.4.6*a* and satisfying the boundary conditions given there.

Solution. Due to the shape of the annulus, we convert to polar coordinates and view $\hat{U}$ as a function of the polar coordinates ρ and σ depicted in Figure 4.4.6*b*. In these coordinates, our boundary value problem becomes

(a) $$\frac{\partial^2\hat{U}}{\partial\rho^2} + \frac{1}{\rho}\frac{\partial\hat{U}}{\partial\rho} + \frac{1}{\rho^2}\frac{\partial^2\hat{U}}{\partial\sigma^2} = 0$$ [3]

(b) $$\hat{U}|_{\rho=R_1} = c_1 \quad \text{and} \quad \hat{U}|_{\rho=R_2} = c_2$$

[3] This is Laplace's equation in polar coordinates and can be derived using the chain rule from advanced calculus.

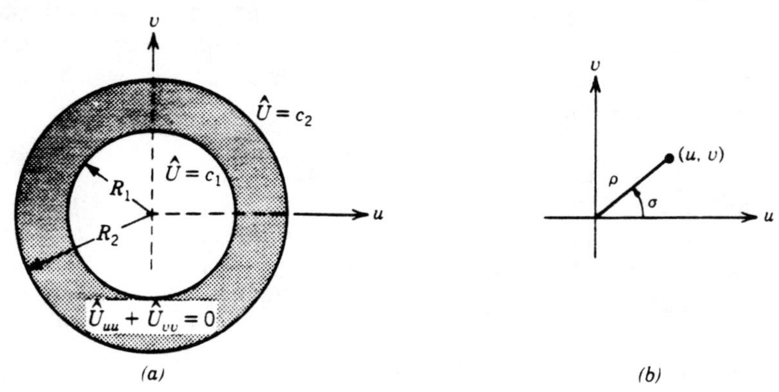

Figure 4.4.6 Model problem for an annulus.

Following the last example, we observe that **(i)** the annulus is independent of the polar angle σ in the sense that each radial line $\sigma =$ constant cuts it in the same cross section, and **(ii)** the (constant) boundary conditions are also independent of σ. Hence, we seek solutions of the above in the form $\hat{U} = \hat{U}(\rho)$, and we get

$$\frac{d^2\hat{U}}{d\rho^2} + \frac{1}{\rho}\frac{d\hat{U}}{d\rho} = 0 \Rightarrow \hat{U}(\rho) = k_1 + k_2 \ln \rho$$

From the boundary conditions at $\rho = R_1$ and $\rho = R_2$, we can evaluate the constants k_1 and k_2 and arrive at

(4.4.18)

$$\hat{U} = k_1 + k_2 \ln \rho$$

$$k_1 = \frac{c_2 \ln R_1 - c_1 \ln R_2}{\ln (R_1/R_2)}$$

$$k_2 = \frac{c_1 - c_2}{\ln (R_1/R_2)}$$

■

We have not yet mentioned anything about uniqueness of solutions to boundary value problems, a subject which is taken up in depth in partial differential equations. Indeed, unless additional conditions relating to continuity and boundedness are prescribed, there may be many solutions. Consider, for example, the function $U(x, y) = a \ln(x^2 + y^2)$. This is harmonic outside the unit disk and, *for all values of* a, vanishes on its boundary. Thus, there are many solutions to the Dirichlet problem $\hat{U}_{xx} + \hat{U}_{yy} = 0$ in $|z| < 1$, with $\hat{U} = 0$ on $|z| = 1$. However, if we demand that the solution be bounded as $x^2 + y^2 \to \infty$, we must choose $a = 0$ in $U(x, y)$. In fact, there is only one bounded solution to this problem, as the following theorem indicates.

(4.4.19) ***Theorem.*** Let D be a bounded, simply connected domain with smooth boundary C. Then there is only one function which is harmonic, continuous, and bounded inside (or outside) D and which takes on prescribed boundary data on C.

■

Riemann Mapping Theorem. As indicated in (4.4.13b), our objective is to find an invertible, analytic mapping of one given region onto another given region. However, though we might desire such a function, it may not always be possible to find one. The following profound result describes a situation which assures us of the existence of such a function. The proof of this would take us too far afield, but it can be found in more advanced complex variables texts.

(4.4.20) ***Theorem* (Riemann Mapping).** Let D_z and D_w be two *simply connected* domains, neither of which is the whole plane. Then there exists an invertible, analytic, conformal function $w = f(z)$ taking D_z onto D_w.

■

The Riemann Mapping Theorem is an existence theorem and is a far cry from providing us with the actual function accomplishing a given mapping task. The actual construction of such a function often requires some ingenuity. We also note that this is a result which holds only for simply connected domains (ones with no *holes* in it). In general, it is not true that you can find an analytic mapping of a multiply connected domain onto another multiply connected domain. For example, it would seem implausible that an analytic function would map a domain with one hole in it onto to one with three holes in it! However, we might ask whether a function can be found which maps, say, a domain with one hole in it onto a similar domain. Even here, the answer is still negative in general. However, a partial answer can be provided. But first a definition.

(4.4.21) ***Definition.*** A **doubly connected domain** is one whose boundary consists of the trace of two *distinct* simple, piecewise smooth, closed curves. (Hence, it has one hole in it.)

■

(4.4.22) ***Theorem.*** There exists an invertible, analytic, conformal function which maps a given doubly connected domain onto *some* annulus.

PROBLEMS (The answers are given on page 432.)

(Problems with an asterisk are more theoretical.)

1.* Here, we offer an alternate proof of Theorem (4.4.1). Thus, let $w = f(z) = u(x, y) + i\,v(x, y)$ be analytic in a domain D_z, with D_w denoting the image of D_z under the mapping. Assume that D_w is simply connected. Let $\hat{U}(u, v)$ be harmonic in D_w. Prove that the function $U(x, y) = \hat{U}(u(x, y), v(x, y))$ is harmonic in D_z. [*Hint*: Let $\hat{V}(u, v)$ be a harmonic conjugate to $\hat{U}(u, v)$. Then, $g(w) = \hat{U}(u, v) + i\,\hat{V}(u, v)$ is analytic in D_w. What is the real part of the composite function $g(f(z))$, and what equation does it satisfy?]

2. Verify that the following functions $U(x, y)$ are harmonic in the given region D_z; and for the given mappings $w = f(z)$, construct the composite function $\hat{U}(u, v)$ and verify that it is harmonic in the image D_w of D_z.

(a) $U = x^2 - y^2$, $\quad D_z$: $|z - 1| = 2$, $\quad f(z) = iz + 3$

(b) $U = e^{2x} \sin(2y)$, $\quad D_z$: $\mathrm{Im}(z) > 0$, $\quad f(z) = \mathrm{Log}\, z$

3. Consider the real function $U(x, y) = x^2 - y^2$ and the complex function $f(z) = (z - i)/(z + i)$.

(a) Verify that the image of the upper half-plane $\mathrm{Im}(z) > 0$ under $w = f(z)$ is the unit disk.

(b) What is the composite function $\hat{U}(u, v)$ under the transformation $w = f(z)$?

(c) What are the values $h(x, y)$ of U on the boundary of the upper half-plane?

(d) Compute the boundary values $\hat{h}(u, v)$ of the composite function $\hat{U}$ on the boundary of the disk, and thus verify part (a) of Theorem (4.4.8).

(e) If $\mathbf{n}_z$ and $\mathbf{n}_w$ denote the unit outer normals to the boundaries of the half-plane and disk, respectively, show that on the respective boundaries, $\partial U/\partial n_z = \partial \hat{U}/\partial n_w = 0$, thus verifying part (c) of Theorem (4.4.8) (use polar coordinates).

4.* Here, we will take the reader through a series of steps which lead to the relationship between normal derivatives under conformal transformations. Thus, let $f(z) = u(x, y) + i\,v(x, y)$ be defined invertible and conformal ($f' \neq 0$) on a simple smooth curve C_z: $z = z(t)$, $t_0 \leq t \leq t_1$, having image C_w: $w = w(t) = f(z(t))$, $t_0 \leq t \leq t_1$ under the mapping. Let $U(x, y)$ be continuously differentiable in a neighborhood of C_z, with the composite function $\hat{U}(u, v) = U(x(u, v), y(u, v))$ being continuously differentiable in a neighborhood of C_w.

(a) Show that $n_z(t) = -i\,z'(t)/|z'|$ and $n_w(t) = -i\,z'(t)f'(z(t))/|z'||f'|$ represent normal vectors to C_z and C_w, respectively. [*Hint:* Show that these are orthogonal to the tangents to C_z and C_w.]

(b) Define the complex quantities $\mathrm{grad}_z(U) = U_x + i\,U_y$ and $\mathrm{grad}_w(\hat{U}) = \hat{U}_u + i\,\hat{U}_v$, and then use the chain rule and the fact that $U(x, y) = \hat{U}(u,(x, y), v(x, y))$ to show that $\mathrm{grad}_z(U) = f'(z) \cdot \mathrm{grad}_w(\hat{U})$

(c) Use the results in (a) and (b) to show that $\partial U/\partial n_z = \mathrm{Re}[\bar{n}_z\, \mathrm{grad}_z(U)]$. Derive a similar expression for the normal derivative $\partial \hat{U}/\partial n_w$. [*Hint:* Use the fact that $\partial U/\partial n_z = \mathbf{n}_z \cdot \nabla U$, where $\mathbf{n}_z$ is the vector representation of the normal and ∇ is the vector gradient.]

(d) Use (a), (b), and (c) to show that $\partial U/\partial n_z = |f'|(\partial U/\partial n_w)$ and thus verify the result of Theorem (4.4.8c).

(e) Verify the result in (d) for Example (4.4.10).

5.* Here, we prove Theorem (4.4.8c) differently than in the previous problem. Thus, let $U(x, y)$ and $V(x, y)$ be harmonic conjugates in a simply connected domain D_z.

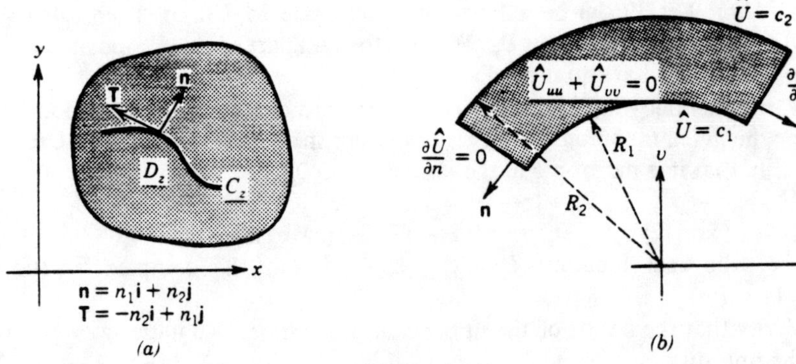

Figure 4.4.7

Let C_z be a smooth curve in D_z with unit normal and tangent fields **n** and **T**, respectively (see Figure 4.4.7a).

(a) Show that on C_z, $\partial U/\partial n_z = \partial V/\partial s$, where $\partial V/\partial s = \mathbf{T} \cdot \nabla V$ is the **tangential derivative** of V.

(b) From (a), show that V is constant on C_z if and only if $\partial U/\partial n_z = 0$ on C_z.

(c) Use (b) to prove Theorem (4.4.8c) for the case when $U(x, y)$ in the theorem is harmonic in a neighborhood of C_z. [*Hint:* Let $V(x, y)$ be a harmonic conjugate of $U(x, y)$. If V = constant on C_z under the mapping $w = f(z)$, what is the value of the composite function $\hat{V}(u, v)$ on C_w?]

6. Show that the solution of Example (4.4.17) also solves the boundary value problem depicted in Figure 4.4.7*b*.

4.5 SOME BOUNDARY VALUE PROBLEMS IN HEAT CONDUCTION AND ELECTROSTATICS

At this point, the reader might wish to read the material in Appendices B and D where the basic equations and subsequent boundary value problems for steady-state temperature distributions and electrostatics are discussed. The special problems to which we will address ourselves here are described below and depicted in Figure 4.5.1. Note how similar the two problems are, even though they describe entirely different phenomena.

	Heat Conduction		Electrostatics	
	(a) $T_{xx} + T_{yy} = 0$	in D	**(a)** $\phi_{xx} + \phi_{yy} = 0$	in D
	(b) $T = T_1$	on C_1	**(b)** $\phi = V_1$	on C_1
(4.5.1)	**(c)** $T = T_2$	on C_2	**(c)** $\phi = V_2$	on C_2
	(d) $\dfrac{\partial T}{\partial n} = 0$	on C_3		

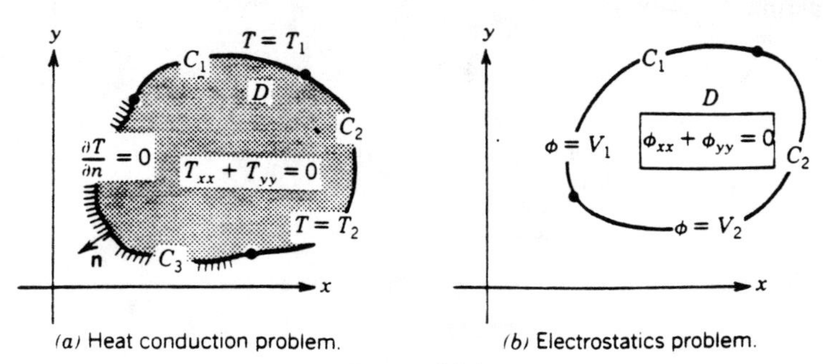

(a) Heat conduction problem.

(b) Electrostatics problem.

Figure 4.5.1

Thus, for the heat conduction problem we seek a harmonic temperature distribution function $T(x, y)$ in a domain D whose boundary curve is partly insulated (the piece C_3). On this insulated piece, the normal derivative $\partial T/\partial n$ of T vanishes (a **Neumann condition**). In addition, on each of the two remaining pieces C_1 and C_2 of the boundary of D, we prescribe constant temperatures T_1 and T_2, respectively (**Dirichlet conditions**). For the electrostatic problem, we seek a harmonic potential function $\phi(x, y)$ in a domain D bounded by a curve which consits of two conductors C_1 and C_2, insulated from each other. On each of these, the potential is a prescribed constant V_1 or V_2, respectively.

These rather special problems will serve to illustrate how conformal mapping techniques can be used to solve boundary value problems for Laplace's equation. **The reader would do well to review the mapping properties of the functions discussed in Chapter 2, in particular the bilinear functions of Section 2.5.**

As noted in the previous section, our objective is to find an appropriate conformal, analytic mapping of the given region in the $z = x + iy$ plane onto a region in the $w = u + iv$ plane which corresponds to a model problem which can be solved explicitly. Two such problems involving strips in the uv-plane have already been solved in Example (4.4.14). For completeness, the problems and their solutions are given below and are depicted in Figure 4.5.2.

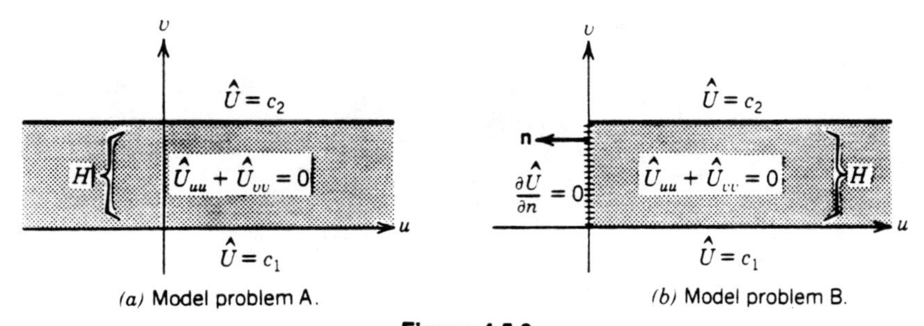

(a) Model problem A.

(b) Model problem B.

Figure 4.5.2

Model Problem A

$$\hat{U}_{uu} + \hat{U}_{vv} = 0 \qquad -\infty < u < \infty, \quad 0 < v < H$$

$$\hat{U}|_{v=0} = c_1 \qquad \text{and} \qquad \hat{U}|_{v=H} = c_2$$

(4.5.2)

$$\boxed{\hat{U} = c_1 + \frac{1}{H}(c_2 - c_1)v}$$

Model Problem B

$$\hat{U}_{uu} + \hat{U}_{vv} = 0 \qquad 0 < u < \infty, \quad 0 < v < H$$

$$\hat{U}|_{v=0} = c_1, \qquad \hat{U}|_{v=H} = c_2 \qquad \text{and} \qquad \left.\frac{\partial \hat{U}}{\partial n}\right|_{u=0} = 0$$

(4.5.3)

$$\boxed{\hat{U} = c_1 + \frac{1}{H}(c_2 - c_1)v}$$

■

(4.5.4) ***Example.*** Solve for the temperature distribution in the upper half-plane subject to the boundary conditions shown in the z-plane of Figure 4.5.3a.

Solution. The objective is to map the upper half-plane onto an infinite strip of the type shown in Model Problem A of Figure 4.5.2. The principal logarithm $w = \operatorname{Log} z$ accomplishes this, and the situation is depicted in

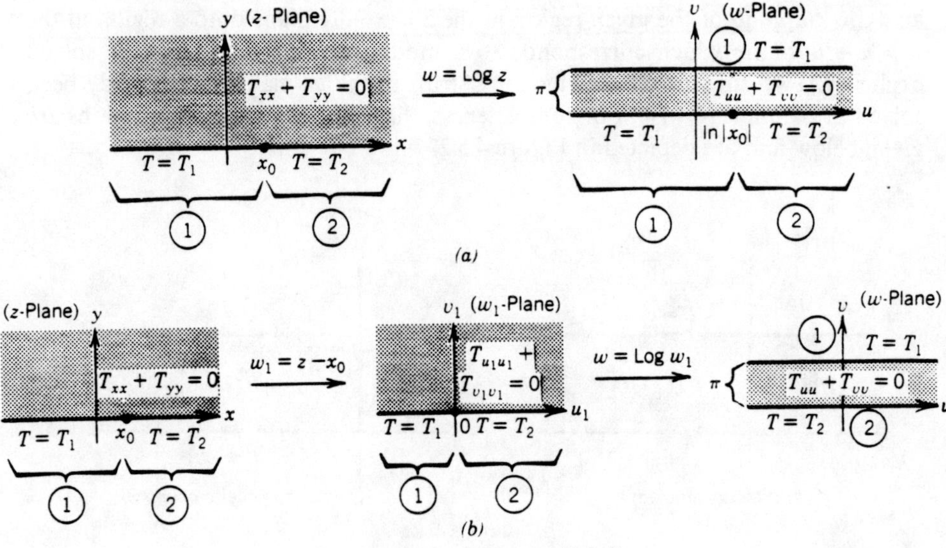

Figure 4.5.3 Example (4.5.4).

Figure 4.5.3*a*. By the results of the previous section, the temperature function T, when viewed as a function of u and v, is harmonic in the strip in the w-plane, and the constant boundary conditions T_1 and T_2 on the boundaries ① and ②, respectively, in the z-plane are transformed into the same constants on the images of these boundaries (also designated by ① and ② in the w-plane). Note that the resulting boundary value problem in the w-plane is *not that of Model Problem A* in (4.5.2), because there we require the same constant on each piece of the boundary of the strip in the w-plane. Thus, we see that, **in general, we not only want to map the domain *D* onto the model domain, but we must also ensure that the pieces of the boundary match up properly.**

To remedy the situation, we first shift the upper half of the z-plane with the mapping

(4.5.5)
$$w_1 = z - x_0$$

Then, the principal logarithm $w = \operatorname{Log} w_1$ will properly match up the boundaries. Note that in each of the w_1- and w-planes, the temperature function is harmonic and satisfies appropriate constant boundary conditions on the images of the boundaries (see Figure 4.5.3*b*)). The composite map $w = \operatorname{Log}(z - x_0)$, that is,

(4.5.6)
$$u + iv = \ln|z - x_0| + i\operatorname{Arg}(z - x_0)$$

produces in the w-plane a model problem of the form (4.5.2) with $H = \pi$, $c_1 = T_2$, and $c_2 = T_1$. Hence, the solution, as a function of u and v, is by (4.5.2)

$$T = T_2 + \frac{1}{\pi}(T_1 - T_2)v$$

From (4.5.6), we can replace v in terms of x and y and find

(4.5.7)
$$T(x, y) = T_2 + \frac{1}{\pi}(T_1 - T_2)\operatorname{Arg}(x - x_0 + iy) \qquad y > 0$$

In Problem 1, the reader is asked to verify that this can be written in terms of the real principal inverse tangent as

$$T(x, y) = T_2 + \frac{1}{\pi}(T_1 - T_2)\cdot\begin{cases} \operatorname{Tan}^{-1}\left(\dfrac{y}{x - x_0}\right) & x > x_0 \\ \dfrac{\pi}{2} & x = x_0 \\ \pi + \operatorname{Tan}^{-1}\left(\dfrac{y}{x - x_0}\right) & x < x_0 \end{cases}$$

■

(4.5.8) ***Example.*** Solve the boundary value problem depicted in Figure 4.5.4 for the temperature distribution in the upper half-plane.

Solution. As in the previous example, $w = \text{Log}\, z$ will map the upper half-plane onto the infinite strip. Also, as before, the final problem in the w-plane will not conform to the Model Problem (4.5.2). What is worse is that, *unlike the previous example*, no simple shift prior to the principal logarithm map will eliminate this difficulty.

The way around this dilemma is to first map the upper half-plane onto another upper half-plane in such a manner that the boundary value problem in the new half-plane has the *same constant temperature on each part of the real axis. Then*, a principal logarithm will do the final job. We know from Section 2.5 that bilinear maps take half-planes onto half-planes. We need only pick three points on the boundaries of each half-plane and find the bilinear function taking the first three points into the second three. [See Example (2.5.30) for a similar problem.] However, we must be careful for two reasons:

(i) We want the image to be the *upper* half-plane (not the lower).
(ii) We want the real interval $0 < x < 1$, on which $T = T_2$ (see Figure 4.5.4) to have as its image one of the two halves of the real axis (why?).

To accomplish (i), we note that the general bilinear function

$$w_1 = \frac{az + b}{cz + d} \tag{4.5.9}$$

is conformal and analytic except at $z = -d/c$, and thus preserves relative directions. Hence, let us start with three points z_1, z_2, and z_3 on the real z-axis having the property that $z_1 < z_2 < z_3$. We now wish to choose three points $\hat{w}_1$, $\hat{w}_2$, and $\hat{w}_3$ on the real w_1-axis, which will be the images of z_1, z_2, and z_3, respectively, and from which we will determine a, b, c, and d in (4.5.9). Now, as we traverse the real z-axis from z_1 to z_2 to z_3, the region to be mapped (the upper half-plane) lies to our left. Since we wish the image in the w_1-plane also to be the upper half-plane, by conformality we must pick $\hat{w}_1$, $\hat{w}_2$, and $\hat{w}_3$ so that $\hat{w}_1 < \hat{w}_2 < \hat{w}_3$ (regions lying to our left will also lie to our left in the image plane).

Now we are left to accomplish (ii) above. To do this, we must choose the images of $z = 0$ and $z = 1$ (in between which $T = T_2$) so that the real interval (0, 1) in the z-plane is "stretched" out onto the positive real

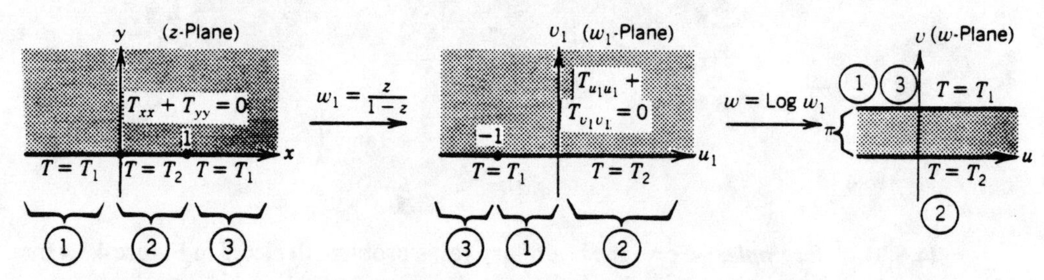

Figure 4.5.4 Example (4.5.8).

axis in the w_1-plane. Hence, we demand that under (4.5.9), $z = 0 \to w_1 = 0$ and $z = 1 \to w_1 = \infty$. For the third pair of points, we can choose *any real* value of z greater than 1 and *any negative* value of w_1. In particular, let us choose $z_1 = \infty \to w_1 = -1$. (Note that since this pair is not unique, *many mapping functions exist which do the job*. However, *there will be only one temperature function* T; see Problem 2.) Now, substituting the pairs $z = 0$ and $w_1 = 0$, $z = 1$ and $w_1 = \infty$, and $z = \infty$ and $w_1 = -1$ into (4.5.9), we find

(4.5.10)
$$w_1 = \frac{z}{1 - z}$$

The reader is urged to verify that this bilinear function accomplishes the mapping shown in the first part of Figure 4.5.4 (the images of the various pieces of the boundaries are shown circled). Finally, the mapping

(4.5.11)
$$w = \operatorname{Log} w_1$$

produces a correct model problem in the w-plane, as shown in the figure, and the solution by (4.5.2) is, with $H = \pi$, $c_1 = T_2$, and $c_2 = T_1$,

$$T = T_2 + \frac{1}{\pi}(T_1 - T_2)v$$

Combining (4.5.10) and (4.5.11), we get the composite map

$$w = \operatorname{Log}\left(\frac{z}{1 - z}\right),$$

from which we deduce that

$$T(x, y) = T_2 + \frac{1}{\pi}(T_1 - T_2)\operatorname{Arg}\left(\frac{z}{1 - z}\right)$$

■

(4.5.12) ***Example.*** Find the electrostatic potential $\phi(x, y)$ in the semiinfinite strip of Figure 4.5.5. The conductors bounding the strip are raised to the potentials shown.

Solution. We know that the sine function maps a strip of width $\pi/2$ onto the first quadrant. A (square) power map will then double this into a half-plane, and a principal logarithm will do the final job. (The appropriate series of mappings is given by (see the figure)

(4.5.13)
$$w_1 = \frac{\pi z}{2}, \qquad w_2 = \sin w_1,$$
$$w_3 = w_2^2, \qquad w = \operatorname{Log} w_3$$

and the composite mapping is

$$w = \operatorname{Log}\left[\sin^2\left(\frac{\pi z}{2}\right)\right]$$

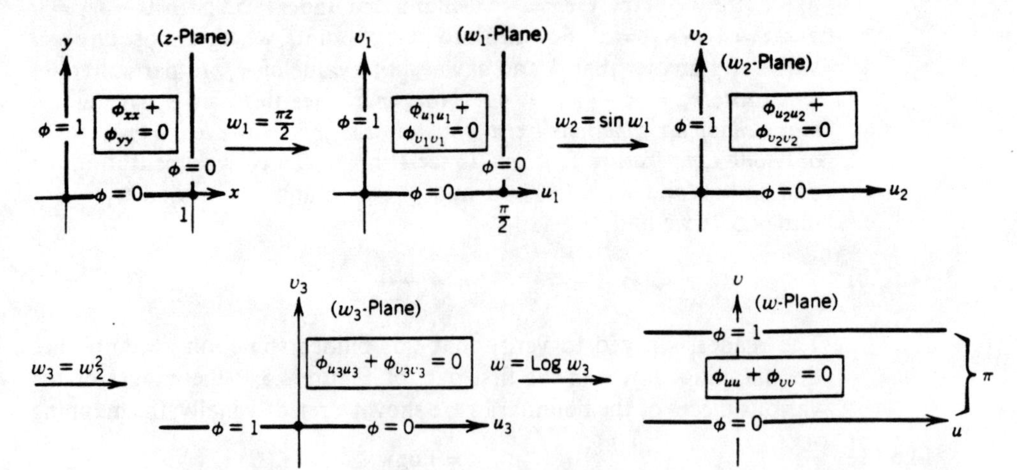

Figure 4.5.5 Example (4.5.12).

Finally, the boundary conditions in the w-plane of the figure, and (4.5.2), yield

$$\phi(x, y) = \frac{1}{\pi} \operatorname{Arg}\left[\sin^2\left(\frac{\pi z}{2}\right)\right]$$

■

(4.5.14) *Example.* Find the electrostatic potential ϕ inside the unit disk if the boundary of the disk consists of two conductors, insulated from each other at the points $z = \pm 1$, and raised to potentials $\phi = 0$ and $\phi = 1$, as shown in the z-plane of Figure 4.5.6a.

Solution. We first look for a bilinear map

$$w_1 = \frac{az + b}{cz + d}$$

taking the unit disk onto the upper half-plane in such a way that the upper half of the unit circle (on which $\phi = 0$) is mapped onto the positive real axis. We determine a, b, c, and d by choosing three points on the unit circle to map into three points on the real axis. By conformality, if we choose the three z values so that we proceed from one to the other, say, counterclockwise, then, to get the upper half of the w-plane, we must choose the three w-values so that we proceed from left to right [see also Example (4.5.8)]. One such choice of pairs of points is $z = 1 \to w = 0$, $z = -1 \to w = \infty$, and $z = -i \to w = -1$. This leads to the bilinear mapping

$$w_1 = i\left(\frac{1 - z}{1 + z}\right)$$

and a final principal logarithm $w = \operatorname{Log} w_1$ yields the model problem

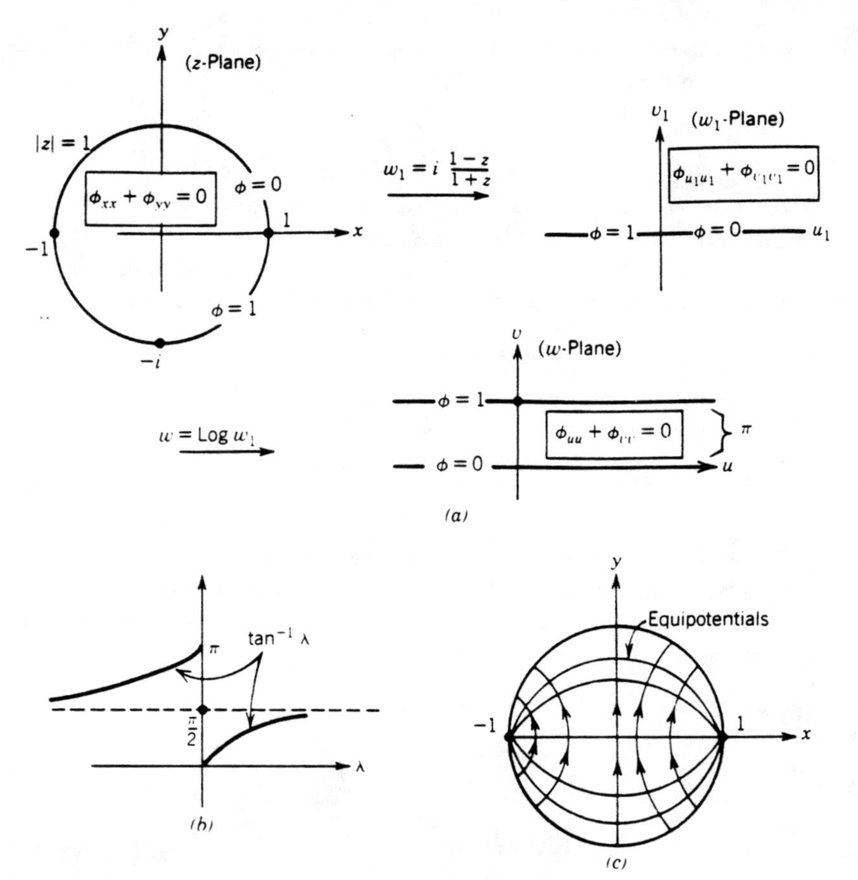

Figure 4.5.6 Example (4.5.14).

shown in the w-plane of Figure 4.5.6a. Hence, with $c_1 = 0$, $c_2 = 1$, and $H = \pi$ in (4.5.2), we get

$$\begin{aligned}\phi(x, y) &= \frac{1}{\pi}\,\text{Arg}\left[i\left(\frac{1-z}{1+z}\right)\right] \\ &= \frac{1}{\pi}\,\text{Arg}\left[\frac{2y + i(1 - x^2 - y^2)}{(x+1)^2 + y^2}\right]\end{aligned} \tag{4.5.15}$$

Note that $x^2 + y^2 < 1$ in the unit disk, and thus

$$\text{Im}\left[\frac{2y + i(1 - x^2 - y^2)}{(x+1)^2 + y^2}\right] > 0$$

Hence, the principal argument above takes on values only in the range $0 \leq \text{Arg} \leq \pi$, and we can write (4.5.15) as

$$\phi(x, y) = \frac{1}{\pi}\tan^{-1}\left(\frac{1 - x^2 - y^2}{2y}\right), \qquad x^2 + y^2 < 1 \tag{4.5.16}$$

where the branch of the inverse tangent is as sketched in Figure 4.5.6*b*. Note from the above, and the figure, that

$$\lim_{x^2+y^2 \uparrow 1} \phi(x, y) = \begin{cases} 0 & y > 0 \\ 1 & y < 0 \end{cases}$$

and thus $\phi(x, y)$ assumes the given boundary values in the limit. Also note from (4.5.15) that $\phi(x, y) = \frac{1}{2}$ for $y = 0$, $-1 < x < 1$. Hence, that part of the real axis lying inside the unit disk is an equipotential. The other equipotentials $\phi(x, y) = k$ are, from (4.5.16), the circles

$$x^2 + [y + \tan(\pi k)]^2 = \sec^2(\pi k) \qquad 0 \le k \le 1, k \ne \tfrac{1}{2}$$

each of which pass through the points $z = \pm 1$. (The case $k = \frac{1}{2}$ yields the equipotential $y = 0$ discussed above.) Some of these equipotentials are plotted in Figure 4.5.6*c*, together with some field lines $\mathbf{E} = -\nabla\phi$.

■

(4.5.17) ***Example.*** Solve the **mixed boundary value problem** for the temperature distribution in the quarter-disk shown in the z-plane of Figure 4.5.7. Here, the temperature is prescribed on two portions of the boundary, while the third piece is insulated.

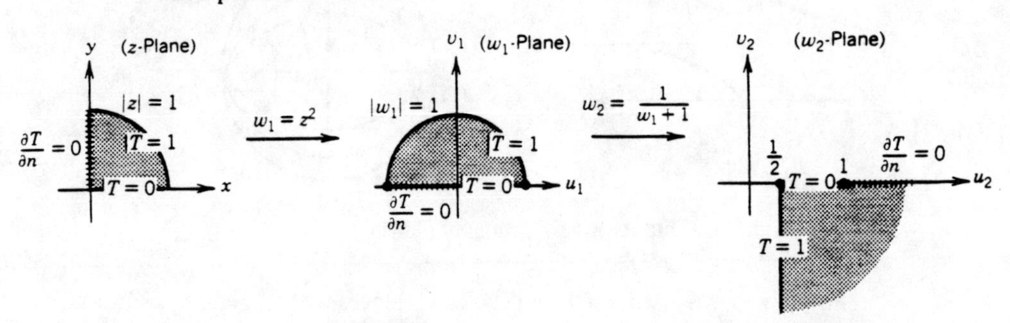

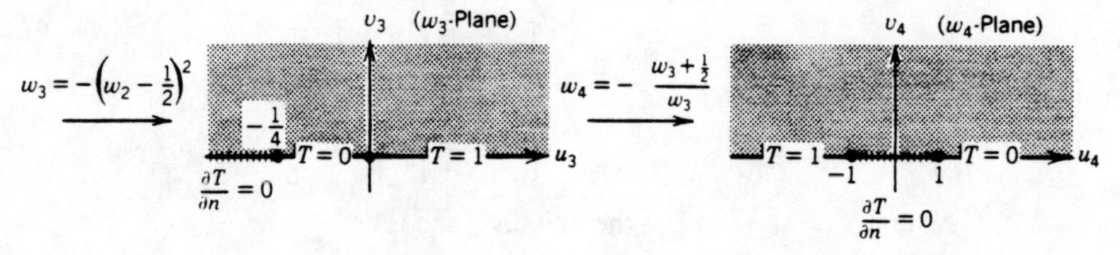

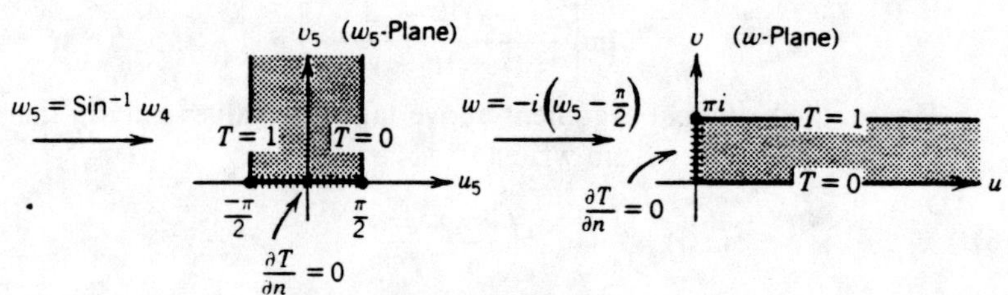

Figure 4.5.7 Example (4.5.17).

Solution. Our objective here will be to map the quarter-disk onto the horizontal, semiinfinite strip shown in Model Problem B of Figure 4.5.2. We must however be sure that in so doing, the insulated piece of the boundary, where $\partial T/\partial n = 0$, that is, the piece of the imaginary axis $x = 0$, $0 \leq y \leq 1$, is mapped onto some interval $u = 0$, $0 \leq v \leq H$ (see Model Problem B). Our game plan will be to

(4.5.18)

(a) Map the quarter-disk onto the upper half-plane so that
  (i) The insulated piece of the boundary maps onto the real interval $(-1, 1)$.
  (ii) The pieces of the boundary on which T is constant map respectively onto the real intervals $(-\infty, -1)$ and $(1, \infty)$.

(b) Then use an inverse sine map to get a semiinfinite strip.

(c) Then finally translate and rotate to produce the configuration of Figure 4.5.2*b*.

This is how it works. The series of mappings to be described can be followed in Figure 4.5.7. First, the power map

$$w_1 = z^2 \tag{4.5.19}$$

produces a half-disk in the w_1-plane. Now, since the boundary of this half-disk consists of pieces of lines and circles, we might try to find a bilinear map taking it onto the upper half-plane. However, since the pieces of the boundary of the half-disk are perpendicular at *two* points ($w_1 = \pm 1$) and bilinear maps are conformal everywhere except at the *one value* where the denominator vanishes, the best we can hope for is a bilinear map taking the half-disk onto a quarter-plane (then another quadratic power map will produce a half-plane). To accomplish this, both pieces of the boundary of the half-disk in the w_1-plane must map onto lines. From the properties of bilinear maps, we can accomplish this using *any* bilinear map which *takes a point in common to the two pieces of the boundary into infinity* (see Example (2.5.26) of Section 2.5). With $w_1 = -1$ as the common point, one such bilinear map is (there are many others)

$$w_2 = \frac{1}{w_1 + 1} \tag{4.5.20}$$

The effect of this is shown in the w_2-plane of Figure 4.5.7. Next the translation, rotation and power maps

$$w_3 = \left[i\left(w_2 - \frac{1}{2}\right)\right]^2 = -\left(w_2 - \frac{1}{2}\right)^2 \tag{4.5.21}$$

produce the upper half of the w_3-plane with the boundary conditions as shown. We must now map this half-plane onto the upper half of the w_4-plane, as was done in Example (4.5.8), in order to properly line up the pieces of the boundary. [Remember, by (4.5.18ai) we want the real axis to eventually be insulated along the interval $(-1, 1)$]. From the figure, the real interval $(-\infty, -\frac{1}{4})$ is insulated in the w_3-plane. Hence, we will seek a bilinear map

$$w_4 = \frac{aw_3 + b}{cw_3 + d}$$

taking $w_3 = \infty \to w_4 = -1$ and $w_3 = -\frac{1}{4} \to w_4 = 1$. Also, note from the w_3-plane in the figure that the temperature is the constant $T = 0$ on the real interval $(-\frac{1}{4}, 0)$. By (4.5.18aii), we should map this interval onto the real interval $(1, \infty)$ in the w_4-plane. Hence, our third pair of points, in order to determine a, b, c, and d above, will be $w_3 = 0 \to w_4 = \infty$. With these, the bilinear map we are looking for is

$$w_4 = \frac{-(w_3 + \frac{1}{2})}{w_3} \tag{4.5.22}$$

Now, the principal inverse sine map (see Section 2.8)

$$w_5 = \mathrm{Sin}^{-1}\, w_4 \tag{4.5.23}$$

produces the vertical strip shown in the w_5-plane of Figure 4.5.7; and finally, the translation and rotation

$$w = -i\left(w_5 - \frac{\pi}{2}\right) \tag{4.5.24}$$

yields the boundary value problem in the w-plane of the figure. Hence, by Model Problem B of (4.5.3), the solution to our problem is (with $c_1 = 0$, $c_2 = 1$, $H = \pi$)

$$T = \frac{1}{\pi}\,\mathrm{Im}(w) \tag{4.5.25}$$

Now combining (4.5.19) through (4.5.24), we get the composite map

$$w = -i\left\{\mathrm{Sin}^{-1}\left[\frac{1 + 6z^2 + z^4}{(1 - z^2)^2}\right] - \frac{\pi}{2}\right\}$$

Hence, from (4.5.25) and Definition (2.8.5) of the principal inverse sine, we have

$$T(x, y) = \frac{1}{2} - \frac{1}{\pi}\,\mathrm{Arg}(i\lambda + \sqrt{1 - \lambda^2}) \tag{4.5.26}$$

$$\lambda = \frac{1 + 6z^2 + z^4}{(1 - z^2)^2}$$

where the principal branch of the square root is to be used.

■

As the reader can now see, sometimes a simple-looking problem can have a rather complicated solution. However, the "magic" of conformal mappings is that each step in our procedure is elementary. All that is required is a thorough knowledge of the mapping properties of the elementary functions.

PROBLEMS (The answers are given on page 432.)

1. Verify the formula after (4.5.7). [*Hint:* for $y > 0$, we have

$$0 < \text{Arg}(x - x_0 + iy) < \pi$$

Now, use the fact that the principal argument can be written as some branch of the real inverse tangent function.]

2. In Example (4.5.8), we sought a bilinear mapping $w_1 = (az + b)/(cz + d)$ taking $z = 0$ into $w_1 = 0$, $z = 1$ into $w_1 = \infty$, and $z = \infty$ into $w_1 = -1$. We mentioned there that this last condition was arbitrary.
   (a) In place of the last condition, let $z = -1$ map into $w_1 = -1$ and find the bilinear map.
   (b) Verify that the model problem in the w_1-plane is the same as that shown in Figure 4.5.4.
   (c) Finish the problem and show that the resulting temperature function $T(x, y)$, though it may look different, actually is the same as the solution given in Example (4.5.8).

3. Solve for the temperature distribution $T(x, y)$ in the domains D shown in each of the parts of Figure 4.5.8 and subject to the boundary conditions given there. [*Hint for part (b):* First, use a bilinear map to get a quarter-plane. Then, a possible shift, square, and logarithm will finish the job.]

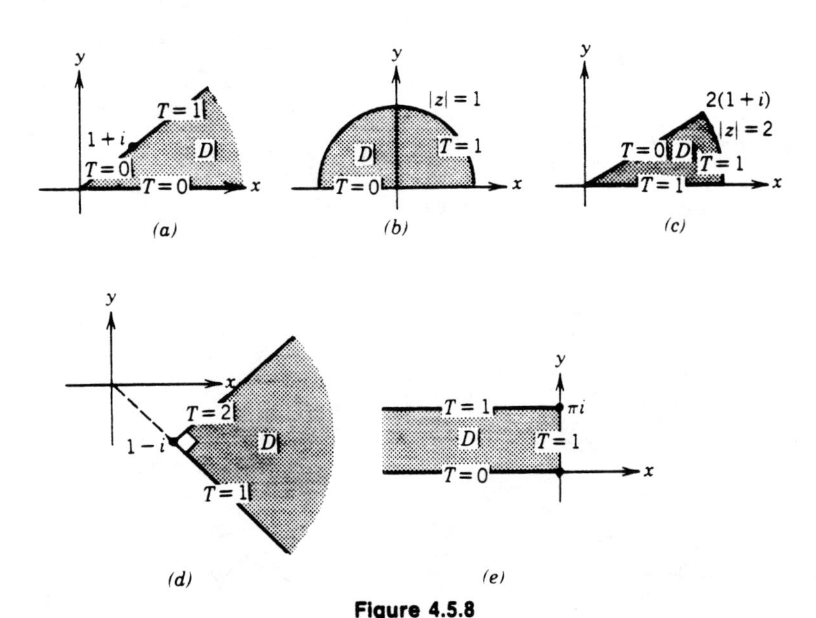

Figure 4.5.8

4. Solve the mixed boundary value problems for the temperature distribution $T(x, y)$ in each of the domains shown in Figure 4.5.9. In each case, the crosshatched piece of the boundary is insulated. You might wish to refer to

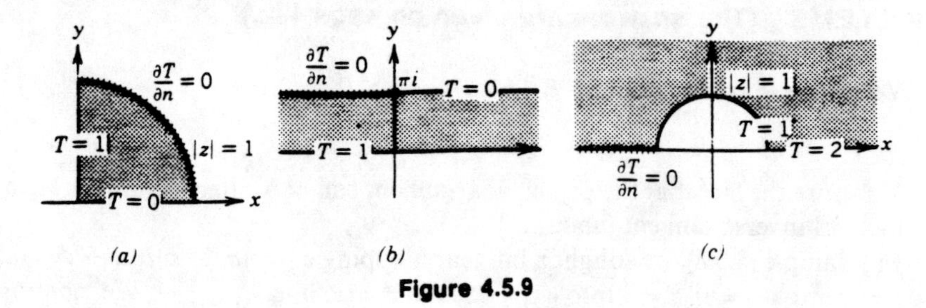

Figure 4.5.9

Example (4.5.17). Remember, the objective is to find an analytic, conformal function mapping the given domain onto that of model problem B in Figure 4.5.2*b*.

5. Find the electrostatic potential $\phi(x, y)$ in each of the domains shown in Figure 4.5.10. The boundary of each domain consists of conductors, insulated at their points of contact and raised to the potentials shown in the Figure. [*Hint*

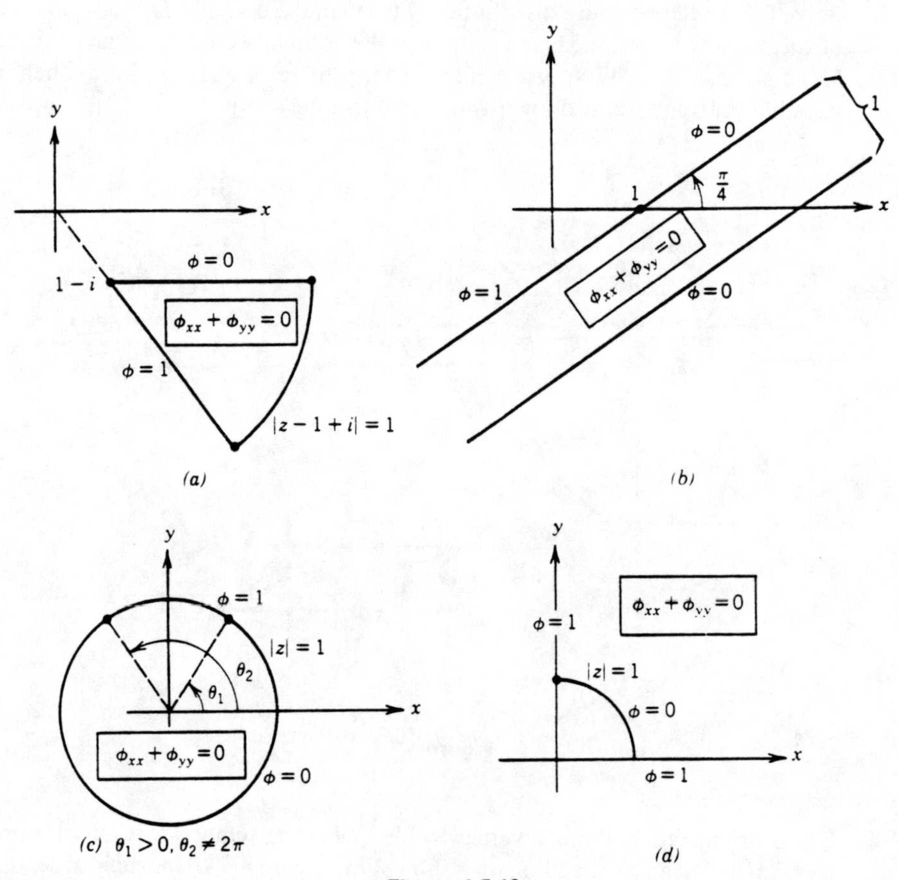

Figure 4.5.10

for part (d): You might try a logarithm, followed by a rotation. Now you are on your own!]

6. Two conducting cylinders whose cross sections are the disks $|z - i| \leq 1$ and $|z + i| \leq 1$ are insulated from each other at their line of contact and are raised to potentials 0 and 1 (see Figure 4.5.11). Find the electrostatic potential outside the cylinders.

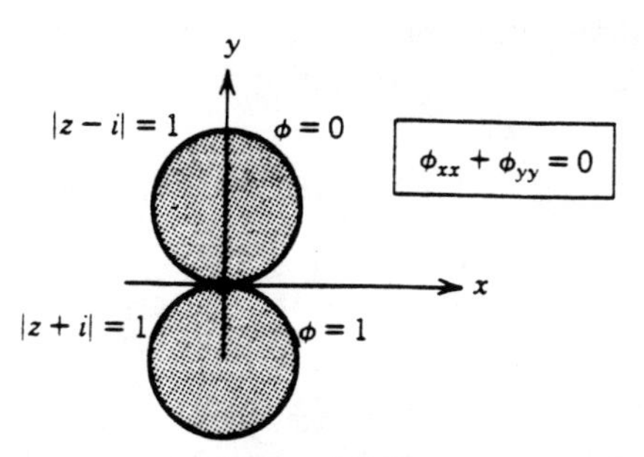

Figure 4.5.11

7. Here, we show how we can treat a half-plane problem where the boundary conditions involve more than two constant values. Thus, consider the temperature distribution $T(x, y)$ in the upper half-plane subject to the boundary conditions shown in Figure 4.5.12*a*.

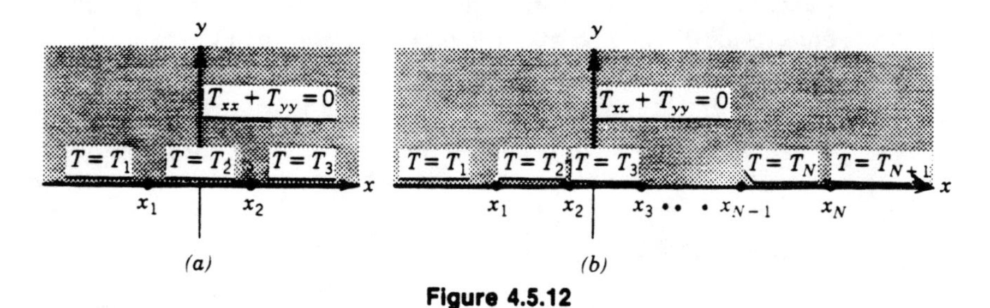

Figure 4.5.12

(a) Let $T^{(1)}(x, y)$, $T^{(2)}(x, y)$, and $T^{(3)}(x, y)$ be temperature distributions in the upper half-plane satisfying the boundary conditions

$$T^{(1)}(x, 0) = \begin{cases} T_1 & x < x_1 \\ 0 & x > x_1 \end{cases}$$

$$T^{(2)}(x, 0) = \begin{cases} 0 & x < x_1 \\ T_2 & x > x_1 \end{cases}$$

$$T^{(3)}(x, 0) = \begin{cases} 0 & x < x_2 \\ T_3 - T_2 & x > x_2 \end{cases}$$

Show that $T(x, y) = T^{(1)} + T^{(2)} + T^{(3)}$.

(b) Note that each of the functions $T^{(1)}$, $T^{(2)}$, and $T^{(3)}$ satisfies boundary conditions with only two different constant boundary temperatures. Compute each of these temperature distributions and show that

$$T(x, y) = T_3 + \frac{1}{\pi}(T_1 - T_2)\operatorname{Arg}(x - x_1 + iy) + (T_2 - T_3)\operatorname{Arg}(x - x_2 + iy)$$

(c) Extend the result of part (b) to the boundary value problem in Figure 4.5.12*b* and show that

$$T(x, y) = \frac{T_1}{\pi}\operatorname{Arg}(z - x_1) + \frac{1}{\pi}\sum_{j=2}^{N} T_j[\operatorname{Arg}(z - x_j) - \operatorname{Arg}(z - x_{j-1})]$$

$$+ \frac{T_{N+1}}{\pi}[\pi - \operatorname{Arg}(z - z_N)]$$

8. All of the problems which we have solved have involved simply connected domains and used model problems for strips. There is another type of model problem, involving an annulus, which is useful for some doubly connected domains. This is shown in Figure 4.4.6, and the solution, in polar coordinates is given in Equation (4.4.18) of Section 4.4. We will use this to find the temperature distribution $T(x, y)$ in the region between the two nonconcentric cylinders shown in Figure 4.5.13 and subject to the boundary conditions there.

(a) From Problem 14*c* of Section 2.5, the function $w = e^{i\psi_0}(z - \alpha)/(\bar{\alpha}z - 1)$, where $|\alpha| < 1$, is the most general bilinear mapping taking the unit disk onto itself. It will also map the inner circle in the figure onto some other circle. Find α so that this other circle has its center at the origin. Thus the "nonconcentric annular region" will be mapped onto a concentric annulus. What is the inner radius of this annulus? [*Hint:* On the inner circle, $z = \frac{1}{4}(2 + e^{i\theta})$. We thus require that $|(z - \alpha)/(\bar{\alpha}z - 1)| = k$, where $k < 1$.]

(b) Use the model problem of Example (4.4.17) to solve the heat conduction problem of Function 4.5.13.

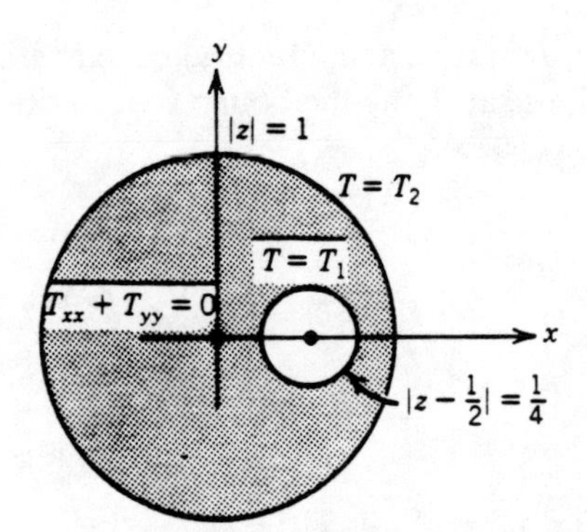

Figure 4.5.13

4.6 APPLICATIONS TO FLUID FLOW AND ELECTROSTATICS

In Sections 3.4 and 3.5, we showed how a given analytic, complex potential function could be viewed as giving rise to fluid flows and electrostatic fields. Here, we will show how we can use conformal mappings to generate additional complex potentials from known ones.

Fluid Flows. We remind the reader that if $\Phi(z) = \phi(x, y) + i\psi(x, y)$ is any analytic function, it gives rise to a fluid flow whose complex velocity is $V(z) = \overline{\Phi'(z)}$. In addition, we can interpret this as a fluid flow around an obstacle whose boundary is the **streamline** $\psi(x, y) =$ constant. The objective here is to *find* a potential which gives rise to a fluid flow around a *given* obstacle. Our procedure will be to find a function $w = f(z)$ which maps the given region of fluid flow onto a region in the w-plane for which a complex potential is known. Then, by going backward, we will generate a complex potential for our given problem. The examples which follow will illustrate these points. First, however, we have the following basic result.

(4.6.1) ***Theorem.*** Let $\Phi(z) = \phi(x, y) + i\psi(x, y)$ be a complex potential function which is analytic in a domain D_z of the z-plane. Let $w = f(z)$ be an invertible, analytic, conformal function mapping D_z onto a domain D_w in the w-plane. Then, any streamline in D_z will map onto a streamline in D_w.

Proof. Since $w = f(z)$ is invertible, it defines a real invertible transformation

$$\begin{matrix} u = u(x, y) \\ v = v(x, y) \end{matrix} \quad \text{If and only if} \quad \begin{matrix} x = x(u, v) \\ y = y(u, v) \end{matrix} \tag{4.6.2}$$

Under this, the potential $\Phi(z)$ becomes the composite function $\hat{\Phi}(w) = \hat{\phi}(u, v) + i\hat{\psi}(u, v)$. This function is analytic in D_w (and thus can serve as a complex potential in the w-plane), with the **stream function** $\hat{\psi}(u, v)$ defined by

$$\hat{\psi}(u, v) = \psi(x(u, v), y(u, v))$$

From this, it is clear that the streamline $\psi(x, y) =$ constant transforms into the streamline $\hat{\psi}(u, v) =$ constant, and vice versa.

■

(4.6.3) ***Example.*** Find a complex potential which gives rise to a fluid flow in the first quadrant as shown in Figure 4.6.1.

Solution. Our objective is to map the first quadrant onto the upper half of the w-plane, and then to use a potential for uniform flow over a rigid half-plane. Now, $w = z^2$ is such a mapping, and the function $\Phi = w$ is a potential for uniform flow over the lower half of the w-plane. Hence, in terms of z,

$$\Phi(z) = z^2 \tag{4.6.4}$$

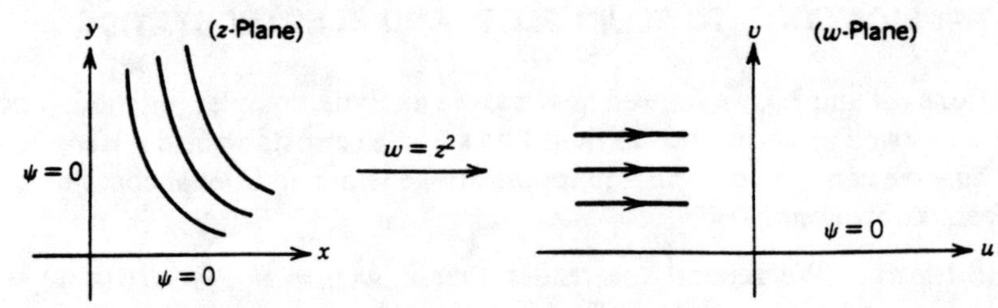

Figure 4.6.1

gives rise to a flow in the first quadrant of the z-plane. As shown in the figure (and from the above theorem), streamlines in the w-plane are "bent around" into streamlines in the z-plane, and the rigid boundaries, on which $\psi = 0$, are mapped onto each other. Note that this is the same as the flow discussed in Example (3.4.14).

■

The function $w = az + (b/z)$ has mapping properties which enable it to be used to solve a wide variety of problems involving flow over certain airfoil shapes. This will be developed in detail in Problems 2 through 6. In the next example, we discuss a special case of this mapping, and in the example which follows we show how it can be used to solve a simple fluid flow problem.

(4.6.5) ***Example.*** Show that

$$w = \frac{1}{2}\left(z + \frac{1}{z}\right)$$

maps the exterior of the unit disk onto the w-plane cut along the real interval $-1 \leq u \leq 1$, and maps the unit circle onto this cut (see Figure 4.6.2).

Solution. The circles $z = r_0 e^{i\theta}, 0 \leq \theta \leq 2\pi$, as r_0 ranges from 1 to ∞, will generate the region exterior to the unit disk, and their images will generate the image of this region. Now for these circles, and with $w = u + iv$, the

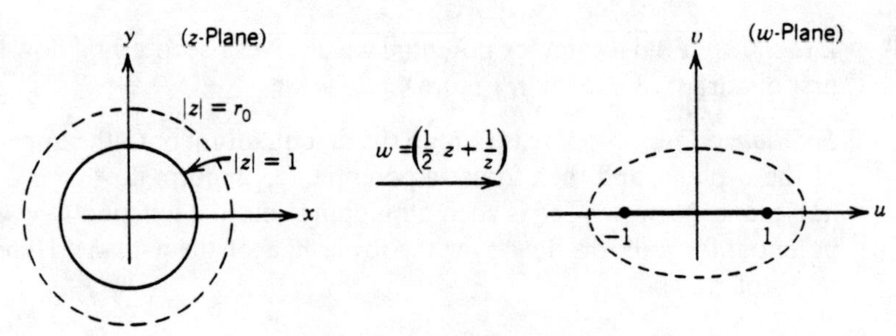

Figure 4.6.2 Example (4.6.5).

mapping yields

(4.6.6)
$$u = \frac{1}{2}\left(r_0 + \frac{1}{r_0}\right)\cos\theta \quad \text{and} \quad v = \frac{1}{2}\left(r_0 - \frac{1}{r_0}\right)\sin\theta$$

From this, the unit circle ($r_0 = 1$) has as its image the curve $v = 0$, $u = \cos\theta, 0 \le \theta \le 2\pi$, which is the interval in question. Now, if $r_0 > 1$, the above curves are the ellipses (shown dashed in the figure)

$$\frac{u^2}{\left(r_0 + \frac{1}{r_0}\right)^2} + \frac{v^2}{\left(r_0 - \frac{1}{r_0}\right)^2} = \frac{1}{4} \qquad r_0 > 1$$

which, as r_0 ranges over the interval $1 < r_0 < \infty$, clearly generate the cut w-plane. Also, note from the mapping that

(4.6.7)
$$\frac{1}{2}\left(z + \frac{1}{z}\right) \approx \frac{1}{2}z \qquad \text{as } z \to \infty$$

■

(4.6.8) ***Example.*** Find a complex fluid potential which describes a uniform fluid flow at infinity impinging from the left at angle α upon a cylinder of radius R_0, as shown in Figure 4.6.3.

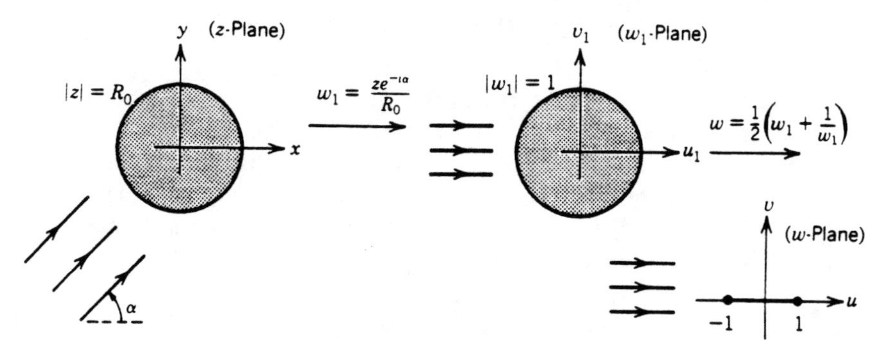

Figure 4.6.3 Example (4.6.8).

Solution. The region of flow is exterior to a disk of radius R_0. We will first map this onto the region exterior to the unit disk by a magnification map (see Section 2.4) and then use the mapping from the previous example to generate a flow in the w-plane over a rigid plate occupying the interval $-1 \le u \le 1$. Note that the magnification map will retain the uniform flow at angle α, while by (4.6.7) the same will be true of the second mapping. Hence, with *only* these two maps, the problem that would result is that of uniform flow *at angle* α toward the plate $v = 0$, $-1 \le u \le 1$. The potential for this problem is much more complicated and will be derived in the next example. What we must first do after the magnification is to rotate so that the uniform flow becomes horizontal. The series of mappings is shown in Figure 4.6.3, with the composite map given by

$$w = \frac{1}{2}\left(\frac{ze^{-i\alpha}}{R_0} + \frac{R_0 e^{i\alpha}}{z}\right)$$

Now, the problem of uniform horizontal flow over the flat plate in the w-plane of the figure leads to the potential $\Phi = w$, which in terms of z yields

$$\Phi(z) = \frac{1}{2}\left(\frac{ze^{-i\alpha}}{R_0} + \frac{R_0 e^{i\alpha}}{z}\right)$$

Note that the circle $|z| = R_0$ is a streamline [$\text{Im}(\Phi) = 0$ here]; and as $z \to \infty$, $\Phi(z) \approx ze^{-i\alpha}/2R_0$, which is just the uniform flow at angle α. Also note that this is the same as the potential discussed in Problem 7 of Section 3.4, but here we derived it.

■

As our final example, we will derive the complex potential for a uniform flow impinging from the left, at angle α, on a finite flat plate. Before reading this, however, the reader is advised to work out Problem 2, which might require some knowledge of the material in Section 3.2 on branch points.

(4.6.9) ***Example.*** A rigid body is in the shape of an infinite, flat plate, and its cross section occupies the interval $-1 \le x \le 1$, as shown in Figure 4.6.4. A uniform flow at angle α impinges on the plate as shown. Find a complex potential.

Solution. By Example (4.6.5), the function $w = \frac{1}{2}(z + 1/z)$ maps the exterior of the unit disk onto the region of flow considered here. Hence, we would expect that there is an inverse function to this which maps our given region back onto the exterior of the unit disk. We can then use the potential developed in the previous example. It is shown in Problem 2 that such a function is given by

$$w = z + \sqrt{z^2 - 1}$$

where the branch of $\sqrt{z^2 - 1}$ is depicted in Figure 4.6.9a. Also, from this problem we have

$$\sqrt{z^2 - 1} \approx z \qquad \text{as } z \to \infty \tag{4.6.10}$$

Now from the previous example, the complex potential for the flow in the w-plane over the cylinder (see Figure 4.6.4) is given by $\Phi = \frac{1}{2}(we^{-i\alpha} + (e^{i\alpha}/w))$. Hence, in terms of z we have

$$\Phi = \frac{1}{2}\left[e^{-i\alpha}(z + \sqrt{z^2 - 1}) + \frac{e^{i\alpha}}{z + \sqrt{z^2 - 1}}\right]$$

$$= \frac{1}{2}[e^{-i\alpha}(z + \sqrt{z^2 - 1}) + e^{i\alpha}(z - \sqrt{z^2 - 1})] \tag{4.6.11}$$

$$= (\cos\alpha)z - i\sin\alpha\sqrt{z^2 - 1}$$

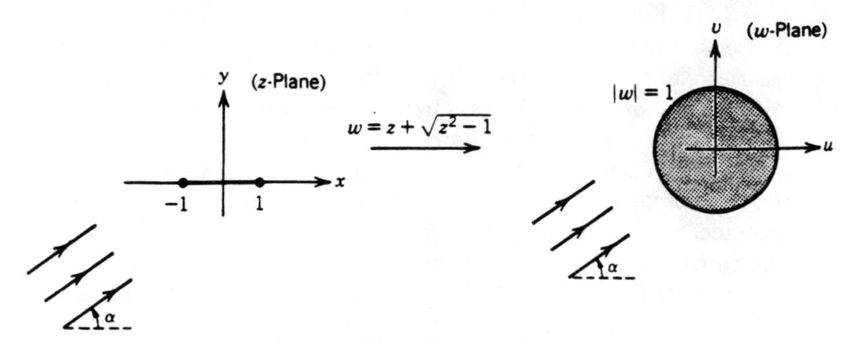

Figure 4.6.4 Example (4.6.9).

where in the second equality we have used the fact that $1/(z + \sqrt{z^2 - 1}) = z - \sqrt{z^2 - 1}$. This is the same potential discussed in Problem 12 of Section 3.4, but here we derived it by conformal mapping techniques. Also note from the above and (4.6.10) that $\Phi(z) \approx e^{-i\alpha} z$ as $z \to \infty$. This is the uniform flow in the z-plane of the problem.

■

Electrostatics. The problems in electrostatics to which we will apply conformal mapping techniques will be those involving point sources in the presence of conductors bounded by lines or circular arcs. The objective will be to first map the regions of interest onto the upper half-plane and then to apply the method of images [see Example (3.5.11) of Section 3.5]. To do so, however, requires that we make use of the following result, which we will state without proof.

(4.6.12) *Under a conformal mapping* $w = f(z)$, *a point charge of strength* q_0 *located at* z_0 *will transform into a source also of strength* q_0 *located at* $w_0 = f(z_0)$.

(4.6.13) ***Example.*** Find the electrostatic potential $\phi(x, y)$ due to a point charge q_0 located at $z = z_0$ and surrounded by grounded conductors as shown in Figure 4.6.5.

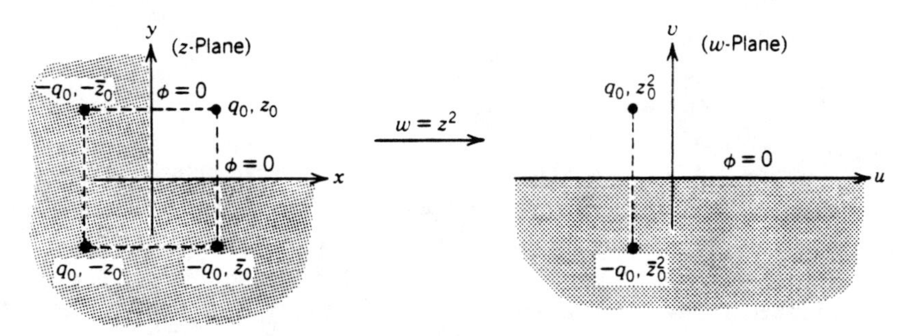

Figure 4.6.5 Example (4.6.13).

Solution. The analytic, conformal mapping $w = z^2$ transforms the quarter-plane, in which the charge is located, onto the upper half-plane. Also, the boundaries of the quarter-plane, on which the potential ϕ is contant ($\phi = 0$), are mapped onto the real axis, on which the transformed potential is also zero. Hence, by the above result, we must solve the problem of a point charge of strength q_0 located at $w = z_0^2$ above a grounded conducting half-plane (see the figure). The complex potential for this problem was derived in Example (3.5.11) by the method of images and is given by $\Phi = -2q_0[\log(w - z_0^2) - \log(w - \bar{z}_0^2)]$. If we replace w by the mapping function $w = z^2$ and take the real part of Φ, we find that the electrostatic potential we are looking for is given by

$$\begin{aligned} \phi(x, y) &= -2q_0 \ln\left|\frac{z^2 - z_0^2}{z^2 - \bar{z}_0^2}\right| \\ &= -2q_0\begin{bmatrix} \ln|z - z_0| + \ln|z + z_0| \\ -\ln|z - \bar{z}_0| - \ln|z + \bar{z}_0| \end{bmatrix} \end{aligned} \tag{4.6.14}$$

Note that this is the same as the potential due to charges of strength q_0 located at $z = z_0$ and $z = -z_0$ and charges of strength $-q_0$ located at $z = \bar{z}_0$ and $z = -\bar{z}_0$. This image system, shown in the z-plane of Figure 4.6.5, is precisely the one discussed in Problem 13*a* of Section 3.5. P. 150

(4.6.15) ***Example.*** Find the electrostatic potential due to a point charge q_0 in the presence of a grounded conducting cylinder $|z| \leq R_0$ (see Figure 4.6.6).

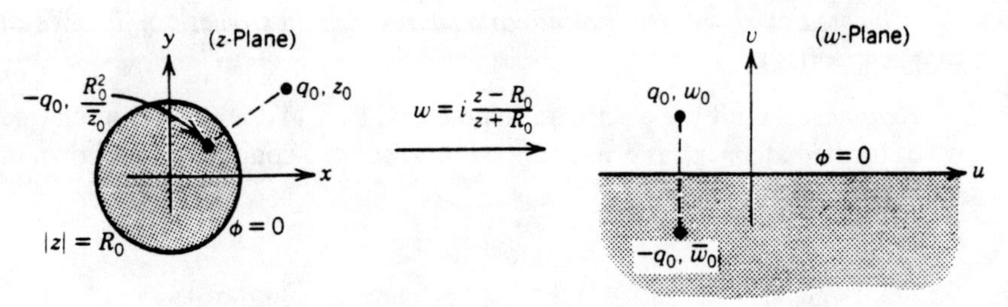

Figure 4.6.6 Example (4.6.15).

Solution. A bilinear map $w = (az + b)/(cz + d)$ can be used to map the exterior of the disk (where the charge is located) onto the upper half-plane. The mapping can be derived by requiring that some point on the circle $|z| = R_0$ be mapped into $w = \infty$, while two additional points be mapped into two real values of w. By conformality, if we pick the points in the proper order, we will be assured of getting the upper half-plane as our image. One such choice of points is to map $z = R_0$, iR_0, and $-R_0$, respectively, onto $w = 0$, -1, and ∞. This leads to the bilinear map

(4.6.16)
$$w = i\left(\frac{z - R_0}{z + R_0}\right)$$

The situation is depicted in the w-plane of Figure 4.6.6, with the point source located at $w_0 = i(z_0 - R_0)/(z_0 + R_0)$. Hence, the complex potential in the w-plane is $\Phi = -2q_0[\log(w - w_0) - \log(w - \bar{w}_0)]$. Now, from (4.6.16) and the expression for w_0, we find

$$w - w_0 = \frac{2i R_0(z - z_0)}{(z_0 + R_0)(z + R_0)}$$

with a similar expression for $w - \bar{w}_0$. From this, the electrostatic potential $\phi = \mathrm{Re}(\Phi)$ can be written as

(4.6.17)
$$\Phi = -2q_0\left[\ln|z - z_0| - \ln\left|z - \frac{R_0^2}{\bar{z}_0}\right|\right] - 2q_0 \ln\left|\frac{R_0}{\bar{z}_0}\right|$$

We leave it to the reader to verify that ϕ vanishes on the cylinder $|z| = R_0$.

Note from the above that the first term in square brackets represents the potential due to the given source q_0 at z_0, while the second term is the potential due to a source $-q_0$ at the point $R_0^2/\bar{z}_0$ which lies inside the conductor. Hence, the second term represents the potential due to the **image of a source in a cylinder**. The situation is depicted in Figure 4.6.6. The third term in (4.6.17) is just a constant which is needed for the potential to vanish on the grounded conductor. (We remind the reader that the electric field, being the gradient of the potential, is not affected by the addition of a constant to the potential.)

■

(4.6.18) ***Example.*** A point charge q_0 is located at a point z_0 between two grounded conducting half-planes. Find the potential between them (see Figure 4.6.7).

Solution. We might be tempted to use the method of images directly. To do this, we would form the image of q_0 in *both* conducting planes. Then, these images would themselves have images in the planes, and so on. In

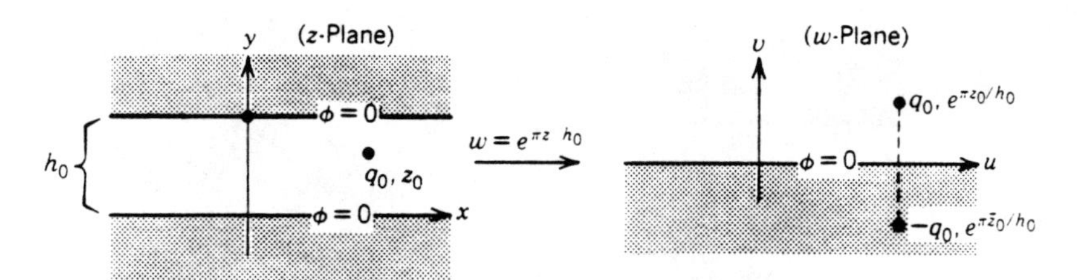

Figure 4.6.7 Example (4.6.18).

this manner, we would construct an infinite number of images, and the final potential would be an infinite series.

Instead, we will map the region between the half-planes onto the upper half of the w-plane by the mapping shown in Figure 4.6.7. Then the complex potential which solves the problem in the w-plane of the figure is given by

$$\phi = -2q_0[\log(w - e^{\pi z_0/h_0}) - \log(w - e^{\pi \bar{z}_0/h_0})]$$

and hence the electrostatic potential is

$$\phi = -2q_0[\ln|e^{\pi z/h_0} - e^{\pi z_0/h_0}| - \ln|e^{\pi z/h_0} - e^{\pi \bar{z}_0/h_0}|]$$

$$= -q_0 \ln\left[\frac{e^{2\pi x/h_0} + e^{2\pi x_0/h_0} - 2e^{\pi(x+x_0)/h_0}\cos\left(\dfrac{\pi(y-y_0)}{h_0}\right)}{e^{2\pi x/h_0} + e^{2\pi x_0/h_0} - 2e^{\pi(x+x_0)/h_0}\cos\left(\dfrac{\pi(y-y_0)}{h_0}\right)}\right] \tag{4.6.19}$$

From the above, it is easily seen that $\phi = 0$ when $y = 0$ and $y = h_0$. Also note that in some miraculous manner, we have "summed" the infinite series due to the infinite image system mentioned at the beginning.

■

PROBLEMS (The answers are given on page 432.)

1. Find a complex potential for a fluid flow for the two regions shown in Figure 4.6.8. The shaded regions are rigid boundaries. [*Hint:* Map these areas onto the upper half-plane.]
2. Here, the reader will be asked to discuss some mapping properties of the function $w = \frac{1}{2}(z + 1/z)$ and its inverse. These results are used in Example (4.6.9).
   (a) It was shown in Example (4.6.5) that under this mapping the exterior of the unit disk is mapped onto the whole w-plane cut along the interval $-1 \leq u \leq 1$. Show that the same is true of the interior of the unit disk. [*Hint:* Consider (4.6.6) with $0 < r_0 < 1$.]

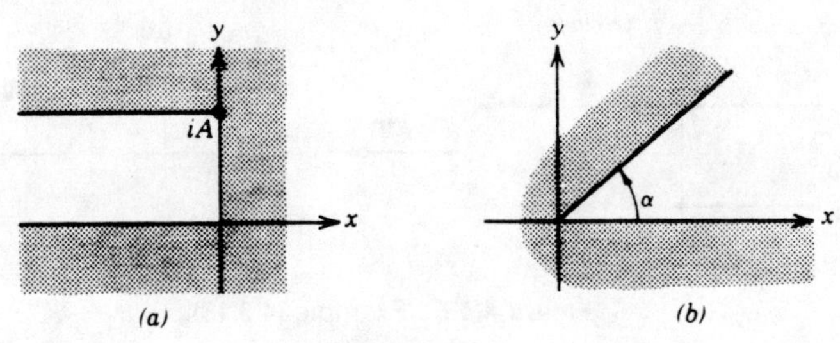

Figure 4.6.8

(b) Solve $w = \frac{1}{2}(z + 1/z)$ for z and show that some branch of $w = z + \sqrt{z^2 - 1}$ maps the z-plane cut along the real interval $-1 \le x \le 1$ onto the region $|w| > 1$.

(c) Consider that branch of $\sqrt{z^2 - 1}$ defined by $\sqrt{z^2 - 1} = |z^2 - 1|^{1/2} \cdot e^{(i/2)(\theta_1 + \theta_2)}$, where $z - 1 = |z - 1|e^{i\theta_1}$, $z + 1 = |z + 1|e^{i\theta_2}$, $-\pi < \theta_1, \theta_2 \le \pi$ (see Figure 4.6.9*a*). Show that

  (i) $\sqrt{z^2 - 1}$ is continuous everywhere except on the interval $-1 < x < 1$.

  (ii) $\sqrt{z^2 - 1}$ is real when $z = x$ is real with $|x| > 1$. Also, it is positive for $x > 1$ and negative for $x < -1$.

  (iii) As $z \to \infty$, $\sqrt{z^2 - 1} \approx z$.

  (iv) The ellipse

$$z = \frac{1}{2}\left(r_0 + \frac{1}{r_0}\right)\cos\theta + \frac{i}{2}\left(r_0 - \frac{1}{r_0}\right)\sin\theta \qquad 0 \le \theta \le 2\pi$$

  where $r_0 > 1$, is mapped by $w = z + \sqrt{z^2 - 1}$ onto the circle $|w| = r_0$. [*Hint:* Show that $z^2 - 1 = \frac{1}{4}[(r_0 - 1/r_0)\cos\theta + i(r_0 + 1/r_0)\sin\theta]^2$ and then use the branch of $\sqrt{z^2 - 1}$ to justify $\sqrt{z^2 - 1} = \frac{1}{2}[(r_0 - 1/r_0)\cos\theta + i(r_0 + 1/r_0)\sin\theta]$.]

  (v) The z-plane cut along the real interval $-1 \le x \le 1$ is mapped by $w = z + \sqrt{z^2 - 1}$ onto the region $|w| > 1$. This then is the inverse mapping of the function in part (a).

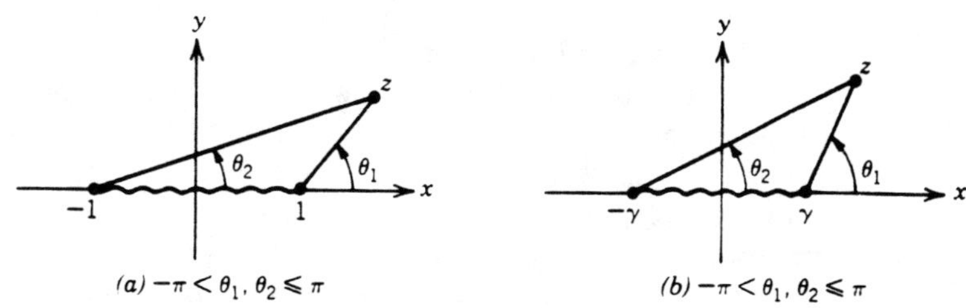

Figure 4.6.9

3. Consider the mapping $w = \frac{1}{2}(z + \gamma^2/z)$, where $\gamma > 0$. The reader is urged to redo both Example (4.6.5) and Problem 2 and show the following:

(a) The function maps the circle $|z| = \gamma$ onto the real interval $-\gamma \le u \le \gamma$.

(b) It maps the circle $|z| = r_0$, where $r_0 > \gamma$, onto an ellipse.

(c) The region $|z| > \gamma$ is mapped onto the w-plane cut along the real interval $-\gamma \le u \le \gamma$.

(d) There is a single-valued inverse defined by $w = z + \sqrt{z^2 - \gamma^2}$, with the branch of the square root given by $\sqrt{z^2 - \gamma^2} = |z^2 - \gamma^2|^{1/2}e^{(i/2)(\theta_1 + \theta_2)}$, where $-\pi < \theta_1, \theta_2 \le \pi$ (see Figure 4.6.9*b*).

(e) The function in part (d) maps the z-plane cut along the interval $-\gamma \le x \le \gamma$ onto $|w| > \gamma$.

(f) If $a > b$, there is a value of γ so that the ellipse $(x^2/a^2) + (y^2/b^2) = 1$ is mapped by the function in part (d) onto a circle, and the region outside of the ellipse is mapped onto the exterior of the circle. In terms of a and b, what is the value of γ? What is the radius of the circle?

4. Consider a uniform flow inclined at angle α against the rigid ellipse $(x^2/4) + y^2 = 1$ (see Figure 4.6.10).

(a) Use the results of Problem 3 and Example (4.6.8) to find the complex potential for the flow.

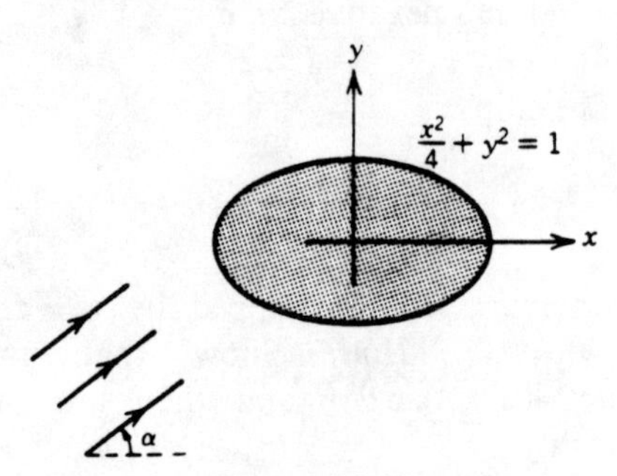

Figure 4.6.10

(b) Use Theorem (3.4.10b) of Section 3.4 and show that there is no lift on the body, that is, there is no component of force perpendicular to the uniform flow.

(c) Refer to Problem 10 of Section 3.4 and indicate how you might arrive at a complex potential which *would* yield a lift on the elliptical body. What is this potential?

5. In Problem 2, we showed that the function $w = \frac{1}{2}(z + 1/z)$ mapped the unit circle onto the real interval $-1 \leq u \leq 1$. (Note that this mapping is not conformal at the points $z = \pm 1$ on the unit circle, since $w' = 0$ there). Here, we investigate how other circles are mapped under this function.

(a) Show that for $z \approx 1$, $w - 1 \approx \frac{1}{2}(z - 1)^2$. [*Hint:* Subtract 1 from each side of the equation $w = \frac{1}{2}(z + 1/z)$].

(b) Use the result in (a) to conclude that the angles between two smooth curves in the z-plane which intersect at $z = 1$ will be doubled under the mapping.

(c) Consider the circle $|z - z_0| = |1 - z_0|$ $(z_0 \neq 0)$ as shown in Figure 4.6.11. This circle passes through $z = 1$ and, if $|z_0 + 1| < |z_0 - 1|$, contains $z = -1$ inside.

(i) If the upper and lower portions of the circle are viewed as separate curves which intersect at $z = 1$, use part (b) to show that in a neighborhood of $w = 1$, the image of the circle must have a cusp-type configuration as shown in the figure.

(ii) Since $z = -1$ (the other point of nonconformality) is inside the circle, the image of the circle must be some smooth curve (with the exception of a neighborhood of $w = 1$). Show that its general shape is as in Figure 4.6.11. This is called a **Joukowski profile**.

(iii) Show that if z_0 is real and negative, the Joukowski profile is symmetric with respect to the real w-axis.

(d) Show that with an appropriate choice of the square root, $w = z + \sqrt{z^2 - 1}$ maps the exterior of a Joukowski profile onto the exterior of some circle.

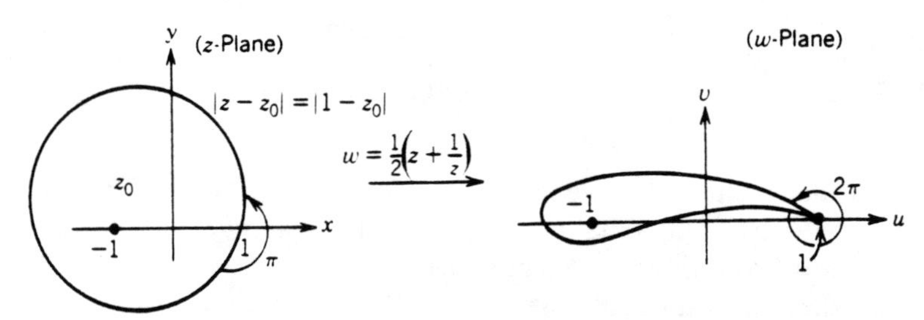

Figure 4.6.11 Joukowski profile.

6. Consider a uniform flow at angle α against a rigid body bounded by the Joukowski profile of the previous problem, (see Figure 4.6.12).
   (a) Find a function $w = f(z)$ which maps the flow region onto the exterior of the unit disk and construct a complex fluid potential function.
   (b) Show that the fluid velocity is singular near the **trailing edge** of the body, that is, near the cusp.
   (c) Introduce a circulation $(i\Gamma/2\pi)\log w$ in the w-plane (see Section 3.4) and derive another complex potential which still has the Joukowski profile as a streamline.
   (d) Show that the vortex parameter Γ in part (c) can be chosen so as to make the fluid velocity finite at the trailing edge. This is called a **Kutta–Joukowski condition**.

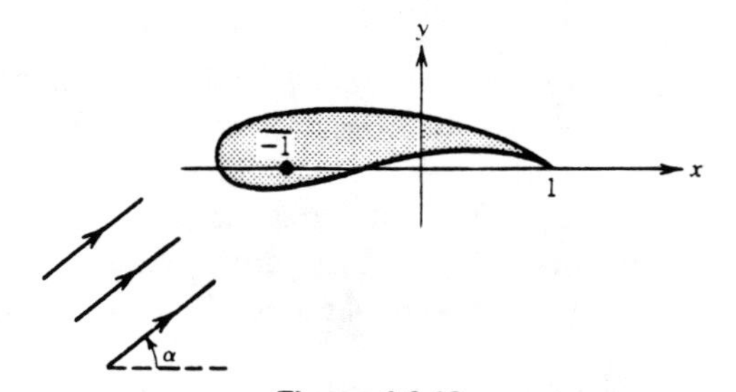

Figure 4.6.12

7. (a) Find the electrostatic potential due to a charge q_0 located in a wedge-shaped region surrounded by grounded conductors as shown in Figure 4.6.13a.

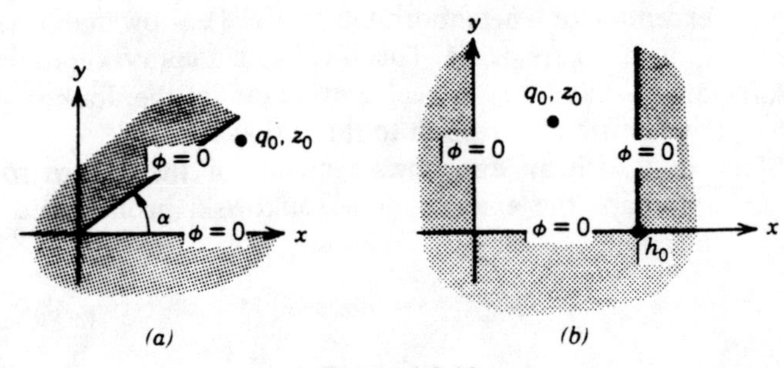

Figure 4.6.13

(b) Show that if the wedge angle α equals π/n ($n = 1, 2, 3, \ldots$), then the potential in (a) is that due to q_0 and a finite number of image charges (see also Problem 13*b* in Section 3.5).

8. Find the electrostatic potential due to a charge q_0 located in a semiinfinite strip surrounded by grounded conductors, as shown in Figure 4.6.13*b*.

9. A grounded cylindrical conductor is in contact with a grounded conducting half-plane as shown in Figure 4.6.14*a*. What is the potential due to a charge q_0 at z_0. [*Hint:* First, use the bilinear mapping $w = 1/z$.]

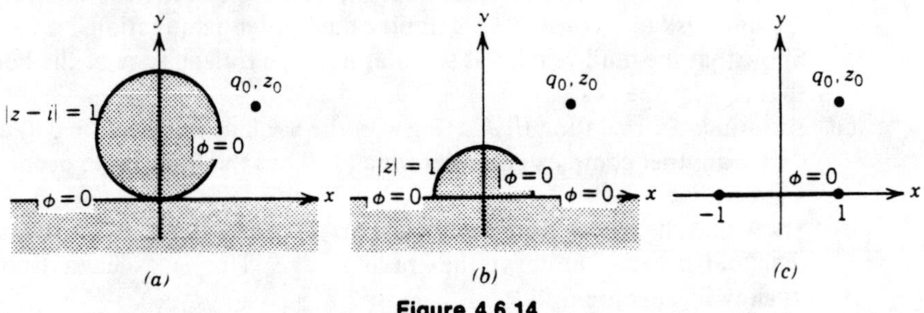

Figure 4.6.14

10. A charge q_0 is located above a grounded conducting half-plane with a "bump," as shown in Figure 4.6.14*b*. Find the electrostatic potential and show that it is due to the charge q_0 and a series of three image charges.

11. Find the electrostatic potential due to a charge q_0 in the presence of a finite grounded, conducting strip whose cross section is the real interval $-1 \leq x \leq 1$, as shown in Figure 4.6.14*c*. [*Hint:* The reader should refer to Problem 2.]

12. If the surface of a conductor is **insulated**, the electrostatic potential satisfies the boundary condition $\partial\phi/\partial n = 0$, where $\partial\phi/\partial n$ is the normal derivative of ϕ.

(a) Show that the electrostatic potential due to a charge q_0 in the first quadrant bounded by conducting walls, one part of which is insulated as shown in Figure 4.6.15*a*, is the same as that due to the system of images shown in this figure, with the signs of the charges as shown.

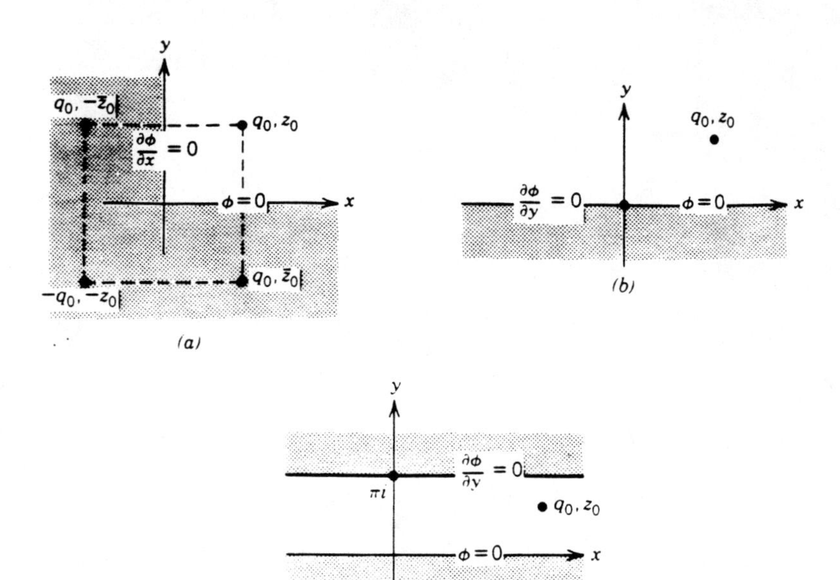

Figure 4.6.15

(b) Use the result of part (a) to find the potential due to the configuration shown in Figure 4.6.15*b*.

(c) Solve for the electrostatic potential due to the configuration of Figure 4.6.15*c*.

5

COMPLEX INTEGRATION

5.1 CONTOUR INTEGRALS[1]

To begin with, we define the **integral of a complex-valued function of a real variable over an interval**.

(5.1.1) ***Definition.*** For the complex-valued function

$$F(t) = U(t) + i\,V(t)$$

we define

$$\int_{t_0}^{t_1} F(t)\,dt \equiv \int_{t_0}^{t_1} U(t)\,dt + i \int_{t_0}^{t_1} V(t)\,dt \tag{5.1.2}$$

if both (real) integrals on the right exist. Then, we say that F is **integrable over the interval $[t_0, t_1]$**. If either t_0 or t_1 (or both) are infinite, or if $F(t)$ is unbounded in the interval, then the integral on the left is still defined by the right-hand side, where these integrals are now considered as (real) **improper integrals** (the reader is urged to review improper integrals from elementary calculus).

■

[1] The reader should thoroughly review the material on curves in Section 4.1. If this section has not yet been covered, then it should be at this time.

(5.1.3) ***Example.*** Compute

$$I = \int_0^1 \left[te^{t^2} + \frac{i}{\sqrt{t}} \right] dt$$

Solution. By (5.1.2),

$$I = \int_0^1 te^{t^2}\, dt + i \int_0^1 \frac{dt}{\sqrt{t}}$$

if both integrals on the right exist. The first, whose integrand is continuous, clearly exists, and its value is $\frac{1}{2}(e - 1)$, while the second is improper with its value given by the limit

(5.1.4

$$\int_0^1 \frac{dt}{\sqrt{t}} = \lim_{\epsilon \downarrow 0} \int_\epsilon^1 \frac{dt}{\sqrt{t}} = 2$$

Thus, $I = \frac{1}{2}(e - 1) + 2i$.

■

(5.1.5) ***Example.*** To compute $I = \int_0^1 (dt/(t - i))$, we first write $1/(t - i) = (t + i)/(1 + t^2)$. Then,

(5.1.6)

$$I = \int_0^1 \frac{t\, dt}{1 + t^2} + i \int_0^1 \frac{dt}{1 + t^2} = \frac{1}{2} \ln 2 + \frac{\pi i}{4}$$

■

The astute reader might have asked why in the above we did not write

$$\int_0^1 \frac{dt}{t - i} = \log(t - i) \Big|_0^1 = \log(1 - i) - \log(-i)$$

In answer, we first observe that this result is dependent upon the branch of the logarithm chosen, and second, as we will see later, not all branches will yield the correct answer.

(5.1.7) ***Theorem***

(a) Let $F(t)$ be defined and piecewise continuous on $[t_0, t_1]$, that is, let there be a finite number of subintervals $\hat{t}_{k-1} < t < \hat{t}_k, k = 1, 2, \ldots, N$, with $\hat{t}_0 = t_0$ and $\hat{t}_N = t_1$, such that $F(t)$ is continuous on each such interval. Then, F is integrable over $[t_0, t_1]$, and

(5.1.8)

$$\int_{t_0}^{t_1} F(t)\, dt = \sum_{k=1}^{N} \int_{\hat{t}_{k-1}}^{\hat{t}_k} F(t)\, dt$$

(b) If $F(t)$ and $G(t)$ are integrable over $[t_0, t_1]$, and α and β are complex constants, then $\alpha F + \beta G$ is also integrable over $[t_0, t_1]$, and

(5.1.9)

$$\int_{t_0}^{t_1} (\alpha F + \beta F)\, dt = \alpha \int_{t_0}^{t_1} F\, dt + \beta \int_{t_0}^{t_1} G\, dt$$

(c) If $F(t)$ is integrable over $[t_0, t_1]$, then so is its magnitude $|F(t)|$, and we have the **triangle inequality for integrals**

$$\left|\int_{t_0}^{t_1} F(t)\,dt\right| \le \int_{t_0}^{t_1} \left|F(t)\right|\,dt \tag{5.1.10}$$

(d) If $G(t)$ is continuous and has a bounded continuous derivative $G'(t)$ on $[t_0, t_1]$, then

$$\int_{t_0}^{t_1} G'(t)\,dt = G(t)\Big|_{t_0}^{t_1} = G(t_1) - G(t_0) \tag{5.1.11}$$

This is a form of the **fundamental theorem of calculus** for complex functions.

Proof. The proofs of (a), (b), and (d) follow from (5.1.2) and the corresponding results from elementary calculus. [In (d), if $G(t) = U(t) + i\,V(t)$, use the fact that $G' = U' + i\,V'$.] The proof of (c) is somewhat more complicated and is given in Problem 3.

■

The length of a curve is a special case of the above type of integral and is defined by the following.

(5.1.12) ***Definition.*** Let C be the differentiable curve

$$C: \quad z = z(t) \qquad t_0 \le t \le t_1$$

We define L, the **length of** C, by

$$L = \int_{t_0}^{t_1} \left|\frac{dz}{dt}\right| dt \tag{5.1.13}$$

when the integral exists.

■

Note that with $z(t) = x(t) + iy(t)$, we have

$$L = \int_{t_0}^{t_1} \{[x'(t)]^2 + [y'(t)]^2\}^{1/2}\,dt \tag{5.1.14}$$

which is just the usual definition of length from elementary calculus. Also, if C is piecewise smooth, then so is the integrand in (5.1.13), and thus the integral exists by the previous theorem.

(5.1.15) ***Example.*** The length of the curve C: $z = (1 + i)t^2$, $-1 \le t \le 1$, is

$$L = \int_{-1}^{1} |2(1 + i)t|\,dt = 2\sqrt{2}\left[\int_{-1}^{0} (-t)\,dt + \int_{0}^{1} t\,dt\right]$$
$$= 2\sqrt{2}$$

Note that the trace of C is the line $y = x, 0 \leq x \leq 1$, which has length $\sqrt{2}$ and not $2\sqrt{2}$. Why do you think there is a difference?

▬

Contour Integrals. Let C be the curve

(5.1.16)
$$C: \quad z = z(t) = x(t) + i\,y(t) \qquad t_0 \leq t \leq t_1$$

and let

(5.1.17)
$$f(z) = u(x, y) + i\,v(x, y)$$

be defined on the trace of C. Then, we have the following.

(5.1.18) ***Definition***

(a) If C is smooth we define the **contour integral of $f(z)$ over C**, denoted by $\int_C f(z)\,dz$, to be

(5.1.19)
$$\int_C f(z)\,dz = \int_{t_0}^{t_1} f(z(t))\frac{dz}{dt}\,dt$$

where the **composite function** $f(z(t))$ is

$$f(z(t)) = u(x(t), y(t)) + iv(x(t), y(t))$$

We say that f **is integrable over C** whenever the integral on the right in (5.1.19) exists.

(b) Let C be piecewise smooth. Hence, there are subintervals $\hat{t}_{k-1} < t < \hat{t}_k$, $k = 1, 2, \ldots, N$, with $\hat{t}_0 = t_0$ and $\hat{t}_N = t_1$ on each of which there is defined a smooth curve C_k given by

$$C_k: \quad z = z(t) \qquad \hat{t}_{k-1} < t < \hat{t}_k,\ k = 1, 2, \ldots, N$$

Then, we define

(5.1.20)
$$\int_C f(z)\,dz = \sum_{k=1}^{N} \int_{C_k} f(z)\,dz$$

If each of the integrals on the right exists.

▬

If z_0 and z_1 are the initial and final points of C, then we will also denote contour integrals by

(5.1.21)
$$\oint_{z_0}^{z_1} f(z)\,dz$$

and will say that $f(z)$ is to be **integrated from z_0 to z_1 along C**. If C is a closed curve, then we will use the notation

(5.1.22)
$$\oint_C f(z)\,dz$$

Also, note that an integral [such as (5.1.2)] of a complex-valued function over a real interval can be viewed as a contour integral over the curve $z(t) = t$, $t_0 \le t \le t_1$. Finally, since $f = u + iv$ and $dz/dt = (dx/dt) + i(dy/dt)$, we have

$$f(z(t))\frac{dz}{dt} = u\frac{dx}{dt} - v\frac{dy}{dt} + i\left(v\frac{dx}{dt} + u\frac{dy}{dt}\right)$$

Hence, a complex contour integral can be written as the following combination of **real line integrals**:

(5.1.23)
$$\int_C f(z)\,dz = \int_C (u\,dx - v\,dy) + i\int_C (v\,dx + u\,dy)$$

(5.1.24) ***Example.*** Evaluate $I = \int_C (x + y^2 + ixy)\,dz$, where

$$C:\quad z = z(t) = \begin{cases} t + 2i & 1 \le t \le 2 \\ 2 + i(4 - t) & 2 < t \le 3 \end{cases}$$

Solution. The curve is piecewise smooth, and

(5.1.25)
$$\frac{dz}{dt} = \begin{cases} 1 & 1 < t < 2 \\ -i & 2 < t < 3 \end{cases}$$

Hence, with $f(z) = x + y^2 + ixy$ in (5.1.19), we have

$$\begin{aligned} I &= \int_1^2 (t + 4 + 2it)\,dt + \int_2^3 [2 + (4 - t)^2 + 2i(4 - t)](-i)\,dt \\ &= \frac{17}{2} - \frac{4i}{3} \end{aligned}$$

■

(5.1.26) ***Example.*** Evaluate $I_k = \oint_C f_k(z)\,dz$, $k = 1, 2, 3$, where $f_1(z) = z$, $f_2(z) = \bar{z}$, $f_3(z) = (z - i)^{-n}$, $n = 0, \pm 1, \pm 2, \ldots$, and C is the closed curve

$$C:\quad z(t) = i + e^{it} \qquad 0 \le t \le 2\pi$$

Solution. Since $z'(t) = ie^{it}$, we have

$$\begin{aligned} I_1 &= \int_0^{2\pi} (i + e^{it})ie^{it}\,dt = \int_0^{2\pi} (-e^{it} + ie^{2it})\,dt \\ &= \left(-\frac{1}{i}e^{it} + \frac{1}{2}e^{2it}\right)\Bigg|_0^{2\pi} \qquad \text{[follows from (5.1.11)]} \\ &= 0 \end{aligned}$$

Similarly, since $\bar{z}(t) = -i + e^{-it}$,

(5.1.27)
$$I_2 = \int_0^{2\pi} (-i + e^{-it})ie^{it}\,dt = \int_0^{2\pi} (e^{it} + i)\,dt = 2\pi i$$

Finally,

$$I_3 = \int_0^{2\pi} ie^{it}e^{-int}\,dt = i\int_0^{2\pi} e^{i(1-n)t}\,dt$$

$$= \begin{cases} 0 & n \neq 1 \\ 2\pi i & n = 1 \end{cases}$$

■

(5.1.28) ***Example.*** Evaluate the integral $I = \oint_C \sqrt{z}\,dz$, where

(5.1.29)
$$C: \quad z = z(t) = e^{it} \qquad 0 \leq t \leq 2\pi$$

and the principal branch of the square root is taken.

Solution. At first glance we might argue that since the initial and final points z_0 and z_1 are the same, and since $\sqrt{z} = \frac{2}{3}(d/dz)(z^{3/2})$, we should have

(5.1.30)
$$I = \frac{2}{3}[z_1^{3/2} - z_0^{3/2}] = 0$$

In fact, this is not the correct answer for reasons which will be discussed in the next theorem. To proceed correctly, we must compute the principal value of $\sqrt{z(t)}$ for each t in the range $0 \leq t \leq 2\pi$, which means that we must take care to represent $z(t) = e^{it}$ in a form whose argument lies in the principal argument range

(5.1.31)
$$-\pi < \arg(e^{it}) \leq \pi \qquad (\text{for } each\ t \text{ in } 0 \leq t \leq 2\pi)$$

Now, in general, $\arg(e^{it}) = t + 2\pi n$, $n = 0, \pm 1, \ldots$. Thus, we must choose n, possibly different for different ranges of t, so that the above is satisfied. This leads to the following representation for $\arg[z(t)]$:

(5.1.32)
$$\arg[z(t)] = \begin{cases} t & 0 \leq t \leq \pi \\ t - 2\pi & \pi < t \leq 2\pi \end{cases}$$

from which it follows that, since $|z(t)| = 1$,

(5.1.33)
$$(\text{Principal branch})\ \sqrt{z(t)} = \begin{cases} e^{it/2} & 0 \leq t \leq \pi \\ e^{i(1/2)(t-2\pi)} & \pi < t \leq 2\pi \end{cases}$$

Hence, since $dz/dt = ie^{it}$, we have

$$I = \int_0^{2\pi} \sqrt{z(t)}\,ie^{it}\,dt$$

$$= \int_0^{\pi} ie^{3it/2}\,dt + \int_{\pi}^{2\pi} ie^{-\pi i}e^{3it/2}\,dt$$

$$= \frac{-4i}{3}$$

■

The results of the next theorem are similar to those from calculus.

(5.1.34) ***Theorem.*** Let C be a piecewise smooth curve with initial point z_0 and final point z_1. Then,

(a) If $f(z)$ is defined and piecewise continuous on the trace of C, it is integrable over C.

(b) If $f(z)$ and $g(z)$ are integrable over C, then

$$\int_C (\alpha f + \beta g)\, dz = \alpha \int_C f\, dz + \beta \int_C g\, dz$$

(c) If $f(z)$ is integrable over C, so is $|f(z)|$ and the **triangle inequality for contour integrals** holds, that is,

(5.1.35)
$$\left| \int_C f(z)\, dz \right| \leq \int_{t_0}^{t_1} |f(z(t))| \left| \frac{dz}{dt} \right| dt$$

Hence, if

(5.1.36)
$$|f(z)| \leq M \qquad \text{for } z \text{ on the trace of } C$$

then

(5.1.37)
$$\left| \int_C f(z)\, dz \right| \leq M \cdot \text{length of } C$$

(d) If $f(z)$ is integrable over C, then it is integrable over $-C$ [see (4.1.19)], and

(5.1.38)
$$\int_{-C} f(z)\, dz = -\int_C f(z)\, dz$$

(e) If $F(z)$ is *continuously differentiable in a domain containing the trace of* C, then

(5.1.39)
$$\oint_{z_0}^{z_1} F'(z)\, dz = F(z) \Big|_{z_0}^{z_1} = F(z_1) - F(z_0)$$

In particular, if C is closed,

(5.1.40)
$$\oint_C F'(z)\, dz = 0$$

Proof. Results (a) and (b) follow from the definition of contour integrals and the corresponding results in elementary calculus, while (c) follows from (5.1.10) and the definition of length. To prove (d), we use Definition

(4.1.19) for $-C$ and write

$$\int_{-C} f(z)\,dz = \int_{-t_1}^{-t_0} f(z(-t))\frac{d}{dt}[z(-t)]\,dt$$

(5.1.41)
$$= \int_{t_1}^{t_0} f(z(\tau))\frac{d}{d\tau}[z(\tau)]\,d\tau \qquad (\text{let } \tau = -t)$$

$$= -\int_C f(z)\,dz$$

In Problem 11, the reader is asked to prove part (e) of the theorem.

∎

If we now return to Example (5.1.28), we can see why, even though $\sqrt{z} = \frac{2}{3}(d/dz)(z^{3/2})$, it was incorrect to use (5.1.30) to evaluate the integral—there is no branch of $z^{3/2}$ which is continuously differentiable in a domain containing the trace (the unit circle) of the curve $z = z(t)$ in (5.1.29). Hence, result (e) is not applicable. The next example also emphasizes this point.

(5.1.42) ***Example.*** Evaluate

$$I_n = \int_C \frac{dz}{z^n} \qquad n = 0, \pm 1, \pm 2, \ldots$$

where C: $z = e^{it}, 0 \le t \le \pi$.

Solution. For $n \ne 1$, $z^{-n} = (d/dz)[z^{1-n}/(1-n)]$. Since $z^{1-n}/(1-n)$ is continuously differentiable in some neighborhood of the trace of C (the upper half of the unit circle), we get from (5.1.39)

(5.1.43)
$$\int_C \frac{dz}{z^n} = \left.\frac{z^{1-n}}{1-n}\right|_1^{-1} = \frac{1}{1-n}[(-1)^{1-n} - 1] \qquad n \ne 1$$

For $n = 1$, we have $1/z = (d/dz)\log z$; and to use (5.1.39), we *must* choose a branch of the logarithm which is continuously differentiable on the upper half of the unit circle. Any branch with a branch cut not intersecting this

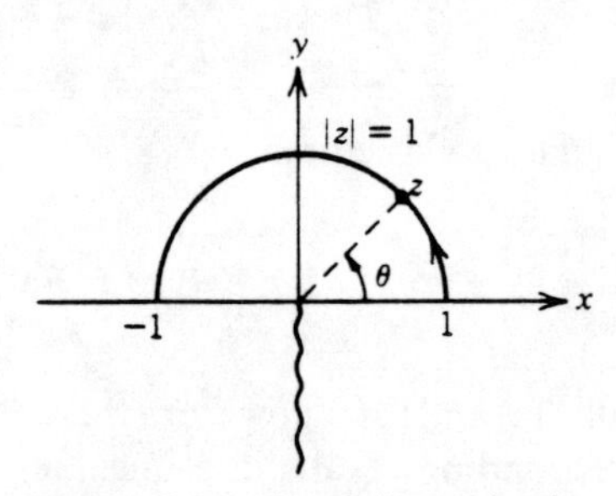

Figure 5.1.1 Branch cut for Example (5.1.42).

semicircle will do, one such being

(5.1.44) $$\log z = \ln|z| + i\theta \qquad -\pi/2 < \theta \le 3\pi/2$$

as shown in Figure 5.1.1. With this, we have

$$\int_C \frac{dz}{z} = \log(-1) - \log 1 = \pi i$$

■

(5.1.45) ***Example.*** Let C: $z = Re^{it}$, $0 \le t \le \pi$, where $R > 1$. In terms of R, compute an upper bound for the magnitude of the integral

(5.1.46) $$I = \int_C \frac{z\,dz}{1+z^4}$$

and show that $\lim_{R\to\infty}(I) = 0$.

Solution. From inequality (1.1.27b), we have $|1 + z^4| \ge |1 - |z|^4|$. Thus,

$$\left|\frac{z}{1+z^4}\right| \le \frac{|z|}{|1-|z|^4|}$$

Now on C, $|z| = R > 1$. Thus, from the above, $|z(1+z^4)^{-1}| \le R(R^4 - 1)^{-1}$. Hence, since the length of C is πR, the triangle inequality for integrals (5.1.36) and (5.1.37) yields

$$\left|\int_C \frac{z}{1+z^4}\,dz\right| \le \frac{\pi R^2}{R^4 - 1}$$

Also, since the magnitude on the left is inherently positive, and since $\lim_{R\to\infty}[R^2/(R^4 - 1)] = 0$, by the *sandwiching theorem from calculus* we must have $\lim_{R\to\infty}(I) = 0$.

■

In Section 4.1, we indicated that two curves which are equivalent [see Definition (4.1.30)] are virtually indistinguishable. Hence, we would expect that integrals over two such curves would yield the same values. This is the content of the next result.

(5.1.47) ***Theorem.*** Let C_1 and C_2 be two simple, smooth curves having the same trace and initial and final points (or the same direction, if they are both closed). Then, if $f(z)$ is continuous in some domain containing this trace, we have

$$\int_{C_1} f(z)\,dz = \int_{C_2} f(z)\,dz$$

Proof. From Theorem (4.1.31) we know that C_1 and C_2 are equivalent, which means that there exists a continuously differentiable function $\phi(\tau)$

defined on an interval $\tau_0 \le \tau \le \tau_1$ and satisfying

$$\text{(5.1.48)} \qquad \begin{aligned} &\text{(a)} \quad \phi' > 0 \\ &\text{(b)} \quad \phi(\tau_0) = t_0, \qquad \phi(\tau_1) = t_1 \end{aligned}$$

and so that C_1 and C_2 have the representations

$$C_1: \quad z = z_1(t) \qquad t_0 \le t \le t_1$$

$$C_2: \quad z = z_1(\phi(\tau)) \qquad \tau_0 \le \tau \le \tau_1$$

Now,

$$\int_{C_2} f(z)\,dz = \int_{\tau_0}^{\tau_1} f(z_1(\phi(\tau)))\frac{d}{d\tau}[z_1(\phi(\tau))]\,d\tau$$

$$= \int_{\tau_0}^{\tau_1} f(z_1(\phi(\tau)))z_1'(\phi(\tau))\frac{d\phi}{d\tau}\,d\tau$$

If in the last integral we change variables from τ to $t = \phi(\tau)$ and use (5.1.48), we get

$$\int_{C_2} f(z)\,dz = \int_{t_0}^{t_1} f(z_1(t))z_1'(t)\,dt = \int_{C_1} f(z)\,dz$$

■

With this last result, we can now adopt the following convention. By the statement

Integrate a continuous function $f(z)$ from z_0 to z_1 over a graph in the z-plane which does not intersect itself

we will mean

Choose any simple, piecewise smooth parametrization of this graph having z_0 and z_1 as initial and final points, and integrate $f(z)$ over the curve.

Also, by the statement

Integrate a continuous function $f(z)$ over a closed graph traversed in a specified direction

we will mean

Choose any simple, piecewise smooth parametrization of the graph having the given direction and integrate $f(z)$ over this curve.

For an integral over a closed curve such as described above, if C denotes either the graph or the curve, we might modify the notation (5.1.22) and write

$$\text{(5.1.49)} \qquad \oint_C f(z)\,dz \qquad \text{or} \qquad \oint_C f(z)\,dz$$

where the arrow indicates the direction of integration.

(5.1.50) ***Example.*** Evaluate the integral of $\bar{z}^2$ over the circle of radius 1, centered at $z = i$, and traversed counterclockwise.

Solution. Let us denote this integral by

$$I = \oint_{|z-i|=1} \bar{z}^2 \, dz$$

Now, one such curve whose trace is this circle is C: $z = i + e^{it}$, $0 \le t \le 2\pi$.
Hence,

$$I = \int_0^{2\pi} (-i + e^{-it})^2 (ie^{it}) \, dt = 4\pi$$

■

Change of Variables. As in elementary calculus, some contour integrals are more easily evaluated after a change of variables is made. However, we must proceed with some caution, since a contour in the z-plane, under a change of variables, will transform in general into an entirely different contour. The following is the analog of the result in calculus.

(5.1.51) ***Theorem.*** Let C_z be a contour in the z-plane. Let $w = g(z)$ be invertible, analytic, and conformal ($g' \neq 0$) in some neighborhood of the trace of C_z, and map C_z onto a contour C_w. Then, if $f(z)$ is integrable over C_z, we have

(5.1.52)
$$\int_{C_z} f(z) \, dz = \int_{C_w} \frac{f(g^{-1}(w)) \, dw}{g'(g^{-1}(w))}$$

where $z = g^{-1}(w)$ is the inverse function.

Proof. If C_z is parametrized by C_z: $z = z(t)$, $t_0 \le t \le t_1$, the image curve C_w is parametrized by C_w: $w = g(z(t))$, $t_0 \le t \le t_1$. Now, by the chain rule, $w' = g'(z(t))\, z'(t)$, and from the definition of contour integral, we have

$$\int_{C_w} \frac{f(g^{-1}(w)) \, dw}{g'(g^{-1}(w))} = \int_{t_0}^{t_1} \frac{f(g^{-1}(g(z(t))))\, g'(z(t))\, z'(t) \, dt}{g'(g^{-1}(g(z(t))))}$$
$$= \int_{t_0}^{t_1} f(z(t))\, z'(t) \, dt$$

where we have used the property that $g^{-1}(g(z(t))) = z(t)$. The reader will recognize the last integral above as the contour integral $\int_{C_z} f(z) \, dz$.

■

(5.1.53) ***Example.*** Evaluate

$$I = \int_{C_z} \frac{e^z \, dz}{1 + e^z}$$

where C_z is that piece of the imaginary axis lying between $y = 0$ and $y = 1$ and traversed upward.

Solution. We make the change of variables

$$w = e^z$$

Under this mapping, the contour C_z is transformed into the arc of the circle

$$C_w: \quad |w| = 1, \qquad 0 \le \arg(w) \le 1$$

traversed counterclockwise, and the integral, by (5.1.52), becomes

$$I = \int_{C_w} \frac{dw}{1 + w}$$

Now, the principal logarithm $\mathrm{Log}(1 + w)$ is continuously differentiable in a neighborhood of C_w, and we have

$$I = \mathrm{Log}(1 + w)\Big|_1^{e^i} = \mathrm{Log}(1 + e^i) - \ln 2$$

where $w = 1$ and $w = e^i$ are, respectively, the initial and final points of C_w.

PROBLEMS (The answers are given on page 433.)

(Problems with an asterisk are more theoretical.)

1. Evaluate $\int_{t_0}^{t_1} F(t)\,dt$, where
   (a) $F = (t + i)^{-1}$ $\quad t_0 = 0, t_1 = 1$ $\qquad$ (b) $F = (t + i)^{-2}$ $\quad t_0 = 0, t_1 = 1$
   (c) $F = e^{(-1+i)t}$ $\quad t_0 = 0, t_1 = \infty$ $\qquad$ (d) $F = te^{it}$ $\quad t_0 = 0, t_1 = \pi$
   (e) $F = (t + i)^{-1/2}$ $\quad t_0 = 0, t_1 = 1$ $\qquad$ (use principal branch)
   (f) $F = e^{int}$ $\quad t_0 = 0, t_1 = 2\pi$ $\qquad$ $(n = 0, \pm 1, \pm 2, \ldots)$
2. Define $\int_0^\infty dt/(1 + t^4)$ by $\lim_{R\to\infty} \int_0^R dt/(1 + t^4)$ and evaluate the integral. [*Hint:* Use partial fraction expansions and appropriate choices of branches of logarithms. Then compute the limit as $R \to \infty$.]
3.* The following steps lead to a proof of the triangle inequality for integrals. Let $F(t)$ be complex-valued and integrable over $[t_0, t_1]$ and define $I = \int_{t_0}^{t_1} F(t)\,dt$.
   (a) Why is $|F(t)|$ integrable over $[t_0, t_1]$?
   (b) Show that $\mathrm{Re}(I) = \int_{t_0}^{t_1} \mathrm{Re}[F(t)]\,dt$.
   (c) Let the polar form of I be $|I|e^{i\theta_0}$. Show that $|I| = \int_{t_0}^{t_1} \mathrm{Re}[e^{-i\theta_0}F(t)]\,dt$.
   (d) Finally, prove the triangle inequality (5.1.10). [*Hint:* Show that $\mathrm{Re}(z) \le |z|$.]

4. The **Laplace transform** $f(s)$ of a function $F(t)$ is defined for complex s by $f(s) = \int_0^\infty e^{-st} F(t)\,dt$ when the integral exists.
   (a) Show that if $F(t)$ has **exponential growth at infinity**, that is, $|F(t)| \le ke^{\gamma t}$ for some real k and γ, then $f(s)$ exists for $\mathrm{Re}(s) > \gamma$.
   (b) Evaluate the following Laplace transforms and indicate for which (complex) values of s they exist.
   (i) e^{at} (ii) t^n $(n = 0, 1, 2, \ldots)$ (iii) $\cos(at)$
   (iv) $\sin(at)$ (v) $\cosh(at)$ (vi) $\sinh(at)$
   (vii) $e^{bt}\sin(at)$
5. Compute the length of the following curves.
   (a) $z(t) = (t-1)^2 + 2i(t-1)^3 \quad 0 \le t \le 2$
   (b) $z(t) = (1+i)e^{it} \quad 0 \le t \le \pi/2$
   (c) $z(t) = t - \sin t + i(1 - \cos t) \quad 0 \le t \le 2\pi$
6. Evaluate $\int_C f(z)\,dz$, where f and C are given below.
   (a) $f = \bar{z} \quad C\colon\ z = e^{it} \quad 0 \le t \le 2\pi$

   (b) $f = x + y + ie^{xy} \quad C\colon\ z = \begin{cases} t + i & 1 \le t \le 2 \\ 2 + i(t-1) & 2 \le t \le 4 \end{cases}$

   (c) $f = e^{\bar{z}} \quad C\colon\ z = (1+i)t \quad 0 \le t \le 1$

   (d) $f = z\,\mathrm{Re}(z) \quad C\colon\ z = \begin{cases} t & 0 \le t \le 1 \\ 1 + i(t-1) & 1 < t \le 2 \end{cases}$

   (e) $f = z^{-2} \quad C\colon\ z = (1+i)t \quad 1 \le t \le 2$
   (f) $f = z|z|^2 \quad C\colon\ z = t + it^3 \quad -1 \le t \le 1$

   (g) $f = (\bar{z} + i)^{-1} \quad C\colon\ z = i + e^{it} \quad \dfrac{-\pi}{2} \le t \le \dfrac{\pi}{2}$

7. Evaluate $\int_C f(z)\,dz$, where $f(z)$ and C are given below. You might wish to refer to Definition (5.1.18) or the result in (5.1.39).

   (a) $f = \dfrac{1}{z} \quad C\colon\ z = t + (1-2t)i \quad 0 \le t \le 1$

   (b) $f = \dfrac{1}{z} \quad C\colon\ z = \dfrac{t}{2} + (1+t)i \quad -2 \le t \le 0$

   [*Hint:* If you use (5.1.39), be careful of the choice of branch of $\log z$.]
   (c) $f = z^3 e^{z^4} \quad C\colon\ z = e^{it} \quad 0 \le t \le \pi$

   (d) $f = \dfrac{e^{2/z}\sin(1/z)}{z^2} \quad C\colon\ z = e^{it} \quad 0 \le t \le \pi$

   (e) $f = z^i$ (*principal branch*) $\quad C\colon\ z = e^{it} \quad \dfrac{-\pi}{2} \le t \le \dfrac{\pi}{2}$

   (f) $f = z\sin z \quad C\colon\ z = \cos t + 2i\sin t \quad 0 \le t \le 2\pi$

(g) $f = \begin{cases} z & \text{Re}(z) > 0 \\ z^2 & \text{Re}(z) < 0 \end{cases}$ $\quad C: \; z = e^{it} \quad 0 \le t \le 2\pi$

(h) $f = \dfrac{e^z}{1 + e^z}$ $\quad C: \; z = \dfrac{1}{2}e^{it} \quad 0 \le t \le 2\pi$

8. The reader is asked to prove each of the following in order to evaluate the integral $I = \oint_C \sqrt{z}\, dz$, where $C: \; z = e^{it}, \; -\pi \le t \le \pi$, and the principal branch of $\sqrt{z}$ is chosen [see also Example (5.1.28)].

(a) $\sqrt{z}$ is bounded and piecewise continuous on the trace of C.

(b) By Theorem (5.1.34a), I exists. Hence, we can write it as

$$I = \lim_{\epsilon \downarrow 0} \left[\oint_{z_1}^{z_2}{}_{C_\epsilon} \sqrt{z}\, dz \right],$$

where z_1, z_2, and C_ϵ are shown in Figure 5.1.2. Now use (5.1.39) to evaluate the integral over C_ϵ and then take the limit.

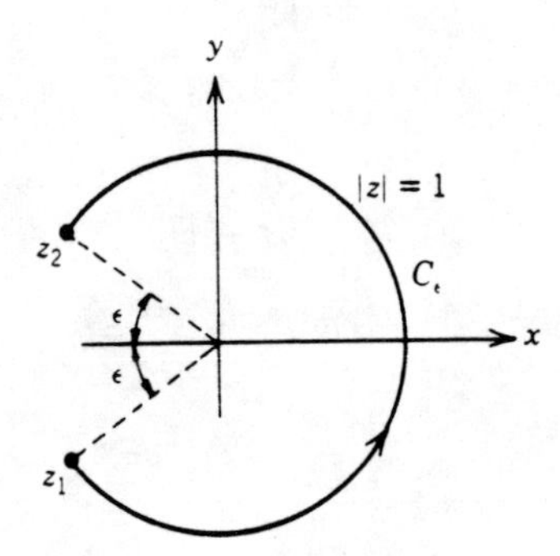

Figure 5.1.2

9. Let C be any simple, smooth, closed curve whose trace is an ellipse and consider $I = \oint_C dz/z$.

(a) Show that $I = 0$ if the origin is not inside of C.

(b) Show that if the origin lies inside of C, the value of I is independent of the ellipse. What is its value? [*Hint:* Use the approach of Problem 8.]

10. (a) Evaluate $\oint dz/(z - z_0)$ over the square having vertices at $\pm 1 \pm i$.

(b) Evaluate $\int_C dz/[z(4 + z^2)]$ over the semicircle $C: \; z = e^{it}, \; -\pi/2 \le t \le \pi/2$. [*Hint:* Use partial fraction expansions.]

(c) Evaluate $\oint dz/(z^3 + 27)$ over the unit circle.

11.* Prove that if $C: \; z = z(t), \; t_0 \le t \le t_1$, is smooth and $F(z)$ is continuously differentiable in a domain containing the trace of C, then $\oint_{z_0}^{z_1} F'(z)\, dz = F(z_1) - F(z_0)$. [*Hint:* Let $F = u + iv$, and $z(t) = x(t) + iy(t)$. Use Definition (5.1.19a) and show by the chain rule that the integrand is the total derivative $(d/dt)[u(x(t), y(t)) + i\, v(x(t), y(t))]$.]

12. Use the triangle inequality for integrals (5.1.35)–(5.1.37) to show that $|\int_C f(z)\,dz| \le K$, where f, C, and K are given below. In some of these you may have need of inequality (1.1.27b).

(a) $f = z/(1+z)$, C is the upper half of the circle $|z| = 2$, $K = 2\pi$.

(b) $f = e^{z^2}$, C is the "broken line" joining $z = 0$ to $z = 1$ to $z = 1 + i$, $K = 2e$.

(c) $f = (\operatorname{Log} z)/z$, C is the unit circle, $K = 2\pi^2$.

(d) $f = e^z/(3 + z^2)$, C is the line joining $z = 0$ to $z = 1 + i$, $K = (e - 1)$. [*Hint:* Show that on C, $|e^z/(3 + z^2)| \le e^x$.]

(e) $f = e^{iz}/(1 + z^2)$, C is the semicircle $z = Re^{it}$, $0 < t < \pi$, with $R > 1$, $K = \pi R/(R^2 - 1)$. [*Hint:* On the semicircle, what is the sign of $\sin t$?]

(f) $f = (z^2 + z + 2)^{-1}$, C is the circle $|z| = R\,(R > 2)$,

$$K = 2\pi R(R^2 - R - 2)^{-1}$$

[*Hint:* On C, $|z^2 + z + 2| \ge |R^2 - |z + 2||$ and $|z + 2| \le R + 2$. Show that $R^2 - (R + 2) \ge 0$, and continue the string of inequalities.]

13.* Prove that if $F(z)$ is continuous differentiable in a domain D with $F'(z) = 0$ there, then F is constant in D. [*Hint:* Fix z_0 and let z be any other point in D. Use (5.1.39) together with the fact that D is connected, to show that $F(z) = F(z_0)$.]

14.* If two contours are equivalent [Definition (4.1.30)], prove that they have the same length.

15.* A complex function $g(t, s)$ of two real variables t and s is said to be **continuous at s_0 uniformly in the interval $t_0 \le t \le t_1$** if, given an $\epsilon > 0$, there is a $\delta(\epsilon)$, *depending only on* ϵ, so that $|g(t, s) - g(t, s_0)| < \epsilon$ when $|s - s_0| < \delta$, for all t in the interval. In Problem 12 of Section 4.1, we defined a continuous deformation of one curve into another. If the function $z(t; s)$ in that problem is such that $(\partial z/\partial t)(t; s)$ is continuous in s, uniformly for $t_0 \le t \le t_1$, prove that the lengths of the deformations vary continuously in s.

16. Make an appropriate change of variables and evaluate the contour integrals $\int_{C_z} f(z)\,dz$, where $f(z)$ and C_z are given below.

(a) $f = \dfrac{z}{1 + z^4}$ $\quad C_z\colon\ z = it \quad 0 \le t \le 1$

(b) $f = \dfrac{e^{1/z}}{z^3}$ $\quad C_z\colon\ z = e^{it} \quad 0 \le t \le 2\pi$

17. If C is a simple, closed contour, show that the area inside $C = -(i/2)\oint_C \bar{z}\,dz$. [*Hint:* Use (5.1.23) together with Green's Theorem from advanced calculus.]

18. Let $\mathbf{F} = F_1\mathbf{i} + F_2\mathbf{j}$ be a **force field** in the plane. The **work** W done by the field on a particle of unit mass traversing the smooth curve C: $x = x(t)$, $y = y(t)$, $t_0 \le t \le t_1$, is the real line integral $\oint \mathbf{F}\cdot\mathbf{T}\,ds$, where $\mathbf{T}$ is the unit tangent $(x'\mathbf{i} + y'\mathbf{j})/(x'^2 + y'^2)^{1/2}$. If $F(z) = F_1 + iF_2$ denotes the complex function corresponding to the vector field $\mathbf{F}$, show that $W = \operatorname{Re}[\oint \bar{F}\,dz]$.

5.2 CAUCHY–GOURSAT THEOREM; INDEFINITE INTEGRALS; MULTIPLY CONNECTED DOMAINS

From (5.1.40), we learn that $\oint_C F'(z)\,dz = 0$ for any function $F(z)$ which is continously differentiable in some domain containing the piecewise smooth, closed curve C. One of the major achievements of complex function theory was to show that $\oint_C f(z)\,dz = 0$ for functions $f(z)$ which are merely analytic inside and on C. The complete result is stated below. However, we will prove it only for a special case using **Green's Theorem** from advanced calculus. The reader is referred to a more advanced text for a more general proof.

(5.2.1) ***Theorem*** **(Cauchy–Goursat).** If $f(z)$ is analytic *inside and on* a simple, closed, piecewise smooth curve C, then

$$\oint_C f(z)\,dz = 0$$

Proof. The special case which we will consider is when $f(z)$ is assumed continuously differentiable.[1] Then, if $f(z) = u(x, y) + i\,v(x, y)$, the real-valued functions u and v are also continuously differentiable. If D denotes the inside of C (See Figure 5.2.1), Green's Theorem states that if $\phi(x, y)$ and $\psi(x, y)$ are any two functions continuously differentiable in D and on C,

$$\oint_C \phi\,dx + \psi\,dy = \iint_D \left(\frac{\partial \psi}{\partial x} - \frac{\partial \phi}{\partial y}\right) dx\,dy \tag{5.2.2}$$

If we now apply this to the two line integrals on the right-hand side of equation (5.1.23) of the previous section, we find

$$\oint_C f(z)\,dz = \iint_D \left[-\left(\frac{\partial v}{\partial x} + \frac{\partial u}{\partial y}\right) + i\left(\frac{\partial u}{\partial x} - \frac{\partial v}{\partial y}\right)\right] dx\,dy \tag{5.2.3}$$

Since f is assumed continuously differentiable in D, the Cauchy–Riemann equations hold. Thus, the integral on the right-hand side above vanishes,

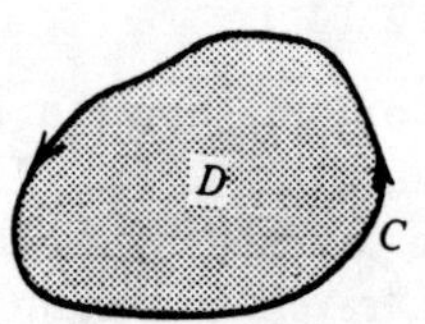

Figure 5.2.1

[1] We remind the reader that analyticity means *only* that the derivative exists—it need not be continuous. In the next section, we will show that in fact analytic functions possess continuous derivatives of all orders.

and we have $_C f(z)\,dz = 0$, which proves the theorem. (The direction of integration does not influence the result, since changing it only changes the sign of the integral and not its zero value.)

■

(5.2.4) ***Example.*** Show that

$$\oint_{|z|=1} \frac{dz}{2+z^4} = 0$$

Solution. The integrand is analytic except at the zeros of $z^4 + 2$, that is, at the four roots $(-2)^{1/4}$. These roots all lie on a circle of radius $2^{1/4}$ and hence outside of the unit circle. Therefore, the integral is zero by the Cauchy–Goursat Theorem.

■

(5.2.5) ***Example.*** Since $\bar{z}$ is nowhere analytic, we *cannot* conclude that $\oint_C \bar{z}\,dz = 0$. In fact, as the reader will easily verify,

$$\oint_{|z|=1} \bar{z}\,dz = 2\pi i$$

■

Let us note that the result of the Cauchy–Goursat Theorem depends not only upon the behavior of $f(z)$ *on* the curve C, but also upon its behavior *inside* of C. For example, even though $1/z$ is analytic *on* the unit circle, and hence $\oint_{|z|=1} (dz/z)$ exists, we cannot conclude that this integral vanishes. This is because $1/z$ is not analytic *inside* of this curve (in fact, the integral is $2\pi i$). However, analyticity, though sufficient, is not necessary for a closed contour integral to vanish. This is evidenced from the result $\oint_{|z|=1} (dz/z^2) = 0$. Here, $1/z^2$ is not analytic inside of the unit circle.

One important corollary of the Cauchy–Goursat Theorem is

(5.2.6) ***Theorem.*** Let $f(z)$ be analytic in some simply connected domain D [see Definition (4.1.28b)].

(a) If C_1 and C_2 are two simple, piecewise smooth curves in D, having the same initial and final points, then

$$\int_{C_1} f(z)\,dz = \int_{C_2} f(z)\,dz \tag{5.2.7}$$

(b) If z_0 and z_1 are any two points in D, then $\oint_{z_0}^{z_1} f(z)\,dz$ is **independent of path** in D, that is, its value is the same no matter what simple, piecewise smooth curve C in D is chosen.

Proof. Note that (b) follows immediately from (a). To prove (a), first assume that, apart from the initial and final points, C_1 and C_2 do not intersect. Then, from Figure 5.2.2*a* the inside of the closed curve consisting of C_1 and $-C_2$ lies in D (since D is simply connected) and the

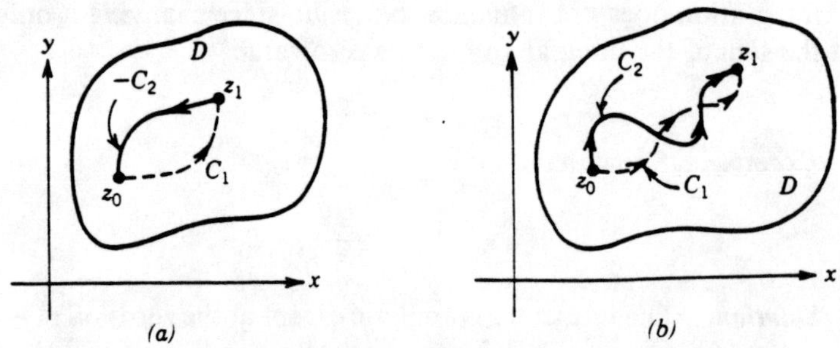

Figure 5.2.2 Curves for Theorem (5.2.6).

hypotheses of the Cauchy–Goursat Theorem are satisfied. Hence,

$$0 = \oint_{C_1 - C_2} f(z)\,dz = \int_{C_1} f(z)\,dz - \int_{C_2} f(z)\,dz$$

from which (5.2.7) follows.

If C_1 and C_2 intersect a finite number of times, then, as in Figure 5.2.2*b*, there will be a finite number of closed curves to which we can apply the above argument. If C_1 and C_2 have an infinite number of intersections, then the analysis is much more subtle and will not be presented here.

■

Let us note the difference between this result and that of Theorem (5.1.47). There, we were also led to the equality of two integrals as in (5.2.7); however, in the previous result, C_1 and C_2 *were to have the same trace.* Here, we merely demand that the integrand be analytic between the two curves.

(5.2.8) ***Example.*** Evaluate $\oint_0^1 z\,dz/(1 + z^2)$, where C is the upper portion of the circle $|z - \frac{1}{2}| = \frac{1}{2}$ as shown in Figure 5.2.3*a*.

Solution. A direct evaluation of this integral, using, for example, the parametrization $z = \frac{1}{2}(1 - e^{it})$, $0 \le t \le \pi$, will lead to some very complicated real integrals. We might also consider using the fact that

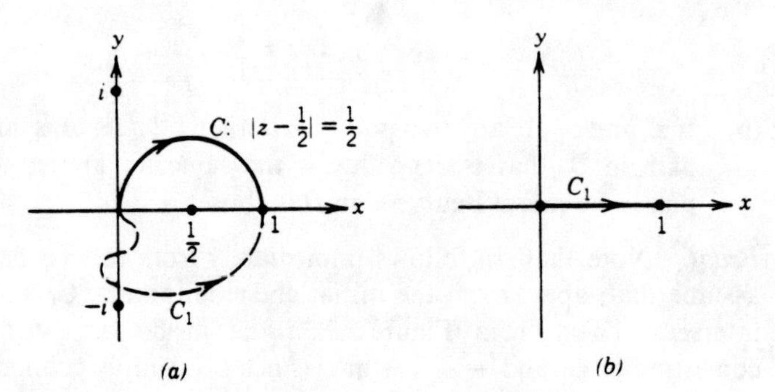

Figure 5.2.3 Example (5.2.8).

$z(1 + z^2)^{-1} = \frac{1}{2}(d/dz)[\log(1 + z^2)]$. However, as we now know, one must be very careful to choose an appropriate branch of the multivalued function $\log(1 + z^2)$.

To use the result of the above theorem, we first note that $z(1 + z^2)^{-1}$ is analytic in any domain not containing $z = \pm i$. Hence, if C_1 is any other curve joining $z = 0$ to $z = 1$ such that $z = \pm i$ is not contained between C_1 and C (see Figure 5.2.3*a*), then, by the above result,

(5.2.9)
$$\oint_0^1 \frac{z\,dz}{1 + z^2} = \oint_{C_1\,0}^{\;\;1} \frac{z\,dz}{1 + z^2}$$

We now pick C_1 (if we can) so that the integral on the right can be easily evaluated. One such choice is the interval $[0, 1]$ of the real axis, as shown in Figure 5.2.3*b*. Then,

$$\oint_0^1 \frac{z\,dz}{1 + z^2} = \int_0^1 \frac{x\,dx}{1 + x^2} = \frac{1}{2}\ln 2$$

where the last integral is an elementary real integral from calculus.

■

The results of the previous theorem have the following very interesting and important interpretation. **If $f(z)$ is analytic on a simple, piecewise smooth curve C joining z_0 to z_1, then we can continuously deform C, keeping its initial and final points fixed, without changing the value of the contour integral $\oint_{z_0}^{z_1} f(z)\,dz$, so long as we do not pass through any singularities of $f(z)$.** The words "continuously deform" can be defined precisely in mathematical terms (see Problem 12 of Section 4.1).

(5.2.10) ***Example.*** To see that the result (5.2.7) need not be true when there are singularities present, consider the two integrals

(5.2.11)
$$\oint_{C_1\,-1}^{\;\;1} \frac{dz}{z} \quad \text{and} \quad \oint_{C_2\,-1}^{\;\;1} \frac{dz}{z}$$

where C_1 and C_2 are the upper and lower portions of the unit circle, respectively, parametrized by

(5.2.12)
$$C_1:\ z = e^{-it} \qquad -\pi \le t \le 0$$
$$C_2:\ z = e^{it} \qquad \pi \le t \le 2\pi$$

Then,

$$\oint_{C_1\,-1}^{\;\;1} \frac{dz}{z} = \int_{-\pi}^{0} \frac{-ie^{-it}}{e^{-it}}\,dt = -\pi i$$

while

$$\oint_{C_2\,-1}^{\;\;1} \frac{dz}{z} = \pi i$$

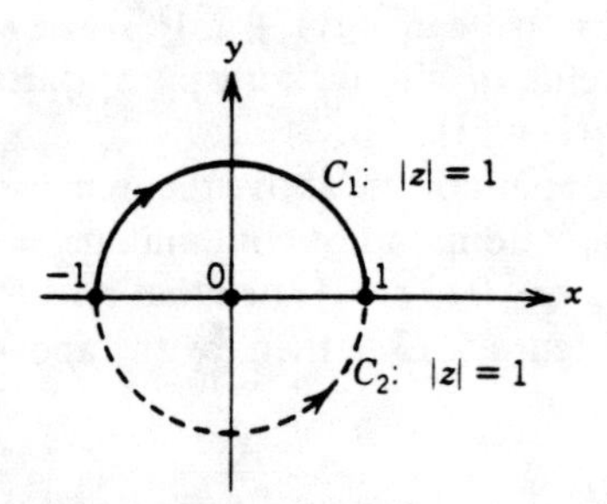

Figure 5.2.4 Example (5.2.10).

Note from Figure 5.2.4 that C_1 cannot be continuously deformed into C_2 without passing through the singularity of the integrand at $z = 0$.

Indefinite Integrals. We will now show that in appropriate regions, every analytic function is the derivative of another analytic function.

(5.2.13) ***Theorem.*** Let D be a simply connected domain and let $f(z)$ be analytic in D. Then,

(a) If z_0 is a fixed point in D, the function

(5.2.14)
$$F(z) = \int_{z_0}^{z} f(\zeta)\, d\zeta$$

is defined and single-valued for all z in D. By the integral we mean to integrate over *any* simple, piecewise smooth curve in D joining z_0 to z_1

(b) $F(z)$ in (5.2.14) is analytic in D and is an **indefinite integral** of $f(z)$, that is,

(5.2.15)
$$F'(z) = f(z)$$

Hence, by (5.1.39), if z_1 is also in D,

(5.2.16)
$$\int_{z_0}^{z_1} f(z)\, dz = F(z_1) - F(z_0)$$

Proof. Part (a) follows from Theorem (5.2.6b), which states that since F is independent of path f is analytic in D. For part (b), we wish to show, using the definition of derivative, that

(5.2.17)
$$\lim_{h \to 0} \left| \frac{F(z + h) - F(z)}{h} - f(z) \right| = 0$$

Now, since D is a domain, and hence an open set (see Section 1.1), if z is in D and $|h|$ is small enough, $z + h$ is also in D. Thus, from (5.2.14),

$$F(z + h) - F(z) = \int_{z_0}^{z+h} f(\zeta)\, d\zeta - \int_{z_0}^{z} f(\zeta)\, d\zeta = \int_{z}^{z+h} f(\zeta)\, d\zeta$$

If we write

$$f(z) = \frac{1}{h}\int_z^{z+h} f(z)\,d\zeta \qquad \text{(the integration variable is } \zeta\text{)}$$

and combine these two results, we get

$$\left|\frac{F(z+h)-F(z)}{h} - f(z)\right| = \frac{1}{|h|}\left|\int_z^{z+h} [f(\zeta) - f(z)]\,d\zeta\right| \tag{5.2.18}$$

Now, if $|h|$ is small enough, we can take as the curve in the integral on the right the line joining z and $z + h$. Since this line has length $|h|$ (see Figure 5.2.5) by the triangle inequality for integrals

$$\left|\frac{F(z+h)-F(z)}{h} - f(z)\right| \le \frac{M_h|h|}{|h|} = M_h \tag{5.2.19}$$

where M_h is the maximum value of $|f(\zeta) - f(z)|$ for ζ in the disk $|\zeta - z| \le |h|$. Hence, if $|h|$ is small enough, M_h can be made arbitrarily small with $|h|$ by the continuity of f. Thus, the limit (5.2.17) is established.

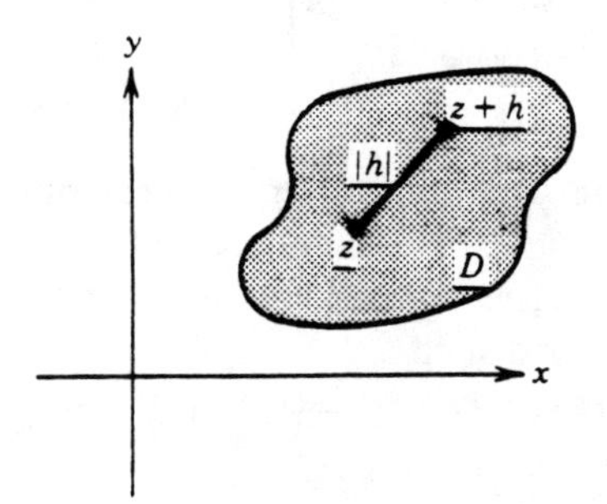

Figure 5.2.5

(5.2.20) ***Example.*** Evaluate the real integral

$$I = \int_0^1 \frac{dx}{x^2+1}$$

Solution. From elementary calculus we can easily find that $I = \pi/4$. However, the point here is to use *complex integration* to evaluate I and therefore indicate how to evaluate more complicated integrals. First, with a partial fraction expansion of the integrand, we can write

$$I = \frac{1}{2i}\oint_0^1 \left(\frac{1}{z-i} - \frac{1}{z+i}\right) dz$$

where the integral is taken over the real interval (0, 1) of the z-plane (see Figure 5.2.6). Since $(z - i)^{-1}$ is analytic in any domain not containing

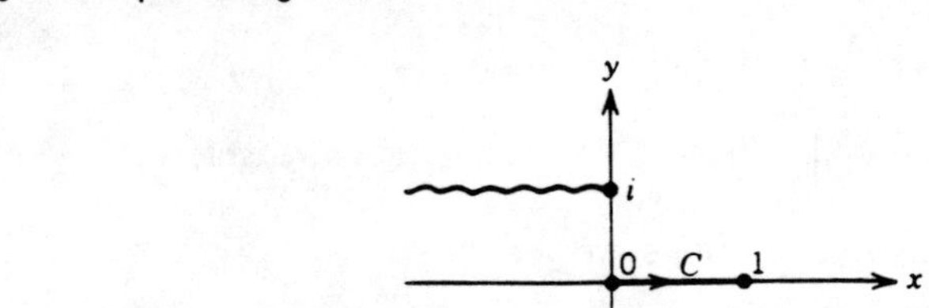

Figure 5.2.6 Branch cuts for Example (5.2.20).

$z = i$, it must, by the previous theorem, have an analytic indefinite integral in some neighborhood of the real axis. *Any* branch of log $(z - i)$ whose branch cut does not intersect the real interval (0, 1) will serve our purposes. A similar argument holds for $(z + i)^{-1}$. Since the principal logarithms have these properties (why?), with the branch cuts shown in Figure 5.2.6, we have

(5.2.21)
$$I = \frac{1}{2i}\left[\operatorname{Log}(z - i) - \operatorname{Log}(z + i)\right]\Bigg|_0^1 = \frac{\pi}{4}$$

■

Multiply Connected Domains. One shortcoming of the Cauchy–Goursat Theorem is that it need not be true for curves in multiply connected domains. This is so because in the hypotheses of that theorem we demand that $f(z)$ be analytic *inside* a simple closed curve, and such a domain is *simply connected.* In spite of this, the Cauchy–Goursat Theorem can be used to prove an important result for multiply connected domains.

(5.2.22) ***Theorem.*** **(Cauchy–Goursat for Multiply Connected Domains).** Let C_0, $C_1, C_2, \ldots, C_N$ be simple, closed, piecewise smooth, nonintersecting curves such that each C_k, $k = 1, 2, \ldots, N$, lies inside of C_0, as shown in Figure 5.2.7*a*. Also, let $f(z)$ be analytic both in the multiply connected

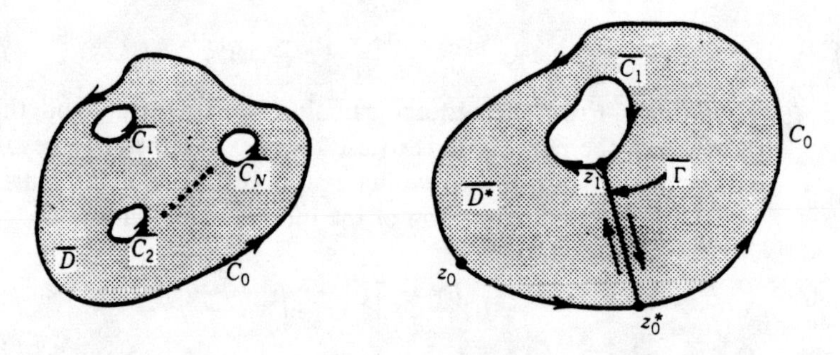

(a) Multiply connected domain *D*. *(b)* Simply connected domain *D** cut out of *D*.

Figure 5.2.7 Cauchy–Goursat Theorem for multiply connected domains.

domain D contained between these curves and in some neighborhood of each of the curves. Then,

(5.2.23)
$$\oint_{C_0} f(z)\,dz = \sum_{k=1}^{N} \oint_{C_k} f(z)\,dz$$

(Note the *same* orientation on all the contour integrals. It could have also been clockwise.)

Proof. We will prove this result only for two curves C_0 and C_1 as shown in Figure 5.2.7*b*. If we connect C_0 and C_1 with a line Γ and denote by D^* the domain D with the points on Γ omitted, then D^* is simply connected and bounded by a curve C which we will traverse as follows (see the figure). First, start at some point z_0 on C_0, proceed along C_0 counterclockwise until z_0^* on Γ is reached. Then, traverse Γ until z_1 on C_1 is reached, go around C_1 *clockwise* until z_1 is again reached, and proceed back along Γ (in the opposite direction) to z_0^*. Finally, continue along C_0 counterclockwise until the initial point z_0 is again reached. By the hypothesis of our theorem, $f(z)$ is analytic in D^* and in some neighborhood of its boundary C. Hence, by the Cauchy–Goursat Theorem,

$$\oint_{C} f(z)\,dz = 0$$

which by the above description of C and the fact that *the integrals along* Γ *cancel*, yields

(5.2.24)
$$\oint_{C_0} f(z)\,dz + \oint_{C_1} f(z)\,dz = 0$$

(Note the *clockwise* orientation on C_1.) Hence,

$$\oint_{C_0} f(z)\,dz = -\oint_{C_1} f(z)\,dz = \oint_{C_1} f(z)\,dz$$

which is our result for $N = 1$. The result for $N > 1$ is proved (in exactly the same way) by joining *each* of the interior curves C_k to C_0 with a curve Γ_k.

■

Just as we noted in Theorem (5.2.6) [see the paragraph preceding Example (5.2.10)], there is an interesting geometric interpretation of this result. Here, **if $f(z)$ is analytic on a closed curve C, we can continuously deform C without changing the value of the integral of f, so long as the deformations do not pass through any singular points of f** (also see Problem 3*a*). In the previous case, we required that all of the deformations have the same initial and final points. That is not necessary here. For example, consider $\oint_C f(z)\,dz$, where f is analytic in the interior of C except for two singular points z_1 and z_2. Now the sequence of deformations is depicted in Figure 5.2.8. In this sequence, C is squeezed over and around z_1 and z_2, keeping its

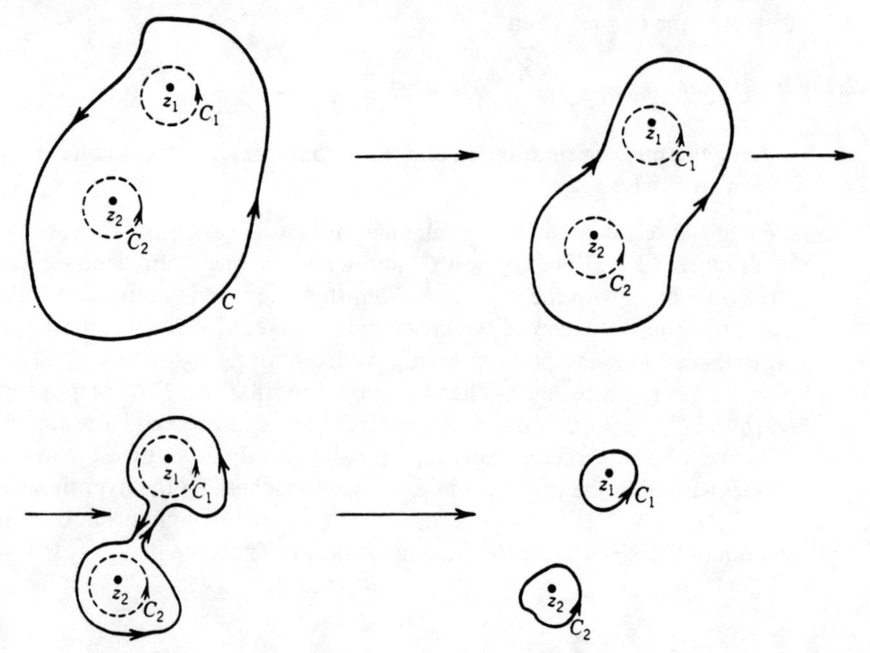

Figure 5.2.8 Deformation of contour interpretation of Theorem (5.2.22).

counterclockwise orientation, until finally the integral over the middle pieces cancel and we are left just with the integrals over C_1 and C_2. This is the result (5.2.23).

(5.2.25) ***Example.*** Evaluate the integral $I = \oint_C (dz/z)$, where C is the ellipse $x^2 + \frac{1}{4}y^2 = 1$.

Solution. Note that the integrand is not analytic inside of the ellipse and so we cannot use the Cauchy–Goursat Theorem. We might try to evaluate I directly by using the parametrization $x = \cos t$, $y = 2\sin t$, $0 \le t \le 2\pi$. We would then arrive at

$$I = 3\int_0^{2\pi} \frac{\sin t \cos t}{\cos^2 t + 4\sin^2 t}\,dt + 2i\int_0^{2\pi} \frac{dt}{\cos^2 t + 4\sin^2 t} \tag{5.2.26}$$

The second of these integrals is not elementary and it is quite difficult to evaluate.

Instead of proceeding in this manner, we note from the theorem that the integral over the ellipse is the same as that over *any* other curve which includes the singular point $z = 0$ in its interior. We now try to choose this new curve so that the integral is easily evaluated. In this case, the choice of the circle $|z| = \epsilon$ will work, for then we have

$$I = \oint_{|z|=\epsilon} \frac{dz}{z} = \int_0^{2\pi} \frac{i\epsilon e^{it}\,dt}{\epsilon e^{it}} = 2\pi i \tag{5.2.27}$$

(We could also have chosen, for example, a rectangle, but the integral would have been more difficult to evaluate.) Note that ϵ, the radius of the circle, does not appear in the final answer. This is as it should be, since the integral is to be independent of the curve chosen. Geometrically, as noted above, we can interpret the circle as a continuous deformation of the ellipse without passing through the singularity at $z = 0$.

■

(5.2.28) ***Example.*** Evaluate $I = \oint_C dz/[z(z-1)]$, where C is the rectangle in Figure 5.2.9.

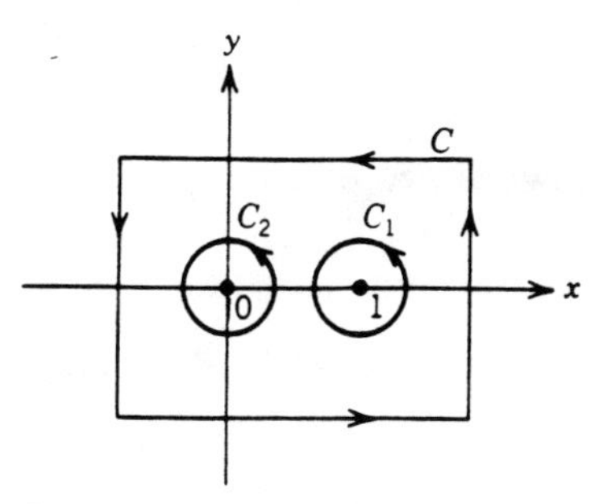

Figure 5.2.9 Example (5.2.28).

Solution. As in the previous example, we could deform the rectangle C into two curves, each surrounding only one of the singular points. However, unlike in the previous example, it is certainly not obvious what special choice of these two curves would permit us to evaluate the resulting integrals. The problem lies in the fact that the *integrand* involves *both* singular points at the same time. To get around this, we first *separate the two singular points* in the integrand by the partial fraction expansion $1/[z(z-1)] = 1/(z-1) - 1/z$. Then,

(5.2.29)
$$I = \oint_C \frac{dz}{z-1} - \oint_C \frac{dz}{z}$$

Note that *now* each integrand has only *one* singular point inside of C. Hence, following the previous example, we deform C in the first integral into a circle C_1 surrounding $z = 1$, and in the second integral into a circle C_2 surrounding $z = 0$ (see the figure). Then,

$$I = \oint_{C_1} \frac{dz}{z-1} - \oint_{C_2} \frac{dz}{z} = 2\pi i - 2\pi i = 0$$

■

Note that the procedure used in the previous examples can also be used to evaluate the integral of any rational function. We first do a partial fraction decomposition thus separating the singular points from each other. Then, in each of the resulting integrals, we deform the contour into a circle surrounding the one singular point of the integrand. However, the results of this section cannot be used for more general integrals. For example, consider

$$I = \oint_C \frac{e^z}{z-1}\,dz \tag{5.2.30}$$

where C is a curve having $z = 1$ in its interior. Due to the exponential e^z, there is no obvious choice of a curve into which we can deform C and be able to evaluate the resulting integral. Cases such as these will be taken up in the next section, where we discuss the remarkable Cauchy integral formula.

PROBLEMS (The answers are given on page 434.)

(Problems with an asterisk are more theoretical.)

1. Use the Cauchy–Goursat Theorem to show that the following integrals are zero.

   (a) $\oint_{|z|=1} e^{\sin(z^2)}\,dz$ (b) $\oint_{|z|=1} (\tan z)\,dz$

   (c) $\oint_C \operatorname{Log}(1-z)\,dz$, C is the square with vertices at $\pm\frac{1}{2} \pm \frac{i}{2}$

   (d) $\oint_{|z|=2} \frac{dz}{\bar{z}}$ (e) $\oint_{|z|=2} \frac{e^z}{z^2+6}\,dz$

   (f) $\oint_C \frac{dz}{z^3+1}$, C is the triangle with vertices at $0, \pm\frac{1}{4} + \frac{i}{2}$

   (g) $\oint_C \csc z\,dz$, C is the ellipse $x^2 + 20\left(y - \frac{1}{4}\right)^2 = 1$

2. For the functions $f(z)$ given below, where in the z-plane is $\int f(z)\,dz$ independent of path, evaluate $\int_C f(z)\,dz$, where C is given below.
   (a) $f(z) = 1/z$, C is the left half of the unit circle joining $-i$ to i
   (b) $f(z) = z\sin(z^2)$, C is the line joining $z = 0$ to $z = 1 + 2i$
   (c) $f(z) = \operatorname{Log} z/z$, C is shown in Figure 5.2.10*a*
   (d) $f(z) = (\text{principal})\ z^i$, C as in Figure 5.2.10*b*

3.* Here, the reader is asked to prove a special case of the Cauchy–Goursat Theorem using the concept of the continuous deformation of a contour.
   (a) Let C_0 and C_1 be two contours which can be continuously deformed into one another in a domain D (see Problem 12 in Section 4.1), and let $f(z)$ be continuously differentiable in D. If the deformation function $z(t, s)$ has continuous second-order partial derivatives with respect to t and s, show

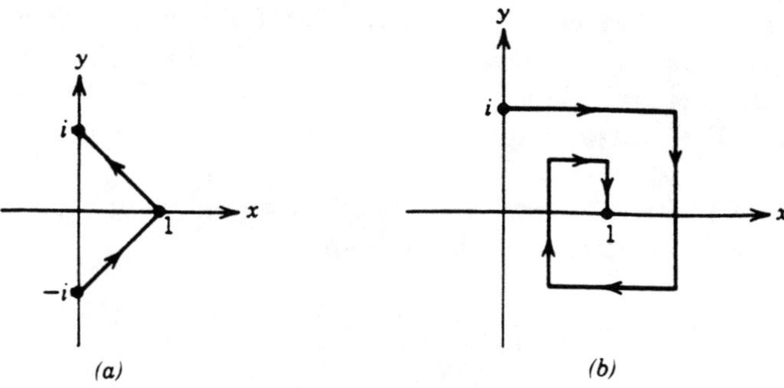

Figure 5.2.10

that $\int_{C_0} f(z)\,dz = \int_{C_1} f(z)\,dz$. [*Hint:* Define the curve C_s by C_s: $z = z(t, s)$, $t_0 \le t \le t_1$, and the function $g(s)$ by $g(s) = \int_{C_s} f(z)\,dz$. Show that

$$\frac{\partial}{\partial s}\left[f(z(t,s))\cdot\frac{\partial z}{\partial t}(t,s)\right] = \frac{\partial}{\partial t}\left[f(z(t,s))\cdot\frac{\partial z}{\partial s}(t,s)\right]$$

Then, compute $g'(s)$ using Leibnitz' rule.]

(b) Prove that if $f(z)$ is continuously differentiable in a simply connected domain D, and if C: $z = z(t)$, $t_0 \le t \le t_1$, is a simple closed contour with $z''(t)$ continuous, then $\oint_C f(z)\,dz = 0$. [*Hint:* Exhibit a function $z(t, s)$ meeting the conditions in part (a) such that C can be deformed to a point.]

4. (a) Let $I = \oint_C \sqrt{z}\,dz$, where the principal branch is used, and let C be as in Figure 5.2.11*a*.

(i) Why is the Cauchy–Goursat Theorem not applicable here?

(ii) Show that $I = 0$ in spite of the result in (i).

(b) Show that $\oint_{\hat{C}} \sqrt{z}\,dz = 0$, where $\hat{C}$ is shown in Figure 5.2.11*b*.

(c) In Problem 8 of Section 5.1, we showed that $\oint_{C'} \sqrt{z}\,dz \ne 0$, where C' is as in Figure 5.2.11*c*. Why do you think the conclusion of the Cauchy–Goursat Theorem was valid for parts (a) and (b) but not here?

5.* The integrand need only be continuous on C, and not necessarily analytic, for the conclusion of the Cauchy–Goursat theorem to hold. We will prove this

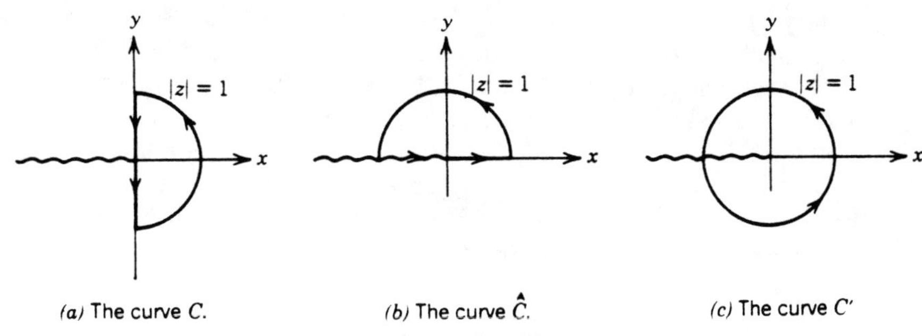

(a) The curve C. (b) The curve $\hat{C}$. (c) The curve C'

Figure 5.2.11

for the special case of a circle. Prove that if $f(z)$ is analytic in $|z| < R$ and *continuous* on $|z| = R$, then $\oint_{|z|=R} f(z)\,dz = 0$. [*Hint:* For $\epsilon > 0$ and small enough, the circle $|z| = R - \epsilon$ is inside and close to $|z| = R$. Use the Cauchy–Goursat Theorem to compute the contour integral over this circle, and then let $\epsilon \to 0$.]

6. Evaluate $\oint_C f(z)\,dz$, where f and C are given below (you may have to do a partial fraction expansion in some cases).

   (a) $f = \dfrac{1}{(z-1)(z-i)^2}$, $\quad C$: $|z| = 2$

   (b) $f = \dfrac{z^n}{(z-1)^k}$, $\quad C$ is the ellipse $\dfrac{x^2}{4} + \dfrac{y^2}{9} = 1$, and n and k are positive integers

   (c) $f = z\,e^{1/(z-2)}$, $\quad C$ is the square with vertices at $z = \pm 1 \pm i$

   (d) $f = \dfrac{1}{1+z^5}$, $\quad C$ is $|z| = 1/2$

   (e) $f = \dfrac{z^2+z+1}{z^3 - iz^2 + 4z - 4i}$, $\quad C$ is $\left|z - \dfrac{i}{2}\right| = 1$

   (f) $f = \dfrac{z+2}{z^3 + 3z^2 + 3z + 1}$, $\quad C$ is $|z| = 10$

   (g) $f = \dfrac{z}{(z^2+4)^3}$, $\quad C$ is $|z - 1| = 5$

7. Let $f(z)$ be analytic everywhere, except *possibly* at $z = 1 \pm i$. Let C_1 be the square with vertices at $z = \pm 3 \pm 3i$ and C_2 the circle $|z - (1+i)| = 1$. If $\oint_{C_1} f(z)\,dz = 2$ and $\oint_{C_2} f(z)\,dz = 3$, why can you conclude that $f(z)$ cannot be analytic at $z = 1 - i$?

8. Let $f(z)$ be analytic everywhere, except *possibly* at $z = \pm i$. Let C_1 be a rectangle with vertices at $1 \pm 2i$ and $-1 \pm 2i$, C_2 be the circle $|z - i/2| = 1$, C_3 the ellipse $4x^2 + 9y^2 = 1$, C_4 the circle $|z + i/2| = 1$, and C_5 the square with vertices at $z = \pm 1$, $1 + 2i$, and $-1 + 2i$. Let the numbers K_j be defined by $K_j = \oint_{C_j} f(z)\,dz$, $j = 1, 2, \ldots, 5$. If $K_1 = 1$ and $K_2 = 2$, find K_3, K_4 and K_5. In fact, what can you conclude about the analyticity of f at $z = \pm i$?

9. Let $f(z)$ be analytic everywhere, except *possibly* at points inside the regions D_1 and D_2 shown shaded in Figure 5.2.12. The curves $C_1, C_2, \ldots, C_6$ are shown in the figure. Let the numbers I_k be defined by $I_k = \oint_{C_k} f(z)\,dz$, $k = 1, 2, \ldots, 6$. If $I_1 = 2$ and $I_2 = 5$, compute I_3, I_4, I_5, and I_6.

10. Let $p(z) = \sum_{k=0}^{N} c_k z^k$ be a polynomial of degree N. Let C be a simple, closed contour and let z_0 be a point inside of C.

    (a) Show that

$$\oint_C \frac{p(z)}{z - z_0}\,dz = 2\pi i\, p(z_0)$$

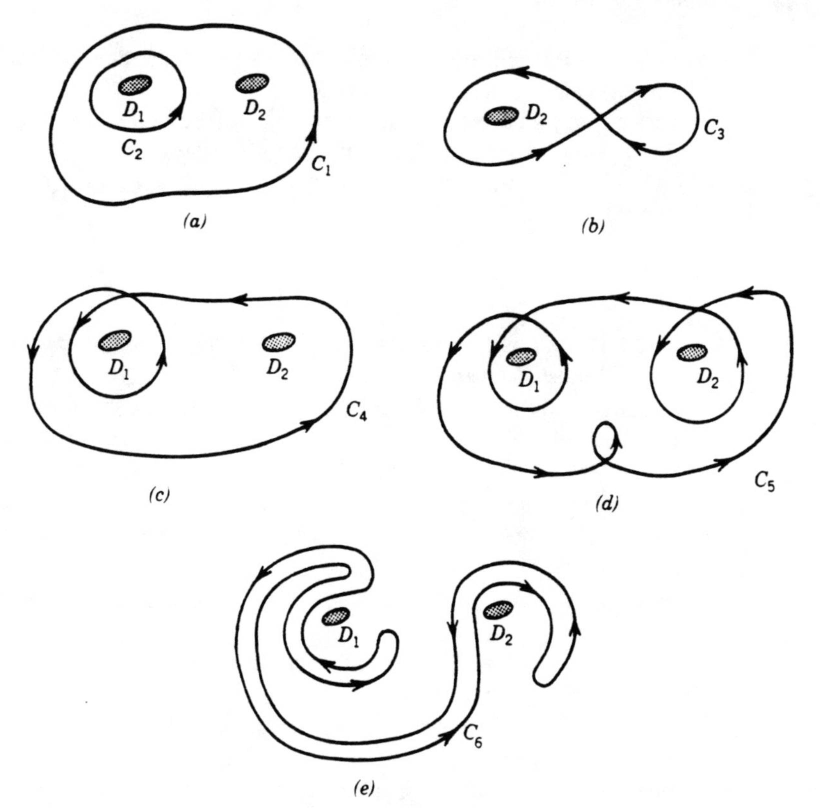

Figure 5.2.12

[*Hint:* First deform C into a circle about $z = z_0$, then write $z = (z - z_0) + z_0$, and finally expand z^k in powers of $z - z_0$ using the **binomial formula**

$$(a + b)^k = \sum_{l=0}^{k} \frac{k!a^l b^{k-l}}{l!(k-l)!}$$

This is a form of the Cauchy integral formula to be discussed in the next section.

(b) Use the hint in (a) to show that

$$\oint_C \frac{p(z)\,dz}{(z - z_0)^{n+1}} = \frac{2\pi i}{n!} p^{(n)}(z_0) \qquad n = 0,1,2,\ldots$$

where $p^{(n)}$ denotes the nth derivative of p.

11.* Here, we will lead the reader through a series of steps which ends with the **fundamental theorem of algebra** which states that every nonconstant polynomial has a complex root. Let $p(z) = \sum_{k=0}^{N} c_k z^k$, with $a_N \neq 0$ and $N \geq 1$.

(a) Show that $\lim_{z \to \infty} [p(z)/z^N] = c_N$.

(b) Use (a) and the definition of limit to show that there are constants K and R so that $|p(z)| \geq K|z|^N$ for $|z| > R$.

(c) Assume that $p(z)$ does not vanish anywhere (*the contradiction of what we wish to prove*). Then, the function $f(z) = 1/p(z)$ is entire. Can you define $[f(z) - f(0)]/z$ at $z = 0$ so that the resulting function is also entire?

(d) From (c),

$$\oint_{|z|=\hat{R}} \frac{f(z) - f(0)}{z}\, dz = 0$$

where $\hat{R}$ is any positive number (why?). Use this to conclude that $f(0) = 0$. [*Hint:* Choose $\hat{R}$ large enough, and solve for $f(0)$. Now use the triangle inequality and (b) to show that $f(0)$ can be made as small as desired.]

(e) Now conclude that since $f(0) \neq 0$, $p(z)$ must vanish somewhere.

12.* Prove that every polynomial $p(z) = \sum_{k=0}^{N} c_k z^k$, $c_N \neq 0$, can be factored into $c_N(z - z_1)(z - z_2)\cdots(z - z_N)$ where the roots $z_1, z_2, \ldots, z_N$ need not be distinct. [*Hint:* Use the result of Problem 11 and the fact that $z^n - \alpha^n = (z - \alpha)(z^{n-1} + \alpha z^{n-2} + \cdots + \alpha^{n-1})$.]

13.* (a) Prove that if $f(z)$ is analytic outside of a simple, closed, piecewise smooth curve C, and if for $|z|$ large enough we have $|f(z)| \leq K/|z|^2$, then $\oint_C f(z)\, dz = 0$. [*Hint:* Deform C to a circle of very large radius and use the bound on $f(z)$ to show that the integral can be made arbitrarily small.]

(b) If C is a simple, closed contour and $p(z)$ and $q(z)$ are polynomials, what are sufficient conditions so that $\int_C [p(z)/q(z)]\, dz = 0$?

14.* Let $p(z)$ be the polynomial $(z - z_1)^{n_1}(z - z_2)^{n_2}\cdots(z - z_N)^{n_N}$, with distinct roots $z_1, z_2, \ldots, z_N$, where $n_1, n_2, \ldots, n_N$ are positive integers. Let C be a simple, closed contour not passing through any of the roots $z_1, z_2, \ldots, z_N$. Show that $\int_C [p'(z)/p(z)]\, dz = 2\pi i \sum_k n_k$, where the the sum is taken over those roots of p lying inside of C. [*Hint:* Compute the general logarithm $\log[p(z)]$ and then differentiate both sides to find a simple expression for the **logarithmic derivative** p'/p. Now use the Cauchy–Goursat Theorem.]

15. This problem shows that, as in the interpretation of Theorem (5.2.22), we can also **deform infinite contours** continuously without changing the value of the integral. We must, however, be careful of the behavior of the integrand at $z = \infty$. Hence, consider $\int_C e^{-z^2}\, dz$, where C is the infinite contour shown in Figure 5.2.13*a*. We would like to deform C to real axis, since from calculus we know that $\int_{-\infty}^{\infty} e^{-x^2}\, dx = \sqrt{\pi}$.

(a) Show that the improper integral $\int_C e^{-z^2}\, dz$ exists. [*Hint:* Bound $|e^{-z^2}|$ on C by an integrable function.]

(b) Consider the closed contour C_R in Figure 5.2.13*b*. Show that $\lim_{R\to\infty} [\int e^{-z^2}\, dz] = 0$, where this integral is taken over the two vertical lines in the figure. Hence, use the Cauchy–Goursat Theorem to conclude that the value of $\int_C e^{-z^2}\, dz$ is the same as when C is deformed to the real axis.

(c) Consider the dotted contour $\hat{C}$ in Figure 5.2.13*c*. Show that there are values of θ_0 for which $\int_{\hat{C}} e^{-z^2}\, dz$ does not exist. Hence, even though C can

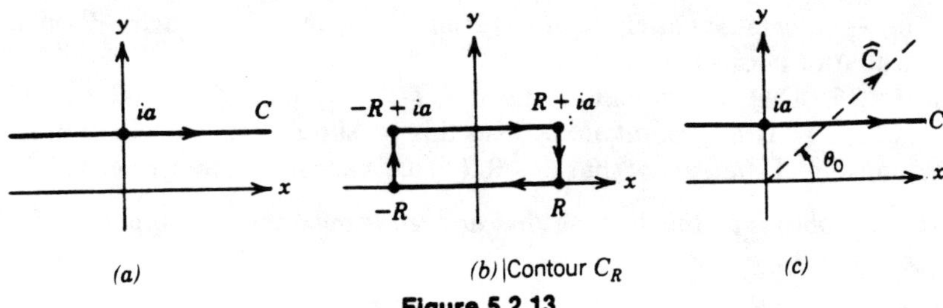

Figure 5.2.13

be deformed continuously into $\hat{C}$, the value of the integral does not remain the same.

16. Use the results of Problems 15a, b to show that $\int_{-\infty}^{\infty} e^{-x^2} \cdot \cos(ax)\,dx = \sqrt{\pi}e^{-a^2/4}$, where a is real. [*Hint:* Write $\cos(ax)$ in terms of exponentials, complete the squares, and change variables.]

17. The **Laplace transform** of $t^{-1/2}$ is defined by $f(s) = \int_0^{\infty} e^{-st}/\sqrt{t}\,dt$ where s is complex.
    (a) Show that $f(s)$ is defined only for $\text{Re}(s) > 0$.
    (b) Letting $t = x^2$, we get $f(s) = 2\int_0^{\infty} e^{-sx^2}\,dx$. A natural complex change of variables would now be $w = \sqrt{s}x$, where a branch $\sqrt{s}$ must be chosen. Show that if the principal branch is picked, then $f(s) = (2/\sqrt{s})\int_C e^{-w^2}\,dw$, where C is the ray $\arg(w) = \frac{1}{2}\text{Arg}(s)$.
    (c) Show that for $\text{Re}(s) > 0$, the contour C above can be deformed to the real axis and $\int_C e^{-w^2}\,dw = \int_0^{\infty} e^{-x^2}\,dx$. Hence, the integral is independent of $\text{Arg}(s)$ and, from calculus, is given by $\sqrt{\pi}/2$. Thus, for $\text{Re}(s) > 0$, $f(s) = \sqrt{\pi}/\sqrt{s}$, where the principal branch of $s^{1/2}$ is used. [*Hint:* Consider the closed contour in Figure 5.2.14 and let $R \to \infty$. You should also consider the two cases $\text{Arg}(s) > 0$ and $\text{Arg}(s) < 0$.]

18.* Let $u(x, y)$ be a real harmonic function ($u_{xx} + u_{yy} = 0$) which possesses continuous second partial derivatives in a simply connected domain D. We will prove that, *up to an additive constant*, u is the real part of an analytic function and thus possesses a harmonic conjugate (see Section 3.3).
    (a) If u were the real part of an analytic function, the derivative of that function would be $u_x - iu_y$. Show that the complex function $g(z) =$

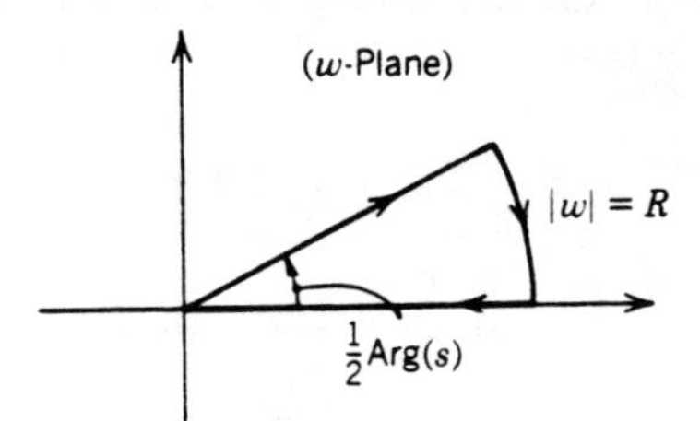

Figure 5.2.14

$u_x - iu_y$ is in fact analytic in D. [*Hint:* Verify that the Cauchy–Riemann equations hold.]

(b) By (5.2.14b), the indefinite integral $f(z) = \int_{z_0}^{z} g(\zeta)\, d\zeta$ is analytic in D, where z_0 is any point in D. Use this to show that $u(x, y) = \text{Re}(f) +$ constant. [*Hint:* show that $u - \text{Re}(f)$ has vanishing x and y derivatives.]

(The next two problems pertain to fluid flow and relate to material in Appendix C and Section 3.4.)

19. Let C: $x = x(t)$, $y = y(t)$, $t_0 \le t \le t_1$ be a simple, piecewise smooth closed curve with unit tangent and normal vectors $\mathbf{T} = (x'^2 + y'^2)^{-1/2}(x'\mathbf{i} + y'\mathbf{j})$ and $\mathbf{n} = (x'^2 + y'^2)^{-1/2}(y'\mathbf{i} - x'\mathbf{j})$, respectively. If $\mathbf{v}$ is a fluid velocity vector (see Appendix C), the **circulation** Γ around C and the **flow** Ω across C are defined by the line integrals

$$\Gamma = \oint_C \mathbf{v} \cdot \mathbf{T}\, ds \quad \text{and} \quad \Omega = \oint_C \mathbf{v} \cdot \mathbf{n}\, ds$$

(a) If $\Phi(z)$ is the **complex potential** for the flow ($V = \overline{\Phi'}$), verify that

$$\oint_C \Phi'(z)\, dz = \Gamma + i\Omega$$

(b) If Φ' is analytic inside and on C, why must the circulation and flow both vanish?

(c) Physically, when do you think $\Phi(z)$ will satisfy the condition in (b)?

(d) If C_1 and C_2 are two simple, closed, piecewise smooth curves, one being inside the other, and if $\Phi'(z)$ is analytic in the region between them, why are the circulation and flow the same for these curves?

20. Consider a rigid body bounded by the curve C of the previous problem and immersed in a fluid flow having complex potential $\Phi(z) = \phi(x, y) + i\psi(x, y)$. By Theorem (3.4.10), the net force on the body is given by the line integral

$$\mathbf{F} = -\rho \oint_C [k - \tfrac{1}{2}|\Phi'|^2]\mathbf{n}\, ds$$

where $\mathbf{n}$ is the unit outer normal to C and k is a constant.

(a) Let F be the complex number represented by the vector $\mathbf{F}$. Use the expression for $\mathbf{n}$ in the previous problem, and the definition of contour integral, to show that

$$F = i\rho \oint_C [k - \tfrac{1}{2}|\Phi'(z)|^2]\, dz$$

(b) Use the Cauchy–Goursat Theorem to show that the above becomes

$$F = \frac{-i\rho}{2} \oint_C |\Phi'(z)|^2\, dz$$

(c) Assume that $\Phi'(z)$ is analytic in some neighborhood of C parametrized by C: $z = x(t) + i\,y(t)$, $t_0 \le t \le t_1$. Show that, on C,

$$\Phi'(z) \cdot \frac{dz}{dt} = \frac{d}{dt}[\phi(x(t), y(t))] + i\frac{d}{dt}[\psi(x(t), y(t))]$$

[*Hint:* Use the Cauchy–Riemann equations, the chain rule, and the fact that $\Phi'(z) = \phi_x + i\psi_x$].

(d) Use the fact that C is a **streamline**, and the result in (c), to show that $\Phi'(z) \cdot (dz/dt)$ is real on C.

(e) Use the above and the result in (b) to prove the **Theorem of Blasius**

$$F = \overline{\frac{i\rho}{2}\oint_C (\Phi')^2\,dz}$$

(f) Show that if $\Phi(z)$ is analytic inside and on C, then there is no net force on the body.

21. Here, we will take the reader through a series of steps which will lead to the **Goursat representation for solutions to the biharmonic equation**. This equation is $\nabla^2(\nabla^2 u) = 0$, where ∇^2 is the Laplacian (see Section 3.3), and arises in the **theory of elasticity**.

(a) From the biharmonic equation, conclude that $\nabla^2 u$ is harmonic and, hence, that

$$\nabla^2 u = \mathrm{Re}[H(z)]$$

where $H(z)$ is analytic.

(b) From the results of this section, the indefinite integral $G(z) = \int^z H(\zeta)\,d\zeta$ is also analytic. Show that

$$\nabla^2[\bar{z}G(z)] = 4H(z)$$

[*Hint:* A "quickie" way is to use the results of Problem 5 of Section 3.3 where it is shown that $\nabla^2 = 4\partial^2/\partial z\,\partial\bar{z}$.]

(c) Use the results in (a) and (b) to conclude that the general solution to the biharmonic is given in terms of two analytic functions $F(z)$ and $G(z)$ by $u(x, y) = \mathrm{Re}[F(z) + \frac{1}{4}\bar{z}\,G(z)]$. [*Hint:* Show that $u - \mathrm{Re}(\frac{1}{4}\bar{z}G)$ is harmonic.]

5.3 THE CAUCHY INTEGRAL FORMULA

From the Cauchy–Goursat Theorem, we know that if $f(z)$ is analytic inside and on a closed curve C, then $\oint_C f(z)\,dz = 0$. Here, we will show that $\oint_C f(z)\,dz$ can be easily evaluated even when f has a certain type of singularity inside of C, and we will discuss some very remarkable consequences of these results.

(5.3.1) ***Theorem* (Cauchy Integral Formula).** Let $g(z)$ be analytic in a simply connected domain D, and let C be a simple, closed, piecewise smooth curve in D. Then,

(5.3.2)
$$\frac{1}{2\pi i}\oint_C \frac{g(z)\,dz}{z-z_0} = \begin{cases} g(z_0) & \text{if } z_0 \text{ is inside } C \\ 0 & \text{if } z_0 \text{ is outside } C \end{cases}$$

(*Note the counterclockwise orientation of* C.) If z_0 is on C, then the integral is improper and may not even exist.

Proof. First, if z_0 is outside C, then the integrand is analytic, and by the Cauchy–Goursat Theorem the integral is zero. If z_0 is inside C, we can no longer apply this result since the integrand is not analytic at z_0. However, by the Cauchy–Goursat Theorem for multiply connected domains, we can deform C to some circle C_0: $|z - z_0| = R_0$ inside of C without changing the value of the integral. Hence,

(5.3.3)
$$\begin{aligned}\oint_C \frac{g(z)\,dz}{z-z_0} &= \oint_{C_0} \frac{g(z)\,dz}{z-z_0} \\ &= \oint_{C_0} \frac{[g(z)-g(z_0)]\,dz}{z-z_0} + g(z_0)\oint_{C_0}\frac{dz}{z-z_0}\end{aligned}$$

The last integral is simply $2\pi i$, and so we have

(5.3.4)
$$\left|\oint_C \frac{g(z)\,dz}{z-z_0} - 2\pi i\,g(z_0)\right| = \left|\oint_{C_0} \frac{[g(z)-g(z_0)]\,dz}{z-z_0}\right|$$

By assumption, $g(z)$ is analytic, and hence continuous, inside and on C_0. Thus, *given any* ϵ *there is a* δ, *so that*

(5.3.5)
$$|z - z_0| < \delta \Rightarrow |g(z) - g(z_0)| < \epsilon$$

On the right-hand side of (5.3.4), the variable of integration z lies on C_0, which is a circle of radius R_0 centered at z_0. Hence, if we choose $R_0 < \delta$, we will have $|z - z_0| < \delta$ for all z on C_0, and thus by the above, $|g(z) - g(z_0)| < \epsilon$ for z on C_0. Now, if we apply the triangle inequality for integrals to the right-hand side of (5.3.4), we get, since $|z - z_0| = R_0$, and the length of C_0 is $2\pi R_0$,

(5.3.6)
$$\left|\oint_C \frac{g(z)\,dz}{z-z_0} - 2\pi i\,g(z_0)\right| \le 2\pi\epsilon$$

Since ϵ can be chosen arbitrarily small, we have shown that the left side above can also be made arbitrarily small. Hence, this term must vanish, thus proving our result when z_0 is inside of C.

■

Note the following remarkable consequence of this result. From (5.3.2), we see that **we need only know the values of an analytic function on the boundary of a simply connected domain in order to compute its values at all points inside.** Also, we can easily evaluate $\oint_C f(z)\,dz$ when f has only one isolated singular point z_0 in the interior of C, so long as $(z - z_0) f(z)$ is analytic there. To do so, we first split $f(z)$ into a "*bad*" (singular) part $(z - z_0)^{-1}$, multiplied by a "*nice*" (analytic) part $g(z) = (z - z_0) f(z)$. Then, **the value of $(1/2\pi i) \oint_C f(z)\, dz$ is simply the nice part of $f(z)$ evaluated at the bad point.**

(5.3.7) ***Example.*** Evaluate

$$I = \oint_{|z - 1/2| = 1} \frac{e^z}{z - 1}\, dz$$

Solution. Note that since the integrand is not a rational function, we cannot evaluate the integral using the deformation of contours approach of the previous section. Instead, we appeal to (5.3.2) and identify the "nice" part and "bad" point, respectively, as

$$g(z) = e^z \quad \text{and} \quad z_0 = 1 \tag{5.3.8}$$

Then,

$$I = (2\pi i) g(1) = 2\pi i e$$

■

(5.3.9) ***Example.*** Evaluate $I = \oint_C [dz/z(z - 1)]$, where C is the rectangle shown in Figure 5.3.1.

Solution. Here we cannot apply the Cauchy integral formula directly since the integrand has two singularities both of which lie inside of C. We must first deform the rectangle into the curves C_1 and C_2 shown in the figure, each of which encloses *only one* of the singular points (why can we do this?). Hence,

$$I = \oint_{C_1} \frac{dz}{z(z - 1)} + \oint_{C_2} \frac{dz}{z(z - 1)} \tag{5.3.10}$$

In the first integral, we identify $g(z) = 1/(z - 1)$ as the "nice" part and $z_0 = 0$ as the "bad" point to get from the Cauchy integral formula

$$\oint_{C_1} \frac{dz}{z(z - 1)} = -2\pi i \tag{5.3.11}$$

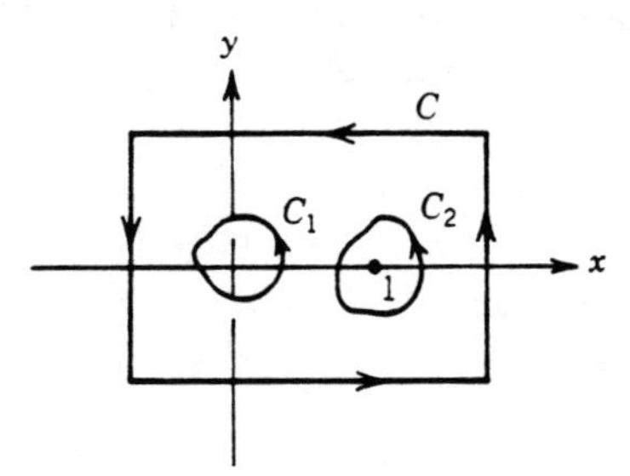

Figure 5.3.1 Example (5.3.9).

while in the second integral we identify $g(z) = 1/z$ and $z_0 = 1$ to get

(5.3.12)
$$\oint_{C_2} \frac{dz}{z(z-1)} = 2\pi i$$

Hence, $I = 0$.

Note that this is the same problem as in Example (5.2.28). There, we first had to do a partial fraction expansion in order to separate the effect of the two singular points, and then we deformed the rectangle into circles about $z = 0$ and $z = 1$. The additional partial fraction expansion is not needed once we have the Cauchy integral formula.

■

(5.3.13) *Example.* The integral $I = \oint_{|z|=1} (e^z dz/z^2)$ cannot be evaluated using the Cauchy integral formula since the integrand is "too singular" at $z = 0$. By this we mean that if we tried to write it in a form which conforms to that in (5.3.2), we would have to identify $g(z) = e^z/z$, which is not analytic at $z = 0$. Below, we will indicate how the Cauchy integral formula can be extended so as to include this type of integral.

■

(5.3.14) *Example.* Consider the function $f(z)$ defined by the integral

$$f(z) = \oint_{|\zeta|=1} \frac{e^{\zeta^2} - 1}{\zeta - z} d\zeta$$

Note that f is not defined for z on the unit circle, since in this case the integrand will be singular at some point on the contour of integration.

For $|z| > 1$, the integrand (as a function of the integration variable ζ) is analytic inside and on the unit circle; and hence by the Cauchy–Goursat Theorem, the integral is zero. For $|z| < 1$, we can use the Cauchy integral formula to evaluate the integral [identify $g(\zeta) = e^{\zeta^2} - 1$ and $\zeta_0 = z$]. Combining these results leads to

(5.3.15)
$$f(z) = \begin{cases} 0 & |z| > 1 \\ 2\pi i(e^{z^2} - 1), & |z| < 1 \end{cases}$$

Note that we cannot define f on the unit circle so as to make the resulting function continuous.

■

Cauchy Integral Formula for Derivatives. The next result provides us with a remarkable property of analytic functions and extends the Cauchy integral formula to cases where the integrand is more singular than that in (5.3.2). To motivate this, consider

(5.3.16)
$$g(z_0) = \frac{1}{2\pi i} \oint_C \frac{g(z)\, dz}{z - z_0}$$

which holds for any z_0 interior to C and for any $g(z)$ analytic inside and on C. If we "formally differentiate" with respect to z_0 inside of the integral and use the fact that

$$\frac{d^n}{dz_0^n}\left(\frac{1}{z-z_0}\right) = n!(z-z_0)^{-(n+1)}$$

we get

$$\frac{d^n g(z_0)}{dz_0^n} = \frac{n!}{2\pi i}\oint_C \frac{g(z)\,dz}{(z-z_0)^{n+1}} \qquad n = 0, 1, 2 \ldots$$

Hence, we are led to the expectation that analytic functions possess derivatives of all orders and that the above formula holds. That this is true is proved below, where we show that the formal interchange of differentiation and integration above is in fact valid.

(5.3.17) ***Theorem.*** Let $g(z)$ be analytic in a simply connected domain D, and let C be a simple, closed, piecewise smooth curve in D. Then, g(z) *possesses derivatives of all orders in* D and, for z_0 inside of C, we have **Cauchy's integral formula for derivatives**

(5.3.18)
$$\oint_C \frac{g(z)\,dz}{(z-z_0)^{n+1}} = \frac{2\pi i}{n!}\, g^{(n)}(z_0) \qquad n = 0, 1, 2, \ldots$$

where $g^{(n)}(z_0)$ denotes the nth derivative of $g(z)$ evaluated at $z = z_0$. By convention, the zero derivative means the function itself, and $0! = 1$. Thus, the case $n = 0$ reduces to the Cauchy integral formula (5.3.2).

Proof. We will first prove (5.3.18) for $n = 1$, that is, we will show

$$g'(z_0) = \frac{1}{2\pi i}\oint_C \frac{g(z)\,dz}{(z-z_0)^2}$$

To do this, we replace $g'(z_0)$ by its definition and prove that

(5.3.19)
$$\lim_{h\to 0}\left|\frac{g(z_0+h)-g(z_0)}{h} - \frac{1}{2\pi i}\oint_C \frac{g(z)\,dz}{(z-z_0)^2}\right| = 0$$

Now, if z_0 is inside C, then so is $z_0 + h$ for $|h|$ small enough; and from the Cauchy integral formula,

$$\begin{aligned}\frac{g(z_0+h)-g(z_0)}{h} &= \frac{1}{2\pi i}\oint_C \frac{g(z)}{h}\left[\frac{1}{z-(z_0+h)} - \frac{1}{z-z_0}\right]dz \\ &= \frac{1}{2\pi i}\oint_C \frac{g(z)\,dz}{(z-z_0)(z-z_0-h)}\end{aligned}$$

Hence,

(5.3.20)
$$\frac{g(z_0+h)-g(z_0)}{h} - \frac{1}{2\pi i}\oint_C \frac{g(z)\,dz}{(z-z_0)^2} = \frac{h}{2\pi i}\oint_C \frac{g(z)\,dz}{(z-z_0)^2(z-z_0-h)}$$

Now, since $g(z)$ is analytic (and hence continuous) on C, it is bounded there, and thus there is an M so that

$$|g(z)| \leq M \qquad (\text{for } z \text{ on } C)$$

Also, since z_0 is inside C, there is a positive number d, the minimum distance from z_0 to C (see Figure 5.3.2), so that

$$|z - z_0| \geq d \qquad (\text{for any } z \text{ on } C) \tag{5.3.21}$$

Hence, with this and the triangle inequalities, we have

$$|z - z_0 - h| \geq ||z - z_0| - |h|| \geq d - |h|$$

for $|h|$ small enough. These results, and the triangle inequality for integrals applied to (5.3.20), leads to

$$\left|\frac{g(z_0 + h) - g(z_0)}{h} - \frac{1}{2\pi i}\oint_C \frac{g(z)\,dz}{(z - z_0)^2}\right| \leq \frac{ML|h|}{2\pi d^2(d - |h|)} \tag{5.3.22}$$

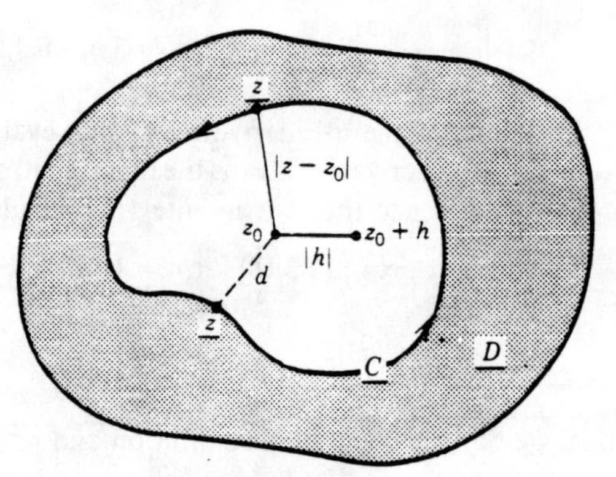

Figure 5.3.2

where L is the length of C. Since the right side above is small with $|h|$, so is the left side, thus proving (5.3.19).

Having proved our result for $n = 1$, we can easily demonstrate it for any n and thus show that $g(z)$ possesses derivatives of all order. For example, if $n = 2$ in (5.3.18), we must show that

$$\lim_{h \to 0}\left|\frac{g'(z_0 + h) - g'(z_0)}{h} - \frac{2}{2\pi i}\oint_C \frac{g(z)\,dz}{(z - z_0)^3}\right| = 0$$

To do this, we proceed in exactly the same way as in the previous proof, but now we replace the derivatives $g'(z_0 + h)$ and $g'(z_0)$ by the result of (5.3.18) for $n = 1$. We leave the details to the reader.

■

Let us note how remarkable this result really is. From it, we learn that **a function which is analytic at a point, that is, possesses at least one derivative in some neighborhood of that point, really possesses derivatives of all order**. That this is certainly not true in elementary calculus can be seen from the function $f(x) = x^3 \sin(1/x)$, with $f(0) = 0$. (The reader is asked to show that $f'(0) = 0$ but that $f''(0)$ does not exist.) The reason why this result is true for analytic functions is deeply rooted in the fact that two arbitrary real functions u and v cannot be combined to produce an analytic function $u + iv$—they must satisfy the Cauchy–Riemann equations.

Finally, since analytic functions possess derivatives of all orders, our result also shows that **the real and imaginary parts of an analytic function are infinitely differentiable**.

(5.3.23) ***Example.*** Evaluate the integrals

$$I_1 = \oint_{|z|=2} \frac{e^z}{z^2}\,dz \qquad \text{and} \qquad I_2 = \oint_{|z|=2} \frac{\sin z}{z^2(2z - i)^3}\,dz$$

Solution. For I_1, if we identify in (5.3.18) $g(z) = e^z$, $z_0 = 0$, and $n = 1$, we get

$$I_1 = 2\pi i \left.\frac{d}{dz}(e^z)\right|_{z=0} = 2\pi i$$

For I_2, we cannot apply (5.3.18) directly because both singular points $z = 0$ and $z = i/2$ lie inside $|z| = 2$, thus making it impossible to split the integrand into a product of a "*nice*" analytic function and a "*bad*" singular function of the form $(z - z_0)^{-n-1}$. Instead, we first use the Cauchy–Goursat theorem for multiply connected domains and write

$$I_2 = \oint_{C_1} \frac{\sin z\,dz}{z^2(2z - i)^3} + \oint_{C_2} \frac{\sin z\,dz}{z^2(2z - i)^3}$$

where C_1 and C_2 are shown in Figure 5.3.3. In the integral over C_1, if we

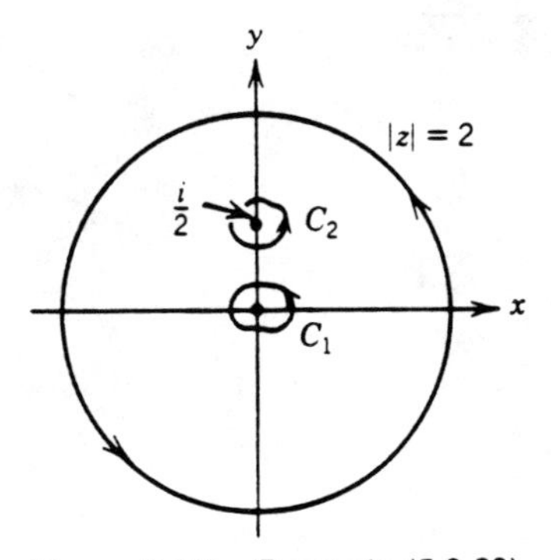

Figure 5.3.3 Example (5.3.23).

identify the analytic part $g(z) = \sin z/(2z - i)^3$, together with $z_0 = 0$ and $n = 1$, we get

$$\oint_{C_1} \frac{\sin z\, dz}{z^2(2z - i)^3} = 2\pi i \frac{d}{dz}\left[\frac{\sin z}{(2z - i)^3}\right]_{z=0} = 2\pi \tag{5.3.24}$$

To evaluate the integral over C_2, we must first write it as $\oint_{C_2} \sin z\, dz/[8z^2(z - i/2)^3]$ (why?). Then, if we identify $g(z) = \sin z/8z^2$, $z_0 = i/2$ and $n = 2$ in (5.3.18), we get

$$\oint_{C_2} \frac{\sin z\, dz}{z^2(2z - i)^3} = \frac{2\pi i}{2} \frac{d^2}{dz^2}\left[\frac{\sin z}{8z^2}\right]_{z=i/2} = \frac{\pi}{2}\left(-25 \sinh \frac{1}{2} + 8 \cosh \frac{1}{2}\right)$$

Hence, adding this to (5.3.24) yields the value for I_2.

■

Finally, we note that even the Cauchy integral formulas will not permit us, at this stage, to evaluate integrals such as

$$\oint_{|z|=1} e^{1/z}\, dz \quad \text{or} \quad \oint_{|z|=1} \frac{dz}{\sin z}$$

For these, we will have to wait until Chapter 6, where infinite series and residues are discussed.

We end this section with a deformation of contour result similar to that of Section 5.2 [see the paragraph after equation (5.2.24)]. There, we learned that if one curve could be continuously deformed into another without passing through a singularity of the (otherwise analytic) integrand, then the integrals would be the same. Here, we extend this result as follows.

(5.3.25) ***Theorem.*** Let C_1 and C_2 be two simple, closed, nonintersecting contours, and let $g(z)$ be analytic on both C_1 and C_2 and in the multiply connected domain D contained between them (see Figure 5.3.4*a*). If z_0 is a point in D, then

$$\oint_{C_2} \frac{g(z)\, dz}{(z - z_0)^{n+1}} = \oint_{C_1} \frac{g(z)\, dz}{(z - z_0)^{n+1}} + \frac{2\pi i}{n!} g^{(n)}(z_0) \tag{5.3.26}$$

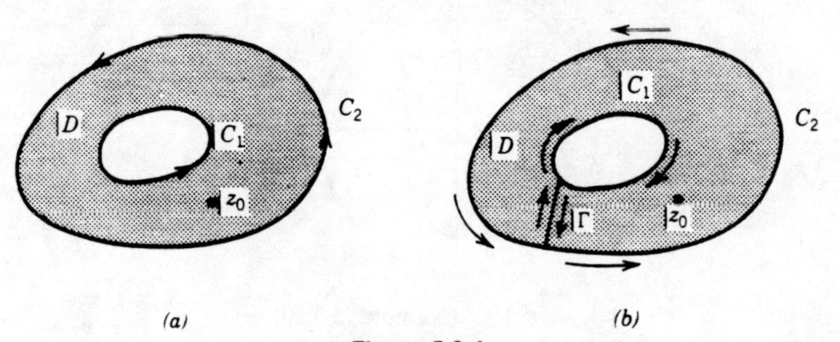

Figure 5.3.4

Proof. The proof proceeds precisely as it did for the Cauchy–Goursat Theorem for multiply connected domains (5.2.22). We join C_1 and C_2 by a curve Γ, as shown in part (*b*) of the figure. This produces a simply connected domain whose boundary is the simple closed contour consisting of C_2, $-C_1$, Γ, and $-\Gamma$. The integral of $g(z)/(z - z_0)^{n+1}$ over this contour is $(2\pi i/n!)\, g^{(n)}(z_0)$ by the Cauchy integral formula for derivatives. The result (5.3.26) then follows, since the integrals over Γ and $-\Gamma$ cancel.

■

Note from Figure 5.3.4*a* that, since any continuous deformation of C_2 into C_1 must pass through the singular point $z = z_0$ of the integrand $g(z)/(z - z_0)^{n+1}$, we cannot invoke the Cauchy–Goursat theorem for multiply connected domains to conclude that the integrals over C_1 and C_2 are equal. Instead, (5.3.26) above tells us that **we can deform C_2 continuously into C_1 but, since we pass through the singularity at z_0, we pick up a contribution from the "nice part" of the integrand.**

PROBLEMS (The answers are given on page 435.)

(Problems with an asterisk are more theoretical.)

1. Evaluate the following integrals:

(a) $\displaystyle\oint_{|z|=2} \frac{(e^z - 1)\,dz}{z}$ (b) $\displaystyle\oint_{|z|=4} \frac{\sin^2 z\,dz}{\left(z - \frac{\pi}{2}\right)^2}$

(c) $\displaystyle\oint_{|z|=1} \frac{\tan z\,dz}{z}$ (d) $\displaystyle\oint_{|z-1/2|=1/4} \frac{\operatorname{Log} z\,dz}{z + 1}$

(e) $\displaystyle\oint_{|z|=1} \frac{\sin z\,dz}{z^n}$ $n = 0, \pm 1, \pm 2, \ldots$ (f) $\displaystyle\oint_{|z-1|=1/2} \frac{\operatorname{Log} z\,dz}{(z - 1)^3}$

(g) $\displaystyle\oint_C \left[\frac{e^z}{z^2 - 1} + \frac{\cos z}{z - \pi} + \sin(iz)\right] dz,$ C is a rectangle with vertices at $\frac{\pi}{4} \pm \pi i$ and $-\frac{\pi}{4} \pm \pi i$.

(h) $\displaystyle\oint_C \left[\frac{z}{(z + 1)^2(z - 3)} + \frac{e^{2z}}{z - 3i}\right] dz,$ C: $x^2 - x + y^2 = \frac{15}{4}$

(i) $\displaystyle\oint_{|z|=2} \frac{(e^{z^2} - e)\,dz}{(z - 1)(z - 4)^3(z - 5i)}$

(j) $\displaystyle\oint_C \frac{(e^z + 5z)}{z^5 + 8iz^2}\,dz,$ C is the square with vertices at $\pm 1 \pm i$

(k) $\oint_{|z|=1} \frac{dz}{e^z - e^2}$ (l) $\oint_{|z|=1} \frac{dz}{2 + \bar{z}}$ *(be careful here)*

(m) $\oint_{|z-5/2|=2} \frac{\operatorname{Log}(z+i)}{z(z-1)^2}\,dz$ (n) $\oint_{|z|=1} \frac{e^z\,dz}{z^2 - 5iz - 6}$

(o) $\oint_{|z|=\ln 2} \frac{dz}{z^2(e^z - 2i)}$

(p) $\oint_C \left(\bar{z} + \frac{1}{z}\right) dz,\quad C\colon\ x = 2 + \cos t,\quad y = \sin t\quad 0 \le t \le 2\pi$

(q) $\oint_{|z|=1} \frac{dz}{(z - z_0)\left(z - \frac{1}{\bar{z}_0}\right)}$, where $z_0 \neq 0$ and $|z_0| \neq 1$

2. Make the change of variables $w = 1/\zeta$ and then evaluate the following integrals.

(a) $\oint_{|\zeta|=2} \frac{e^{1/\zeta}}{\zeta^2 + 1}\,d\zeta$ (b) $\oint_{|\zeta|=1} \zeta^4 \sin\left(\frac{1}{\zeta}\right) d\zeta$

(c) $\oint_{|\zeta|=1} \frac{g(\zeta)\,d\zeta}{\zeta(\zeta - z)}$, where $|z| \neq 1$ and $g(\zeta)$ is analytic for $|\zeta| \geq 1$

3. (a) Let $g(z)$ be analytic inside and on a simple closed contour C, and let $p(z)$ be the polynomial $p(z) = (z - z_1)(z - z_2)\ldots(z - z_N)$. If no two of the roots $z_1, z_2, \ldots, z_N$ are equal and none lies on C, show that

$$\oint_C \frac{g(z)\,dz}{p(z)} = 2\pi i \sum \frac{g(z_k)}{p'(z_k)}$$

where the sum is taken over those roots lying inside C.

4. For which z is the function $f(z) = \oint_{|\zeta|=1} [e^{\zeta}/(\zeta^2 - z^2)]\,d\zeta$ defined? Compute all of its values.

5. The contours C_1 and C_2 are shown in Figure 5.3.5. If we are given that $\int_{C_1} e^z\,dz/[(z-1)^2(z+i)] = 2$, what is the value of $\int_{C_2} e^z\,dz/[(z-1)^2(z+i)]$

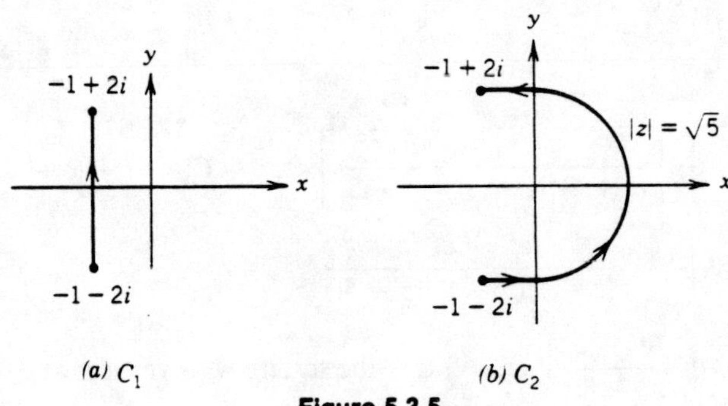

Figure 5.3.5

6. Compute $f(0)$, $f(1)$, $f(i)$, and $f(5i)$, where

$$f(z) = \oint_C \frac{(\zeta - 2z)e^{\zeta}}{(\zeta - z)^2}\, d\zeta$$

and C is the ellipse $4x^2 + y^2 = 16$.

7. Consider the integral

$$I = \oint_C \frac{\operatorname{Log} z\, dz}{(z - i)^2(z + i)}$$

where C is shown in Figure 5.3.6*a*. We cannot apply the Cauchy integral formula directly since the principal logarithm is not analytic on C (why?).

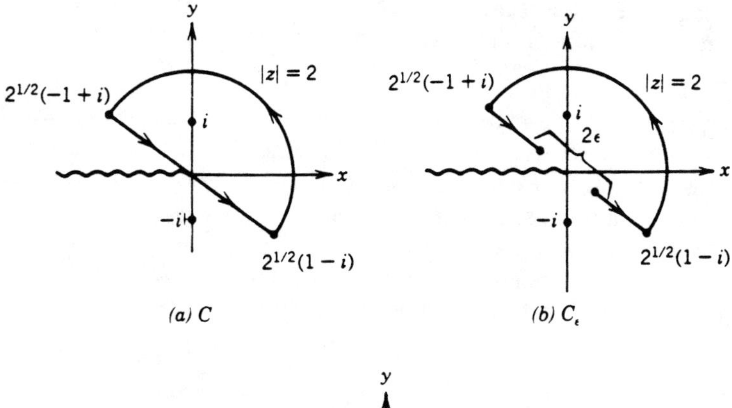

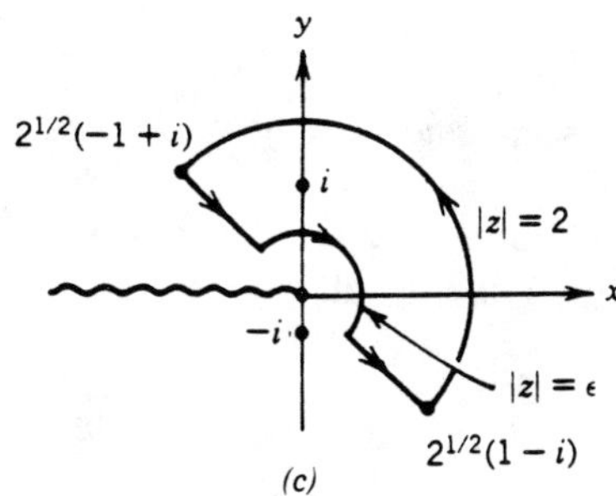

Figure 5.3.6

(a) By parametrizing C, show that the integral exists as an improper integral and is given by

$$\lim_{\epsilon \to 0} \int_{C_\epsilon} \frac{\operatorname{Log} z\, dz}{(z - i)^2(z + i)}$$

where C_ϵ is shown in part (b) of the figure.

(b) Use the closed contour in Figure 5.3.6c to show that $I = -\pi/2(\pi/2 + 2i)$. [*Hint:* Use the Cauchy integral formula and part (a) and show that the integral over the cirlce $|z| = \epsilon$ vanishes as $\epsilon \to 0$.]

8. If $g(z)$ is analytic on and inside a simple closed contour C, show that if z_0 is inside C,

(a) $$\oint_C \frac{g(z)\,dz}{(z - z_0)^{n+1}} = \frac{1}{n!}\oint_C \frac{g^{(n)}(z)\,dz}{z - z_0}$$

(b) $$g^{(n)}(z_0) = \frac{k!}{2\pi i}\oint_C \frac{g^{(n-k)}(z)\,dz}{(z - z_0)^{k+1}} \qquad k \le n$$

9.* Prove that the real-valued harmonic function $u(x, y)$ possesses continuous second partial derivatives in a simply connected domain D, then u possesses continuous partial derivatives of any order in D. [*Hint:* Does u possess a harmonic conjugate? If so, what can be said about $u + iv$?]

10.* Here, we lead the reader through a proof that, under certain conditions, the integral of a function involving a parameter is analytic in the parameter.

Theorem. Let $h(z, \zeta)$ be a complex-valued function of the two complex variables z and ζ, and be defined for z in a simply connected domain D and for ζ on the simple, closed contour C bounding D. Assume that **(i)** $|h| \le K$ (h is bounded), **(ii)** for each ζ on C, $h(z, \zeta)$ is analytic in z, and **(iii)** for each z in D, $h(z, \zeta)$ is continuous in ζ. Then, the function $f(z) = \oint h(z, \zeta)\,d\zeta$ is analytic for z in D and $f'(z) = \oint \partial h(z, \zeta)/\partial z\,d\zeta$.

(a) Why does the integral defining $f(z)$ exist for each z in D?

(b) Use the Cauchy integral formula to show that for each z in D,

$$f(z) = \frac{1}{2\pi i}\oint_C d\zeta\left[\oint_{C_z} \frac{h(\mu, \zeta)\,d\mu}{\mu - z}\right]$$

where C_z is some *small enough* circle of radius ρ in D with center at z (see Figure 5.3.7.)

(c) Choose Δz small enough so that $z + \Delta z$ lies inside the circle C_z. Then use Cauchy's integral formula for derivatives to show that

$$\frac{f(z + \Delta z) - f(z)}{\Delta z} - \oint_C \frac{\partial h(z, \zeta)}{\partial z}\,d\zeta = \frac{\Delta z}{2\pi i}\oint_C d\zeta \oint_{C_z} \frac{h(\mu, \zeta)\,d\mu}{(\mu - z - \Delta z)(\mu - z)^2}$$

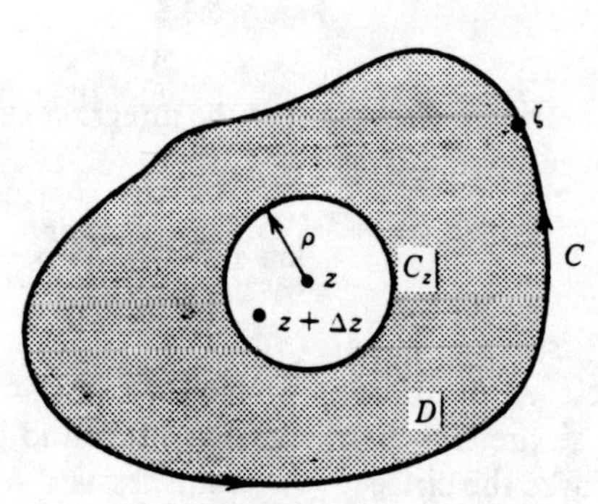

Figure 5.3.7

(d) Finally, use the triangle inequality for integrals and the boundedness of h to conclude that $f'(z)$ exists and equals $\oint \partial h(z,\zeta)/\partial z\, d\zeta$.

11.* Verify the hypothesis of Problem 10 and show that if $g(\zeta)$ is continuous on a simple, closed contour C, then $f(z) = \oint g(\zeta)\, d\zeta/(\zeta - z)$ is analytic for z not on C.

12. Here, we will show how solutions of **ordinary differential equations with constant coefficients** can be written as contour integrals. Thus, consider the nth order ordinary differential equation

$$c_n u^{(n)} + c_{n-1} u^{(n-1)} + \cdots + c_1 u' + c_0 u = 0 \qquad (c_n \neq 0) \tag{5.3.27}$$

for a function $u(z)$, where $u^{(k)}$ denotes the kth derivative and c_k, $k = 0, 1, \ldots, n$, are constants.

(a) Let $p(\zeta)$ be the polynomial $p(\zeta) = \sum_{k=0}^{n} c_k \zeta^k$, and C be any simple, closed contour containing within it *all* of the roots of $p(\zeta) = 0$. If $g(\zeta)$ is *any* function which is analytic inside and on C, show that

$$u(z) = \frac{1}{2\pi i} \oint_C \frac{g(\zeta)\, e^{\zeta z}\, d\zeta}{p(\zeta)}$$

is a solution of (5.3.27). The polynomial $p(\zeta)$ is called the **characteristic polynomial**. [*Hint:* Use the result of Problem 10 to justify differentiating under the integral sign.]

(b) If $r_1, r_2, \ldots, r_N$ are the *distinct* roots of $p(\zeta) = 0$, then we can factor p into $p(\zeta) = c_n(\zeta - r_1)^{n_1}(\zeta - r_2)^{n_2} \ldots (\zeta - r_N)^{n_N}$, where the positive integers n_l, $l = 1, 2, \ldots, N$, are the orders of the roots. Use part (*a*) and the arbitrainess of $g(\zeta)$ to show that the general solution of equation (5.3.27) is of the form $u(z) = \sum_{k=1}^{N} q_k(z) e^{r_k z}$ where $q_k(z)$ is a polynomial of order $n_k - 1$. This is well known from differential equations. [*Hint:* Do a partial fraction expansion for $1/p(\zeta)$ and then use the Cauchy integral formula in part (a).]

(The next problem refers to fluid flow and relates to the material in Appendix C and in Sections 3.4 and 4.6.)

13. Consider a fluid flow with complex potential $\Phi(z)$ (see Section 3.4). In Problems 19 and 20 of Section 5.2, it was shown that if C is a simple, piecewise smooth, closed curve in the flow, then

$$\Gamma + i\Omega = \oint_C \Phi'\, dz \qquad \text{and} \qquad F = \overline{\frac{i\rho}{2} \oint_C (\Phi')^2\, dz}$$

where Γ and Ω are the **circulation** around C and the **flow** across C, respectively, while F is the net **force** acting on a rigid body immersed in the flow and bounded by C.

(a) If $\Phi(z) = A \log z$, find the circulation and flow in terms of A for a curve C surrounding the origin. What are your conclusions if A is real? If A is imaginary? Compare these results with Examples (3.4.15) and (3.4.19). What is the net force on the rigid body immersed in such a flow?

(b) If $\Phi(z) = g(z) + B/z$, where g is entire, what are the circulation and flow for a curve surrounding the origin? What is the net force on a rigid body? Compare these results with Example (3.4.22) and Problem 7 of that section.

(c) Compute the circulation, flow, and force for the potential $\Phi(z) = a_0(z + 1/z) + B \log z$ when the curve C surrounds the origin. Compare these results with those in Problem 10 of Section 3.4.

5.4 EVALUATION OF SOME REAL INTEGRALS (Optional at this Time)

In this section, we will show how the Cauchy integral formula can be used to evaluate some real integrals which might otherwise be quite difficult to compute using the usual methods of elementary calculus. In Section 6.7, we will return to this subject and show how these results can be interpreted in terms of residue theory and how they can be extended to include other types of integrals. Indeed, if the reader wishes, this material can be delayed until that time.

Trigonometric Integrals. The first class of integrals to be evaluated are those which can be put into the form

(5.4.1)
$$I = \int_0^{2\pi} F(\sin x, \cos x)\, dx$$

where the integrand is a ratio of polynomials in $\sin x$ and $\cos x$.

(5.4.2) ***Example.*** Evaluate

$$I = \int_0^{2\pi} \frac{dx}{2 + \cos x}$$

Solution. First, observe that since $|\cos x| \leq 1$, the denominator never vanishes and hence the integral exists. Now, in elementary calculus, when confronted with a difficult integral we often try to find a proper substitution. There is in fact a rather obscure trigonometric substitution which is useful here. However, we shall not pursue this since the complex variables approach which we shall develop is much more direct, simple, and elegant.

First, since $\cos x = \frac{1}{2}(e^{ix} + e^{-ix})$, we see that the integrand involves only the complex function e^{ix}. Thus, we are naturally led to the *complex change of variables* $w = e^{ix}$. However, before proceeding with this, we first interpret I as the contour integral

(5.4.3)
$$I = \int_{C_x} \frac{dz}{2 + \frac{1}{2}(e^{iz} + e^{-iz})}$$

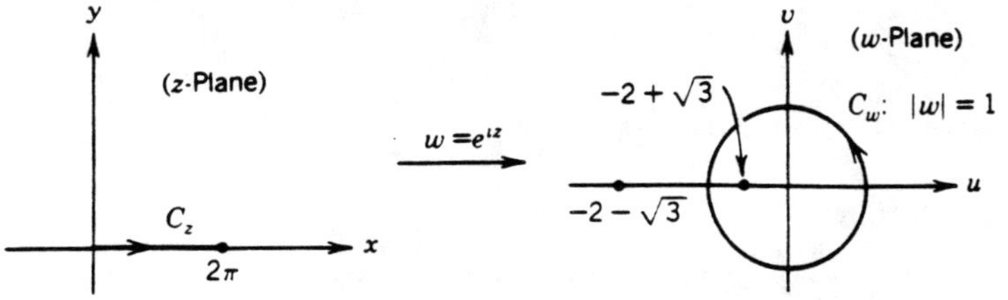

Figure 5.4.1 Example (5.4.2).

where C_z is the real interval $[0, 2\pi]$ as shown in Figure 5.4.1. Then, under the change of variables $w = e^{iz}$, in which case $dw = ie^{iz}\,dz = iw\,dz$, we get

$$I = \int_{C_w} \frac{dw}{iw\left[2 + \frac{1}{2}\left(w + \frac{1}{w}\right)\right]}$$

where C_w is the image in the w-plane of the real interval $[0, 2\pi]$ under the mapping $w = e^{iz}$. This image is the unit circle traversed counterclockwise. Hence, the above becomes

$$I = \frac{2}{i} \oint_{|w|=1} \frac{dw}{w^2 + 4w + 1} \tag{5.4.4}$$

Note that the integrand is a rational function of the new integration variable w and the contour of integration is a closed curve in the w-plane. If we factor the denominator into

$$w^2 + 4w + 1 = [w - (-2 - \sqrt{3})][w - (-2 + \sqrt{3})]$$

and note that the only root lying *inside* the unit circle is $w = -2 + \sqrt{3}$, then, by the Cauchy integral formula,

$$I = \frac{2\pi}{\sqrt{3}} \tag{5.4.5}$$

■

Observe that the above procedure will work for any integral of the form (5.4.1). This is so because any rational function of $\sin x$ and $\cos x$ is also a rational function of e^{ix}, and hence under the substitution $w = e^{iz}$ the integrand becomes a rational function of w and the real interval $[0, 2\pi]$ transforms into the unit circle in the w-plane. This new integral, in turn, can be evaluated using the Cauchy integral formula, *where we pick up contributions due only to those singularities inside the unit circle.* For example, to evaluate $\int_0^{2\pi} [\cos(2x)\,dx]/[2 + \sin^2(3x)]$, we first write

$$\cos(2x) = \frac{1}{2}(e^{2ix} + e^{-2ix}) = \frac{1}{2}(w^2 + w^{-2})$$

and

$$\sin(3x) = \frac{1}{2i}(e^{3ix} - e^{-3ix}) = \frac{1}{2i}(w^3 - w^{-3})$$

where $w = e^{ix}$. Hence,

$$\int_0^{2\pi} \frac{\cos(2x)\,dx}{2 + \sin^2(3x)} = \frac{-2}{i}\oint_{|w|=1} \frac{w^3(1 + w^4)\,dw}{w^{12} - 10w^6 + 1}$$

and this can be evaluated by the Cauchy integral formula.

Also note that the interval of integration in (5.4.1) can be *any interval of length* 2π, not necessarily $[0, 2\pi]$. In addition, sometimes because of symmetries in the integrand in (5.4.1), integrals not over a 2π-interval might be able to be converted into one. For example,

$$\int_0^{\pi} \frac{dx}{2 + \sin^2 x} = \frac{1}{2}\int_{-\pi}^{\pi} \frac{dx}{2 + \sin^2 x}$$

and

$$\int_0^{\pi/2} \frac{dx}{2 + \cos^2 x} = \frac{1}{4}\int_0^{2\pi} \frac{dx}{2 + \cos^2 x}$$

Finally, as the next example shows, it is possible to use these same ideas, coupled with the results of Section 5.2, to evaluate an integral which cannot be converted to one over a 2π-interval.

(5.4.6) ***Example.*** Evaluate $I = \int_0^{\theta_0} \sec x\,dx$, $0 \le \theta_0 < \pi/2$.

Solution. First, note that since

$$\sec x = \frac{1}{\cos x} = \frac{2}{e^{ix} + e^{-ix}}$$

the given range on the upper limit of integration θ_0 ensures that the integral exists. Also, for arbitrary θ_0, I cannot be converted into an integral over a 2π-range. However, keeping faith, we proceed as before and make the substitution $w = e^{iz}$. Thus, I becomes

$$I = \frac{2}{i}\int_{C_w} \frac{dw}{1 + w^2} \tag{5.4.7}$$

However, now the contour C_w is not the whole (closed) unit circle but just that part shown in Figure 5.4.2*a*. Thus, we cannot use the Cauchy integral formula. Instead we search for an indefinite integral using the partial fraction expansion

$$\frac{1}{1 + w^2} = \frac{1}{2i}\left(\frac{1}{w - i} - \frac{1}{w + i}\right)$$

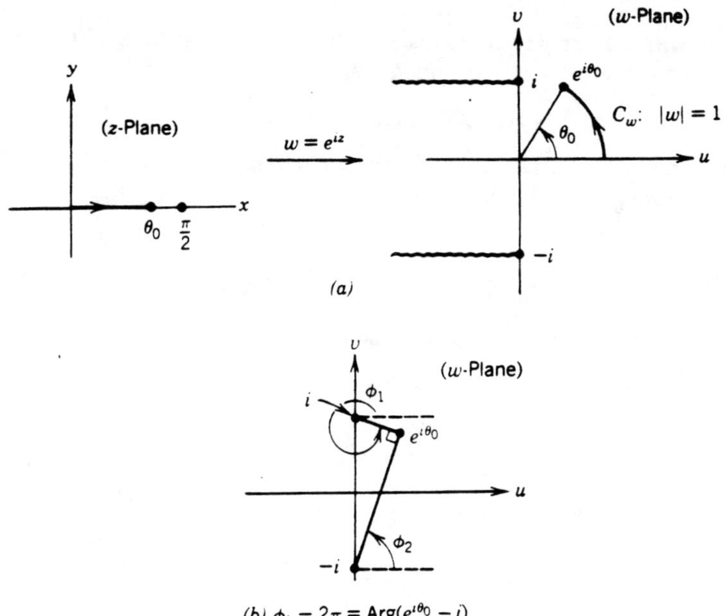

(b) $\phi_1 - 2\pi = \operatorname{Arg}(e^{i\theta_0} - i)$
$\phi_2 = \operatorname{Arg}(e^{i\theta_0} + i)$

Figure 5.4.2 Example (5.4.6).

Hence, we can perform the integration in (5.4.7) to get

$$I = -[\log(w - i) - \log(w + i)]\Big|_{w=1}^{w=e^{i\theta_0}} \tag{5.4.8}$$

where the two logarithms must be chosen in such a way that they are analytic in some domain containing the arc C_w. In particular, we can pick the principal logarithms since these have branch cuts emanating from $w = \pm i$, as shown in the figure. Hence from the above, with Arg denoting the principal argument, we get

$$I = \ln\left|\frac{e^{i\theta_0} + i}{e^{i\theta_0} - i}\right| + i\left[\operatorname{Arg}(e^{i\theta_0} + i) - \operatorname{Arg}(e^{i\theta_0} - i) - \frac{\pi}{2}\right] \tag{5.4.9}$$

Now,

$$\left|\frac{e^{i\theta_0} + i}{e^{i\theta_0} - i}\right|^2 = \frac{1 + \sin\theta_0}{1 - \sin\theta_0} = (\sec\theta_0 + \tan\theta_0)^2$$

and thus, because $\sec(\theta_0) + \tan(\theta_0) > 0$, we have

$$\ln\left|\frac{e^{i\theta_0} + i}{e^{i\theta_0} - i}\right| = \ln(\sec\theta_0 + \tan\theta_0) \tag{5.4.10}$$

Now, with the angles ϕ_1 and ϕ_2 defined as in Figure 5.4.2*b*, we have (remember the principal argument range is $-\pi < \operatorname{Arg}(z) \leq \pi$)

$$\operatorname{Arg}(e^{i\theta_0} + i) = \phi_2 \quad \text{and} \quad \operatorname{Arg}(e^{i\theta_0} - i) = \phi_1 - 2\pi$$

Also, since i, $-i$, and $e^{i\theta_0}$ all lie on the unit circle, the lines joining i and $-i$ with $e^{i\theta_0}$ are perpendicular, and hence

$$\left(\phi_1 - \frac{3\pi}{2}\right) + \left(\frac{\pi}{2} - \phi_2\right) = \frac{\pi}{2}$$

Combining this with the above yields

(5.4.11)
$$\operatorname{Arg}(e^{i\theta_0} + i) - \operatorname{Arg}(e^{i\theta_0} - i) = \phi_2 - \phi_1 + 2\pi = \frac{\pi}{2}$$

Thus, by (5.4.10), (5.4.11), and (5.4.9), we have

$$\int_0^{\theta_0} \sec x \, dx = \ln(\sec\theta_0 + \tan\theta_0)$$

This of course is a well-known result from calculus.

■

Improper Integrals of Rational Functions. Here, we will show how to evaluate **improper real integrals** of the form

(5.4.12)
$$I = \int_{-\infty}^{\infty} \frac{p(x)}{q(x)} \, dx$$

where $p(x)$ and $q(x)$ are polynomials in x. From calculus[1] we know that such integrals exist if $q(x)$ has no (real) zeros and if the integrand approaches zero fast enough at $x = \pm\infty$. This will be ensured if

(5.4.13)
$$\text{degree}(q) - \text{degree}(p) \geq 2$$

With these conditions, the integral may be expressed as the limit

(5.4.14)
$$I = \lim_{R\to\infty} \left[\int_{-R}^{R} \frac{p(x)}{q(x)} \, dx \right]$$

This is called the **Cauchy principal value.**

(5.4.15) ***Example.*** Evaluate

$$I = \int_{-\infty}^{\infty} \frac{dx}{(x^2+1)^2}$$

[1] The reader might wish to review improper integrals at this time.

Solution. First, note that the integral exists since the denominator has no real zeros and the integrand behaves like x^{-4} as $x \to \pm\infty$. Hence,

$$I = \lim_{R \to \infty} \left[\int_{-R}^{R} \frac{dx}{(x^2 + 1)^2} \right] \tag{5.4.16}$$

We now make the following observations.

(a) The (real) integral in (5.4.16) can be viewed as a contour integral of the complex function $(z^2 + 1)^{-2}$ over the real interval $[-R, R]$.

(b) The function $(z^2 + 1)^{-2}$ is continuous on the contour in (a) but has singularities *in the complex plane* at $z = \pm i$.

(c) The function $(z^2 + 1)^{-2}$ "behaves like" z^{-4} for $|z|$ large (we will show this below.)

(d) If the curve in (a) was not the interval $[-R, R]$ but was a closed curve not passing through $z = \pm i$, then the Cauchy integral formula could be used to evaluate the integral in (5.4.16).

Observation (d) is pivotal in determining how to proceed. What we do is to consider a new (complex) contour integral

$$I_R = \oint_{\Gamma_R} \frac{dz}{(z^2 + 1)^2} \tag{5.4.17}$$

where the curve Γ_R is to have the following properties:

(i) It is to be closed.

(ii) Part of it is to consist of the real interval $[-R, R]$.

(iii) The limit of the integral as $R \to \infty$ can easily be computed on the remaining part.

One such curve, as we shall soon see, is that shown in Figure 5.4.3*a* and consists of the real interval $[-R, R]$ together with the semicircle

$$C_R: \quad z = Re^{i\theta} \qquad 0 \le \theta \le \pi \tag{5.4.18}$$

First, note that for R large enough, the two singularities $z = \pm i$ of the integrand cannot lie on Γ_R, and hence the integral in (5.4.17) exists. Also,

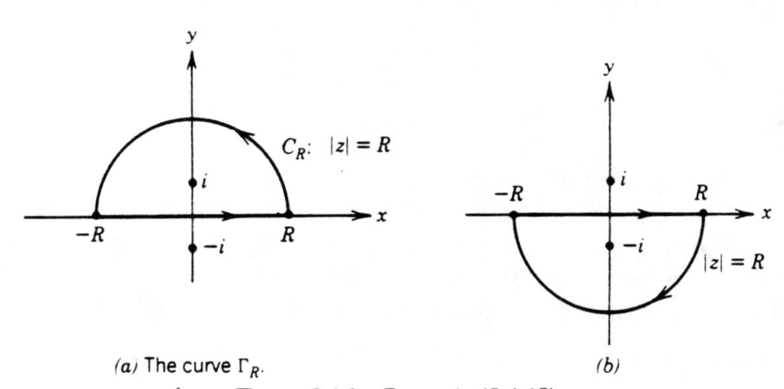

(a) The curve Γ_R. *(b)*

Figure 5.4.3 Example (5.4.15).

since Γ_R is closed, the Cauchy integral formula yields

(5.4.19)
$$I_R = \frac{\pi}{2}$$

However, this is not yet the value of the integral we wish to evaluate; we must first compute the limit in (5.4.16). To do this, we note that the curve Γ_R consists of two pieces—the real interval $[-R, R]$, which can be parametrized by

$$z = x \qquad -R \leq x \leq R$$

and the semicircle C_R given by (5.4.18). Hence, from (5.4.17) and (5.4.19), we have

$$\frac{\pi}{2} = \int_{-R}^{R} \frac{dx}{(x^2+1)^2} + \int_{C_R} \frac{dz}{(z^2+1)^2}$$

Taking the limit $R \to \infty$ yields

(5.4.20)
$$\int_{-\infty}^{\infty} \frac{dx}{(x^2+1)^2} = \frac{\pi}{2} - \lim_{R\to\infty} \int_{C_R} \frac{dz}{(z^2+1)^2}$$

To evaluate the limit on the right, we first observe that

(5.4.21)
$$\begin{aligned} |(z^2+1)^2| &\geq (|z|^2 - 1)^2 \\ &= (R^2-1)^2 \qquad \text{for } z \text{ on } C_R \text{ and } R > 1 \end{aligned}$$

Hence, for z on C_R, $|(z^2+1)^{-2}| \leq (R^2-1)^{-2}$. This result, coupled with the triangle inequality for integrals and the fact that the length of C_R is πR, yields

$$\left| \int_{C_R} \frac{dz}{(z^2+1)^2} \right| \leq \frac{\pi R}{(R^2-1)^2} \xrightarrow[R\to\infty]{} 0$$

Thus, the limit in (5.4.20) is zero, and we have

$$\int_{-\infty}^{\infty} \frac{dx}{(x^2+1)^2} = \frac{\pi}{2}$$

■

Let us note the salient features of the above procedure. One was our ability to evaluate the closed contour integral (5.4.17) using Cauchy's integral formula, and another was our ability to show that the integral over the semicircle in (5.4.20) was zero as $R \to \infty$. The following result, which is proved later as a special case of Theorem (5.4.37), shows that such limits are zero under very general conditions.

(5.4.22) ***Theorem.*** Let $p(z)$ and $q(z)$ be polynomials such that $q(z)$ has no real zeros and degree(q) − degree$(p) \geq 2$. Consider the circular arc

$$C_R\colon \quad z = Re^{i\theta} \qquad \theta_0 \leq \theta \leq \theta_1 \; (\theta_0 > 0 \quad \text{and} \quad \theta_1 < 2\pi)$$

where R is large enough so that the complex zeros of $q(z)$ all lie inside the circle $|z| = R$. Then,

$$\lim_{R\to\infty}\left[\int_{C_R}\frac{p(z)}{q(z)}\right] = 0$$

■

We point out that *this result hinges entirely on the fact that* p(z)/q(z) *approaches* 0 *fast enough at infinity*, and it also does not depend upon the location of the circular arc C_R. Hence, in the previous example, the reader is urged to show that if we had considered instead of Γ_R the contour shown in Figure 5.4.3*b*, we would have arrived at the same answer. (Note that for this contour, the Cauchy integral formula will pick up a contribution from the singularity lying in the *lower half-plane*, together with a change of sign due to the *clockwise orientation* of the curve.) Also, we can use this same procedure to evaluate integrals of the form $\int_0^\infty [p(x)/q(x)]\,dx$ if p/q contains only even powers of x (thus, the integrand is an even function). For in this case, we have

(5.4.23)
$$\int_0^\infty \frac{p(x)}{q(x)}\,dx = \frac{1}{2}\int_{-\infty}^\infty \frac{p(x)}{q(x)}\,dx \qquad \left(\text{for } \frac{p}{q} \text{ even}\right)$$

For integrands which are not even, we can sometimes use a modification of the above ideas, as the next example shows.

(5.4.24) ***Example.*** Evaluate the integral

$$I = \int_0^\infty \frac{dx}{1 + x^3}.$$

Solution. This improper integral exists and is given by

(5.4.25)
$$I = \lim_{R\to\infty}\int_0^R \frac{dx}{1 + x^3}$$

First note that we can factor $1 + z^3$ into

$$(1 + z^3) = (z + 1)(z - e^{\pi i/3})(z - e^{-\pi i/3})$$

Now, as in the previous example, we seek a closed contour having none of the singularities of the integrand on it and part of which is the real interval $[0, R]$. Since the singularity at $z = 1$ is located on the contour Γ_R of Figure 5.4.3*a*, we cannot use this contour here. Instead, we will use the contour C shown in Figure 5.4.4 and which consists of the interval $[0, R]$, an arc of the circle $|z| = R$, and the line L_R. As we will shortly see, the choice of L_R is not arbitrary, for something "nice" will happen to the integrand on it.

First, since C encloses the singularity at $z = e^{\pi i/3}$, the Cauchy integral formula yields

$$\oint_C \frac{dz}{1 + z^3} = \frac{2\pi i}{(e^{\pi i/3} + 1)(e^{\pi i/3} - e^{-\pi i/3})} = \frac{2\pi}{\sqrt{3}(1 + e^{\pi i/3})}$$

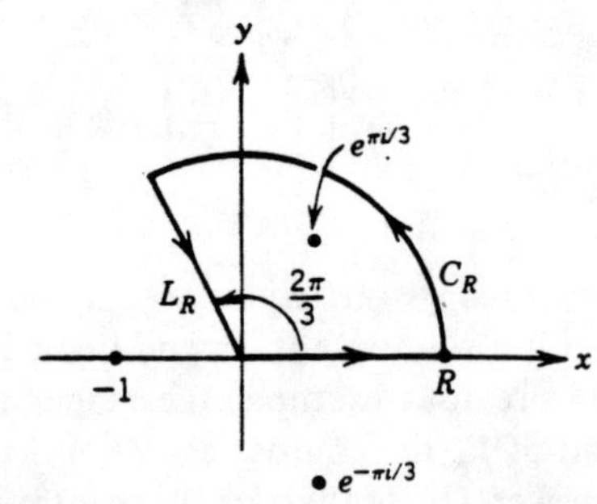

Figure 5.4.4 Contour C for Example (5.4.24).

Equating this to the sum of the integrals over the three pieces of C leads to

(5.4.26)
$$\frac{2\pi}{\sqrt{3}(1 + e^{\pi i/3})} = \int_0^R \frac{dx}{1 + x^3} + \int_{L_R} \frac{dz}{1 + z^3} + \int_{C_R} \frac{dz}{1 + z^3}$$

Now, the line L_R can be parametrized by L_R: $z = -te^{2\pi i/3}$, $-R \le t \le 0$, and hence,

(5.4.27)
$$\int_{L_R} \frac{dz}{1 + z^3} = \int_{-R}^{0} \frac{-e^{2\pi i/3}}{1 - t^3}\, dt \qquad (\text{since } e^{2\pi i} = 1)$$
$$= -e^{2\pi i/3} \int_0^R \frac{dx}{1 + x^3} \qquad (\text{let } x = -t \text{ above})$$

Note that the last term is simply proportional to the first integral on the right-hand side in (5.4.26), whose limit is the integral we want. Indeed, *this was precisely the reason why the line* $\mathrm{L_R}$ *was chosen to complete the contour* C. Finally, from Theorem (5.4.22), we have

$$\lim_{R \to \infty} \int_{C_R} \frac{dz}{1 + z^3} = 0$$

Hence, combining all of the above and taking the limit, we get

(5.4.28)
$$I = \frac{2\pi}{\sqrt{3}(1 + e^{\pi i/3})(1 - e^{2\pi i/3})} = \frac{2\pi}{3\sqrt{3}}$$

■

Improper Integrals Involving Trigonometric Functions. Here, we will discuss the evaluation of real integrals which can be put into the form

(5.4.29)
$$I = \int_{-\infty}^{\infty} \frac{p(x)}{q(x)} \{\sin(ax) \quad \text{or} \quad \cos(ax)\}\, dx$$

where p and q are polynomials. As we will see in the examples, the presence of the trigonometric functions dictates that we proceed more carefully than before.

(5.4.30) ***Example.*** Evaluate

$$I = \int_{-\infty}^{\infty} \frac{\cos x}{1 + x^2}\, dx$$

Solution. Since the integrand behaves like $1/x^2$ for $x \to \pm\infty$, this improper integral exists and is given by

$$I = \lim_{R\to\infty} \left[\int_{-R}^{R} \frac{\cos x}{1 + x^2}\, dx \right] \tag{5.4.31}$$

Following the previous procedure, we might be inclined to consider the integral

$$\hat{I}_R = \oint_{\Gamma_R} \frac{\cos z}{1 + z^2}\, dz$$

where Γ_R is our old friend the contour of Figure 5.4.5*a*. Then, since $1 + z^2 = (z - i)(z + i)$, the Cauchy integral formula would yield

$$\hat{I}_R = \frac{2\pi i \cos i}{2i} = \pi \cosh 1$$

As before, we equate this to the sum of the integrals over the two parts of Γ_R and take the limit $R \to \infty$ to get

$$I = \pi \cosh 1 - \lim_{R\to\infty} \left[\int_{C_R} \frac{\cos z\, dz}{1 + z^2} \right] \tag{5.4.32}$$

In the previous examples, the above limit was zero since the integrand decayed rapidly enough at infinity. However, since the trigonometric functions are not necessarily bounded for large values of their arguments (see Section 2.6), this limit may not be zero here. Indeed, since $\cos z = \frac{1}{2}(e^{iz} + e^{-iz})$, we have on C_R: $z = Re^{i\theta}, 0 \le \theta \le \pi$,

$$\cos z = \frac{1}{2} [e^{iR\cos\theta} \cdot e^{-R\sin\theta} + e^{-iR\cos\theta} \cdot e^{R\sin\theta}], \tag{5.4.33}$$

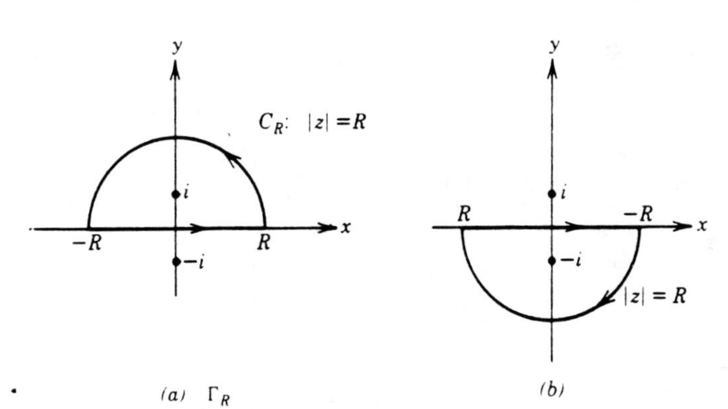

Figure 5.4.5 Example (5.4.30).

Note that since $\sin\theta > 0$ for θ in the range $0 < \theta < \pi$, the term $e^{R\sin\theta}$ is unbounded exponentially as $R \to \infty$, and so is the integrand in (5.4.32). Hence, we cannot easily compute the limit. Nevertheless, our procedure is sound, but we must start at a different place. To see where, first note from (5.4.33) that on the semicircle the source of the difficulty was with the term $e^{R\sin\theta}$, which is unbounded as $R \to \infty$ for $0 < \theta < \pi$. This in turn arose from the e^{-iz} term in the expression for $\cos z$. If we had no need for this term, then we would have been able to proceed as before. Now, we can eliminate the e^{-iz} term by first noting that for real x, $\cos x = \text{Re}(e^{ix})$. Hence, we first return to our original real integral $I = \int_{-\infty}^{\infty} [\cos x/(1 + x^2)]\, dx$ and write

$$
\begin{aligned}
I &= \text{Re}\left[\int_{-\infty}^{\infty} \frac{e^{ix}}{1 + x^2}\, dx\right] \\
&= \text{Re}\left[\lim_{R\to\infty} \int_{-R}^{R} \frac{e^{ix}}{1 + x^2}\, dx\right]
\end{aligned}
\tag{5.4.34}
$$

Now we can follow the previous procedure and consider the integral

$$I_R = \oint_{\Gamma_R} \frac{e^{iz}}{1 + z^2}\, dx$$

where Γ_R is the contour of Figure 5.4.5*a*. The Cauchy integral formula yields

$$I_R = \frac{\pi}{e}$$

which we equate to the sum of the integrals over the interval $[-R, R]$ and the semicircle C_R. We then take the limit $R \to \infty$ to get

$$I = \frac{\pi}{e} - \text{Re}\left[\lim_{R\to\infty} \int_{C_R} \frac{e^{iz}}{1 + z^2}\, dz\right] \tag{5.4.35}$$

On C_R, $|e^{iz}| = e^{-R\sin\theta} \le 1$ (since $0 \le \theta \le \pi$), while, by the usual inequalities,

$$\left|\frac{1}{1 + z^2}\right| \le \frac{1}{R^2 - 1} \qquad \text{for } z \text{ on } C_R$$

Hence, since the length of C_R is πR,

$$\left|\int_{C_R} \frac{e^{iz}}{1 + z^2}\, dz\right| \le \frac{\pi R}{R^2 - 1} \tag{5.4.36}$$

Thus, the limit in (5.4.35) is zero, and we have

$$\int_{-\infty}^{\infty} \frac{\cos x}{1 + x^2}\, dx = \frac{\pi}{e}$$

■

Note in this example that if we had started with the semicircle located in the lower half-plane (Figure 5.4.5*b*), then the term $e^{-R\sin\theta}$ in (5.4.33) would be unbounded, and we would have had to modify (5.4.34) to read

$$\int_{-\infty}^{\infty} \frac{\cos x}{1+x^2}\,dx = \mathrm{Re}\left[\int_{-\infty}^{\infty} \frac{e^{-ix}}{1+x^2}\,dx\right] \qquad \text{(why?)}$$

Finally, the above procedure worked since we were able to show that the limit in (5.4.35) was zero. This in turn depended upon inequality (5.4.36), where we used the "rough estimate" $|e^{iz}| \le 1$ for z on the semicircle C_R. However, consider the integral

$$\int_{-\infty}^{\infty} \frac{x\sin x}{1+x^2}\,dx = \mathrm{Im}\left[\lim_{R\to\infty}\int_{-R}^{R} \frac{xe^{ix}}{1+x^2}\,dx\right]$$

Proceeding as above, and again using the rough estimate $|e^{iz}| \le 1$ on the semicircle C_R: $z = Re^{i\theta}$, $0 \le \theta \le \pi$, we would find

$$\left|\int_{C_R} \frac{ze^{iz}}{1+z^2}\,dz\right| \le \frac{\pi R^2}{R^2-1}$$

which does not imply that the term on the left is zero in the limit $R \to \infty$. Nevertheless, this limit can still be shown to be zero, but the proof is more subtle. It is given by the following general result, which provides an extension of Theorem (5.4.22) and is known as Jordan's lemma.

(5.4.37) ***Theorem* (Jordan's Lemma).** Let $p(z)$ and $q(z)$ be polynomials and C_R the semicircle

$$C_R\colon\ z = Re^{i\theta} \qquad 0 \le \theta \le \pi$$

If the zeros of $q(z)$ all lie inside of the circle $|z| = R$, then

(5.4.38)
$$\lim_{R\to\infty}\left[\int_{C_R} \frac{p(z)\,e^{iaz}}{q(z)}\,dz\right] = 0 \qquad (a \text{ is real})$$

if either

(a) $a > 0$ and degree (q) − degree $(p) \ge 1$, or

(b) $a = 0$ and degree (q) − degree $(p) \ge 2$

Proof. From the parametrization of C_R, we have

$$|e^{iaz}| = e^{-aR\sin\theta} \qquad (z \text{ on } C_R)$$

Let n_p = degree (p) and n_q = degree (q). Now in Problem 23g of Section 1.1, it is shown that for R large enough,

$$\left|\frac{p(z)}{q(z)}\right| \le kR^{n_p - n_q} \qquad (K \text{ is a constant})$$

Hence, the triangle inequality for integrals and the above yields

(5.4.39)
$$\left|\int_{C_R} \frac{p(z)e^{iaz}}{q(z)}\,dz\right| \le KR^{n_p - n_q + 1}\int_0^{\pi} e^{-aR\sin\theta}\,d\theta$$

We would now like to show that the right-hand side approaches zero as $R \to \infty$.

For $a = 0$, if $n_p - n_q + 1 < 0$, then the right-hand side approaches zero. This proves part (b) of the theorem, since there we assume that $n_p - n_q \le -2$. This in turn is a proof of Theorem (5.4.22). Now, for $a > 0$, since $\sin\theta \ge 0$ in this θ range, $e^{-aR\sin\theta} \le 1$ and the right-hand side above is certainly bounded by $\pi KR^{n_p - n_q + 1}$. However, our assumption in (a) is that degree(q) − degree$(p) \ge 1$ and this yields $n_p - n_q + 1 \le 0$. Hence, it need not be true that in all cases

$$R^{n_p - n_q + 1} \xrightarrow[R\to\infty]{} 0.$$

Thus, we need a more delicate estimate of the exponential $e^{-aR\sin\theta}$ for $0 \le \theta \le \pi$.

Consider first the range $0 \le \theta \le \pi/2$. In Figure 5.4.6*a*, we plot $\sin\theta$ for this range and we see that its graph lies above the line shown there, that is,

$$\sin\theta \ge \frac{2\theta}{\pi} \qquad \text{for } 0 \le \theta \le \frac{\pi}{2}$$

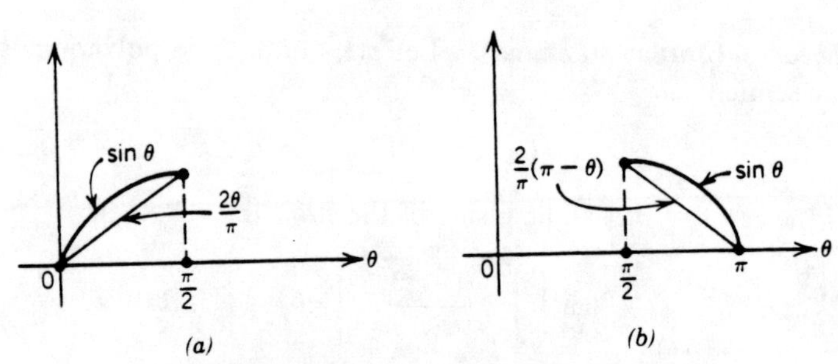

Figure 5.4.6 Jordan's lemma.

Hence, since a and R are positive,

$$e^{-aR\sin\theta} \le e^{-2aR\theta/\pi}$$

and thus,

$$\int_0^{\pi/2} e^{-aR\sin\theta}\,d\theta \le \frac{\pi}{2aR}(1 - e^{-aR})$$

A similar analysis for $\pi/2 \le \theta \le \pi$ will also yield (see Figure 5.4.6*b*)

$$\int_{\pi/2}^{\pi} e^{-aR\sin\theta}\,d\theta \le \frac{\pi}{2aR}(1 - e^{-aR})$$

Substituting these results into (5.4.39) leads to

$$\left|\int_{C_R} \frac{p(z)e^{iaz}}{q(z)}\,dz\right| \leq \frac{\pi K}{a}(1 - e^{-aR})R^{n_p - n_q}$$

Finally, since $a > 0$ and $n_p - n_q \leq -1$, we have that the limit of the right-hand side is zero as $R \to \infty$.

■

(5.4.40) ***Example.*** Evaluate

$$I = \int_{-\infty}^{\infty} \frac{\sin x}{x}\,dx$$

Solution. Note that there are two sources of difficulty here. First, the integrand is singular at $x = 0$. However, since $\sin x \approx x$ for small x, this singularity is *removable*. Second, the integral is improper due to the infinite limits of integration. In elementary calculus, it is shown that even though the integrand is bounded only by $|x|^{-1}$, which is *not integrable* at $x = \pm\infty$, the improper integral still exists due to the oscillatory nature of $\sin x$ and is given by the Cauchy principal value

$$I = \lim_{R \to \infty} \int_{-R}^{R} \frac{\sin x}{x}\,x \tag{5.4.41}$$

Now, following example (5.4.30), we might write

$$I = \text{Im}\left[\lim_{R \to \infty} \int_{-R}^{R} \frac{e^{ix}}{x}\,dx\right]$$

However, we have not eliminated all of the difficulties since the integral $\int_{-R}^{R} (e^{ix}/x)\,dx$ does not exist due to the singularity at $x = 0$ (e^{ix} does not vanish at $x = 0$). To get around this, we first rewrite (5.4.41) as [remember $(\sin x)/x$ is integrable at $x = 0$]

$$I = \lim_{R \to \infty}\left\{\lim_{\epsilon \downarrow 0}\left[\int_{-R}^{-\epsilon} \frac{\sin x}{x}\,dx + \int_{\epsilon}^{R} \frac{\sin x}{x}\,dx\right]\right\}$$

We now replace $\sin x$ by $Im(e^{ix})$ in the integrals above, which we can do since $x = 0$ is no longer in the intervals of integration, and we finally get

$$I = \text{Im}\left\{\lim_{R \to \infty}\left[\lim_{\epsilon \downarrow 0}\left(\int_{-R}^{-\epsilon} \frac{e^{ix}}{x}\,dx + \int_{\epsilon}^{R} \frac{e^{ix}}{x}\,dx\right)\right]\right\} \tag{5.4.42}$$

We now seek a closed contour in the complex plane, part of which consists of the two real intervals $[-R, -\epsilon]$ and $[\epsilon, R]$, and such that the limits above are easily computed on the remaining parts. Here, we choose the **indented contour** Γ of Figure 5.4.7a, which consists of the two intervals in question, the usual semicircle C_R of radius R, and an

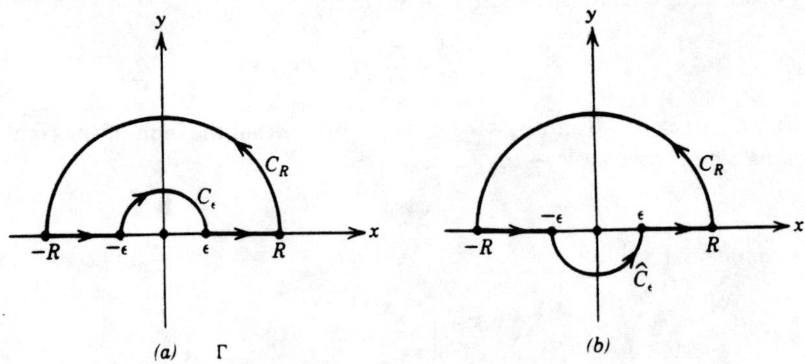

Figure 5.4.7 Example (5.4.40).

additional semicircle C_ϵ of radius ϵ given by:

(5.4.43)
$$C_\epsilon: \quad z = \epsilon e^{-i\theta} \qquad -\pi \le \theta \le 0$$

We now consider the contour integral of e^{iz}/z over Γ. First, since there are no singularities inside, we have

$$\oint_\Gamma \frac{e^{iz}}{z}\,dz = 0$$

We now equate this result to the sum of the integrals over the four parts of Γ to get

$$0 = \int_{-R}^{-\epsilon} \frac{e^{ix}}{x}\,dx + \int_{\epsilon}^{R} \frac{e^{ix}}{x}\,dx + \int_{C_R} \frac{e^{iz}}{z}\,dz + \int_{C_\epsilon} \frac{e^{iz}}{z}\,dz$$

From this and (5.4.42), we have

$$I = -\operatorname{Im}\left\{ \lim_{R\to\infty} \left[\int_{C_R} \frac{e^{iz}}{z}\,dz \right] + \lim_{\epsilon \downarrow 0} \left[\int_{C_\epsilon} \frac{e^{iz}}{z}\,dz \right] \right\}$$

By Jordan's lemma, the first limit is zero, and from the parametrization (5.4.43) we get

$$I = -\operatorname{Im}\left[\lim_{\epsilon\downarrow 0} \int_{-\pi}^{0} -ie^{i\epsilon e^{-i\theta}}\,d\theta \right]$$

$$= \lim_{\epsilon\downarrow 0} \left[\int_{-\pi}^{0} e^{\epsilon \sin\theta} \cos(\epsilon\cos\theta)\,d\theta \right]$$

Since $\sin\theta$ and $\cos\theta$ are bounded for all θ, we can expand $e^{\epsilon\sin\theta}\cos(\epsilon\cos\theta)$ in a Taylor series in ϵ (for small ϵ) and get

$$e^{\epsilon\sin\theta}\cos(\epsilon\cos\theta) = 1 + (\text{higher powers of } \epsilon)$$

Hence, the limit above is π, and we have

$$\int_{-\infty}^{\infty} \frac{\sin x}{x}\,dx = \pi \tag{5.4.44}$$

The reader is urged to redo this problem using the contour in Figure 5.4.7*b* which has the small semicircle $\hat{C}_\epsilon$ located in the lower half-plane. Here, there *is* a singularity inside Γ.

PROBLEMS (The answers are given on page 435.)

1. Evaluate the following trigonometric integrals:

(a) $\displaystyle\int_0^{2\pi} \frac{dx}{2+\cos(2x)}$ (b) $\displaystyle\int_{-\pi}^{\pi} \frac{dx}{3+2\sin x}$ (c) $\displaystyle\int_0^{\pi} \frac{dx}{(2+\cos x)^2}$

(d) $\displaystyle\int_0^{2\pi} \frac{\sin x\,dx}{3+\cos x}$ (e) $\displaystyle\int_0^{2\pi} \cos^3 x\,dx$ (f) $\displaystyle\int_0^{\pi} \frac{dx}{\cos^2 x + 4\sin^2 x}$

(g) $\displaystyle\int_{-\pi}^{\pi} \frac{dx}{a+\cos x}$ $a^2 > 1$ (h) $\displaystyle\int_0^{2\pi} \frac{dx}{a+\cos^2 x}$ $a > 1$

(i) $\displaystyle\int_{-\pi}^{\pi} \frac{dx}{1-(2a\cos x)+a^2}$ $a \neq \pm 1$

2. Use contour integration to verify the following:

(a) $\displaystyle\int_0^{2\pi} \sin^{2n} x\,dx = \frac{\pi(2n)!}{2^{2n-1}(n!)^2}$ $n = 1, 2, \ldots$

(b) $\displaystyle\int_0^{2\pi} \frac{dx}{a^2\sin^2 x + b^2\cos^2 x} = \frac{2\pi}{ab}$ $a, b > 0$

(c) $\displaystyle\int_{-\pi}^{\pi} (1-\cos x)^n \cos(nx)\,dx = \frac{\pi(-1)^n}{2^{n-1}}$ $n = 0, 1, 2, \ldots$

(d) $\displaystyle\int_0^{\pi} \frac{\cos(nx)\,dx}{1+k\cos x} = \frac{\pi}{\sqrt{1-k^2}}\left(\frac{\sqrt{1-k^2}-1}{k}\right)^n$ $k^2 < 1$

(e) $\displaystyle\int_0^{2\pi} \frac{\cos(n\theta)-\cos(n\phi)}{\cos\theta-\cos\phi}\,d\phi = \frac{\pi\sin(n\theta)}{\sin\theta}$ $n = 1, 2, \ldots$

3. If $f(z)$ is analytic inside and on the circle $|z| = R$, show that for $r < R$,

$$\frac{R^2-r^2}{2\pi}\int_0^{2\pi} \frac{f(Re^{i\phi})\,d\phi}{R^2+r^2-2rR\cos(\phi-\theta)} = f(re^{i\theta})$$

[*Hint:* Let $z = Re^{i\phi}$.]

4. For $0 < \theta < 2\pi$, evaluate in terms of θ and n, $n = 2, 3, \ldots$.

(a) $\displaystyle\int_0^\theta \sin^n x\, dx$ (b) $\displaystyle\int_0^\theta \cos^n x\, dx$

[*Hint:* See Example (5.4.6).]

5. Consider the integral $I = \int_0^{2\pi} dx/[2 + \cos(2x)]$ of Problem 1*a*. This was evaluated using the change of variables $w = e^{iz}$. However, since $\cos(2x) = \frac{1}{2}(e^{2ix} + e^{-2ix})$, it would seem that the change of variables $w = e^{2iz}$ would also be appropriate. Make this change of variables and evaluate I again. Do you get the same answer as before? If you were off by a factor of 2, do you know why?

6. (a) Consider the integral $I = \int_0^{2\pi} dx/[2 + \cos(x/2)]$.
  (i) Why won't the usual substitution $w = e^{iz}$ lead to an integral which can be evaluated by the Cauchy integral formula?
  (ii) First make the real substitution $y = x/2$, and then evaluate I.
(b) Evaluate $\int_0^{2\pi} dx/[2 + \cos(x/3)]$. [*Hint:* Let $y = x/3$ and then evaluate the resulting integral using the approach of Example (5.4.6).]

7. Evaluate the following integrals:

(a) $\displaystyle\int_{-\infty}^{\infty} \frac{x^2\, dx}{1 + x^4}$ (b) $\displaystyle\int_{-\infty}^{\infty} \frac{x\, dx}{(x^2 + x + 13)^2}$

(c) $\displaystyle\int_{-\infty}^{\infty} \frac{(x + 2)\, dx}{(x^2 + 1)(x^2 + 4)}$ (d) $\displaystyle\int_0^{\infty} \frac{(x^2 + 12)\, dx}{x^4 + 8x^2 + 16}$

(e) $\displaystyle\int_{-\infty}^{\infty} \frac{dx}{2x^2 + 2x + 5}$ (f) $\displaystyle\int_0^{\infty} \frac{x^2\, dx}{(1 + x^2)^2}$

8. (a) Use the approach of Example (5.4.24) to show that

$$\int_0^{\infty} \frac{dx}{1 + x^n} = \frac{\pi}{n \sin(\pi/n)} \qquad n = 2, 3, \ldots$$

(b) Evaluate $\int_0^{\infty} x^m\, dx/(1 + x^n)$, where $n - m \geq 2$. [*Hint:* See part (*a*).]
(c) Evaluate $\int_0^{\infty} x^2\, dx/(x^6 + 2x^3 + 1)$.

9. The procedure for evaluating integrals of the type (5.4.12) used the semicircle C_R of Figure 5.4.3*a*. Except for the nice symmetry of circles, which makes the calculations somewhat easier, pretty much any curve all of whose parts go off to infinity in the upper half-plane will do. Redo Example (5.4.15) using the closed rectangle shown in Figure 5.4.8*a*. [*Hint:* You might have need of the inequality $|R^2(1 + i\lambda)^2 + 1|^2 \geq R^4 - 4R^2 + 1$, where λ is real and $|\lambda| \leq 1$.]

10. Evaluate the following integrals:

(a) $\displaystyle\int_0^{\infty} \frac{\cos(5x)}{x^2 + 1}\, dx$ (b) $\displaystyle\int_{-\infty}^{\infty} \frac{x \sin x\, dx}{x^2 + 4x + 5}$

(c) $\displaystyle\int_{-\infty}^{\infty} \frac{\cos^2 x}{(1 + x^2)^2}\, dx$ [*Hint:* Write $\cos^2 x = \frac{1}{4}(e^{2ix} + 2 + e^{-2ix}).] = \frac{1}{2}\text{Re}(e^{2ix} + 1)$

(a)

(b)

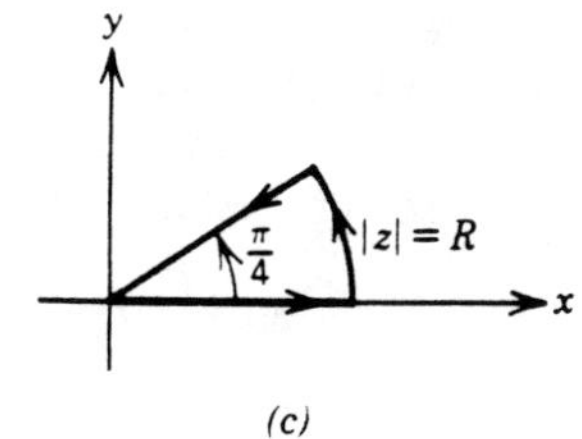

(c)

Figure 5.4.8

11. (a) Show that if $a, b > 0$, then

$$\int_{-\infty}^{\infty} \frac{\cos^2 x \, dx}{(x^2 + a^2)(x^2 + b^2)} = \frac{\pi}{2(b^2 - a^2)} \left[\frac{1 + e^{-2a}}{a} - \frac{1 + e^{-2b}}{b} \right]$$

[*Hint:* See the hint in Problem 10c.] Also, rederive the answer of Problem 10c by letting $a, b \to 1$.

(b) For $a, b > 0$, and in terms of a and b, what is the value of $\int_{-\infty}^{\infty} \sin^2 x \, dx/[(x^2 + a^2)(x^2 + b^2)]$?

(c) For $a, b > 0$, evaluate $\int_{-\infty}^{\infty} \sin^4 x \, dx/[(x^2 + a^2)(x^2 + b^2)]$.

12. Use the approach of Example (5.4.40) to evaluate

(a) $\displaystyle\int_{-\infty}^{\infty} \frac{\sin^2 x}{x^2} dx$ (b) $\displaystyle\int_{-\infty}^{\infty} \frac{(\sin x)^{2n}}{x^2} dx$ $n = 2, 3, \ldots$

(c) $\displaystyle\int_{0}^{\infty} \frac{(1 - x^n)x^p}{1 - x^m} dx$ where $n, m,$ and p are all positive integers with $m - n - p \geq 2$ [*Hint:* See Figure 5.4.8*b*.]

13. (a) Evaluate $\int_0^{\infty} \cos(x^2) dx$. This is called a **Fresnel integral** and arises in optics. [*Hint:* Integrate e^{iz^2} around the contour of Figure 5.4.8*c*. For the limit over the circular arc, use the same procedure as in Theorem (5.4.37). Also, we know that $\int_0^{\infty} e^{-x^2} dx = \sqrt{\pi}/2$.]

(b) Evaluate $\int_0^{\infty} \sin(x^2) dx$.

5.5 THE SCHWARZ–CHRISTOFFEL TRANSFORMATION AND ITS APPLICATIONS

The solution of a variety of interesting physical problems using conformal mappings has already been taken up at length in Chapter 4. There, we used the mapping properties of the elementary functions (e^z, $\sin z$, $\log z$, etc.) to develop other (more complicated) mappings. Here, we will show how additional complex functions, defined by integrals, can be derived which map regions bounded by polygons onto the upper half-plane.[1] We will begin by considering the problem of finding a function which maps the upper half-plane onto a given polygonal region. The inverse of this will then be the desired mapping.

The general situation is depicted later on in Figure 5.5.2*a*. There, the upper half-plane is to be mapped by $w = f(z)$ onto a region bounded by a polygon having vertices at $w_1, w_2, \ldots, w_n$. The points $x_1, x_2, \ldots, x_n$ are those real values of z whose images under the mapping are these vertices and which have the property that

> *The image of the real interval (x_j, x_{j+1}) is the line segment connecting the vertices w_j and w_{j+1}.*

Thus, the function we are looking for must have the property that the *slope of the image of each real interval between neighboring points of the set $x_1, x_2, \ldots, x_n$ is constant*. Now we know (see Section 4.3) that if C_z: $z = z(t)$, $t_0 \le t \le t_1$, is a curve in the z-plane, with image C_w: $w = w(t) = f(z(t))$, $t_0 \le t \le t_1$, then $\arg[w'(t)]$ is the local polar angle made by the tangent line to the curve C_w. Hence, by the above criteria

(5.5.1) *The mapping must have the property that $\arg(w')$ is constant on the image of each of the real intervals between neighboring pairs of the set of points $x_1, x_2, \ldots, x_n$.*

To motivate the general result, we will consider two simple, and special, problems whose solutions we can easily develop. The "polygons" in the w-plane will be a wedge and a semiinfinite strip (note that these can be considered as polygons with one vertex at $w = \infty$). The first problem is to find a mapping of the upper half-plane onto a wedge of angle ϕ_0, as shown in Figure 5.5.1*a*. The game plan is to

(i) First map the half-plane onto a similar wedge using the power map $\zeta = z^{(1/\pi)\phi_0}$.

(ii) Then find constants a and b so that the linear map $w = a\zeta + b$ will properly align the wedge.

Hence, we wind up with the composite function $w = az^{(1/\pi)\phi_0} + b$, which has the property that

[1] We remind the reader that oftimes we can easily develop the solution to problems whose physical domain is the upper half-plane [heat conduction with the temperature prescribed on the real axis, Example (4.5.4); fluid flow over a half-plane, Example (3.4.11)—an electrostatic field due to a point charge above a conducting half-plane Example (3.5.11); etc.]

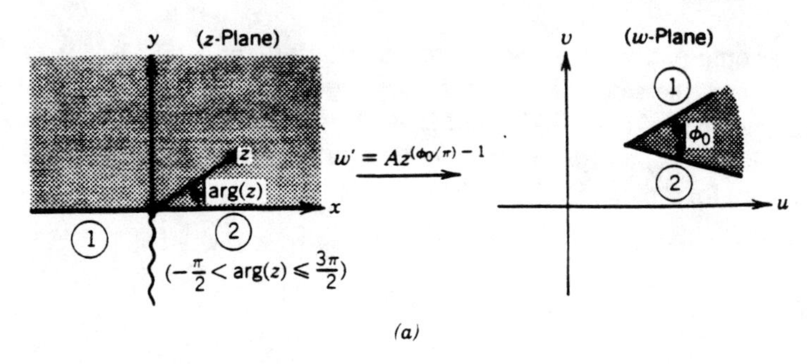

(a)

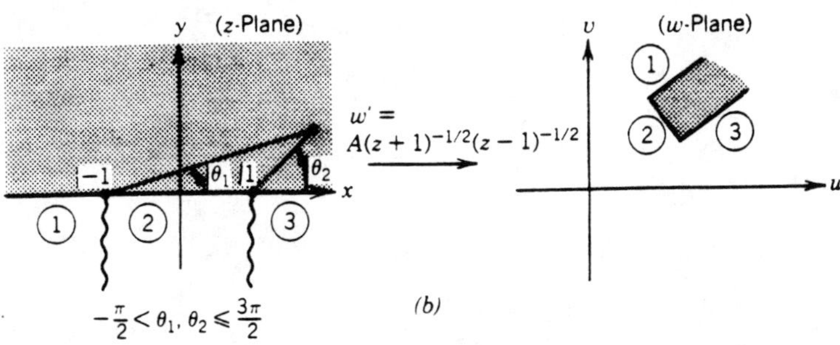

(b)

Figure 5.5.1

(5.5.2) $$w'(z) = Az^{(1/\pi)\phi_0 - 1} \qquad A = a\phi_0/\pi$$

Note from Figure 5.5.1*a* that if we choose the branch of the power function $z^{(1/\pi)\phi_0}$ so that $-\pi/2 < \arg(z) \leq 3\pi/2$, then from the above, and for z real,

$$\arg[w'(x)] = \arg(A) + \begin{cases} \phi_0 - \pi & x < 0 \\ 0 & x > 0 \end{cases}$$

Hence, $\arg(w')$ is constant on each of the real intervals $x < 0$ and $x > 0$.

As our second example, consider the problem of finding a mapping of the upper half-plane onto a semi-infinite strip (see Figure 5.5.1*b*). Our game plan here is to

(i) First, use an inverse sine map $\zeta = \sin^{-1} z$ to produce a *similar* strip in the ζ-plane (see Section 2.8). The points $z = -1$ and $z = 1$ are to be mapped onto the vertices of the strip, and the real intervals $(-\infty, -1)$ $(-1, 1)$, and $(1, \infty)$ are to be mapped onto the sides.

(ii) Then, use a magnification $\eta = k\zeta (k > 0)$, which will produce the proper size of the base of the strip.

(iii) Finally, use a linear map $w = a\zeta + b$ to properly align the strip.

Putting these together, the composite map is $w = ak\sin^{-1}(z) + b$. Now (without concerning ourselves with choosing a proper branch), since $(d/dz)(\sin^{-1} z) = (1 - z^2)^{-1/2} = i(z + 1)^{-1/2} \cdot (z - 1)^{-1/2}$, the mapping has the property that

(5.5.3) $$w'(z) = A(z + 1)^{-1/2}(z - 1)^{-1/2}$$

where A is a complex constant and the real intervals $(-\infty, -1)$, $(-1, 1)$, and $(1, \infty)$ are mapped onto the sides of the strip (see Figure 5.5.1*b*). If we now, as before, choose the branches of $(z+1)^{-1/2}$ and $(z-1)^{-1/2}$ as shown in the figure, with the angles $\theta_1 = \arg(z+1)$ and $\theta_2 = \arg(z-1)$ both chosen in the range $-\pi/2 < \theta_1$, $\theta_2 \le 3\pi/2$, then for z real, (5.5.3) yields

$$\arg[w'(x)] = \arg A + \begin{cases} -\pi & -\infty < x < -1 \\ -\dfrac{\pi}{2} & -1 < x < 1 \\ 0 & 1 < x < \infty \end{cases}$$

Hence, $\arg(w')$ is again constant on each of the three real intervals which map onto the sides of the half-strip.

We now generalize the mappings in (5.5.2) and (5.5.3) and consider a function $w = f(z)$ with the property that

(5.5.4)
$$\frac{dw}{dz} = A(z - x_1)^{\beta_1}(z - x_2)^{\beta_2}\cdots(z - x_n)^{\beta_n}$$

where $x_1, \ldots, x_n$ and $\beta_1, \ldots, \beta_n$ are real constants and A is a complex constant. We will make the following assumptions about the mapping

(5.5.5) ***Assumptions***

(a) It is analytic and conformal in the upper half-plane.

(b) It maps the real axis onto a polygon, with the real quantities $x_1, x_2, \ldots, x_n$ mapping onto the vertices $w_1, w_2, \ldots, w_n$ of the polygon and the intervals between the x's mapping onto the respective line segments between the w's (see Figure 5.5.2*a*).

(c) The upper half-plane is mapped onto the region bounded by the polygon.

■

First note that the mapping in (5.5.4) will be analytic in the upper half-plane if the branch of each of the power functions $(z - x_k)^{\beta_k}$, $k = 1, 2, \ldots, n$ is chosen so that the branch cut does not lie in the upper half-plane. One such choice is shown in Figure 5.5.2*a*, where, for each k,

(5.5.6)
$$(z - x_k)^{\beta_k} = |z - x_k|^{\beta_k} e^{i\beta_k\theta_k} \qquad -\frac{\pi}{2} < \theta_k \le \frac{3\pi}{2}$$

This function is conformal since from (5.5.4), $w' = 0$ in the upper half-plane. Now, in order that assumption (c) in (5.5.5) be satisfied, the points $w_1, w_2, \ldots, w_n$ must be such that as you proceed from one to the other, you progress counterclockwise. This is so since conformal maps preserve relative orientation, and the upper half-plane lies to your left as you proceed from x_1 *to* x_2 *to* ... *to* x_n. We now investigate those conditions which are needed so that the real axis maps onto a closed polygon.

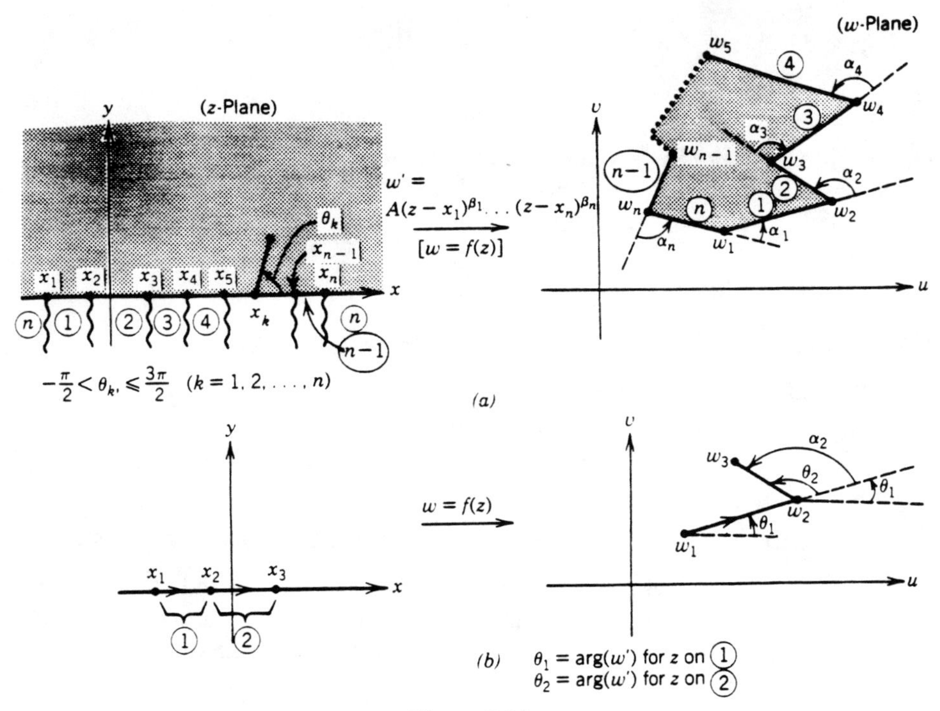

w = f(z)

(b) $\theta_1 = \arg(w')$ for z on (1)
$\theta_2 = \arg(w')$ for z on (2)

Figure 5.5.2

First, note from Figure 5.5.2*a* and equation (5.5.6) that for z real,

$$\text{(5.5.7)} \qquad \arg[w'(x)] = \arg(A) + \begin{cases} \pi \sum_{k=1}^{n} \beta_k & \text{for } -\infty < x < x_1 \\ \pi \sum_{k=l+1}^{n} \beta_k & \text{for } x_l < x < x_{l+1} \qquad (l = 1, \ldots, n-1) \\ 0 & \text{for } x_n < x < \infty \end{cases}$$

Hence, $\arg(w')$ is constant for z on each interval of the real axis shown in the figure. From this result, we also find that for each $j = 1, 2, \ldots, n$,

(5.5.8) $\quad -\pi\beta_j =$ change in $\arg(w')$ as x crosses the point x_j from left to right on the real axis

Now, consider a neighborhood of, say, the vertex w_2, as shown in Figure 5.5.2*b*. As x passes from left to right through x_2, its image will jump from the line segment between w_1 and w_2 to the line segment between w_2 and w_3. Hence, from the figure, the argument change is also given by the angle α_2 shown. Note that α_2 *is the exterior angle of the polygon at* $w = w_2$. We can argue the same way for each vertex

$w = w_k$, and hence we have, from this and from (5.5.8),

$$\beta_k = -\frac{1}{\pi}\alpha_k \qquad k = 1, 2, \ldots, n \tag{5.5.9}$$

where **α_k is the exterior angle of the polygon at the vertex w_k**. This is not all, however, for we have from geometry that the sum of the exterior angles of a polygon is 2π. Hence, the $\{\beta_k\}$ in (5.5.9) must also satisfy

$$\sum_{k=1}^{n} \beta_k = -2$$

What we have accomplished so far can be summed up as follows.

(5.5.10) ***Theorem.*** The mapping

$$w = f(z) = A\int^{z} (z - x_1)^{-\alpha_1/\pi} \cdots (z - x_n)^{-\alpha_n/\pi}\, dz + B$$

where $\sum_{k=1}^{n} \alpha_k = 2\pi$, maps the upper half-plane onto *some* region bounded by a polygon with vertices at $w_k = f(x_k)$, $k = 1, \ldots, n$, and with exterior angles respectively given by $\alpha_1, \alpha_2, \ldots, \alpha_n$. The branches of the powers are chosen as in (5.5.6), and the integrals is to be taken over any contour not intersecting these cuts.

■

Now, the problem which we posed at the beginning of this section was to *find a mapping of a region bounded by a given polygon onto the upper half-plane*. The next result, which we will not prove, tells us that the inverse of the function given in the previous theorem provides the answer. It is arrived at by interchanging z and w in the equation $dw/dz = A(z - x_1)^{-\alpha_1/\pi} \cdots (z - x_n)^{-\alpha_n/\pi}$.

(5.5.11) ***Theorem*** **(Schwarz–Christoffel)**

(a) There are values of the real constants $u_1, u_2, \ldots, u_n$ and the complex constants A and B so that the function $w = f(z)$, defined implicitly by

$$z = A\int^{w} (w - u_1)^{-\alpha_1/\pi} \cdots (w - u_n)^{-\alpha_n/\pi} dw + B$$

$$f(z_k) = u_k \qquad k = 1, 2, \ldots, n$$

maps a region bounded by a polygon with vertices at $z = z_1, z_2, \ldots, z_n$ and respective exterior angles $\alpha_1, \alpha_2, \ldots, \alpha_n$ onto the upper half-plane (see Figure 5.5.3). *Three of the real quantities* $u_1, u_2, \ldots, u_n$ *can be* chosen arbitrarily. The branches of each power function are such that

$$(w - u_k)^{-\alpha_k/\pi} = |w - u_k|^{-\alpha_k/\pi} e^{-i\phi_k\alpha_k/\pi} \qquad k = 1, \ldots, n$$

$$\frac{-\pi}{2} < \phi_k \le \frac{3\pi}{2} \tag{5.5.12}$$

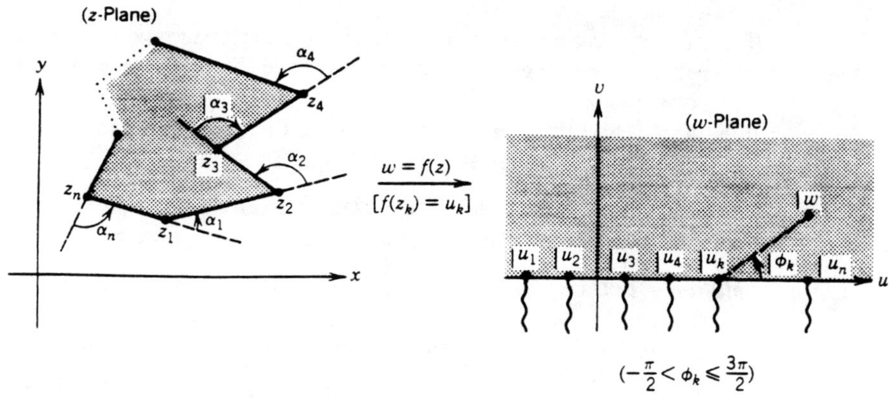

Figure 5.5.3 Schwarz–Christoffel Theorem.

and the integration is over any contour which does not intersect the branch cuts (see the figure).

(b) If u_n (or u_1) in the above is chosen to be the point at infinity, then with an appropriate choice of $u_1, u_2, \ldots, u_{n-1}$ (or $u_2, \ldots, u_n$) A and B, the function $w = f(z)$ defined implicitly by

$$z = A \int^{w} (w - u_1)^{-\alpha_1/\pi} \cdots (w - u_{n-1})^{-\alpha_{n-1}/\pi} dw + B$$

(5.5.13)
$$\begin{cases} f(z_k) = u_k & k = 1, \ldots, n-1 \\ f(z_n) = \infty \end{cases}$$

accomplishes the same task as in part (a), except that now *only two of the* u_k*'s can be chosen arbitrarily.*

■

We offer the following observations concerning the above results and then proceed to some applications.

(i) The formula in part (a) contains $2n$ independent real constants (nu_k's, $n\alpha_k$'s, and four *real* quantities in A and B, from which we subtract the three arbitrarily chosen u values and the one constraint $\sum_{k=1}^{n} \alpha_k = 2\pi$). Note that this is the same count as the $2n$ independent coordinates of the n vertices of an arbitrary polygon.

(ii) It is not really surprising that *three* of the u_k's can be arbitrarily assigned. This is so since we know from Section 2.5 that we can find a bilinear map of the upper half-plane *onto itself* which takes three arbitrarily assigned real values onto three other arbitrarily assigned real values.

(iii) The remaining $n - 3$ values of the u_k's and the magnitude of A will adjust the polygon so that the sides are the proper size (this is not obvious).

(iv) $\arg(A)$ and the constant B produce a final rotation and translation so that the polygon is properly aligned.

(5.5.14) ***Example.*** Find a function which maps a triangle onto the upper half-plane. Discuss the special case of an isoceles triangle (see Figure 5.5.4).

Solution. Let us assume that the triangle has two real vertices at $z_1 = -x_0$ and $z_2 = x_0$, and the third vertex is at $z = z_3$. (A more general triangle can be brought into this configuration by an appropriate translation and rotation.) Since there are only three vertices, in (5.5.13) (with $u_3 = \infty$) we can arbitrarily choose u_1 and u_2 to be, say, the symmetric points $u_1 = -1$ and $u_2 = 1$. Then, we get

$$z = A \int_0^w (w + 1)^{-\alpha_1/\pi}(w - 1)^{-\alpha_2/\pi} dw + B \tag{5.5.15}$$

where the following points correspond to each other:

z	$-x_0$	x_0	z_3
w	-1	1	∞

We have, for definiteness, chosen the lower limit of integration in (5.5.13) to be zero. Now, to evaluate A and B, we use the conditions stated above and get

$$\begin{aligned} -x_0 &= A \int_0^{-1} (w + 1)^{-\alpha_1/\pi}(w - 1)^{-\alpha_2/\pi}\, dw + B \\ x_0 &= A \int_0^{1} (w + 1)^{-\alpha_1/\pi}(w - 1)^{-\alpha_2/\pi}\, dw + B \end{aligned} \tag{5.5.16}$$

where the integrations are to be taken over any contours not crossing the branch cuts shown in Figure 5.5.4. In particular, we choose in the first case

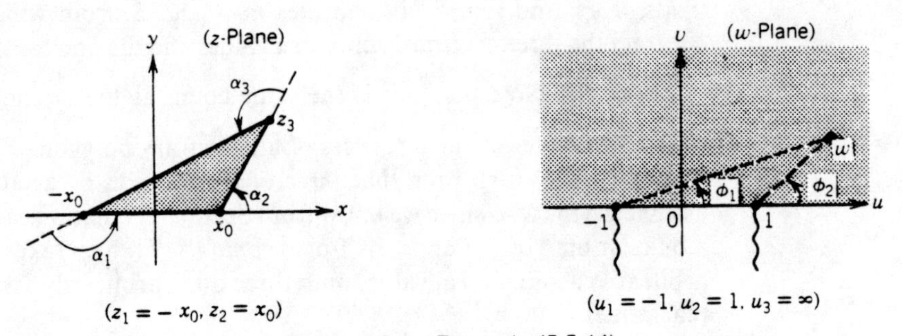

Figure 5.5.4 Example (5.5.14).

the real interval between $w = 0$ and $w = -1$, and in the second the interval between $w = 0$ and $w = 1$. For each of these, and from the w-plane in the figure, we have $\phi_1 = 0$ and $\phi_2 = \pi$. Hence, from (5.5.12),

$$(w+1)^{-\alpha_1/\pi} = |w+1|^{-\alpha_1/\pi}$$

$$(w-1)^{-\alpha_2/\pi} = |w-1|^{-\alpha_2/\pi}e^{-i\alpha_2}$$

Now, the interval between $w = 0$ and $w = -1$ can be parametrized by $w = -t$, $0 \leq t \leq 1$, while the interval between $w = 0$ and $w = 1$ is parametrized by $w = t$, $0 \leq t \leq 1$. Hence, (5.5.16) becomes

$$-x_0 = -Ae^{i\alpha_2}\int_0^1 (1-t)^{-\alpha_1/\pi}(1+t)^{-\alpha_2/\pi}\,dt + B$$

$$x_0 = Ae^{-i\alpha_2}\int_0^1 (1+t)^{-\alpha_1/\pi}(1-t)^{-\alpha_2/\pi}\,dt + B$$

We would now, in general, solve these two equations for A and B and substitute back into (5.5.15). However, let us consider the special case of an isoceles triangle. From Figure 5.5.4, this means that $\alpha_2 = \alpha_1$. Hence, both integrals in the above equations for A and B are the same, and after solving these, (5.5.15) yields

$$z = \left[\frac{x_0 e^{i\alpha_1}}{\int_0^1 (1-t^2)^{-\alpha_1/\pi}\,dt}\right]\int_0^w (w+1)^{-\alpha_1/\pi}(w-1)^{-\alpha_1/\pi}\,dw \tag{5.5.17}$$

Note that the actual function $w = f(z)$ which maps the triangle onto the upper half-plane is the inverse of the above relationship. This means that we must first perform the integration (a rather difficult task) and then solve for w in terms of z (an even more difficult task, in general). However, we sometimes can gather some interesting information without ever performing the integration. For example, we have not really verified the last of the conditions following (5.5.15), namely, that $z = z_3$ is mapped into $w = \infty$. To do so, we set $w = \infty$ in the integral in (5.5.17) and choose the path of integration to be the imaginary axis, that is,

$$w = it, \qquad 0 \leq t < \infty$$

Now, from Figure 5.5.4, for w on the imaginary axis we have $\phi_2 = \pi - \phi_1$, and by the above and (5.5.12) (remember $\alpha_1 = \alpha_2$),

$$(w+1)^{-\alpha_1/\pi}(w-1)^{-\alpha_1/\pi} = (1+t^2)^{-\alpha_1/\pi}e^{-i\alpha_1}$$

Hence, with $z = z_3$ and $w = \infty$ in (5.5.17), we get (since $dw = i\,dt$)

$$z_3 = \left[\frac{ix_0}{\int_0^1 (1-t^2)^{-\alpha_1/\pi}\,dt}\right]\int_0^\infty (1+t^2)^{-\alpha_1/\pi}\,dt \tag{5.5.18}$$

Note that z_3 lies on the imaginary axis, *as it should for the isoceles triangle.* Also, as $\alpha_1 \to \pi/2$, $z_3 \to \infty$.

In the above example, it was noted that the function which did the actual mapping of a triangle onto a half-plane could be determined only after an integration and a function inversion [see (5.5.17)]. In general, this is rather difficult to accomplish. However, often times the problems of interest involve **degenerate polygons**. By this we mean figures whose boundaries are in some sense the limit of the boundaries of appropriate polygons. For some of these situations, we *can* carry out the integration and inversion, as the next example shows.

(5.5.19) ***Example.*** Find a function which maps the horizontal strip $-\infty < \text{Re}(z) < \infty, 0 < \text{Im}(z) < \pi$ onto the upper half-plane.

Solution. From Figure 5.5.5, we see that even though the strip is not bounded by a polygon, it can be considered as the limit of the shaded triangular region shown in the z-plane, as z_1 and z_2 approach infinity in the directions shown. In these limits, it is clear from the figure that the exterior angles α_1, α_2, and α_3 approach

$$\alpha_1 = \pi, \qquad \alpha_2 = \pi, \qquad \alpha_3 = 0$$

Hence, we might venture an educated guess that the Schwarz–Christoffel formula

$$z = A \int^{w} (w - u_1)^{-\pi/\pi}(w - u_2)^{-\pi/\pi}(w - u_3)^{0/\pi}\, dw + B$$

will provide the correct answer, where u_1, u_2, u_3 are arbitrary real quantities in the w-plane. First, note that the term $(w - u_3)^0$ is 1 and does not appear in the integrand. Also, we have the option [see Theorem (5.5.11b)] of choosing u_1 to be infinity, and thus it, too, will not appear in the integrand. Finally, we can *still* arbitrarily choose u_2 and u_3, and our

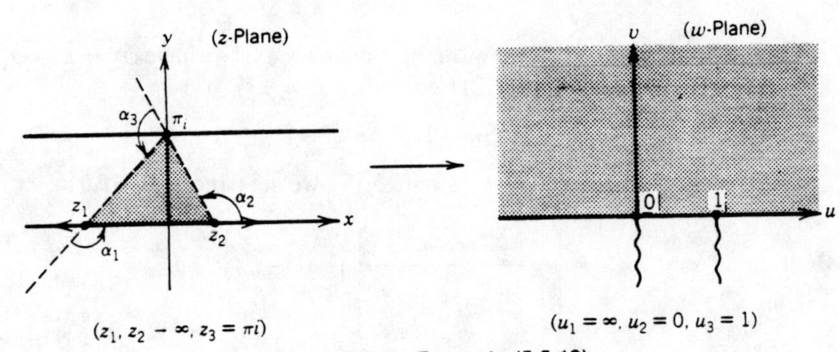

Figure 5.5.5 Example (5.5.19).

choice will be $u_2 = 0$, $u_3 = 1$. Hence, the above becomes

(5.5.20) $$z = A \int^{w} \frac{dw}{w} + B$$

From the figure, since $z_1 = \infty$, $z_2 = \infty$ in the limit, and $z_3 = \pi i$ are to be mapped respectively onto $u_1 = \infty$, $u_2 = 0$, and $u_3 = 1$, we must have in the above that

(5.5.21)
(a) $z = \infty$ corresponds to $w = 0$ or $w = \infty$
(b) $z = \pi i$ corresponds to $w = 1$.
(c) The real z-axis is mapped onto the negative real axis in the w-plane.
(d) The line $z = x + \pi i$, $-\infty < x < \infty$, is mapped onto the

$$\begin{cases} \text{Real interval } 0 < u < 1 & \text{for } x > 0 \\ \text{Real interval } 1 < u < \infty & \text{for } x < 0 \end{cases}$$

In this case, we can easily integrate (5.5.20) to get

$$z = A \log w + B$$

where we choose the branch

$$\log w = \ln|w| + i\phi \qquad \frac{-\pi}{2} < \phi \leq \frac{3\pi}{2}$$

Note from this expression for z that *both* conditions in (5.5.21a) are automatically satisfied, and condition (b) there yields

$$\begin{aligned} i\pi &= A \log 1 + B \\ &= B \end{aligned}$$

since $\log 1 = 0$ for our choice of branch. Hence, we have so far

(5.5.22) $$z = A \log w + \pi i$$

Ordinarily at this stage, there would be one remaining condition to be used to evaluate the constant A; see Example (5.5.14). That this is not the case here is due to the fact that two vertices of our triangle, z_1 and z_2, *both went "off to infinity."* The only conditions which are left are (5.5.2c, d). As we will now show, either of these will determine A uniquely. To see this, note that with w real and negative ($w = u$, $u < 0$), we have $\log w = \ln|u| + \pi i$, and condition (5.5.2c) and the above yield

$$x = A(\ln|u| + i\pi) + i\pi \qquad u < 0, \quad x \text{ is real}$$

Setting $A = A_1 + iA_2$, we get

(5.5.23)
(a) $A_1 \ln|u| - \pi A_2 = x \qquad u < 0, \quad x \text{ is real}$
(b) $\pi A_1 + A_2 \ln|u| = -\pi \qquad u < 0$

Now (b) must hold for all $u < 0$, and this can only be if $A_2 = 0$ and $A_1 = -1$, and thus $A = -1$. Substituting this into (a) above yields

$$\ln|u| = -x \qquad u < 0, \quad x \text{ is real}$$

from which we have

$$x < 0 \Rightarrow u < -1$$

$$x > 0 \Rightarrow -1 < u < 0$$

This tells us that the negative real axis is mapped onto the real interval $-\infty < u < -1$ and the positive real axis onto the interval $-1 < u < 0$.

Now, with $A = -1$, (5.5.22) yields

$$z = -\log w + \pi i$$

Note that the mapping is now defined completely, and hence the remaining condition (5.5.21d) *must automatically be satisfied.* To see this, set $z = x + \pi i$ and $w = u\,(>0)$ above to get

$$x + \pi i = \pi i - \ln u \Rightarrow x = -\ln u$$

This simply verifies (5.5.21d), since it shows that if $x > 0$, then $0 < u < 1$, while if $x < 0$, $u > 1$. Finally, note that we can, in this simple case, invert the mapping to get

$$w = -e^{-z}$$

As the reader should have known all along, the exponential indeed maps an infinite horizontal strip onto the upper half-plane.

■

The next result, which we also state without proof, shows how to find a function which maps an *infinite* region, whose boundary consists of linear segments, onto the upper half-plane.

(5.5.24) ***Theorem.*** There are values of the real constants $u_1, u_2, \ldots, u_n$ and the complex constants A and B so that the function $w = f(z)$ *defined implicitly* by

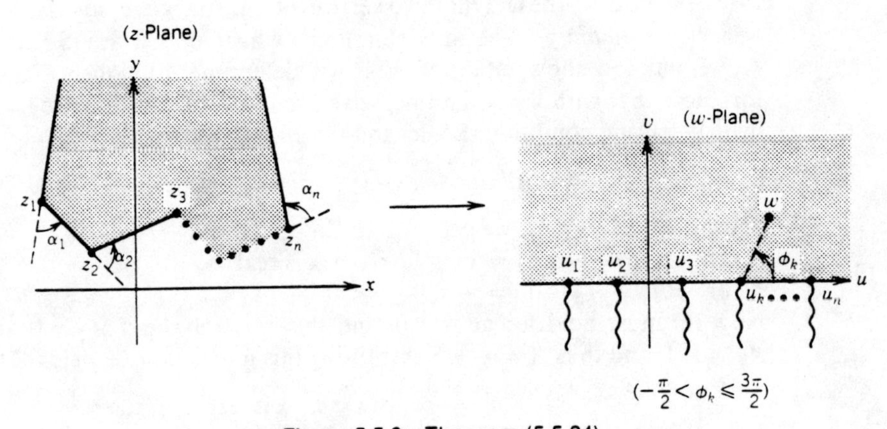

Figure 5.5.6 Theorem (5.5.24).

$$z = A\int^{w}(w-u_1)^{-\alpha_1/\pi}\cdots(w-u_n)^{-\alpha_n/\pi}dw + B$$

(5.5.25)
$$\begin{cases} f(z_k) = u_k & k = 1, 2, \ldots, n \\ f(\infty) = \infty \end{cases}$$

maps the region shown in Figure 5.5.6 onto the upper half-plane. *Two of the real quantities* $u_1, u_2, \ldots, u_n$ can be arbitrarily chosen, and the branches are (as usual) defined by

$$(w-u_k)^{-\alpha_k/\pi} = |w-u_k|^{-\alpha_k/\pi}e^{-i\alpha_k\phi_k/\pi} \qquad -\frac{\pi}{2} < \phi_k \le \frac{3\pi}{2}$$

■

We will conclude this section with two examples. These show how a clever choice of appropriate polygonal boundaries, and the above result, can be used to solve problems more complicated than those before, and how the mappings can be applied to some physically interesting situations. The reader might find it helpful to review (or read for the first time, as the case might be) the material on branch cuts in Section 3.2.

(5.5.26) ***Example.*** Find a function $w = f(z)$ which maps the upper half-plane, cut along the horizontal line $y = 1$, $-\infty < x \le 0$, onto the upper half of the w-plane. If the boundaries of this region of the z-plane are conductors raised to potentials $\phi = 0$ and $\phi = 1$ as shown in Figure 5.5.7b, find parametric equations for the **equipotentials** corresponding to the electric field set up in this region.

Solution. Consider the open-ended polygon shown shaded in the z-plane of Figure 5.5.7a. The exterior angles α_1 and α_2 at the two vertices approach the limits (as $z_2 \to \infty$)

$$\alpha_1 \to -\pi \qquad \text{and} \qquad \alpha_2 \to \pi$$

Since there are only two vertices, we can choose both u_1 and u_2 in (5.5.25) arbitrarily. Hence, with $u_1 = -1$ and $u_2 = 0$, we get

(5.5.27)
$$z = A\int^{w}\frac{(w+1)}{w}dw + B = A(w + \log w) + B$$

We must choose A and B so that

$$z = i \ \text{ when } w = -1 \qquad \text{and} \qquad z = \infty \ \text{ when } w = 0, \infty$$

Note that the second condition is automatically satisfied, while the first condition yields (since $\log(-1) = \pi i$)

(5.5.28)
$$A(-1 + \pi i) + B = i$$

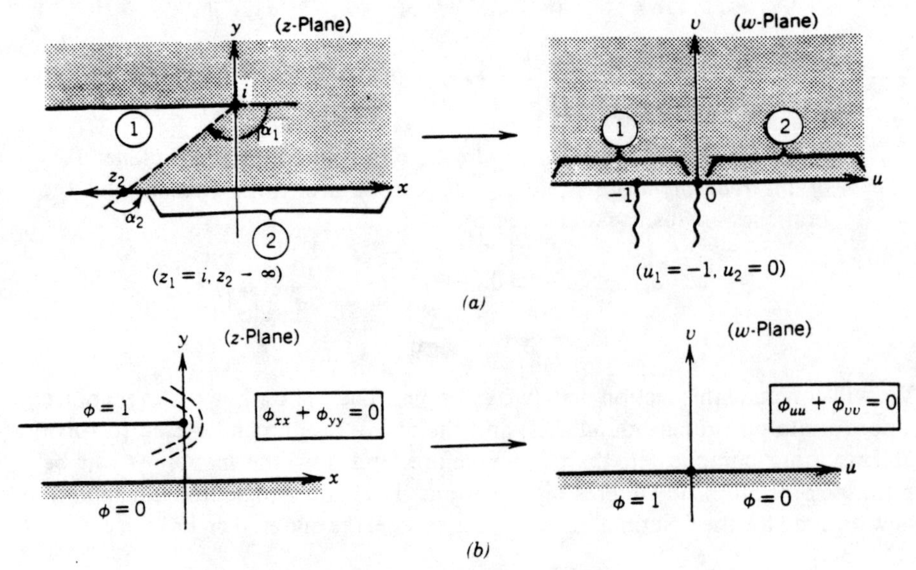

Figure 5.5.7 Example (5.5.26).

We can generate further conditions which will determine A and B by noting from part (a) of Figure 5.5.7 that

$$z \text{ real} \Rightarrow w \text{ is real and positive}$$

Hence, setting $z = x$ and $w = u(>0)$ in (5.5.27) yields

$$x = A(u + \ln u) + B$$

Thus, with $A = A_1 + iA_2$ and $B = B_1 + iB_2$, we get

$$A_2(u + \ln u) + B_2 = 0 \qquad \text{for all } u > 0$$

$$A_1(u + \ln u) + B_1 = x$$

The first of these, since it is to hold for all $u > 0$, yields

$$A_2 = B_2 = 0$$

Using this in (5.5.28) and equating real and imaginary parts there leads to

$$A_1 = B_1 = \frac{1}{\pi}$$

Hence, we find

$$z = \frac{1}{\pi}(w + \log w + 1) \tag{5.5.29}$$

The inverse of this relationship (which cannot be written in terms of elementary functions) is the mapping function which we desire.

Clearly, from the figure, the upper line (labeled ①) is mapped onto the negative real axis, while the real axis in the z-plane (labeled ②) is mapped onto the real positive w-axis. Hence, under this map, the electrostatic problem in the z-plane is mapped into the problem in the w-plane of Figure 5.5.7b. Thus, we seek a harmonic potential function in the upper half of the w-plane whose values on the real axis are as shown in part (b) of the figure. The solution to this boundary value problem is from, Section 4.5,

$$\phi = \frac{1}{\pi}\operatorname{Arg}(w)$$

Now, under conformal, analytic maps, equipotentials are preserved. From the above, the equipotentials are $\operatorname{Arg}(w) = k$, $0 \le k \le \pi$. Hence, from (5.5.29), after setting $w = te^{ik}$, $0 < t < \infty$ (since $\operatorname{Arg}(w) = k$), we find the following parametric equations, in terms of t, for the equipotentials in the (physical) z-plane:

(5.5.30)
$$z = \frac{1}{\pi}(te^{ik} + \ln t + 1 + ik) \qquad 0 < t < \infty, \qquad 0 \le k \le \pi$$

In Problem 7, the reader is asked to show that near the sharp edge $z = i$, the equipotentials are approximately parabolas, as depicted by the dashed curves in Figure 5.5.7b.

■

(5.5.31) ***Example.*** Find a function mapping the upper half-plane, cut along the portion of the imaginary axis $x = 0, 0 \le y \le 1$, onto the upper half of the w-plane. Use this to find the **fluid velocity potential** corresponding to uniform flow from the left over a rigid half-plane (the lower half of the z-plane) having a protrusion in the shape of the line $x = 0, 0 \le y \le 1$.

Solution. First note from Figure 5.5.8b that we can consider the region of interest as the limit of the region in this figure as both z_1 and $z_3 \to 0$. In this limit, we have that the exterior angles α_1 and α_2 satisfy

$$\alpha_1 \to \frac{\pi}{2}, \qquad \alpha_2 \to -\pi, \qquad \alpha_3 \to \frac{\pi}{2}$$

There are three vertices, z_1, z_2, and z_3, which are to have real images u_1, u_2, and u_3. By Theorem (5.5.24), we can choose two of the u's arbitrarily. If we pick $u_1 = -1$ and $u_3 = 1$, then (5.5.25), and the above yield

(5.5.32)
$$z = A\int^{w} \frac{(w - u_2)\,dw}{(w+1)^{1/2}(w-1)^{1/2}} + B$$

Now it seems clear, from symmetry in the figure, that u_2 should be zero (the reader is asked to show this in Problem 4). With this value of u_2, we

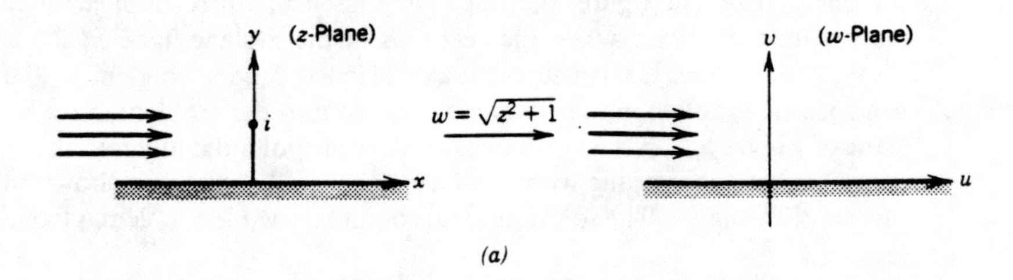

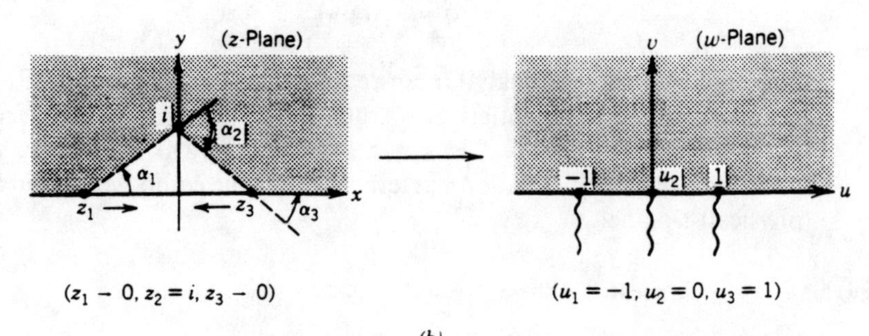

Figure 5.5.8 Example (5.5.31).

can perform the above integration to get

$$z = A(w+1)^{1/2}(w-1)^{1/2} + B$$

To determine A and B, we invoke the conditions that

$$z = 0 \quad \text{when } w = \pm 1 \qquad \text{and} \qquad z = i \quad \text{when } w = 0$$

These lead to $B = 0$ and $A = 1$. Thus, $z = (w+1)^{1/2}(w-1)^{1/2}$, which can easily be solved for w to yield

$$w = (z^2 + 1)^{1/2} \tag{5.5.33}$$

where that branch of the square root is chosen so that the resulting function is analytic in the upper half of the z-plane cut along $x = 0$, $0 \leq y \leq 1$, and has a positive imaginary part here. This final choice of branch often does not easily follow from the previous branch cut in the w-plane. Here, the reader is asked to verify that if we choose the square root function so that

$$\sqrt{\zeta} = |\zeta|^{1/2} e^{i\psi/2} \qquad 0 < \psi \leq 2\pi \tag{5.5.34}$$

then $(z^2+1)^{1/2}$ will have the required properties (see Problem 8). Now, a complex fluid potential for a uniform flow from the left over the lower half of the w-plane is $\Phi = w$ (see Section 3.4). Hence, by (5.5.33), the complex fluid potential for the flow in the z-plane is

$$\Phi(z) = (z^2 + 1)^{1/2} \tag{5.5.35}$$

The reader should note that this is exactly the same as the potential given in Problem 11 of Section 3.4, except that here it was *derived* using the Schwarz–Christoffel transformation.

PROBLEMS (The answers are given on page 436.)

1. Use Theorem (5.5.11a) and the fact that the bilinear function $w = (\zeta - i)/(\zeta + i)$ maps the upper half of the ζ-plane onto the unit disk to show that the inverse of the relationship

$$z = K\int^{w}(w - w_1)^{-\alpha_1/\pi}\cdots(w - w_n)^{-\alpha_n/\pi}\,dw + B$$

maps a polygon with exterior angles $\alpha_1, \ldots, \alpha_n$ onto the unit disk, with the vertices $z_1, z_2, \ldots, z_n$ of the polygon mapping onto the points w_k on the unit circle ($|w_k| = 1$).

2. Show that the function $w = f(z)$ defined implicitly by

$$z = \int^{w}\frac{dw}{(w-1)^{1/2}(w+1)^{1/2}\left(w - \dfrac{1}{k}\right)^{1/2}\left(w + \dfrac{1}{k}\right)^{1/2}} \qquad 0 < k < 1$$

maps a rectangle onto the upper half-plane.

3. (a) Derive an expression, similar to (5.5.17), for a function which maps an isoceles triangle with vertices at $z_1 = -x_0$, $z_2 = x_0$, and $z_3 = ib (b > 0)$, and respective exterior angles α_1, α_2, and α_3, onto the upper half-plane, with the images of the vertices being, respectively, $w = u_1$, u_2, and ∞, with $u_1 \neq 0$ and $u_1 < u_2$. (In Example (5.5.14), we choose $u_1 = -1$ and $u_2 = 1$. Here, they are arbitrary.)

   (b) If $b \to +\infty$ in part (a), the triangle will generate the semiinfinite strip $-x_0 < x < x_0, 0 < y < \infty$, and the exterior angle α_1 will approach $\pi/2$. Show that the result in part (a) can be integrated and will lead to a sine map. Then, verify that this will map the strip onto the upper half-plane no matter what the values of u_1 and u_2.

4. In Example (5.5.31), we argued from symmetry that the real constant u_2 in equation (5.5.32) had to be zero. Here, we will show that it is zero. Remember, the problem is to map the upper half-plane cut along $x = 0, 0 \leq y \leq 1$, onto the upper half of the w-plane, with $z = 0$ mapping into $w = \pm 1$ and $z = i$ mapping into $w = 0$.

   (a) Show that the integration in (5.5.32) can be performed to yield

$$z = A\{(w+1)^{1/2}(w-1)^{1/2} - u_2\,\mathrm{Log}[w + (w+1)^{1/2}(w-1)^{1/2}]\} + B$$

   where the logarithm is the principle logarithm.

   (b) Show that u_2 must be zero if the mapping conditions on z and w mentioned above are to be satisfied.

5. Consider the infinite strip $-h < \text{Im}(z) < h$ with the positive real axis removed (see Figure 5.5.9).
   (a) Consider this region as the limit of the open polygon shown in the figure as z_1 and $z_3 \to \infty$. Choose the image of z_1 to be $u_1 = -1$, the image of $z_2 = 0$ to be $u_2 = 0$, and the image of z_3 to be $w = u_3$, with $u_3 > 0$, and show that the function which maps the slit strip onto the upper half-plane is given implicitly by

$$z = K[\log(w+1) + u_3 \log(w - u_3)] + B$$

   where the branches of the logarithms are

$$\log(w+1) = \ln|w+1| + i\theta_1 \qquad \text{and} \qquad \log(w-u_3) = \ln|w-u_3| + i\theta_2$$

$$\frac{-\pi}{2} < \theta_1, \theta_2 \leq \frac{3\pi}{2}$$

   This ensures that the mapping is analytic in the upper half of the w-plane.

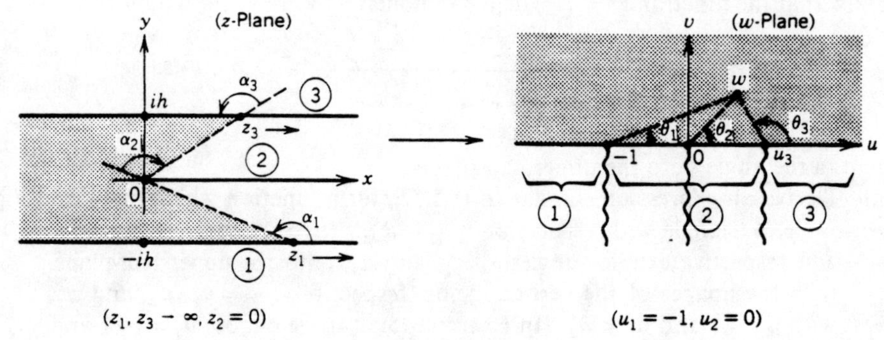

Figure 5.5.9

   (b) Show that the condition which requires $z = 0$ to map into $w = 0$ leads to $B = -Ku_3(\ln u_3 + \pi i)$.
   (c) By symmetry, we might expect u_3 to be 1. Use the fact that the lines marked ①, ②, and ③ in the z-plane map onto the respective lines in the w-plane to show that $u_3 = 1$ and $K = -h/\pi$. Hence, derive the mapping

$$z = -\frac{h}{\pi}[\log(w+1) + \log(w-1) - \pi i]$$

   (d) Invert the result in (c) and verify that the function $w = (1 - e^{-\pi z/h})^{1/2}$, where the branch of the square root is taken to be $\xi^{1/2} = |\xi|^{1/2} e^{i\theta/2}$, $0 < \theta \leq 2\pi$, maps the strip onto the upper half-plane. In the process, show that with this choice of square root the positive real z-axis is a branch cut for $[1 - e^{-\pi z/h}]^{1/2}$. [*Hint:* View the function as a composite of simpler maps.]

6. Consider the problem of mapping the region shown in Figure 5.5.10 onto the upper half of the w-plane. The result of such a mapping might be used to find

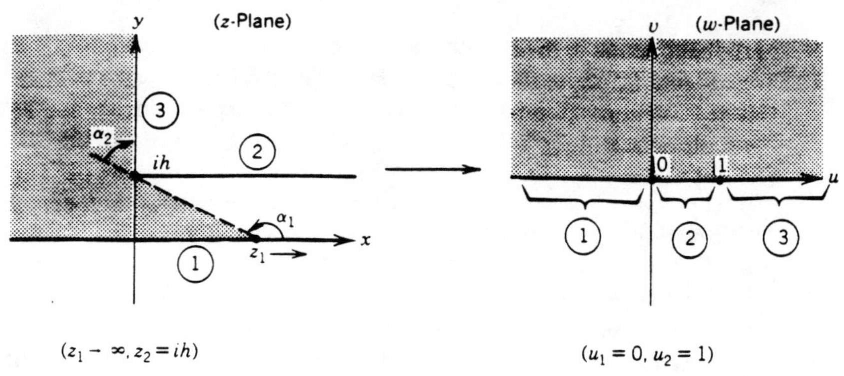

Figure 5.5.10

the fluid potential for a flow which enters a narrow channel (or leaves a narrow channel).

(a) Choose a point z_1 on the real axis and consider the (shaded) open polygon shown in the z-plane. Show that with $u_1 = 0$ and $u_2 = 1$ in (5.5.25), we have, in the limit as $z_1 \to \infty$,

$$z = A \int^{w} \frac{\sqrt{w-1}}{w}\, dw + B$$

(b) Perform the integration above to get

$$z = 2A\left[\sqrt{w-1} + \frac{i}{2}\log(i - \sqrt{w-1}) - \frac{i}{2}\log(i + \sqrt{w-1})\right] + B$$

(c) In the above, the branch of $\sqrt{w-1}$ is the usual one $\sqrt{w-1} = |w-1|^{1/2}e^{i\theta/2}$, $-\pi/2 < \theta \le 3\pi/2$. However, the branch of the logarithms must be chosen so that the resulting function is analytic in the upper half of the w-plane. Show that if we choose $\log \xi = \ln|\xi| + i\phi$, $0 < \phi \le 2\pi$, then the *only possible* place of nonanalyticity of the function in part (b) is along that piece of the parabola $u = v^2/4$, $v \le 0$. [*Hint:* For which w is $i \pm \sqrt{w-1}$ real and positive?]

(d) From Figure 5.5.10, we see that $z = \infty$ is mapped into $w = 0$ and $w = \infty$, and $z = ih$ is mapped into $w = 1$. Show that this first condition is automatically satisfied, while, with our choice of branches, the second yields $B = ih$ in part (b).

(e) Use the fact that the real axis in the z-plane is to map onto the negative real axis is the w-plane to determine the remaining constant A in part (b). [*Hint:* For $w = u < 0$, $\sqrt{w-1} = i\sqrt{1-u}$. Find similar expressions for $\log(i \pm \sqrt{w-1})$. You might want to refer to Example (5.5.26).]

7. Consider equations (5.5.30), which are parametric equations for the equipotentials of example (5.5.26).

(a) Show that the real axis and the line $y = 1, x \leq 0$, are equipotentials. These are the conductors in the example. [*Hint:* Set $k = 0$ and $k = \pi$ in (5.5.30).]

(b) From Figure 5.5.7*b*, the top conductor has a sharp edge at $z = i$, which from (5.5.30) corresponds to $k = \pi$ and $t = 1$. To construct the equipotentials near this tip, we proceed as follows.

(i) Let $k = \pi - \delta$, where δ is positive and very small. Using the Taylor series for $\cos \delta$ and $\sin \delta$, show that $e^{ik} \approx -1 + i\delta + \frac{1}{2}\delta^2$ for small δ.

(ii) Expand $\ln t$ in a Taylor series about the point $t = 1$ to get $\ln t \approx (t-1) - \frac{1}{2}(t-1)^2$.

(iii) Use (i) and (ii) in (5.5.30) to show that the equipotentials in a neighborhood of the sharp edge are approximately the parabolas $x = (\delta^2/2\pi) - [\pi(y-1)^2]/2\delta^2$.

8. Consider Example (5.5.31). With the branch of the square root as in (5.5.34), show that the sequence of mappings $w_1 = z^2$, $w_2 = 1 + w_1$, $w = w_2^{1/2}$, the composite of which is (5.5.33), accomplishes the mapping shown in Figure 5.5.8*b*.

9. A fluid is flowing over a river bottom which has a sharp drop. The topography is as shown in the z-plane of Figure 5.5.11, where the drop is of height h and occurs at $z = ih$.

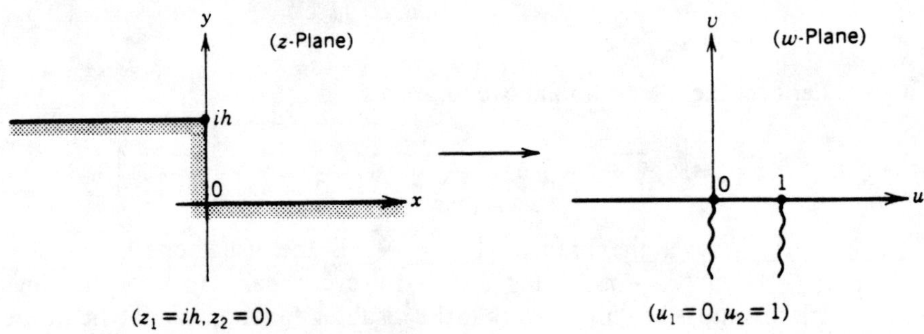

Figure 5.5.11

(a) Show that with the usual choice of the branches of the square roots, the inverse of the relationship $z = \int^w w^{1/2}(w-1)^{-1/2}\,dw + B$, with appropriate choices of A and B, will map the region above the river bottom onto the upper half of the w-plane.

(b) Integrate the above to get

$$z = A[w^{1/2}(w-1)^{1/2} + \text{Log}(w^{1/2} + (w-1)^{1/2})] + B$$

and find A and B so that $z = ih$ maps into $w = 0$ and $z = 0$ maps into $w = 1$.

(c) Show that with the usual branches of $w^{1/2}$ and $(w-1)^{1/2}$, we get

$$z \approx \left(\frac{2h}{\pi}\right) w \qquad \text{for } w \to \infty \text{ and } \text{Im}(z) > 0$$

(d) Use the result in (c) to show that a complex fluid potential $\Phi(z)$ which gives rise to a uniform flow $V_0 z$ at infinity flowing horizontally over the river bottom is given by $\Phi = (2hV_0/\pi)w(z)$, where $w(z)$ is the inverse of the function in part (b).

10. (a) Find a function which maps the upper half-plane, cut along the line $x = 0$, $1 \le y < \infty$, onto the upper half of the w-plane. [*Hint:* First, find a map which yields the region of Example (5.5.31).] (See Figure 5.5.8*a*.)

(b) Find the electrostatic potential $\phi(x, y)$ in the upper half-plane when the real axis and the cut in part (a) are conductors raised to potentials $\phi = 0$ and $\phi = 1$, respectively (see Figure 5.5.12). [*Hint:* To find the boundary value problem in the w-plane, find the images of each conductor as it is approached from the appropriate direction.]

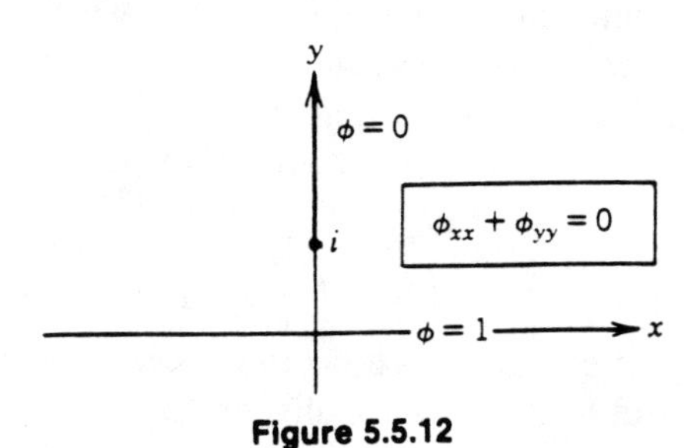

Figure 5.5.12

5.6 SOME CONSEQUENCES OF THE CAUCHY INTEGRAL FORMULA: MOREA'S THEOREM; MEAN VALUE THEOREM; LIOUVILLE'S THEOREM; POISSON'S INTEGRAL FORMULA; MAXIMUM PRINCIPLE

Much of the material in this section will be theoretical in nature and will exhibit some of the more remarkable properties of analytic functions. In the problems, the reader is asked to extend some of these results. We will start with Morea's theorem, which is converse to the Cauchy–Goursat Theorem.

(5.6.1) ***Theorem*** **(Morea).** If $f(z)$ is continuous in a simply connected domain D, and if for *every* simple, closed contour C in D we have

$$\int_C f(\zeta)\, d\zeta = 0$$

then $f(z)$ is analytic in D.

Proof. Let z be an arbitrary point in D and consider a fixed point z_0, also in D. By the above hypothesis, the function $F(z)$, given by the indefinite integral

$$F(z) = \int_{z_0}^{z} f(\zeta)\, d\zeta$$

is defined and single-valued for all z in D, where the integral is taken along any simple contour in D joining z_0 to z. From this, we find that

$$\frac{F(z+h)-F(z)}{h}-f(z)=\frac{1}{h}\int_{z}^{z+h}[f(\zeta)-f(z)]\,d\zeta \tag{5.6.2}$$

Now, if $|h|$ is small enough, the line joining z and $z+h$ lies entirely in D (since D is a domain) and has length $|h|$. With this as the contour in (5.6.2), we get, by the triangle inequality for integrals,

$$\left|\frac{F(z+h)-F(z)}{h}-f(z)\right|\leq M_h$$

where M_h is the maximum value of $|f(\zeta)-f(z)|$ for ζ on the line joining z to $z+h$ [this exists since $f(\zeta)$ is continuous in D]. Again, by the continuity of f, as $h\to 0$, ζ approaches z (remember ζ lies on the line joining z to $z+h$) and hence $M_h\to 0$. Thus, from the above

$$\lim_{h\to 0}\left[\frac{F(z+h)-F(z)}{h}\right]=f(z)$$

for all z in D. Therefore, $F(z)$ is differentiable at all points in D, and hence analytic here, and $F'(z)=f(z)$. Finally, by the Cauchy integral formula for derivatives, Theorem (5.3.17), $F'(z)$ is also analytic, hence so is $f(z)$.

■

The next result shows that an analytic function (and a real harmonic function) is related to its average value over a circle.

(5.6.3) ***Theorem*** **(Mean Value Theorem)**

(a) Let $f(z)$ be analytic in a simply connected domain D. If z_0 is in D, then $f(z_0)$ is the average of the integral of f around any circle in D centered at z_0. That is, *for any* ρ small enough,

$$f(z_0)=\frac{1}{2\pi}\int_0^{2\pi}f(z_0+\rho e^{i\phi})\,d\phi$$

We say that $f(z)$ satisfies the **mean value property**.

(b) If the real-valued function $u(x,y)$ possesses continuous second partial derivatives and is harmonic ($u_{xx}+u_{yy}=0$) in D, then for any point (x_0,y_0) in D we have the **mean value property for harmonic functions**

$$u(x_0,y_0)=\frac{1}{2\pi}\int_0^{2\pi}u(x_0+\rho\cos\phi,\,y_0+\rho\sin\phi)\,d\phi$$

Proof. By the Cauchy integral formula

$$f(z_0)=\frac{1}{2\pi i}\oint_{|\zeta-z_0|=\rho}\frac{f(\zeta)}{\zeta-z_0}\,d\zeta$$

If we parametrize the circle $|\zeta - z_0| = \rho$ by $\zeta = z_0 + \rho e^{i\phi}$, $0 \leq \phi \leq 2\pi$, then the above immediately leads to the result of part (a).

Now, if $u(x, y)$ is harmonic with continuous second partial derivatives, then it has a harmonic conjugate $v(x, y)$ in D (see Problem 18 of Section 5.2). Thus, $u(x, y)$ is the real part of the analytic function $f(z) = u + iv$. If we apply the result of part (a) to f and take real parts of both sides of the equation, then we get the mean value theorem for u.

■

We all take for granted that an nth degree polynomial has n complex roots. This rather profound result will be shown to follow easily from the next theorem.

(5.6.4) ***Theorem* (Liouville).** The only function which is *both entire and bounded* is the constant function.

Proof. If $f(z)$ is entire, then the Cauchy integral formula for derivatives yields

$$f'(z) = \frac{1}{2\pi i} \oint_{|\zeta - z| = R} \frac{f(\zeta)}{(\zeta - z)^2} \, d\zeta$$

where this is to hold for any z and for any value of R. If f is bounded, that is, $|f(\zeta)| \leq M$ for all ζ, then the triangle inequality for integrals yields

$$|f'(z)| \leq \frac{M}{R}$$

If we choose R large enough, we can make $|f'(z)|$ as small as we please. Thus, we can conclude that $f'(z) = 0$ for all z, and therefore $f(z)$ must be constant.

■

The reader should note that the above result implies that **an entire nonconstant function cannot be bounded at infinity**. This is valid for all of the entire elementary functions (polynomials, e^z, $\sin z$, etc.), as the reader is asked to show.

(5.6.5) ***Theorem* (Fundamental Theorem of Algebra).** Any polynomial

$$p(z) = z^n + a_{n-1}z^{n-1} + a_{n-2}z^{n-2} + \cdots + a_1 z + a_0 \qquad n \geq 1$$

has precisely n complex roots (some of these may be equal, or real).

Proof. If we can prove the existence of one root, say z_1, then we would factor $p(z)$ into

$$p(z) = (z - z_1)\, q(z)$$

where $q(z)$ is a polynomial of degree $n - 1$, and apply the argument again to $q(z)$. In this way, we would show the existence of n roots.

Now, assume that $p(z)$ never vanishes. Then, the function $f(z)$ defined by

$$f(z) = \frac{1}{p(z)} = \frac{1}{z^n}\left[\frac{1}{1 + a_{n-1}z^{-1} + \cdots + a_0 z^{-n}}\right]$$

is entire and satisfies $\lim_{z\to\infty} [f(z)] = 0$. This last condition implies that for any $\epsilon > 0$, there is an R so that

(5.6.6) $$|z| > R \Rightarrow |f(z)| < \epsilon$$

Hence, $f(z)$ is bounded *outside* of some disk. That f is also bounded inside and on this disk follows from the fact that it is continuous here, and the disk is a closed bounded set (see Problem 21b of Section 2.1). Hence, f is bounded and entire and, by Liouville's Theorem, it must be constant. Since it clearly is not a constant, the assumption that $p(z)$ has no roots must be false.

■

Poisson's Integral Formula. In Section 4.5, we discussed at length how conformal mappings could be used to solve boundary value problems for Laplace's equation. In order to accomplish this, we had to construct a simple "model problem" to map into. In all cases considered there, the model problem involved boundary conditions which were *piecewise constant*. That is, the boundary of the region in which Laplace's equation was to be solved was split into pieces, on each of which the value of the function was a given constant. Here, we will show how the Cauchy integral formulas can be used to express the values of a function harmonic inside a disk in terms of *general boundary conditions* on the circular boundary of the disk. In Problems 12 and 13, the reader is asked to show how conformal mappings can be used to derive from this the solution to the half-plane problem.

Now, let $u(x, y)$ be a real function which is harmonic and has continuous second partial derivatives in a simply connected domain D which contains the origin. Then, there is a function $f(z)$, analytic in D, so that

(5.6.7) $$u(x, y) = \text{Re}[f(z)]$$

Consider the disk $|\zeta| \le r_0$ in D (see Figure 5.6.1). If z is inside this disk, then by the

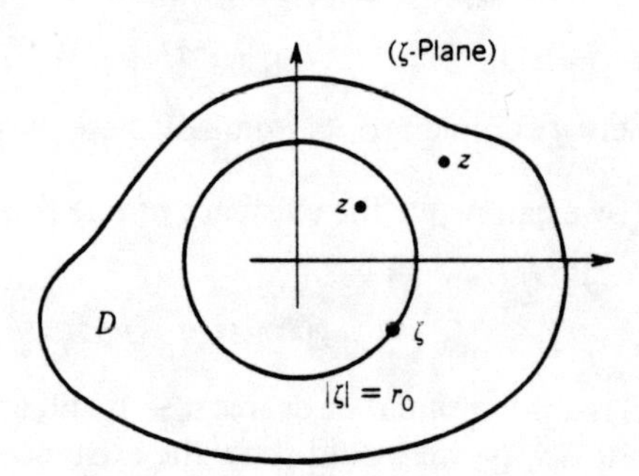

Figure 5.6.1

Cauchy integral formula

$$f(z) = \frac{1}{2\pi i} \oint_{|\zeta| = r_0} \frac{f(\zeta)}{\zeta - z}\, d\zeta \qquad |z| < r_0$$

while if $\hat{z}$ is a point *outside* the disk, the Cauchy–Goursat Theorem yields

$$0 = \frac{1}{2\pi i} \oint_{|\zeta| = r_0} \frac{f(\zeta)}{\zeta - \hat{z}}\, d\zeta \qquad |\hat{z}| > r_0$$

If we subtract these two equations, we get

(5.6.8)
$$f(z) = \frac{1}{2\pi i} \oint_{|\zeta| = r_0} \frac{(z - \hat{z}) f(\zeta)\, d\zeta}{(\zeta - z)(\zeta - \hat{z})}$$

Now on the circle, $\zeta = r_0 e^{i\phi}$, $0 \le \phi \le 2\pi$, and hence

$$\begin{aligned} f(z) &= \frac{1}{2\pi} \int_0^{2\pi} \frac{r_0 e^{i\phi}(z - \hat{z}) f(r_0 e^{i\phi})\, d\phi}{(r_0 e^{i\phi} - z)(r_0 e^{i\phi} - \hat{z})} \\ &= \frac{1}{2\pi} \int_0^{2\pi} \frac{(\hat{z} - z) f(r_0 e^{i\phi})\, d\phi}{(z - r_0 e^{i\phi})\left(1 - \dfrac{\hat{z}}{r_0} e^{-i\phi}\right)} \end{aligned}$$

where in the second equation we have divided numerator and denominator by $-r_0 e^{i\phi}$. So far, this result is valid for any $\hat{z}$ for which $|\hat{z}| > r_0$. We now try to choose a $\hat{z}$ so that that part of the integrand above which multiplies $f(r_0 e^{i\phi})$ is real. We will then be able to equate real parts of both sides and find an expression for $u(x, y) = \mathrm{Re}[f(z)]$ in terms of its values on the circle. To do this, first multiply numerator and denominator of the integrand by the complex conjugate $\bar{z}$ to get

$$f(z) = \frac{1}{2\pi} \int_0^{2\pi} \frac{(\bar{z}\hat{z} - |z|^2) f(r_0 e^{i\phi})\, d\phi}{(z - r_0 e^{i\phi})\left(\bar{z} - \dfrac{\bar{z}\hat{z}}{r_0} e^{-i\phi}\right)}$$

Note that the second term of the denominator will be the conjugate of the first if $\bar{z}\hat{z}/r_0 = r_0$, that is, if we choose

$$\hat{z} = \frac{r_0^2}{\bar{z}}$$

[The astute reader will recognize z and $\hat{z}$ as being *images of each other in a circle of radius* r_0 (see Example (4.6.15) of Section 4.6).] With this value of $\hat{z}$, $f(z)$ above becomes

$$f(z) = \frac{1}{2\pi} \int_0^{2\pi} \frac{(r_0^2 - |z|^2) f(r_0 e^{i\phi})\, d\phi}{|z - r_0 e^{i\phi}|^2}$$

With $z = re^{i\theta}$, we have $|z - r_0 e^{i\phi}|^2 = |re^{i\theta} - r_0 e^{i\phi}|^2 = r^2 - 2rr_0 \cos(\theta - \phi) + r_0^2$.

Hence, equating real parts of the above, we get **Poisson's integral formula**

(5.6.9)
$$u(r,\theta) = \frac{1}{2\pi}\int_0^{2\pi} \frac{(r_0^2 - r^2)u(r_0,\phi)\,d\phi}{r^2 - 2rr_0\cos(\theta-\phi) + r_0^2} \qquad r < r_0$$

Here, $u(r,\theta)$ designates the value of the harmonic function u at $z = re^{i\theta}$, and the result expresses u in terms of its boundary values $u(r_0,\phi)$ on the circle $|z| = r_0$. The quantity

(5.6.10)
$$K(r,\theta;r_0,\phi) = \frac{(r_0^2 - r^2)}{r^2 - 2rr_0\cos(\theta-\phi) + r_0^2}$$

is called the **Poisson kernel**.

The above result was derived under the assumption that u was also harmonic on the circle. The next theorem, which we present without proof, shows that this same basic formula provides a solution to the Dirichlet problem for Laplace's equation in a disk when the boundary data is merely piecewise continuous.

(5.6.11) ***Theorem.*** If $h(\theta)$ is piecewise continuous in the interval $0 \le \theta \le 2\pi$, and $h(0) = h(2\pi)$, then the solution to the **Dirichlet problem for the unit disk**

$$\nabla^2 u = 0 \qquad r < r_0$$

$$u(r_0,\theta) = h(\theta) \qquad 0 \le \theta \le 2\pi$$

is given by

$$u(r,\theta) = \frac{(r_0^2 - r^2)}{2\pi}\int_0^{2\pi} \frac{h(\phi)\,d\phi}{r^2 - 2rr_0\cos(\theta-\phi) + r_0^2} \qquad r < r_0$$

At every point of continuity of $h(\theta)$, we have

$$\lim_{r\uparrow r_0} u(r,\theta) = h(\theta)$$

The function $u(r,\theta)$ possesses continuous partial derivatives of all orders in $r < r_0$.

■

The Maximum Principle. This very interesting result, which has several important physical consequences, can be proved in many ways. Here, we will begin with a result which holds for a disk.

(5.6.12) ***Lemma.*** Let $f(z)$ be analytic inside and on the boundary of the disk $|z - z_0| < \rho_0$ and satisfy

$$|f(z)| \le M_0 \qquad \text{for } |z - z_0| \le \rho_0$$

If the maximum value M_0 is taken on at the center of the disk, that is, if $|f(z_0)| = M_0$, then $|f(z)| = M_0$ for all z in the disk.

Proof. Consider any circle $|z - z_0| = \rho \le \rho_0$ lying inside the disk. We will show that $|f(z)| = M_0$ for all z on this circle. Since ρ is arbitrary, we can then conclude that $|f(z)| = M_0$ in the whole disk $|z - z_0| \le \rho_0$.

Now, by the Cauchy integral formula,

(5.6.13)
$$f(z_0) = \frac{1}{2\pi i} \oint_{|\zeta - z_0| = \rho} \frac{f(\zeta)\, d\zeta}{\zeta - z_0}$$

Since $|f(z)| \le M_0$ for all z in the disk $|z - z_0| \le \rho_0$, then for each ζ on the circle $|\zeta - z_0| = \rho$, either $|f(\zeta)| < M_0$ or $|f(\zeta)| = M_0$. Assume there is a ζ_1 on this circle for which $|f(\zeta_1)| < M_0$. Then, since $|f(\zeta)|$ is continuous, there must be a whole arc C_1 of the circle on which (see Figure 5.6.2)

$$|f(\zeta)| < M_0 \qquad \text{for } \zeta \text{ on } C_1$$

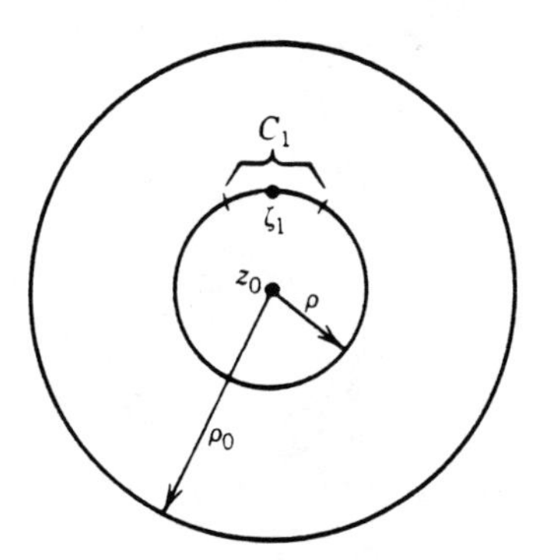

Figure 5.6.2

Also, $|f(\zeta)| \le M_0$ on the rest of the circle. Now, we can write (5.6.13) as

$$f(z_0) = \frac{1}{2\pi i}\left[\int_{C_1} \frac{f(\zeta)\, d\zeta}{\zeta - z_0} + \int_{\text{rest}} \frac{f(\zeta)}{\zeta - z_0}\, d\zeta\right]$$

If we use the triangle inequality, and the above inequalities on $|f(\zeta)|$, we get

$$|f(z_0)| < M_0$$

This contradicts the hypothesis that $|f(z_0)| = M_0$, and hence we conclude that *there is no point* ζ_1 on the circle $|\zeta - z_0| = \rho$ for which $|f(\zeta_1)| < M_0$. Thus, $|f(\zeta)| = M_0$ for all ζ on the circle, and hence our theorem is proved.

■

(5.6.14) ***Theorem* (Maximum Principle)**

(a) If $f(z)$ is analytic and bounded inside a domain D, and if $|f(z)|$ attains its maximum at an interior point, then $f(z)$ must be constant in D.

(b) If D is a bounded domain, and if $f(z)$ is a nonconstant function which is analytic in D and continuous in D and on its boundary, then $|f(z)|$ *attains its maximum on the boundary.*

Proof. We will show, for part (a), that $|f(z)|$ is constant in D. Then, the Cauchy–Reimann equations will imply that the same is true for $f(z)$ (see, for example, Problem 14 of Section 2.2).

To begin with, $|f(z)|$ is bounded by its maximum value M_0. That is,

$$|f(z)| \leq M_0 \qquad \text{for } z \text{ in } D$$

By the assumption in (a), there is a point z_0 in D at which

$$|f(z_0)| = M_0 \qquad \text{for } z_0 \text{ in } D$$

Since D is an open set (it is a domain), there is some disk $|z - z_0| < \rho_0$ contained in D (see Figure 5.6.3). By the above lemma, $|f(z)| = M_0$ for all z in this disk.

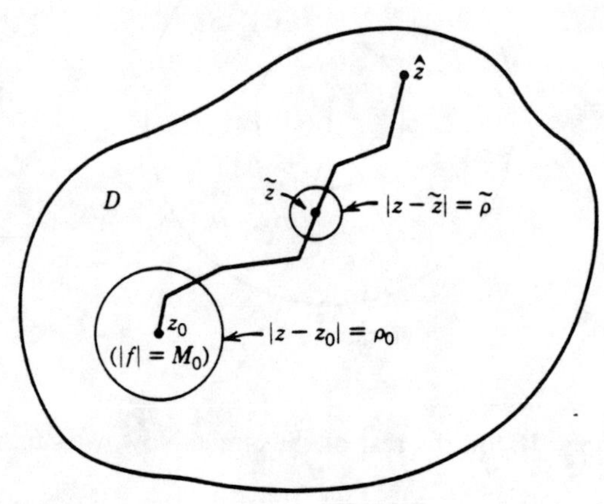

Figure 5.6.3

Now assume there is a $\hat{z}$ in D for which $|f(\hat{z})| \neq M_0$. Then, we must have strict inequality

$$|f(\hat{z})| < M_0 \tag{5.6.15}$$

Since D is connected, we can join $\hat{z}$ to z_0 with a broken line, as shown in the figure. Now proceed along this curve starting from $\hat{z}$. By the continuity of $|f(z)|$ and by (5.6.15), there must be points on this curve *near* $\hat{z}$ at which $|f| < M_0$. Eventually, we must arrive at a first point $\tilde{z}$ for which $|f(\tilde{z})| = M_0$. (One such point, for example, is where the circle $|z - z_0| = \rho_0$ meets the broken line.) Now, consider a circle $|z - \tilde{z}| = \tilde{\rho}$ lying inside of D. Again, by the lemma [since $|f(\tilde{z})| = M_0$], we must have that $|f(z)| = M_0$ for all z in the disk $|z - \tilde{z}| < \tilde{\rho}$. However, there are points inside this disk which *also* lie on the broken line and for which $|f(z)| < M_0$. Hence, we have a contradiction, and (5.6.15) cannot hold.

For part (b), since $|f(z)|$ is continuous, it must attain its maximum somewhere on the closed bounded set consisting of D and its boundary. Now, if this maximum is attained at an interior point of D, then by part (a),

$f(z)$ is constant in D. Since f is continuous up to and including the boundary of D, it must also be constant on the boundary. Hence, $f(z)$ is constant everywhere, which contradicts the hypothesis. Thus, the maximum is attained on the boundary.

■

(5.6.16) ***Example.*** Verify that the maximum principle holds for the functions

$$f_1(z) = e^z \quad \text{and} \quad f_2(z) = z^2 + 1$$

where D is the unit disk.

Solution. For $f_1(z)$, $|f_1(z)| = e^x$. The range of x is $-1 \leq x \leq 1$, and clearly e^x is maximum when $x = 1$. But for the unit disk, x is 1 only at the point $z = 1$ on the boundary.

Now, for f_2,

$$|f_2(z)|^2 = (x^2 - y^2 + 1)^2 + 4x^2y^2 \qquad x^2 + y^2 \leq 1 \tag{5.6.17}$$

From calculus, the function on the right attains a local *interior* maximum at only those points where the x and y derivatives both vanish. An easy calculation shows that this can happen only at $z = 0$, in which case

$$|f_2(0)| = 1 \qquad \text{(possible interior maximum)}$$

(Note that we are maximizing $|f(z)|^2$ in order to eliminate the square root.) To determine any boundary maxima, we substitute $y^2 = 1 - x^2$ into (5.6.17) and seek to maximize

$$h(x) = 4x^2 \qquad -1 \leq x \leq 1$$

This clearly has a maximum on the boundary of the interval at $x = \pm 1$ (which yields $y = 0$). Thus, on the boundary of the disk we have maxima at $z = \pm 1$, for which

$$|f_2(\pm 1)| = 2 \qquad \text{(boundary maximum)}$$

Hence, the maximum of $|f_2(z)|$ also occurs on the boundary of the unit disk.

■

Finally, we have a maximum principle for real harmonic functions.

(5.6.18) ***Theorem.*** Let $u(x, y)$ be harmonic, possess continuous second partial derivatives in a bounded domain D, and be continuous up to and on its boundary. Then, u attains its maximum on the boundary of D.

Proof. By the hypotheses on u, there is a complex conjugate $v(x, y)$ so that $u + iv$ is analytic in D and continuous up to and on the boundary of D. Hence, the same is true of the composite function

$$f(z) = e^{u+iv}$$

Since $|f| = e^u$, by the previous theorem, $e^{u(x,y)}$ attains its maximum on the boundary of D. However, since the real exponential is an increasing function of its argument, $u(x, y)$ itself must attain its maximum on the boundary.

∎

In the above, if $u(x, y)$ represents a harmonic, steady-state temperature distribution in a region D (see Section 4.5), then we learn that the temperature attains its maximum on the boundary. That this must be so, physically, is clear, since heat is dissipated as you enter a heat conducting body.

PROBLEMS (The answers are given on page 437.)

(Problems with an asterisk are more theoretical.)

1.* Use Morea's theorem to prove that if $g(\zeta)$ is analytic inside and on a simple, closed contour C, then $f(z) = \oint_C [g(\zeta)/(\zeta - z)]\, d\zeta$ is also analytic for z inside C. [*Hint:* Integrate $f(z)$ about any closed contour interior to C and interchange the order of integration.]

2. Let $u(x, y)$ be harmonic and possess continuous second partial derivatives in a domain D. If the disk $|z - z_0| \le r_0$ lies in D, use the result of the mean value Theorem (5.6.3b) to prove the alternate **mean value theorem for areas**

$$u(x_0, y_0) = \frac{1}{\pi r_0^2} \iint_{(x-x_0)^2+(y-y_0)^2 \le r_0^2} u(x, y)\, dx\, dy$$

3. Use Poisson's integral formula (5.6.9) to verify the mean value theorem for a circle with center at the origin. [*Hint:* Set $r = 0$ in (5.6.9).]

4. Directly verify the mean value theorem for the functions $f(z)$ given below.
(a) z^n (b) e^z (c) $\sin z$

5.* A function is said to have at worst **algebraic growth at infinity** if $|f(z)| \le K|z|^n$ for $|z|$ large, where K is a constant and n is a nonnegative integer. For such a function, the following is an **extension of Liouville's Theorem**. Prove that an entire function with at worst algebraic growth at infinity must be a polynomial. [*Hint:* Fix z and consider $\oint_{|\zeta - z| = R} f(\zeta)\, d\zeta/(\zeta - z)^{n+2}$. Now use the Cauchy integral formula for derivatives to show that

$$|f^{n+1}(z)| \le \frac{K(n+1)!(R + |z|)^n}{R^{n+1}}$$

for any R.]

6.* Prove that if $u(x, y)$ is harmonic and bounded and possesses continuous second partial derivatives everywhere, then it must be constant. [*Hint:* Show that e^u is the magnitude of an entire function, and then use Liouville's Theorem.]

7.* Prove that if $f(z)$ is entire and $\lim_{z \to \infty} f(z) = \infty$, then $f(z)$ must vanish somewhere. [*Hint:* Apply Liouville's Theorem to $1/f(z)$.]

8.* A function is said to have at worst **logarithmic growth at infinity** if $|f(z)| \le K(\ln|z|)^n$ for some $n \ge 0$ and $|z|$ large. Prove that an entire function with at worst logarithmic growth must be constant. (Hence, logarithmic growth is rather weak compared to, say, algebraic growth in Problem 5). [*Hint:* Show that the above inequality implies that $|f| \le K|z|$ for $|z|$ large, and then use Problem 5 to show that $f = az + b$. Now conclude that $a = 0$.]

9. With the Poisson kernel $K(r, \theta; r_0, \phi)$ defined as in (5.6.10), show that $\int_0^{2\pi} K(r, \theta; r_0, \phi)\, d\phi = 2\pi$ by

(a) Direct integration using the methods of Section 5.4.

(b) Using equation (5.6.9), with an appropriate choice of $u(r, \theta)$.

10. If u and v are harmonic conjugates in the disk $|z| < r_0$, show that

$$v(r, \theta) = k - \frac{rr_0}{\pi} \int_0^{2\pi} \frac{u(r_0, \phi) \sin(\phi - \theta)\, d\phi}{r_0^2 + r^2 - 2rr_0 \cos(\phi - \theta)}$$

where k is the value of v at the center of the disk. [*Hint:* Add the two equations after (5.6.7) with $\hat{z} = r_0^2/\bar{z}$. Then, take imaginary parts and use the mean value theorem. Also note that $\zeta = r_0^2/\bar{\zeta}$.]

11. Consider the Dirichlet problem for the unit disk

$$\nabla^2 u = 0 \qquad \text{in } |z| < 1$$

$$u|_{|z|=1} = h(\theta) \qquad 0 \le \theta \le 2\pi$$

If $h(\theta) = \text{Re}[f(e^{i\theta})]$, where $f(z)$ is analytic inside and on the unit disk, what is $u(x, y)$ in terms of $f(z)$?

12. In the next two problems, we will take the reader through a series of steps which lead to a representation of the solution of the **Dirichlet problem for a half-plane.**

(a) Show that the bilinear function $w = (z - i)/(z + i)$ maps the upper half-plane onto the unit disk $|w| < 1$.

(b) Let $f(z)$ be analytic in the upper half-plane. Then, under the mapping in (*a*), the composite function

$$g(w) = f\left[i\left(\frac{w + 1}{1 - w}\right)\right]$$

is analytic inside the unit disk. Use equation (5.6.8), applied to $g(w)$, together with $r_0 = 1$, $\hat{z} = 1/\bar{w}$, and the fact that

$$f(z) = g\left(\frac{z - i}{z + i}\right)$$

to show that

$$f(z) = \frac{1}{2\pi i} \oint_{|\zeta|=1} \frac{\left(\dfrac{z - i}{z + i} - \dfrac{\bar{z} - i}{\bar{z} + i}\right) g(\zeta)\, d\zeta}{\left(\zeta - \dfrac{z - i}{z + 1}\right)\left(\zeta - \dfrac{\bar{z} - i}{\bar{z} + i}\right)} \qquad \text{Im}(z) > 0$$

(c) In the above integral, make the change of variables

$$\eta = i\left(\frac{\zeta + 1}{1 - \zeta}\right)$$

which takes the unit disk back onto the upper half-plane, and show that

$$f(z) = \frac{y}{\pi}\int_C \frac{f(\eta)\,d\eta}{|\eta - z|^2}$$

where the contour is the real η-axis traversed from $-\infty$ to $+\infty$.

(d) Use the above to show that if $u(x, y)$ is a real harmonic function in the upper half-plane, then in terms of its values $u(x, 0)$ on the real axis, we have

$$u(x, y) = \frac{y}{\pi}\int_{-\infty}^{\infty} \frac{u(\xi, 0)\,d\xi}{(\xi - x)^2 + y^2}$$

13.* With Problem 12*d* as motivation, we are led to the following result, which the reader is then asked to prove.

Theorem: If $h(x)$ is bounded and piecewise continuous on the interval $-\infty < x < \infty$, then the function

$$u(x, y) = \frac{y}{\pi}\int_{-\infty}^{\infty} \frac{h(\xi)\,d\xi}{(\xi - x)^2 + y^2}$$

is harmonic is the upper half-plane and satisfies $\lim_{y \downarrow 0} [u(x, y)] = h(x)$ at the points of continuity of h.

(a) For $y > 0$, show that the integral above converges uniformly, as do the integrals of the partial derivatives of the integrand (see an advanced calculus book for a definition of uniform convergence). Hence, show that $u_{xx} + u_{yy} = 0$ for $y > 0$.

(b) For $y > 0$, rewrite $u(x, y)$ as

$$u(x, y) = \frac{1}{\pi}\int_{-\infty}^{\infty} \frac{h(x + y\eta)\,d\eta}{1 + \eta^2}$$

Show that $\lim_{y \downarrow 0} [u(x, y)] = h(x)$ for those values of x where h is continuous. [*Hint:*

$$u(x, y) - h(x) = \frac{1}{\pi}\int_{-\infty}^{\infty} \frac{[h(x + y\eta) - h(x)]\,d\eta}{1 + \eta^2}$$

$$= \frac{1}{\pi}\left[\int_{-\infty}^{-N} + \int_{-N}^{N} + \int_{N}^{\infty}\right]\left(\frac{h(x + y\eta) - h(x)}{1 + \eta^2}\right) d\eta$$

Now, choose N so that the first and last integrals are small (why can you do this?). Then, use the continuity of h to show that the second integral is small for y small.]

(c) If $h(x)$ is the piecewise constant function

$$h(x) = \begin{cases} c_1 & x < 0 \\ c_2 & x > 0 \end{cases}$$

show that $u(x, y)$ reduces to the solution of the Dirichlet problem solved in Example (4.5.4) of Section 4.5.

14. Find the maximum value of $|f(z)|$ in the region D when $f(z)$ and D are given by

(a) $f = 1 + z$, D: $0 \leq x \leq 1$, $0 \leq y \leq 1$

(b) $f = e^{z^2}$, D: $|z| \leq 1$

15.* Prove the following **minimum principles**.

(a) If $f(z)$ is analytic in a bounded domain D, continuous up to and on the boundary of D, and does not vanish in D, then $|f(z)|$ attains its minimum value on the boundary. [*Hint:* Consider $1/f(z)$ and assume that $|f|$ attains its minimum inside D.]

(b) If $u(x, y)$ is harmonic and possesses continuous second partial derivatives in a bounded domain D, and if u is continuous up to and on the boundary of D, then u attains its minimum value on the boundary. (*Note that, unlike part* (*a*), *we need not assume that* u *does not vanish in* D.)

(c) If $f(z)$ is as in part (a) and if $|f(z)|$ is constant on the boundary of D, then either f = constant in D or f vanishes somewhere in D.

16. Let **E** be the **electrostatic field** in a bounded domain D with potential ϕ which is constant on the boundary of D. Use the maximum principle for harmonic functions and the fact that $\mathbf{E} = -\nabla\phi$ to show that at some point on the boundary of D, **E** must point into the interior. [*Hint:* If **n** is a unit normal vector on the boundary of D which points into D, then $\mathbf{n} \cdot \nabla\phi$ measures the rate of change of ϕ in the direction **n**.]

17.* Prove that if both $f(z)$ and $g(z)$ are analytic in a bounded domain D, and are continuous up to and on the boundary of D, and if $g(z) \neq 0$ in D and on the boundary, then

$$|f| \leq |g| \quad \text{on boundary} \Rightarrow |f| \leq |g| \text{ in } D$$

18.* (a) Prove that if $u(x, y)$ is harmonic with continuous second partial derivatives in a bounded domain D and continuous up to and on the boundary of D, and if $u = 0$ on the boundary, then $u(x, y) \equiv 0$ in D. [*Hint:* Use the maximum principle, *and* the minimum principle of Problem 15.]

(b) Prove that in the class of functions with continuous second partial derivatives, the solution of the Dirichlet problem in a bounded domain is unique. [*Hint:* Assume there are two solutions, $u_1(x, y)$ and $u_2(x, y)$, and consider $u_1 - u_2$.]

6

INFINITE SERIES

6.1 CONVERGENCE OF SEQUENCES AND SERIES; GEOMETRIC SERIES[1]

To begin with, we have the following basic definitions as in calculus.

(6.1.1) ***Definition***

(a) An ordered set of complex numbers $z_1, z_2, z_3, \ldots$ is called an **infinite sequence** and is denoted by $\{z_n\}$.

(b) The infinite sequence $\{z_n\}$ is said to **converge to A**, written as

$$\lim_{n \to \infty} z_n = A \qquad \text{or} \qquad z_n \to A$$

(*sometimes we will omit the symbol* $n \to \infty$) if, given any $\epsilon > 0$, there is an integer $N(\epsilon)$, which may depend upon ϵ, so that

(6.1.2) $$\text{if } n \geq N(\epsilon), \qquad \text{then } |z_n - A| < \epsilon$$

The complex number A is called the **limit of the sequence**. If no such A exists, we say that the sequence is **divergent**.

▬

[1] The reader who is familiar with the corresponding concepts from real calculus can skip over this material. It is suggested, however, that a brief review, in particular of the material on geometric series, might be worthwhile.

(6.1.3) ***Example.*** The sequence $\{[(1-i)/3]^n\}$ converges to zero. To show this, we note that since $\sqrt{2}/3 < 1$,

$$\left|\left(\frac{1-i}{3}\right)^n - 0\right| = \left(\frac{\sqrt{2}}{3}\right)^n \to 0.$$

Hence, $|[(1-i)/3]^n - 0|$ can be made arbitrarily small with a large enough choice of n, and thus certainly less than any preassigned ϵ.

On the other hand, the sequence $\{i^n\}$ diverges. To see this, write it explicitly as

$$\{i, -1, -i, 1, i, -1, -i, \ldots\} \tag{6.1.4}$$

By the definition, if the sequence *did* converge to a number A, then *all the terms far enough out in the sequence should cluster about* A. This is not the case, since the sequence consists of four numbers which repeat adinfinitum.

■

The proof of the next theorem follows along the same lines as in calculus. Parts (a) and (c) are proved in Problem 2.

(6.1.5) ***Theorem***

(a) If $\{z_n\}$ converges, then the limit is unique.

(b) $z_n \to A$ if and only if $|z_n - A| \to 0$.

(c) Let $z_n = x_n + iy_n$ and $A = A_1 + iA_2$. Then, $z_n \to A$ if and only if $x_n \to A$ and $y_n \to A_2$. If so, we have

$$\lim z_n = (\lim x_n) + i(\lim y_n) \tag{6.1.6}$$

Hence, the convergence of complex sequences is directly related to the convergence of the real sequences consisting of the real and imaginary parts of z_n.

(d) If $\{z_n\}$ and $\{w_n\}$ are convergent sequences, then

(i) $\{\alpha z_n + \beta w_n\}$ converges and

$$\lim(\alpha z_n + \beta w_n) = \alpha \lim z_n + \beta \lim w_n$$

(ii) $\{z_n \cdot w_n\}$ converges and

$$\lim(z_n \cdot w_n) = (\lim z_n) \cdot (\lim w_n)$$

(iii) If $\lim w_n \neq 0$, then for those $w_n \neq 0$ the sequence $\{z_n/w_n)$ converges and

$$\lim\left(\frac{z_n}{w_n}\right) = \frac{\lim z_n}{\lim w_n}$$

■

(6.1.7) ***Example.*** To compute

$$A = \lim_{n\to\infty}\left[\frac{n + i\sqrt{1+n^2}}{1+2in}\right],$$

we first note that both numerator and denominator get large in magnitude with n, and also that $\sqrt{1+n^2} \approx n$, for n large. Hence, dividing numerator and denominator by n, we get

$$A = \lim_{n\to\infty}\left(\frac{1 + i\left(1+\dfrac{1}{n^2}\right)^{1/2}}{\dfrac{1}{n} + 2i}\right) = \frac{1+i}{2i}$$

■

(6.1.8) ***Example.*** Without evaluating the integral, compute the limit of the sequence $\{z_n\}$ defined by

$$z_n = \oint_0^{(1+i)/\sqrt{2}} \frac{z^n}{n+z}\,dz$$

where C is the joining $z = 0$ to $z = (1+i)/\sqrt{2}$.

Solution. On C, we have that $|z| \leq 1$. Hence,

$$\begin{aligned} |n+z| &\geq |n - |z|| \\ &= n - |z| \\ &\geq n - 1 \qquad \text{since } |z| \geq 1 \end{aligned}$$

where the equality above holds for n large enough. Thus, $|z^n(n+z)^{-1}| \leq (n-1)^{-1}$ for z on C, and by the triangle inequality for integrals

$$|z_n| \leq \frac{1}{n-1} \qquad \text{since the length of } C \text{ is } 1$$

Thus, $\lim\limits_{n\to\infty} z_n = 0$.

■

The next result, part of which is proved in Problem 6, provides a test for the convergence of sequences which does not depend upon knowing the actual limit of the sequence.

(6.1.9) ***Theorem.* (Cauchy Test for Convergence).** The sequence $\{z_n\}$ converges if and only if

$$\lim_{m,n\to\infty} |z_n - z_m| = 0$$

By this limit notation we mean that m *and* n *are to approach infinity independently.*

■

Any sequence satisfying the above condition is called a **Cauchy sequence.**

(6.1.10) ***Example.*** That the sequence $\{(n + i)/n\}$ converges to 1 is rather clear. However, to illustrate the above result, note that

$$\left|\frac{n+i}{n} - \frac{m+i}{m}\right| = \left|\frac{i}{n} - \frac{i}{m}\right| \leq \frac{1}{n} + \frac{1}{m} \xrightarrow[m,n\to\infty]{} 0$$

Hence, the sequence converges by the Cauchy test.

■

Infinite Series. As in elementary calculus, we have the following definitions and results.

(6.1.11) ***Definition***

(a) The sum $z_1 + z_2 + z_3 + \cdots$ of the terms of the infinite sequence $\{z_n\}$ is an **infinite series** and is denoted by

$$\sum_{n=1}^{\infty} z_n \quad \text{or} \quad \sum_{1}^{\infty} z_n$$

(b) An infinite series is said to **converge** if the sequence of ***N*th partial sums** (the sum of the first N terms in the sequence) converges. That is, if there is a number L so that

$$\lim_{N\to\infty}\left[\sum_{1}^{N} z_n\right] = L \qquad (6.1.12)$$

Otherwise, the series is said to **diverge.**

(c) The series $\sum_1^\infty z_n$ **converges absolutely** if the real series of magnitudes $\sum_1^\infty |z_n|$ converges.

■

In these definitions, the **dummy index of summation** n need not necessarily start at $n = 1$.

(6.1.13) ***Theorem***

(a) If $\sum_1^\infty z_n$ converges, then

(i) $\lim_{n\to\infty} z_n = 0$

(ii) There is an M so that $|z_n| \leq M$ (for all n).

(b) If $z_n = x_n + iy_n$, then $\sum_1^\infty z_n$ converges if and only if both real series $\sum_1^\infty x_n$ and $\sum_1^\infty y_n$ converge, and then,

$$\sum_1^\infty z_n = \sum_1^\infty x_n + i\sum_1^\infty y_n$$

(c) If $\sum_1^\infty z_n$ and $\sum_1^\infty w_n$ both converge, then so does $\sum_1^\infty (\alpha z_n + \beta w_n)$, and

$$\sum_1^\infty (\alpha z_n + \beta w_n) = \alpha\sum_1^\infty z_n + \beta\sum_1^\infty w_n$$

(d) **(Cauchy Test).** $\sum_1^\infty z_n$ converges if and only if

$$\lim_{n,p\to\infty}\left(\sum_{k=n}^{n+p} z_k\right) = 0 \tag{6.1.14}$$

(e) If $\sum_1^\infty z_n$ converges absolutely, then it converges.

Proof. To prove (a), we write, for $n \geq 2$,

$$z_n = \sum_{k=1}^{n} z_k - \sum_{k=1}^{n-1} z_k$$

The sums on the right are the nth and $(n-1)$st partial sums for the series $\sum_1^\infty z_k$. Since these both have the same limit, their difference, which is z_n, converges to zero, thus proving (a*i*). Using this and (6.1.2) (with $A = 0$), we know that for any ϵ, there exists an $N(\epsilon)$, so that

$$n \geq N(\epsilon) \Rightarrow |z_n| < \epsilon$$

Thus, if we define M to be the largest member of the finite set of numbers $\{|z_1|, |z_2|, \ldots, |z_{N-1}|, \epsilon\}$ then $|z_n| \leq M$ for all n, thus proving (a*ii*).

Part (d) follows by first recognizing $\sum_{k=n}^{n+p} z_k$ as the difference of the $(n+p)$th and $(n-1)$st partial sums. Then, by the Cauchy test for convergence (6.1.9), this difference approaches zero as $n, p \to \infty$. This result in turn can be used to prove (e), since by the triangle inequality

$$\left|\sum_{k=n}^{n+p} z_k\right| \leq \sum_{k=n}^{n+p} |z_k| \xrightarrow[n,p\to\infty]{} 0$$

where the last limit holds by the assumption of absolute convergence and the Cauchy test.

Parts (b) and (c) are proved in Problem 8.

■

The following two simple convergence tests will supply additional ammunition to use in order to determine whether or not an infinite series converges.

(6.1.15) ***Theorem***

(a) **(Comparison Test).** Let $M_n \geq 0$, $n = 1, 2, \ldots$, and consider the series $\sum_{n=1}^\infty z_n$. If for all $n \geq N_0$, $|z_n| \leq M_n$, then,

$$\sum_1^\infty M_n \text{ converges} \Rightarrow \sum_1^\infty z_n \text{ converges (also absolutely)}$$

(b) **(Ratio Test).** Assume that the sequence $\left\{\left|\dfrac{z_{n+1}}{z_n}\right|\right\}$ has a limit, that is,

$$\lim_{n\to\infty}\left|\frac{z_{n+1}}{z_n}\right| = L \qquad (L \text{ can be infinite})$$

Then,

$$\sum_{1}^{\infty} z_n \begin{cases} \text{converges} & \text{if } L < 1 \\ \text{diverges} & \text{if } L > 1 \\ \text{no conclusion} & \text{if } L = 1. \end{cases} \tag{6.1.16}$$

The convergence is also absolute.

Proof. The comparison test follows simply from the Cauchy test (6.1.14), since for $n, p \geq N_0$,

$$\left|\sum_{k=n}^{n+p} z_k\right| \leq \sum_{k=n}^{n+p} |z_k| \leq \sum_{k=n}^{n+p} M_k \underset{n,p\to\infty}{\longrightarrow} 0$$

where the limit follows because $\sum_1^\infty M_k$ converges.

To prove the convergence part of the ratio test, assume that

$$\left|\frac{z_{n+1}}{z_n}\right| \to L < 1$$

Since $L < 1$, we can find an $\epsilon_0 > 0$ so that

$$L + \epsilon_0 < 1$$

Then, for $\epsilon = \epsilon_0$, we conclude from (6.1.2) that there is an N_0 so that, for $n \geq N_0$,

$$\left|\left|\frac{z_{n+1}}{z_n}\right| - L\right| < \epsilon_0 \Rightarrow \left|\frac{z_{n+1}}{z_n}\right| < L + \epsilon_0$$

Then, from the above,

$$\begin{aligned} |z_{N_0+1}| &\leq (L+\epsilon_0)|z_{N_0}| \\ |z_{N_0+2}| &\leq (L+\epsilon_0)|z_{N_0+1}| \leq (L+\epsilon_0)^2|z_{N_0}| \\ &\ \vdots \\ |z_{N_0+k}| &\leq (L+\epsilon_0)^k|z_{N_0}| \qquad k \geq 1 \end{aligned}$$

Hence, if we define

$$M_n = \begin{cases} |z_n| & n = 1, 2, \ldots, N_0 \\ (L+\epsilon_0)^{n-N_0}|z_{N_0}| & n = N_0+1, N_0+2, \ldots \end{cases}$$

then

$$|z_n| \leq M_n \qquad \text{(for all } n\text{)}$$

Also, since $L + \epsilon_0 < 1$, $\sum_1^\infty M_n$ converges (it is basically a real geometric series). Thus, by the comparison test, $\sum_1^\infty z_n$ also converges.

In the problem 9, the reader is asked to prove the divergence of the ratio test.

▬

(6.1.17) ***Example.*** Consider the series $\sum_{n=1}^{\infty} [\sin(i+n)]/n^2$. Since $\sin(n+i) = (1/2i)(e^{in}e^{-1} - e^{-in}e)$, we have

$$\left|\frac{\sin(i+n)}{n^2}\right| \leq \frac{e+e^{-1}}{2n^2}$$

By the integral test from elementary calculus, $\sum_{n=1}^{\infty} 1/n^2$ converges, and hence by our comparison test the above complex series converges (also absolutely). The reader should note that the ratio test is inoperable here since

$$\left|\frac{z_{n+1}}{z_n}\right| = \left(\frac{n}{n+1}\right)^2 \cdot \left|\frac{\sin(i+n+1)}{\sin(i+n)}\right|$$

has no limit as $n \to \infty$.

■

(6.1.18) ***Example.*** For which values of z does the series $\sum_{n=0}^{\infty} e^{-nz}$ converge?

Solution.

$$\left|\frac{e^{-(n+1)z}}{e^{-nz}}\right| = |e^{-z}| = e^{-x}$$

Since $e^{-x} < 1$ for $x > 0$ and $e^{-x} > 1$ for $x < 0$, by the ratio test the series

$$\begin{cases} \text{Converges} & \text{for } \operatorname{Re}(z) > 0 \\ \text{Diverges} & \text{for } \operatorname{Re}(z) < 0 \end{cases}$$

The ratio test is not definitive for $\operatorname{Re}(z) = 0$ since then the limit L in (6.1.16) is 1. However, for $z = iy$, the general term in the series is e^{-iny}, and this has no limit as $n \to \infty$ unless y is a multiple of 2π, in which case the limit is 1. Hence, result (a*i*) of Theorem (6.1.13) does not hold, and the series $\sum_{n=0}^{\infty} e^{-iny}$ cannot converge.

■

Geometric Series. The next section will be devoted to a full discussion of power series. Here, we will examine the special case of the **geometric series**

$$\sum_{n=0}^{\infty} z^n \tag{6.1.19}$$

(6.1.20) ***Theorem.*** The geometric series converges absolutely within the unit circle and diverges outside and on the unit circle. More specifically,

$$\sum_{n=0}^{\infty} z^n = \begin{cases} \dfrac{1}{1-z} & \text{for } |z| < 1 \\ \text{diverges} & \text{for } |z| \geq 1 \end{cases} \tag{6.1.21}$$

Proof. To sum this series, which is one of the few which can be summed explicitly, we consider the N-th, partial sum $S_N(z)$. This will depend upon the value of z and is given by

$$S_N(z) = \sum_{n=0}^{N} z^n = 1 + z + \cdots + z^N$$

We first multiply both sides of the above by z, subtract S_N from both sides, and then divide by $1 - z$ (assuming $z \neq 1$) to get

$$\sum_{n=0}^{N} z^n = \frac{1 - z^{N+1}}{1 - z} \qquad z \neq 1 \tag{6.1.22}$$

(If $z = 1$, the infinite series in (6.1.21) cannot converge since the general term does not approach zero.) Now, the series above converges if and only if the *sequence* on the right converges as $N \to \infty$. For $|z| < 1$, $\lim_{N\to\infty} z^{N+1} = 0$, and the first part of (6.1.21) is proved. For $|z| > 1$, $\lim_{N\to\infty} z^{N+1} = \infty$, and the series does not converge. For $|z| = 1$, we can write $z = \cos\theta + i\sin\theta$. Then, $z^{N+1} = \cos(N+1)\theta + i\sin(N+1)\theta$, and this has no limit as $N \to \infty$.

That the convergence is absolute follows from (6.1.22) upon replacing z by $|z|$.

■

(6.1.23) ***Example.*** Sum the series

$$\sum_{n=1}^{\infty} \left(\frac{1+i}{3}\right)^n$$

Solution. First note that, save for the lower limit of summation, the series is a geometric series (6.1.19) with $z = (1 + i)/3$. Hence, since $(1 + i/3)^0 = 1$, we first write

$$\sum_{n=1}^{\infty} \left(\frac{1+i}{3}\right)^n = \sum_{n=0}^{\infty} \left(\frac{1+i}{3}\right)^n - 1$$

Now, since $|(1 + i/3)| < 1$, the series on the right converges and sums to $1/(1 - (1 + i)/3) = 3/(2 - i)$. Hence,

$$\sum_{n=1}^{\infty} \left(\frac{1+i}{3}\right)^n = \frac{1+i}{2-i}$$

■

(6.1.24) ***Example.*** Determine for which z the two series

$$\sum_{n=0}^{\infty} (-1)^n z^n \qquad \text{and} \qquad \sum_{n=0}^{\infty} z^{2n}$$

converge, and sum them.

Solution. We recognize the first series as the geometric series $\sum_{n=0}^{\infty} (-z)^n$ with z in (6.1.19) replaced by $-z$. Hence, the series converges for $|-z| < 1$ and sums to

(6.1.25)
$$\sum_{n=0}^{\infty} (-1)^n z^n = \frac{1}{1+z} \qquad \text{for } |z| < 1$$

Similarly, the second series can be written as $\sum_{n=0}^{\infty} (z^2)^n$. Thus, by (6.1.21), if we replace z by z^2, we have

$$\sum_{n=0}^{\infty} z^{2n} = \frac{1}{1-z^2} \qquad \text{for } |z| < 1$$

where the range on z follows since $|z|^2 < 1$ if and only if $|z| < 1$.

■

(6.1.26) ***Example.*** For which z does the series $\sum_{n=1}^{\infty} nz^n$ converge? Sum it!

Solution. By the ratio test, the series converges for $|z| < 1$ and diverges for $|z| > 1$. On the unit circle, $\lim_{n\to\infty} (nz^n) \neq 0$, and so the series diverges here.

To sum the series for $|z| < 1$, let $S_N(z)$ denote the Nth partial sum

$$S_N(z) = \sum_{n=1}^{N} nz^n \qquad |z| < 1$$

Now $S_N(\zeta)/\zeta$ is a polynomial in ζ. Hence, it is analytic for $|\zeta| < 1$ and can be integrated over any simple contour C joining $\zeta = 0$ to $\zeta = z$. Thus, from the above

$$\oint_0^z \frac{S_N(\zeta)}{\zeta}\, d\zeta = \sum_{n=1}^{N} \oint_0^z n\zeta^{n-1}\, d\zeta = \sum_{n=1}^{N} z^n$$

Now, this last sum is just $\sum_{n=0}^{N} z^n - 1$, which by (6.1.22) yields

$$\oint_0^z \frac{S_N(\zeta)}{\zeta}\, d\zeta = \frac{1-z^{N+1}}{1-z} - 1 \qquad |z| < 1$$

The integral above is an analytic function of z. (Why?) Thus, we can differentiate both sides with respect to z and find

$$S_N(z) = \frac{z}{(1-z)^2} - z\frac{d}{dz}\left(\frac{z^{N+1}}{1-z}\right)$$

The reader is asked to show that for $|z| < 1$, the second term vanishes in the limit as $N \to \infty$, and hence

$$\sum_{n=1}^{\infty} nz^n = \frac{z}{(1-z)^2}$$

■

PROBLEMS (The answers are given on page 437.)

(Problems with an asterisk are more theoretical.)

1. Discuss the convergence or divergence of the following sequences. In the case of convergence, give the limit.

(a) $\left\{\dfrac{n^2+n+1}{in^2+2}\right\}$ (b) $\left\{\dfrac{\sin n}{n}\right\}$ (c) $\left\{\dfrac{\sin(in)}{n}\right\}$

(d) $\{e^{-n\alpha}\}$ (e) $\left\{\dfrac{\alpha^n}{\alpha^n+\beta^n}\right\}$

(f) $\left\{\dfrac{\alpha}{n^2+\alpha}\right\}$ $(\alpha \neq -1, -4, -9, \ldots$

(g) $\{(1+n\alpha)^{-1}\}$ $\alpha \neq -1, -\frac{1}{2}, \ldots$ (h) $\left\{\left(\dfrac{\alpha-i}{\alpha+i}\right)^n\right\}$ $\alpha \neq -i$

(i) $\left\{\mathrm{Log}\left(\alpha+\dfrac{1}{n}\right)\right\}$ $\alpha \neq -1/n$ (j) $\{\mathrm{Arg}(i+n^2)\}$

(k) $\{n-(i+n^2)^{1/2}\}$ where the principle branch is used

(l) $\{n^\alpha\}$

2.* Use the following hints to prove parts (a) and (c) of Theorem (6.1.5).

(a) To prove (a) of the theorem, assume $z_n \to A$ and $z_n \to B$, with $A \neq B$. Then, show that $|A-B|$ can be made as small as desired.

(b) To prove one part of (c) of the theorem, note that $|z_n - A|^2 = (x_n - A_1)^2 + (y_n - A_2)^2$. One implication of this is that $|x_n - A_1| \leq |z_n - A|$ and $|y_n - A| \leq |z_n - A|$. (Why?) Use this to prove the second part of (c) of Theorem (6.1.5).

3. The following sequences $\{z_n\}$ are defined by integrals. Compute the limits of those which converge.

(a) $z_n = \oint_0^2 dz/(z^5 + n^2)$, where C is the upper half of the circle $|z-1| = 1$. [*Hint:* For z on C, show that $|z| \leq 2$. Then show that if n is large enough, $|1 + z^5 n^{-2}| \geq 1 - (32/n^2)$.]

(b) $z_n = \oint_0^2 n\, dz/(1 + n^2 z^2)$, where C is as part (a).

4.* (a) Prove that if $z_n \to A$, then $|z_n| \to |A|$. [*Hint:* Use inequality (1.1.27b).]

(b) Is the converse of (a) true? That is, does $|z_n| \to |A|$ imply that $z_n \to A$?

(c) Prove that $z_n \to A$ if and only if $\bar{z}_n \to \bar{A}$.

(d) Is it true that if $z_n \to A$, then $\mathrm{Arg}(z_n) \to \mathrm{Arg}(A)$, where $A \neq 0$?

(e) Prove that if $z_n \to A$, then the sequence $\{z_{n+n_0}\}$, where n_0 is some nonnegative integer, also converges to A.

5.* Prove that if $z_n \to A$, and if $f(z)$ is defined at each z_n and is continuous in some neighborhood of $z = A$, then $f(z_n) \to f(A)$. [*Hint:* For any ϵ, there is a δ so that $|z - A| < \delta \Rightarrow |f(z) - f(A)| < \epsilon$. (Why?) Now, use $z_n \to A$ to show that for this ϵ there is an N so that $n \geq N$ implies that $|f(z_n) - f(A)| < \epsilon$.]

6.* Prove that if $z_n \to A$, then $\lim\limits_{n,m\to\infty} |z_n - z_m| = 0$. This is half of the Cauchy test for convergence, Theorem (6.1.9). [*Hint:* Write $z_n - z_m = (z_n - A) - (z_m - A)$ and use the triangle inequality.]

7. Discuss the convergence or divergence of the following series.

(a) $\sum_{n=1}^{\infty} \frac{e^{\alpha n}}{n^2}$ [*Hint:* For $\mathrm{Re}(\alpha) = 0$, use the comparison test.]

(b) $\sum_{n=0}^{\infty} \left(\frac{1}{n+i} - \frac{1}{n+i+1}\right)$

(c) $\sum_{n=0}^{\infty} \frac{1}{(\alpha+n)^2}$ $\alpha \neq 0, -1, -2, \ldots$

(d) $\sum_{n=1}^{\infty} \frac{1}{\alpha^n(1-\alpha^n)}$ $|\alpha| \neq 1$

8.* Prove parts (b) and (c) of Theorem (6.1.13). [*Hint:* For part (b), write the Nth partial sum of $\sum z_n$ in terms of the Nth partial sums of $\sum x_n$ and $\sum y_n$. Then, use the corresponding results on limits.]

9.* Prove the divergence part of the ratio test (6.1.16b). [*Hint:* If $|z_{n+1}/z_n| \to L$, then show that for any ϵ, $|z_{n+1}| > (L - \epsilon)|z_n|$ for n large enough. (Why?) Now, if $L > 1$, choose ϵ so that $L - \epsilon > 1$ and show that $|z_{n+1}| > (L - \epsilon)^n |z_1|$. This should now imply, from (6.1.13a), that $\sum_1^{\infty} z_n$ diverges.]

10. For which z do the following series converge? Sum them.

(a) $\sum_{n=0}^{\infty} \left(\frac{z-i}{z+i}\right)^n$ (b) $\sum_{n=0}^{\infty} e^{n/z}$ (c) $\sum_{n=0}^{\infty} z^{-n}$

11. Here, we will use the approach of Example (6.1.26) to sum the series $\sum_{n=1}^{\infty} z^n/n$.
 (a) Show that the series converges for $|z| < 1$.
 (b) If $S_N(z) = \sum_{n=1}^{N} z^n/n$ is the Nth partial sum, show that $S_N'(z) = (1 - z^N)/(1 - z)$.
 (c) If $|z| < 1$ and C is the line joining $\zeta = 0$ to $\zeta = z$, integrate $S_N'(\zeta)$ over C and show that $S_N(z) = -\mathrm{Log}(1 - z) - \oint_0^z (\zeta^N/(1 - \zeta)\, d\zeta$.
 (d) Show that $|\zeta^{N-1}/(1 - \zeta)| \leq |z|^N/(1 - |z|)$, and hence $\lim_{N\to\infty} [S_N(z)] = -\mathrm{Log}(1 - z)$. Thus, $\sum_{n=1}^{\infty} (z^n/n) = -\mathrm{Log}(1 - z)$ for $|z| < 1$.

12. Show that the sequence $\{z_n\}$ defined recursively by $z_0 = 1, z_n = (i/2)z_{n-1} + i$, $n = 1, 2, \ldots$, converges, and find the limit. [*Hint:* Show that $z_{n-1} = (i/2) \cdot z_{n-2} + i$, and use this in the above to find z_n in terms of z_{n-2}. Do the same for z_{n-2}, and finally arrive at $z_n = (i/2)^n + i\sum_{k=0}^{n-1} (i/2)^k$, $n \geq 1$. Now take the limit.]

13. Show that

$$\frac{1}{z-\alpha} = \begin{cases} -\sum_{n=0}^{\infty} z^n \alpha^{-n-1} & \text{if } |z| < |\alpha| \\ \sum_{n=0}^{\infty} \alpha^n z^{-n-1} & |z| > |\alpha| \end{cases}$$

14. Consider the real-valued function $u(\theta) = (2 + \cos\theta)^{-1}$.

(a) What is the relationship between $u(\theta)$ and the complex function $f(z) = 1/[2 + \frac{1}{2}(z + z^{-1})]$?

(b) From (a), $f(z) = 2z(z^2 + 4z + 1)^{-1}$. Now, do a partial fraction expansion for $f(z)$ and use the result of Problem 13 to show that

$$f(z) = 3^{-1/2}\left[1 + \sum_{n=1}^{\infty} (-1)^n(2 - \sqrt{3})^n(z^n + z^{-n})\right]$$

for z in the annulus $-2 - \sqrt{3} < |z| < -2 + \sqrt{3}$.

(c) Use the result in (b) to derive the following **Fourier series** for $u(\theta)$:

$$u(\theta) = \frac{1}{\sqrt{3}}\left[1 + 2\sum_{n=1}^{\infty} (-1)^n(2 - \sqrt{3})^n \cos(n\theta)\right]$$

6.2 POWER SERIES; UNIFORM CONVERGENCE AND ANALYTIC FUNCTIONS[1]

A **power series about** $z = z_0$ is an infinite series of the form

(6.2.1)
$$\sum_{n=0}^{\infty} a_n(z - z_0)^n = a_0 + a_1 z + a_2 z^2 + \cdots$$

where the a_n are complex constants. The geometric series will be recognized as a special case of this with $z_0 = 0$ and $a_n = 1$ for all n. The power series is said to **converge at a point** $z = \hat{z}$ if the infinite series of numbers $\sum_{n=0}^{\infty} a_n(\hat{z} - z_0)^n$ converges. It **converges in a region** D if it converges at each point of D.

Note that a *power series always converges at* $z = z_0$, *the point about which it is being expanded*, to the number a_0. The set of all points in the z-plane for which the power series converges constitutes the domain of some function which is represented by this power series. The objective in this section is to determine this domain and some properties of this function. Our first result states something about the domain.

(6.2.2) ***Theorem.*** Consider the power series $\sum_{n=0}^{\infty} a_n(z - z_0)^n$.

(a) If the series converges at $\hat{z} \neq z_0$, then it converges for all z in the disk $|z - z_0| < |\hat{z} - z_0|$.

(b) One of the following must hold:

(i) The series converges only at $z = z_0$.

(ii) There is an $R > 0$ (R could be infinite), called the **radius of convergence** of the power series, such that the series converges

[1] Much of this material is theoretical in nature and is used to justify some of the manipulations in the next two sections. It would be advantageous to include this material in a more thorough mathematics course. In any case, it advisable to cover the first part of this section, where an elementary discussion of power series is provided.

for $|z - z_0| < R$ and diverges for $|z - z_0| > R$. The circle $|z - z_0| = R$, on which there may or may not be convergence, is called the **circle of convergence**.

(c) The radius of convergence is given by

(6.2.3)
$$R = \frac{1}{\lim_{n\to\infty} \left|\frac{a_{n+1}}{a_n}\right|}$$

if the limit exists.[2] If the limit is infinite, then we have case (b*i*) above.

Proof. The proof of parts (a) and (b) will be given in Problem 2. To prove (c), we first identify $z_n = a_n(z - z_0)^n$. Then, from the ratio test (6.1.15b), we have convergence for all those z such that

$$\lim_{n\to\infty} \left|\frac{a_{n+1}(z - z_0)^{n+1}}{a_n(z - z_0)^n}\right| = |z - z_0| \cdot \lim_{n\to\infty} \left|\frac{a_{n+1}}{a_n}\right| < 1$$

and divergence when the limit is greater than 1. Hence, the result follows.

■

Note that the radius of convergence in (6.2.3) can be zero or infinite, as the next example shows.

(6.2.4) ***Example.*** For the geometric series $\sum_{n=0}^{\infty} z_n$, we have $a_n = 1$ for all n, and thus by (6.2.3), $R = 1$ and this series converges for $|z| < 1$ (as we already know).

Now consider the power series

(6.2.5)
$$\sum_{n=0}^{\infty} \frac{(z - z_0)^n}{n!}$$

Here, $a_n = 1/n!$ and we have

$$\lim_{n\to\infty} \left|\frac{a_{n+1}}{a_n}\right| = \lim_{n\to\infty} \left[\frac{n!}{(n+1)!}\right] = 0$$

Thus, the series converges for all z.

Finally, consider the series $\sum_{n=0}^{\infty} n!(z - z_0)^n$.

Here, $a_n = n!$ and we have

$$\left|\frac{a_{n+1}}{a_n}\right| = n + 1 \xrightarrow[n\to\infty]{} \infty$$

Hence, the radius of convergence is zero and the series converges only at $z = z_0$.

■

[2] If the limit does not exist, the radius of convergence can still be defined. We refer the reader to a more advanced textbook.

Uniform Convergence. The concept of uniform convergence is one of the most important in mathematics. It is through this property that one can easily prove many of the outstanding results in complex function theory.

(6.2.6) ***Definition.*** If the power series $\sum_{n=0}^{\infty} a_n(z-z_0)^n$ converges for all z in some region D, it defines a function

$$f(z) = \sum_{n=0}^{\infty} a_n(z-z_0)^n \qquad \text{for } z \text{ in } D$$

We say that the **power series converges uniformly in D to $f(z)$** when, given an $\epsilon > 0$, there is an integer $N(\epsilon)$, *depending only on* ϵ, so that *for all z in* D, if $n \geq N(\epsilon)$, then

$$\left| \sum_{k=0}^{n} a_k(z-z_0)^k - f(z) \right| < \epsilon. \tag{6.2.7}$$

■

Note the following implication of this definition. First, whenever we terminate a convergent infinite series at some finite point, we will incur an error which becomes smaller as we take more and more terms in the series. If we have uniform convergence, then there is a single cutoff point $n = N(\epsilon)$, which does not depend upon the value of z in the power series, so that the error ϵ is the same for all z.

(6.2.8) ***Example.*** Show that the geometric series $\sum_{n=0}^{\infty} z^n$ does not converge uniformly inside of its whole circle of convergence $|z| = 1$, but does converge uniformly inside of any smaller circle $|z| = r_1 < 1$.

Solution. The difference between the nth partial sum of the geometric series and its sum $(1-z)^{-1}$ is

$$\left| \sum_{k=0}^{n} z^k - \frac{1}{1-z} \right| = \left| \frac{1-z^{n+1}}{1-z} - \frac{1}{1-z} \right| = \frac{|z|^{n+1}}{|1-z|} \tag{6.2.9}$$

We now wish to determine whether

for any ϵ there is an $N(\epsilon)$ so that if $n \geq N(\epsilon)$, then $(|z|^{n+1})/(|1-z|) < \epsilon$ for all $|z| < 1$?

First, consider the case when $|z| \leq r_1 < 1$. Then,

$$|1-z| \geq |1-|z|| = 1-|z| \geq 1-r_1$$

Hence,

$$\frac{|z|^{n+1}}{|1-z|} \leq \frac{r_1^{n+1}}{1-r_1} \tag{6.2.10}$$

Since $r_1 < 1$ implies that $r_1^{n+1} \xrightarrow[n\to\infty]{} 0$, we can find an $N(\epsilon)$ so that

$$n \geq N(\epsilon) \Rightarrow \frac{r_1^{n+1}}{1-r_1} < \epsilon$$

From this and from (6.2.9) and (6.2.10), we can conclude that the series converges uniformly for $|z| \le r_1 < 1$.

Now consider the full region of convergence $|z| < 1$. The difference between this case and the above is that now $|z|$ can get arbitrarily close to 1. To show that the series does not converge uniformly here, we will use a **proof by contradiction** procedure. Thus, we first assume that the geometric series does converge uniformly for $|z| < 1$, that is, for any $\epsilon > 0$, we assume there is an $N(\epsilon)$, so that [from (6.2.9)]

$$n \ge N(\epsilon) \Rightarrow \frac{|z|^{n+1}}{|1 - z|} < \epsilon \qquad \text{for all } |z| < 1$$

Now, in the above fix, $n = n_0$, where $n_0 \ge N(\epsilon)$. Then,

$$\frac{|z|^{n_0+1}}{|1 - z|} < \epsilon$$

This must hold for all $|z| < 1$. However, the left side of the above can be made as large as desired, and certainly greater than the given ϵ, by simply choosing a z inside the unit circle and close to $z = 1$. Hence, we have a contradiction, and the geometric series does not converge uniformly for $|z| < 1$.

■

The next result provides a criterion for uniform convergence.

(6.2.11) ***Theorem* (Weierstrass *M*-test).** For all z in a region D, let there be a sequence of positive numbers $\{M_n\}$ so that

$$|a_n(z - z_0)^n| \le M_n$$

Then, if $\sum_{n=0}^{\infty} M_n$ converges, the power series $\sum_{n=0}^{\infty} a_n(z - z_0)^n$ converges uniformly and absolutely in D.

■

The proof of this uses the Cauchy criterion for convergence (6.1.14) and is given in Problem 6.

(6.2.12) ***Example.*** In Example (6.2.8), we showed that the geometric series $\sum_{n=0}^{\infty} z^n$ converged uniformly for $|z| \le r_1 < 1$. The Weierstrass test yields this result immediately, for when $|z| \le r_1$, we have $|z^n| \le r_1^n$. For $r_1 < 1$, the geometric series of numbers $\sum_{n=0}^{\infty} r_1^n$ converges, and hence $\sum_{n=0}^{\infty} z^n$ converges uniformly. Here, we have identified $M_n = r_1^n$.

■

The next theorem shows that the uniform convergence result for the geometric series is also true for the general power series.

(6.2.13) ***Theorem.*** Let $|z - z_0| = R (R \neq 0)$ be the circle of convergence for the power series $\sum_{n=0}^{\infty} a_n(z - z_0)^n$. Then this series **converges uniformly and absolutely within and on any circle interior to the circle of convergence.**

Proof. Consider Figure 6.2.1. There, we show the circle of convergence of radius R, an interior circle of radius $r_1 < R$, and another (dotted) circle, lying between these two and of radius $\frac{1}{2}(R + r_1)$. Also shown is an arbitrary point z which may lie inside or on the circle of radius r_1, and a point z_1 on the dotted circle. From the figure, it is clear that

$$|z - z_0| \leq r_1 \quad \text{and} \quad |z_1 - z_0| = \tfrac{1}{2}(R + r_1)$$

Now, since z_1 lies inside the circle of convergence, the series of numbers $\sum_{n=0}^{\infty} a_n(z_1 - z_0)^n$ converges. Hence, the general term is bounded, that is,

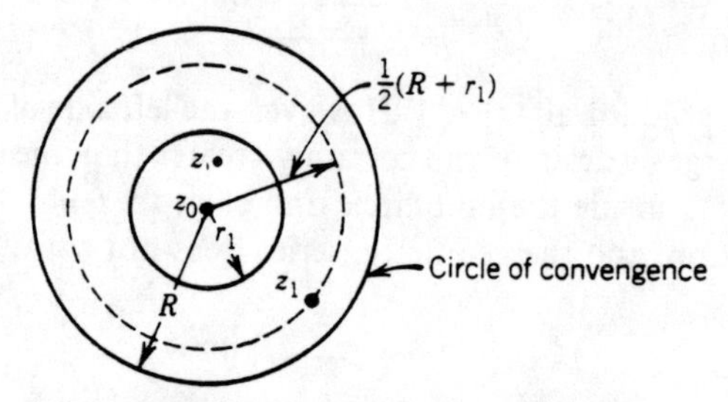

Figure 6.2.1 Theorem (6.2.13).

there is an M so that for all n, $|a_n(z_1 - z_0)^n| \leq M$. From this and the above inequalities,

$$|a_n(z - z_0)^n| = |a_n(z_1 - z_0)^n| \cdot \left|\frac{z - z_0}{z_1 - z_0}\right|^n \leq M\left(\frac{2r_1}{R + r_1}\right)^n$$

Now, $2r_1/(R + r_1) < 1$, since $r_1 < R$. Thus, the geometric series

$$\sum_{n=0}^{\infty} M\left(\frac{2r_1}{r_1 + R}\right)^n$$

converges, and by the Weierstrass M-test, the power series converges uniformly and absolutely in $|z - z_0| \leq r_1 < R$.

■

Power Series and Analytic Functions. The next few results, which are the highlights of this section, show that power series define analytic functions and can be differentiated and integrated term by term. For ease of exposition, and without loss of generality, *we will deal with power series about the point* $z_0 = 0$.

(6.2.14) ***Theorem.*** Let $f(z)$ be defined by the power series

$$f(z) = \sum_{n=0}^{\infty} a_n z^n \qquad |z| < R \qquad (R \neq 0)$$

Then, $f(z)$ is continuous within the circle of convergence. In terms of the power series, this means that

(6.2.15)
$$\lim_{z \to \hat{z}} \left[\sum_{n=0}^{\infty} a_n z^n \right] = \sum_{n=0}^{\infty} a_n \hat{z}^n \qquad \text{for } |\hat{z}| < R$$

Proof. Let $\hat{z}$ lie inside the circle of convergence, as shown in Figure 6.2.2, and choose $|\Delta z|$ small enough so that $\hat{z} + \Delta z$ also lies inside this circle. To prove continuity at $\hat{z}$, we must show that $\lim_{\Delta z \to 0} |f(\hat{z} + \Delta z) - f(\hat{z})| = 0$. Now, for any choice of N,

(6.2.16)
$$\begin{aligned} |f(\hat{z} + \Delta z) - f(\hat{z})| = \Bigg| & f(\hat{z} + \Delta z) - \sum_{n=0}^{N} a_n (\hat{z} + \Delta z)^n \\ & + \sum_{n=0}^{N} a_n [(\hat{z} + \Delta z)^n - \hat{z}^n] \\ & + \sum_{n=0}^{\infty} a_n \hat{z}^n - f(\hat{z}) \Bigg| \\ \le \Bigg| f(\hat{z} + \Delta z) - \sum_{n=0}^{N} & a_n (\hat{z} + \Delta z)^n \Bigg| + \Bigg| f(\hat{z}) - \sum_{n=0}^{N} a_n \hat{z}^n \Bigg| \\ & + \sum_{n=0}^{N} |a_n| |(\hat{z} + \Delta z)^n - \hat{z}^n| \end{aligned}$$

Choose r_1 so that $|\hat{z}| < r_1 < R$. Then, by the previous theorem, $\sum_{n=0}^{\infty} a_n z^n$ converges uniformly to $f(z)$ for $|z| \le r_1$. (The circle $|z| = r_1$ is shown dashed in the figure.) Since $\hat{z}$ and, for $|\Delta z|$ small enough, $\hat{z} + \Delta z$, both lie within this dashed circle, we have from the definition of uniform convergence that for any ϵ there is an $N_0(\epsilon)$ so that

$$\left| f(\hat{z}) - \sum_{n=0}^{N_0} a_n \hat{z}^n \right| < \epsilon/3 \qquad \text{and} \qquad \left| f(\hat{z} + \Delta z) - \sum_{n=0}^{N_0} a_n (\hat{z} + \Delta z)^n \right|$$

$< \epsilon/3$, for $|\Delta z|$ small enough. These results provide bounds for the first two terms on the right-hand side of the inequality in (6.2.16). To bound the last term there, we choose $N = N_0$ and note from the binomial theorem that for $n \ge 1$,

$$(\hat{z} + \Delta z)^n - \hat{z}^n = n(\Delta z)\hat{z}^{n-1} + \cdots + (\Delta z)^n$$

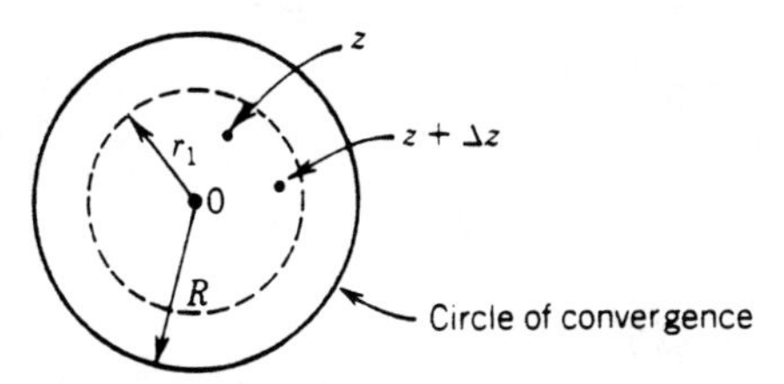

Figure 6.2.2 Theorem (6.2.14).

Hence, every term in the last summation on the right-hand side of (6.2.16) has a factor of Δz. Thus, with Δz small enough, this term can be made less than $\epsilon/3$ also. Hence, combining these results we have for any ϵ, and with $|\Delta z|$ small enough, $|f(\hat{z} + \Delta z) - f(\hat{z})| < \epsilon$. Thus, $f(z)$ is continuous at any $\hat{z}$ inside of the circle of convergence.

■

We now show that we can integrate and differentiate power series term by term.

(6.2.17) ***Theorem.*** Consider the power series

$$f(z) = \sum_{n=0}^{\infty} a_n z^n \qquad |z| < R \qquad (R \neq 0)$$

Let C be any simple piecewise smooth curve whose trace lies inside the circle of convergence. Then we can **integrate the power series term by term,** that is,

$$\int_C \left(\sum_{n=0}^{\infty} a_n z^n \right) dz = \sum_{n=0}^{\infty} a_n \int_C z^n \, dz \tag{6.2.18}$$

Proof. By the previous theorem, the function $f(z)$ defined by the power series is continuous on C, and hence the integrals above are well defined. From (6.2.18) and the definition of $f(z)$, we would like to show that

$$\lim_{n \to \infty} \left| \int_C \left[f(z) - \sum_{k=0}^{n} a_k z^k \right] dz \right| = 0 \tag{6.2.19}$$

Also, since the trace of C lies inside the circle of convergence of the power series, the series converges uniformly on C to $f(z)$. Hence, for any ϵ, there is an $N(\epsilon)$ so that, *for all z on* C,

$$n \geq N(\epsilon) \Rightarrow \left| f(z) - \sum_{k=0}^{n} a_k z^k \right| < \epsilon$$

By the triangle inequality for integrals and the above inequalities, we have, *for* $n \geq N$,

$$\left| \int_C \left[f(z) - \sum_{k=0}^{n} a_k z^k \right] dz \right| \leq \epsilon \cdot (\text{length of } C)$$

This proves that the limit in (6.2.19) is zero, since ϵ is arbitrary.

■

(6.2.20) ***Theorem.*** Let

$$f(z) = \sum_{n=0}^{\infty} a_n z^n \qquad |z| < R \qquad (R \neq 0)$$

Then,

(a) The *formally derived series* $\sum_{n=1}^{\infty} n a_n z^{n-1}$ has the same radius of convergence as does $f(z)$. The same is true for the higher-order, formally derived series.

(b) $f(z)$ is analytic inside the circle of convergence, and its derivatives can be computed by term-wise differentiation of the power series; that is, for $n = 1, 2, \ldots,$

(6.2.21)
$$f^{(n)}(z) = \sum_{k=n}^{\infty} k(k-1)\cdots(k-n+1)a_k z^{k-n} \qquad \text{for } |z| < R$$

Proof. We will leave the proof of part (a) to Problem 8 and will prove (b) using Morera's Theorem from Section 5.6.

First, let C be any closed contour lying inside the circle of convergence. Then, by the previous theorem

$$\oint_C f(z)\,dz = \sum_{n=0}^{\infty} a_n \oint_C z^n\,dz = 0$$

since z^n is entire. Hence, since $f(z)$ is continuous and satisfies the hypotheses of Morera's Theorem, it is analytic for $|z| < R$.

With $n = 1$ in (6.2.21), we now wish to show that for $|\hat{z}| < R$,

$$f'(\hat{z}) = \lim_{n\to\infty}\left(\sum_{k=1}^{n} k a_k \hat{z}^{k-1}\right)$$

If $S_n(z) = \sum_{k=0}^{n} a_k z^k$ is the nth partial sum of the series for $f(z)$, then $S_n'(\hat{z}) = \sum_{k=1}^{n} k a_k \hat{z}^{k-1}$, and by the Cauchy integral formula

(6.2.22)
$$f'(\hat{z}) - S_n'(\hat{z}) = \frac{1}{2\pi i}\oint_{|\zeta-\hat{z}|=R_1} \frac{[f(\zeta) - S_n(\zeta)]\,d\zeta}{(\zeta-\hat{z})^2}$$

where $R_1 < R$, and thus the integration variable ζ lies inside the circle of convergence of the power series. Because of this, $S_n(\zeta)$ converges uniformly to $f(\zeta)$ for $|\zeta - \hat{z}| = R_1$, and thus for any ϵ, there is an $N(\epsilon)$ so that, for $|\zeta - \hat{z}| = R_1$,

$$|f(\zeta) - S_n(\zeta)| < \epsilon \qquad \text{for } n \geq N(\epsilon)$$

From (6.2.22), this implies that

$$|f'(\hat{z}) - S_n'(\hat{z})| \leq \frac{\epsilon}{2\pi R_1}$$

and hence $S_n'(\hat{z}) \xrightarrow[n\to\infty]{} f'(\hat{z})$ pointwise for any $\hat{z}$ such that $|\hat{z}| < R$. The results for the higher-order derivatives follow along the same lines.

■

The implications of the last two results are that **power series define analytic functions and can basically be manipulated at will within their circle of convergence**, as the next example shows.

(6.2.23) ***Example.*** Sum the series $f(z) = \sum_1^{\infty} z^n/n, |z| < 1$ (also see Problem 11 in Section 6.1).

Solution. If there were no n in the denominator, the series would simply be a geometric series. Thus, we are led to first differentiate f to get

$$f'(z) = \sum_{n=1}^{\infty} z^{n-1} = \frac{1}{1-z}$$

Then, integration along any contour inside of $|z| = 1$ yields

$$f(z) = k - \log(z-1) \qquad (k = \text{constant}) \tag{6.2.24}$$

Now, since f is analytic within $|z| = 1$, we must choose a branch of the logarithm which is also analytic here. The constant k will then depend upon the choice of branch. One such branch is given by

$$\log(z-1) = \ln|z-1| + i\theta \qquad 0 < \theta \leq 2\pi$$

where the branch cut is shown in Figure 6.2.3. Now, we must compute k in (6.2.24). From the original power series, we have $f(0) = 0$. Thus, for our choice of branch, and (6.2.24),

$$0 = k - \log(-1) \Rightarrow k = \pi i$$

Hence,

$$\sum_{n=1}^{\infty} \frac{z^n}{n} = \pi i - \log(z-1) \qquad |z| < 1$$

Note the branch point singularity at $z = 1$ on the circle of convergence.

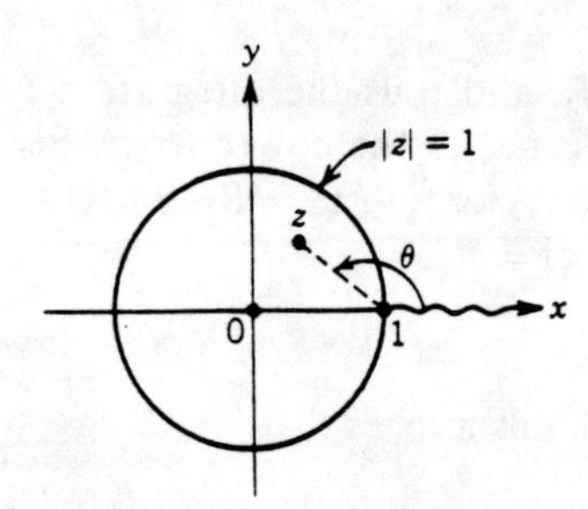

Figure 6.2.3 Example (6.2.23).

Finally, *returning to the general power series about* z_0, we have the following.

(6.2.25) ***Theorem.*** Let $f(z)$ have a power series representation

$$f(z) = \sum_{n=0}^{\infty} a_n(z-z_0)^n \qquad |z-z_0| < R \qquad (R \neq 0)$$

Then,

(a) $f(z)$ is analytic for $|z - z_0| < R$.

(b) The coefficients a_n are given in terms of the derivatives of $f(z)$ by

(6.2.26)
$$a_n = \frac{f^{(n)}(z_0)}{n!}$$

(c) This power series representation for f is unique.

Proof. Part (a) is just a repetition of Theorem (6.2.20b). From this same theorem and the analog of (6.2.21) in powers of $(z - z_0)$, we get

$$f^{(k)}(z_0) = k!a_k$$

thus proving (b). Finally, (c) follows from this result, since if there were two power series representations, both sets of coefficients must be given by (6.2.26).

■

PROBLEMS (The answers are given on page 438.)

(Problems with an asterisk are more theoretical.)

1. What are the radii of convergence for the following power series? What are the circles of convergence?

(a) $\displaystyle\sum_{n=1}^{\infty} \frac{(z-i)^n}{n(n+1)}$ (b) $\displaystyle\sum_{n=0}^{\infty} \frac{n!(z-1+i)^n}{(2n)!}$

(c) $\displaystyle\sum_{n=0}^{\infty} (n+1+2^n)z^n$ (d) $\displaystyle\sum_{n=0}^{\infty} \frac{z^{2n}}{a^n}$ $(a \neq 0)$

(e) $\displaystyle\sum_{n=0}^{\infty} \frac{z^n}{n!n^n}$ (f) $\displaystyle\sum_{n=0}^{\infty} \frac{n!z^n}{n^n}$

2.* The following series of steps leads to a proof of Theorem (6.2.2a, b). For (a), we are assuming that the power series $\sum_{n=0}^{\infty} a_n(z-z_0)^n$ converges at $z = \hat{z}$, $\hat{z} \neq z_0$.
(a) Show that there is an $M > 0$ so that for all n, $|a_n(\hat{z} - z_0)^n| \leq M$. [*Hint:* Use Theorem (6.1.13b).]
(b) If $\tilde{z}$ is such that $|\tilde{z} - z_0| < |\hat{z} - z_0|$, use part (*a*) to show that $|a_n(\tilde{z} - z_0)^n| \leq M\delta^n$, where δ is a number less than 1.
(c) Now use the comparison test (6.1.15a) to show that $\sum_{n=0}^{\infty} a_n(z-z_0)^n$ converges for z in the disk $|z - z_0| < |\hat{z} - z_0|$. Thus, we have proved part (a) of the theorem
(d) Finally, prove part (b) of the theorem.

3.* Prove that if the power series $\sum_{n=0}^{\infty} a_n(z-z_0)^n$ *diverges* at $z = \hat{z}$, then it diverges for $|z - z_0| > |\hat{z} - z_0|$. [*Hint:* Assume it converges at a point $\tilde{z}$ such that $|\tilde{z} - z_0| > |\hat{z} - z_0|$. Then, use the approach of Problem 2.]

4.* Prove that if $|a_n| \leq M_n$, and if $\sum_{n=0}^{\infty} M_n(z-z_0)^n$ converges absolutely for z in a region D, then $\sum_{n=0}^{\infty} a_n(z-z_0)^n$ converges for z in D.

5. Where do the following power series converge uniformly?

(a) $\sum_{n=0}^{\infty} (a^n + b^n)z^n \quad a, b > 0, a \neq b$ (b) $\sum_{n=2}^{\infty} \frac{(-1)^n(2z+1)^n}{\sqrt{n}}$

(c) $\sum_{n=1}^{\infty} \frac{(z-i)^n \ln n}{ne^n}$ (d) $\sum_{n=0}^{\infty} \frac{(iz-1)^n e^{-n}}{n!}$

(e) $1 + \sum_{n=0}^{\infty} \frac{\alpha(\alpha+1)\cdots(\alpha+n)\beta(\beta+1)\cdots(\beta+n)z^{n+1}}{\gamma(\gamma+1)\cdots(\gamma+n)(n+1)!}$, where α, β, and γ are neither negative integers nor zero. This is called a **hypergeometric series.**

6.* The following series of steps leads to a proof of the Weierstrass M-test of Theorem (6.2.11). We are given that $|a_n(z-z_0)^n| \leq M_n (n = 0, 1, 2, \ldots)$ uniformly for z in a region D, and that the series of constants $\sum_{n=0}^{\infty} M_n$ converges.

(a) Use the Cauchy test for convergence (6.1.14d) to show that

$$\lim_{n,p\to\infty} \left(\sum_{k=n}^{n+p} M_k\right) = 0$$

(b) Use the triangle inequality to show that $\lim_{n,p\to\infty} \left[\sum_{k=n}^{n+p} a_k(z-z_0)^k\right] = 0$ for *each* z in D, and hence conclude from the Cauchy test that $\sum_{n=0}^{\infty} a_n(z-z_0)^n$ converges in D to some function $f(z)$.

(c) Use the proof of part (*b*) and the definition of limit to show that given an $\epsilon > 0$, there is an $N(\epsilon)$ (independent of z) so that

$$n, p > N \Rightarrow \left|\sum_{k=n}^{n+p} a_k(z-z_0)^k\right| < \epsilon$$

for all z in D.

(d) Now take the limit above as $p \to \infty$ and show that given an $\epsilon > 0$, there is an $N(\epsilon)$ so that

$$n > N \Rightarrow \left|\sum_{k=0}^{n} a_k(z-z_0)^k - f(z)\right| < \epsilon$$

Hence, conclude that the series converges uniformly.

7.* Prove that if $\sum_{n=0}^{\infty} a_n(z-z_0)^n$ converges uniformly on a set D, then the convergence is also uniform on any subset of D.

8.* The following series of steps leads to a proof of part (a) of Theorem (6.2.20). We are given that the radius of convergence of $\sum_{n=0}^{\infty} a_n z^n$ is R.

(a) Show that $n\delta^n \xrightarrow[n\to\infty]{} 0$, where $0 < \delta < R$.

(b) Assume that $\sum_{n=0}^{\infty} a_n z^n$ converges at $\hat{z}$. Prove that the derived series $\sum_{n=0}^{\infty} na_n z^{n-1}$ converges for each z such that $|z| < |\hat{z}|$. Thus, conclude that the radius of convergence of the derived series is at least as large as that of the original series. [*Hint:* For $z \neq 0$, write

$$na_n z^{n-1} = a_n \hat{z}^n \left[\frac{n}{z}\left(\frac{z}{\hat{z}}\right)^n\right]$$

and then use part (a) and the Weierstrass M-test.

(c) Now show that if the derived series $\sum_{n=0}^{\infty} na_n z^{n-1}$ converges at $z = z^*$, then the original series converges for $|z| < |z^*|$. Hence, use part (b) to conclude that the radii of convergence of the two series are the same. [*Hint:* $a_n z^n = [na_n(z^*)^{n-1}](z^*/n)(z/z^*)^n$.]

(d) Prove these results for all higher-order derivatives.

9. Use the procedure of Example (6.2.23) to sum the following series:

(a) $\displaystyle\sum_{n=1}^{\infty} \frac{z^n}{n^2}$ (b) $\displaystyle\sum_{n=1}^{\infty} \frac{z^n}{n(n+1)}$

6.3 TAYLOR SERIES; ZEROS OF ANALYTIC FUNCTIONS

In the previous section,[1] we learned that a power series with a nonzero radius of convergence represents an analytic function. Here, we will prove the important converse of this, namely, that every analytic function must possess a power series expansion.

(6.3.1) ***Definition.*** Let $f(z)$ be analytic at $z = z_0$. Then, the **Taylor series for $f(z)$ about $z = z_0$** is defined to be the power series

$$\sum_{n=0}^{\infty} \frac{f^{(n)}(z_0)}{n!}(z - z_0)^n$$

The numbers $f^{(n)}(z_0)/n!$ are called the **Taylor coefficients**. If $z_0 = 0$, the series is often called a **Maclaurin series**.

■

The power series above will have some radius of convergence which, in general, will be quite difficult to determine. However, the next result tells us for *which z the Taylor series actually converges to the number* f(z).

(6.3.2) ***Theorem.*** Let $f(z)$ be analytic at z_0. Then f has a Taylor series representation

$$f(z) = \sum_{n=0}^{\infty} \frac{f^{(n)}(z_0)}{n!}(z - z_0)^n \qquad |z - z_0| < \hat{R}$$

which converges to f(z) *for z in the above disk*, where **$\hat{R}$ is the distance from z_0 to the singularity of f closest to z_0**. If $f(z)$ is entire, then the above is valid for all z. In addition, the convergence is uniform within and on any circle centered at z_0 and lying inside the above disk.

[1] Though the material in Section 6.2 on uniform convergence is necessary for rigorous proofs of the main results here, the reader can acquire a working knowledge of Taylor series without it. All that is really needed is an acceptance of the fact that when dealing with power series, the order in which operations are carried out may be interchanged.

Proof. **We must exhibit the validity of the above for each z *interior to the largest disk about z_0 inside of which f is analytic.*** In Figure 6.3.1, let z_1 be that singularity of f closest to z_0 and choose z to lie inside the circle centered at z_0 with radius $|z_1 - z_0|$ (the dotted circle in the figure). Now pick a circle C of radius ρ centered at z_0, lying inside of this disk, and containing z in its interior. If ζ designates the running parameter along C (see the figure), we have by the Cauchy integral formula

(6.3.3)
$$f(z) = \frac{1}{2\pi i} \oint_{|\zeta - z_0| = \rho} \frac{f(\zeta)\, d\zeta}{(\zeta - z)}$$

Now, since $\zeta \neq z_0$ on C, we can write

(6.3.4)
$$\begin{aligned} \frac{1}{\zeta - z} &= \frac{1}{\zeta - z_0}\left[\frac{1}{1 - \left(\dfrac{z - z_0}{\zeta - z_0}\right)}\right] \\ &= \frac{1}{\zeta - z_0}\left[\frac{1}{1 - \left(\dfrac{z - z_0}{\zeta - z_0}\right)} - \sum_{k=0}^{n}\left(\frac{z - z_0}{\zeta - z_0}\right)^k + \sum_{k=0}^{n}\left(\frac{z - z_0}{\zeta - z_0}\right)^k\right] \end{aligned}$$

where to the term in brackets on the first line above we have added and subtracted the first n terms of a geometric series. From (6.1.22),

$$\sum_{k=0}^{n}\left(\frac{z - z_0}{\zeta - z_0}\right)^k = \frac{1 - \left(\dfrac{z - z_0}{\zeta - z_0}\right)^{n+1}}{1 - \left(\dfrac{z - z_0}{\zeta - z_0}\right)}$$

Hence, from (6.3.4),

$$\frac{1}{\zeta - z} = \frac{1}{\zeta - z}\left(\frac{z - z_0}{\zeta - z_0}\right)^{n+1} + \sum_{k=0}^{n} \frac{(z - z_0)^k}{(\zeta - z_0)^{k+1}}$$

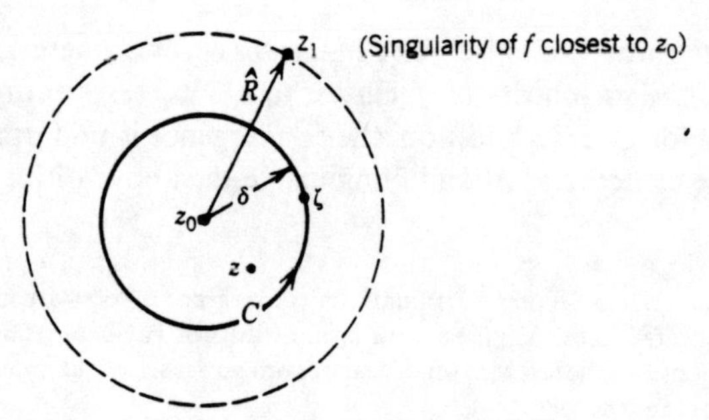

Figure 6.3.1 Theorem (6.3.2).

Substituting this into the integrand in (6.3.3) yields

(6.3.5)
$$f(z) = \sum_{k=0}^{n} (z - z_0)^k \left[\frac{1}{2\pi i} \oint_{|\zeta - z_0| = \rho} \frac{f(\zeta)\, d\zeta}{(\zeta - z_0)^{k+1}}\right] + \frac{1}{2\pi i} \oint_{|\zeta - z_0| = \rho} \frac{f(\zeta)}{\zeta - z} \left(\frac{z - z_0}{\zeta - z_0}\right)^{n+1} d\zeta$$

Since $f(\zeta)$ is analytic on the circle $|\zeta - z_0| = \rho$, it is bounded there, and thus $|f(\zeta)| \le M$. Also, note from Figure 6.3.1 that

$$|z - z_0| < |z_1 - z_0| \quad \text{and} \quad |z - z_0| < \rho$$

Hence,

$$\begin{aligned} |\zeta - z| &\ge \big| |\zeta - z_0| - |z - z_0| \big| \\ &= \rho - |z - z_0| \\ &\ge \rho - |z_1 - z_0| \end{aligned}$$

This, together with the triangle inequality applied to the second integral in (6.3.5), yields

(6.3.6)
$$\left| f(z) - \sum_{k=0}^{n} (z - z_0)^k \left[\frac{1}{2\pi i} \oint_{|\zeta - z_0| = \rho} \frac{f(\zeta)\, d\zeta}{(\zeta - z_0)^{k+1}}\right] \right| \le \frac{\rho M}{\rho - |z_1 - z_0|} \left|\frac{z - z_0}{\rho}\right|^{n+1}$$

Now, since $|(z - z_0)/\rho| < 1$, the term of the right side above approaches zero as $n \to \infty$. Also, by the Cauchy integral theorem for derivatives, the integral on the left will be recognized as $2\pi i\, f^{(k)}(z_0)/k!$ Thus, the Taylor series for f converges pointwise for all z inside the dotted circle in the figure. That the series also converges uniformly inside of any smaller circle is shown in Problem 1.

■

Note that a Taylor series is of course a power series itself. Consequently, this power series has its own radius of convergence. Depending upon the function being expanded in a Taylor series, this radius of convergence may not be the same as the radius $\hat{R}$ of the disk inside of which the Taylor series converges to the function [see Example (6.3.17)]. However, $\hat{R}$ *cannot be greater than this radius of convergence.*

If we now append the present result to those of the previous section, we have the following.

(6.3.7) ***Theorem.*** Let $f(z)$ be analytic at $z = z_0$. Then,

(a) *$f(z)$ possesses a unique power series expansion* about $z = z_0$, and this power series is the Taylor series

$$f(z) = \sum_{n=0}^{\infty} a_n (z - z_0)^n \qquad \text{for } |z - z_0| < \hat{R}$$

where

$$a_n = \frac{f^{(n)}(z_0)}{n!}$$

Furthermore, the radius $\hat{R}$ of the above disk, *in which the power series converges to* f(z), is the distance between z_0 and the nearest singularity of $f(z)$.

(b) Derivatives and integrals of $f(z)$ are given, within the above disk, by term-by-term differentiation and integration of the Taylor series.

(c) If $g(z)$ is analytic at z_0 with Taylor series

(6.3.8)
$$g(z) = \sum_{n=0}^{\infty} b_n(z - z_0)^n \qquad |z - z_0| < \hat{R}_1$$

then

(i) $\alpha f(z) + \beta g(z) = \sum_{n=0}^{\infty} (\alpha a_n + \beta b_n)(z - z_0)^n$ for $|z - z_0| < \min\{\hat{R}, \hat{R}_1\}$.

(ii) The Taylor series for the product $f \cdot g$ is given by the **Cauchy product**

$$f(z) \cdot g(z) = \sum_{n=0}^{\infty} c_n(z - z_0)^n \qquad \text{for } |z - z_0| < \min\{\hat{R}, \hat{R}_1\}$$

where

(6.3.9)
$$c_n = \sum_{k=0}^{n} a_k b_{n-k} = \sum_{k=0}^{n} \frac{f^{(k)}(z_0) g^{(n-k)}(z_0)}{k!(n-k)!}$$

Proof. The results of (a) and (b) are simply a collection of those of the previous section, and the previous theorem. The proof of (c) follows from the fact that the Taylor coefficients for $\alpha f + \beta g$ and $f \cdot g$ are just those given in the respective series. Finally, the radii of the disks where these series converge to their respective functions must be the distances from z_0 to the nearest singularities of $\alpha f + \beta g$ and $f \cdot g$, that is, $\min\{\hat{R}, \hat{R}_1\}$.

■

What we learn from the above results, and what will be exhibited in the examples to follow, is that **if, by whatever means, a convergent power series representation can be computed for an analytic function, this must be its Taylor series.**

(6.3.10) ***Example.*** The exponential function e^z, being entire, possesses a Taylor series *about any point* which converges for all z. Thus, since all derivatives of e^z are 1 at $z = 0$, we have the usual expansion

(6.3.11)
$$e^z = \sum_{n=0}^{\infty} \frac{z^n}{n!} \qquad \text{for } |z| < \infty$$

We can use this to generate the Taylor series of e^z about any other point. For example, for the series about $z_0 = 1$, we would first write $e^z = e \cdot e^{z-1}$ and then expand e^{z-1} in powers of $z - 1$ by simply replacing z in (6.3.11) by $z - 1$. Hence,

$$e^z = e \sum_{n=0}^{\infty} \frac{(z-1)^n}{n!} \qquad \text{for } |z - 1| < \infty$$

To derive the Taylor series for e^{z^2} about $z = 0$, we do not have to evaluate all of the derivatives but can simply replace z in (6.3.11) by z^2 to get

$$e^{z^2} = \sum_{n=0}^{\infty} \frac{z^{2n}}{n!} \qquad \text{for } |z| < \infty$$

Finally, for the Taylor series of e^{z^2} about $z = 1$, we first write

$$e^{z^2} = e^{(z-1)^2} \cdot e^{2(z-1)} \cdot e$$

and then use (6.3.11) and the above result for e^{z^2} to get

$$e^{z^2} = e\left[\sum_{n=0}^{\infty} \frac{(z-1)^{2n}}{n!}\right]\left[\sum_{n=0}^{\infty} \frac{2^n(z-1)^n}{n!}\right] \qquad \text{for } |z-1| < \infty$$

which is, by the Cauchy product (6.3.9),

$$e^{z^2} = e \sum_{n=0}^{\infty} (z-1)^n \left(\sum_{k=0}^{n} a_k \frac{2^{n-k}}{(n-k)!}\right)$$

where

$$a_k = \begin{cases} 0 & k \text{ odd} \\ \dfrac{1}{(k/2)!} & k \text{ even} \end{cases}$$

■

(6.3.12) ***Example.*** Compute the Taylor series for

(a) $f(z) = \dfrac{1}{z}$ about $z_0 = 1$

(b) $g(z) = \dfrac{1}{z^2 + (1-2i)z - 2i}$ about $z_0 = 0$

Solution. For $f(z)$, we first write

$$f(z) = \frac{1}{1 + (z-1)} \tag{6.3.13}$$

and expand this in a geometric series for $|z - 1| < 1$ to get

$$f(z) = \sum_{n=0}^{\infty} (-1)^n (z-1)^n \qquad \text{for } |z-1| < 1$$

By uniqueness of power series, the above *must* be the Taylor series of $f(z)$ about $z_0 = 1$. Note that the radius of convergence of the power series on the right is 1. This also happens to be the distance between $z = 1$ (*the point about which we are expanding*) and $z = 0$ (*the singularity of $f(z)$ closest to $z = 1$*).

For the rational function $g(z)$, with singularities at $z = -1, 2i$, we first perform the partial fraction expansion

$$g(z) = \frac{1}{1+2i}\left[\frac{1}{z-2i} - \frac{1}{z+1}\right]$$

and then perform the geometric series expansions

$$\frac{1}{z-2i} = -\frac{1}{2i}\sum_{n=0}^{\infty}\left(\frac{z}{2i}\right)^n \qquad |z| < 2$$

and

$$\frac{1}{1+z} = \sum_{n=0}^{\infty}(-1)^n z^n \qquad |z| < 1$$

The difference of these series has the common region of convergence $|z| < 1$, and hence,

$$g(z) = \frac{-1}{1+2i}\sum_{n=0}^{\infty}[(2i)^{-n-1} + (-1)^n]z^n \qquad |z| < 1$$

Again, note that the radius of convergence of the above is also the distance between $z = 0$ and the nearest singularity of g.

■

(6.3.14) ***Example.*** Compute the Taylor series for $f(z) = 1/z^2$ about $z_0 = 1$.

Solution. We will do this in two ways. By far the easiest is to write

$$\frac{1}{z^2} = -\frac{d}{dz}\left(\frac{1}{z}\right) = -\frac{d}{dz}\left[\frac{1}{1+(z-1)}\right]$$

$$= -\frac{d}{dz}\sum_{n=0}^{\infty}(-1)^n(z-1)^n \qquad |z-1| < 1$$

$$= \sum_{n=0}^{\infty}(n+1)(-1)^n(z-1)^n \qquad |z-1| < 1$$

We could have also multiplied the Taylor series

$$1/z = \sum_{n=0}^{\infty}(-1)^n(z-1)^n$$

by itself using the Cauchy product. The reader is urged to do this and reproduce the above result.

■

(6.3.15) ***Example.*** Consider the ζ-plane cut along the two lines as shown in Figure 6.3.2. If C is *any* simple, piecewise smooth curve not intersecting the cuts, the function $f(z)$ defined by the integral

(6.3.16)
$$f(z) = \oint_0^z \frac{d\zeta}{1+\zeta^2}$$

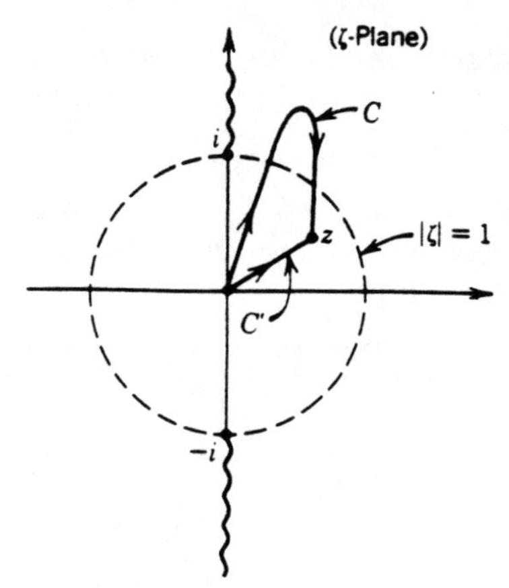

Figure 6.3.2 Example (6.3.15).

is analytic in the z-plane cut the same way. What is its Taylor series about $z = 0$?

Solution. One approach is to do a partial fraction expansion of the integrand and then integrate both terms to get logarithms. Then, being careful of the choice of the branches, a Taylor expansion can be done for each term. A different approach, however, is to first expand the integrand $1/(1 + \zeta^2)$ in a geometric series about $\zeta = 0$ and then integrate term by term. However, to do this, one must first be assured that every value of the variable of integration ζ in (6.3.16), which lies on C, also lies within the unit circle. (Why?) By the analyticity of the integrand in (6.3.16), the integral is unchanged if C is deformed into any other curve joining $\zeta = 0$ to $\zeta = z$ so long as we do not pass through a singularity. Hence,

$$f(z) = \oint_0^{z}{}_{C'} \frac{d\zeta}{1 + \zeta^2}$$

where C' is the line joining $\zeta = 0$ to $\zeta = z$ (see the figure). For z close to zero, $|z| < 1$; hence, $|\zeta| < 1$ for all ζ on C'. Thus,

$$\frac{1}{1 + \zeta^2} = \sum_{n=0}^{\infty} (-1)^n \zeta^{2n} \qquad (\text{for } \zeta \text{ on } C')$$

and

$$\begin{aligned} f(z) &= \sum_{n=0}^{\infty} (-1)^n \oint_0^{z}{}_{C'} \zeta^{2n}\, d\zeta \\ &= \sum_{n=0}^{\infty} \frac{(-1)^n z^{2n+1}}{2n + 1} \end{aligned}$$

■

(6.3.17) ***Example.*** Compute the Taylor series for the principal logarithm $f(z) = \operatorname{Log} z$ about $z_0 = -1 + i$. Discuss the radius of convergence of the resulting power series.

Solution. We will first expand $f'(z) = 1/z$ in a geometric series about $z = -1 + i$ and then integrate to get $f(z)$. Thus,

$$f'(z) = \frac{1}{[z-(-1+i)]+(-1+i)}$$

$$= \frac{1}{-1+i}\left[\frac{1}{1+\left(\dfrac{z-(-1+i)}{-1+i}\right)}\right]$$

$$= \sum_{n=0}^{\infty} \frac{(-1)^n[z-(-1+i)]^n}{(-1+i)^{n+1}}$$

and hence, upon integrating, we get

$$f(z) = \operatorname{Log} z = K + \sum_{n=0}^{\infty} \frac{(-1)^n[z-(-1+i)]^{n+1}}{(n+1)(-1+i)^{n+1}}$$

To evaluate the constant K, set $z = -1 + i$ above to get

$$K = \operatorname{Log}(-1+i) = \ln\sqrt{2} + \frac{3\pi i}{4}$$

Thus, the Taylor series is

$$\operatorname{Log} z = \ln\sqrt{2} + \frac{3\pi i}{4} + \sum_{n=0}^{\infty} \frac{(-1)^n[z-(-1+i)]^{n+1}}{(n+1)(-1+i)^{n+1}} \tag{6.3.18}$$

Since the singularity of $\operatorname{Log} z$ closest to $z = -1 + i$ is at $z = -1$ (see Figure 6.3.3), this formula is valid for

$$|z-(-1+i)| < 1$$

However, by the ratio test, the radius of convergence of the *power series* in (6.3.18) is $\sqrt{2}$, which is greater than 1. Hence, this is a case where the *radius of convergence of a Taylor series is larger than the radius of that circle inside*

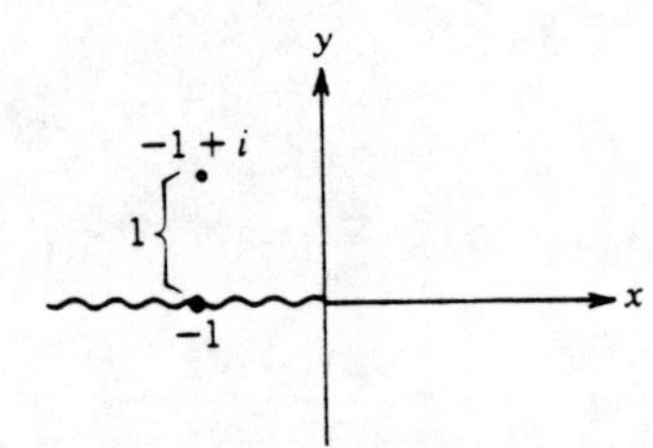

Figure 6.3.3 Example (6.3.17).

of which the Taylor series converges to the function. Indeed, the series cannot converge to $\operatorname{Log} z$ for all z in $|z - (-1 + i)| < \sqrt{2}$. This is so since the power series in (6.3.18) defines an analytic function in this disk and $\operatorname{Log} z$ is not analytic here.

■

The next (intuitively obvious) result, which is proved in Problem 10, provides us with an important computational tool.

(6.3.19) ***Theorem.*** If f is analytic at $z = z_0$, then for any $N = 0, 1, 2, \ldots$ and for those z for which the Taylor series for $f(z)$ converges to $f(z)$, we have

$$f(z) = \sum_{n=0}^{N} \frac{f^{(n)}(z_0)}{n!}(z - z_0)^n + (z - z_0)^{N+1} g_{N+1}(z)$$

where the remainder $g_{N+1}(z)$ is analytic at $z = z_0$ and

$$g_{N+1}(z_0) = \frac{f^{N+1}(z_0)}{(N+1)!}$$

■

The above result conveys the message that **we can terminate a Taylor series at the power $(z - z_0)^N$, and what is left is an analytic function with a zero at least of order $N + 1$.** (See the end of this section for a discussion of the zeros of an analytic function.) The next two examples indicate how this type of result can be used.

(6.3.20) ***Example.*** To compute $\lim_{z \to 0} [(z^2 + \sin^2 z)/(1 - \cos z)]$, we first apply the above result to both the (analytic) numerator and denominator. First, from their respective Taylor series about $z = 0$, we have

$$\sin z = z + z^3 h_1(z)$$

and

$$1 - \cos z = \frac{z^2}{2} + z^4 h_2(z)$$

where h_1 and h_2 are analytic at $z = 0$. Thus,

$$\frac{z^2 + \sin^2 z}{1 - \cos z} = \frac{2z^2 + 2z^4 h_1(z) + z^6 h_1^2(z)}{(z^2/2) + z^4 h_2(z)}$$

$$= \frac{2 + 2z^2 h_1(z) + z^4 h_1^2(z)}{\frac{1}{2} + z^2 h_2(z)} \xrightarrow[z \to 0]{} 4$$

■

In Problem 11, the reader is asked to show that this last result is a special case of **L'Hospital's Rule for Analytic Functions.**

(6.3.21) ***Example.*** Compute the terms, through quadratic powers of z, in the Taylor series for $1/(2 + e^z)$ about $z = 0$.

Solution. Note that the function in the denominator is analytic at $z = 0$ and hence possesses a Taylor series. Thus, from the above theorem, if we terminate this Taylor series at the power z^2, we have

$$2 + e^z = 3 + z + \frac{z^2}{2} + z^3 h(z)$$

where $h(z)$ is analytic at $z = 0$. Hence,

$$\frac{1}{2 + e^z} = \frac{1/3}{1 + p(z)} \tag{6.3.22}$$

where

$$p(z) = \frac{z}{3} + \frac{z^2}{6} + \frac{z^3 h(z)}{3}$$

Since $|p(z)| < 1$ for $|z|$ sufficiently small, we can perform a geometric series expansion in (6.3.22) to get $1/(2 + e^z) = \frac{1}{3}[1 - p(z) + p^2(z) + \cdots]$, where "$+ \cdots$" indicates *powers of p greater than 2*. Thus, from the definition of p,

$$\frac{1}{2 + e^z} = \frac{1}{3}\left[1 - \left(\frac{z}{3} + \frac{z^2}{6} + \frac{z^3 h(z)}{3}\right) + \left(\frac{z}{3} + \frac{z^2}{6} + \frac{z^3 h(z)}{3}\right)^2 + \cdots\right] \tag{6.3.23}$$

where the terms omitted are of the form z^3 + (higher powers of z) and are not to be included in our answer anyway (remember, we only want the Taylor series through quadratic powers). Hence, from the above,

$$\frac{1}{2 + e^z} = \frac{1}{3} - \frac{z}{9} - \frac{z^2}{54} + \text{(higher powers of } z) \tag{6.3.24}$$

The reader should note that it is quite difficult to determine all of the Taylor coefficients for $(2 + e^z)^{-1}$, and it is therefore difficult to determine where the resulting power series would converge. However, from Theorem (6.3.7), we know that it must converge to $(2 + e^z)^{-1}$ up to the nearest singularity, which in this case is that root of $e^z = -2$ closest to $z = 0$. Thus, the series (6.3.24) converges to $(2 + e^z)^{-1}$ for

$$|z| < [\pi^2 + (\ln 2)^2]^{1/2}$$

■

Zeros of an Analytic Function. One reason (out of many) for the importance of this topic is that very often the singularities of an analytic function are located at the zeros of some denominator.

(6.3.25) ***Definition***

(a) Let $f(z)$ be analytic at $z = z_0$. We call z_0 a **zero of order $m \geq 1$**, or **a zero of multiplicity m**, of $f(z)$ if

$$f^{(k)}(z_0) = 0 \qquad \text{for } k = 0, 1, 2, \ldots, m-1$$

while

$$f^{(m)}(z_0) \neq 0$$

If $m = 1$, we say that f has a **simple zero**.

(b) We say that f has a **zero of order m at $z = \infty$** if $f(1/z)$ has a zero of order m at $z = 0$.

■

From this definition, we conclude that **the order of the first nonvanishing derivative of an analytic function is the order of its zero.**

(6.3.26) ***Example.*** The function $\sin z$ has simple zeros at $z = n\pi, n = 0, \pm 1, \ldots$, since it vanishes there but its derivative, $\cos z$, does not.

The function $(1 - e^z)^2$ has a zero of order 2 (a **double zero**) at $z = 2\pi i n$, $n = 0, \pm 1, \ldots$, as the reader will easily verify, while $(1 - e^{1/z})^2$ also has double zeros at $z = 1/2\pi i n$, $n = \pm 1, \pm 2, \ldots$, and at $z = \infty$.

The function $g(z)$ defined by

$$g(z) = \oint_0^z e^{\zeta} \sin^3 \zeta \, d\zeta \tag{6.3.27}$$

is independent of the contour C joining $\zeta = 0$ to $\zeta = z$ (why?) and is analytic at $z = 0$ (in fact, it is entire). It also has a zero at $z = 0$. The reader is asked to show that the order of the zero is four by computing the derivatives of g.

■

(6.3.28) ***Theorem***

(a) If $f(z)$ is analytic at z_0, then it has a zero of order m there if and only if *in some neighborhood of* z_0

$$f(z) = (z - z_0)^m g(z) \qquad g(z_0) \neq 0 \tag{6.3.29}$$

where $g(z)$ is analytic at $z = z_0$.

(b) If $f(z)$ is not identically zero, then its zeros are isolated.

Proof. If $f(z)$ is as in (6.3.29), then a simple computation will show that its first nonvanishing derivative at z_0 is of order m. Hence, f has a zero of order m there. Conversely, if f has a zero of order m, then (6.3.29) follows from Theorem (6.3.19) if we choose $N = m - 1$ there. Thus, part (a) is proved.

To prove part (b), let z_0 be a zero of $f(z)$. Then, from (6.3.29), since $g(z)$ is analytic and $g(z_0) \neq 0$, there is some neighborhood of z_0 in which $g(z) \neq 0$. Hence, in this neighborhood, $f(z)$ can only vanish at z_0.

■

(6.3.30) ***Example.*** The function

$$f(z) = \begin{cases} \sin \dfrac{1}{z} & z \neq 0 \\ 0 & z = 0 \end{cases}$$

has zeros at $z = 0$ and $1/n\pi$, ($n = \pm 1, \pm 2, \ldots$. The zeros at $1/n\pi$ are isolated since f is analytic here. However, note that since $\lim_{n \to \infty} (1/n\pi) = 0$, the zero at $z = 0$ is not isolated. Also note that f is not analytic at $z = 0$. (Why?)

■

As a final application, we will show how some of these results relate to the topic of conformal mappings discussed in Section 4.3. There [see Theorem (4.3.12)],we showed that if $f(z)$ is analytic at $z = z^*$, with $f'(z^*) \neq 0$, then local angle changes at z^* are preserved. In Example (4.3.19), we showed, through the simple function $w = z^2$, that this result is not true at points where the derivative vanishes. The next example illustrates how we can use Taylor series to investigate local mapping properties in such a case.

(6.3.31) ***Example.*** Determine the mapping properties of $w = \sin z$ in a neighborhood of $z = \pi/2$, where the derivative vanishes.

Solution. We first expand $w = \sin z$ in a Taylor series about $z = \pi/2$ to get $w = 1 - \frac{1}{2}(z - \pi/2)^2 +$ (higher-order powers). Now, we ignore the higher-order terms and consider the effect of the *approximate mapping*

$$w = 1 - \frac{1}{2}\left(z - \frac{\pi}{2}\right)^2$$

on a wedge whose vertex is at $z = \pi/2$ and whose wedge angle is $\Delta\theta$ (see Figure 6.3.4). Note from the figure that $\Delta\theta$ is measured from the curve marked ① in the z-plane to the curve marked ②. The effect of the above mapping is to produce another wedge in the w-plane with vertex at $w = 1$ and with ① and ② denoting the images of the corresponding lines in the z-plane. Note that the angle between these two lines in the w-plane is *doubled* to $2\,\Delta\theta$.

If we now have two arbitrary smooth curves in the z-plane intersecting at $z = \pi/2$, then we would approximate these curves in a neighborhood of $\pi/2$ by their tangent lines. Thus, the above result tells us that the angle between two smooth curves is doubled at $z = \pi/2$ under

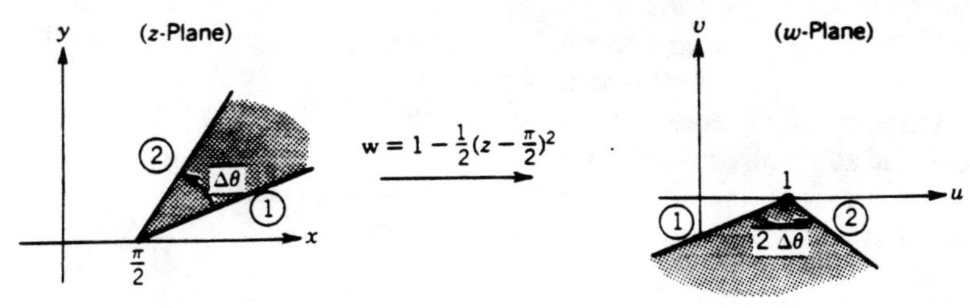

Figure 6.3.4 Example (6.3.31).

the mapping $w = \sin z$. In Problem 18, the reader is asked to generalize this result and show that local mapping properties are related to the order of the derivative of the mapping function.

PROBLEMS (The answers are given on page 438.)

(Problems with an asterisk are more theoretical.)

1.* Prove that if $f(z)$ is analytic at $z = z_0$, then its Taylor series *converges uniformly* to $f(z)$ in any disk whose radius is smaller than the distance from z_0 to the nearest singularity of f. [*Hint:* Use (6.3.6).]

2. Compute the Taylor series for the following functions about the point z_0. Where does the series converge to the function?
   (a) $\sin z \quad z_0 = \pi k$ (b) $\sin z \quad z_0 = (k + \frac{1}{2})\pi$
   (c) $(z - i)^{-1} \quad z_0 = 2$ (d) $(z + i)^{-1}(3z - 1)^{-1} \quad z_0 = 0$
   (e) $\mathrm{Log}(1 + z^2) \quad z_0 = 0$ (f) $(2z + z^2)^{-1} \quad z_0 = 1$
   (g) $z(z - i)^{-3} \quad z_0 = 1$
   (h) (prin.) $(z + \alpha)^\beta \quad z_0 = 0 \quad (\alpha \neq 0)$ (This is an extension of the **binomial theorem.**)
   (i) $e^z \cos z \quad z_0 = 0$ [*Hint:* Write $\cos z$ in terms of exponentials. Also, do this using the Cauchy product.]

3. Compute the Taylor series for the following functions about $z_0 = 0$. The functions are defined by integrals.

   (a) $\displaystyle\int_0^z e^{-\zeta^2}\, d\zeta$ (b) $\displaystyle\int_0^z e^{-z\zeta}\, d\zeta$

   (c) $\displaystyle\oint_0^z \frac{d\zeta}{\sqrt{1 + \zeta^2}}$, where C is the line joining $\zeta = 0$ to $\zeta = z$, and the principal branch of the square root is used.

4. Consider the function $f(z) = \mathrm{Log}(1 + \sqrt{1 - z})$, where the principal branches of both the logarithm and square root are used.
   (a) Show that $f(z)$ is analytic for all z except when $z = x$, $x \geq 1$. [*Hint:* Show that $1 + \sqrt{1 - z}$ can never vanish.]

(b) Compute the terms through z^3 of the Taylor series for f about $z = 0$. For which z does this series converge to f?

5. A function $f(z)$ is **even** if $f(-z) = f(z)$ and **odd** if $f(-z) = -f(z)$. If $f(z)$ is analytic at $z = 0$, prove the following.
(a) f is even if and only if $f^{(2n+1)}(0) = 0$, $n = 0, 1, 2, \ldots$.
(b) f is odd if and only if $f^{(2n)}(0) = 0$, $n = 0, 1, 2, \ldots$.

6. Let

$$f(z) = \begin{cases} z^{-1}\sin z & z \neq 0 \\ 1 & z = 0 \end{cases}$$

(a) Show that f is analytic at $z = 0$.
(b) Compute the Taylor series for f about $z = 0$.

7. Taylor series for ratios of functions are sometimes computed by first doing a multiplication. For example, consider $f(z) = e^z/(1 + e^z)$.
(a) Compute the first three terms in the Taylor series for $f(z)$ about $z = 0$ by direct calculation of the derivatives of f.
(b) From f, we have $(1 + e^z) f(z) = e^z$. Write the Taylor series for e^z about $z = 0$, together with $f(z) = \sum_{n=0}^{\infty} a_n z^n$, and compute the first few of the a_n by multiplying the series. Compare your answers with those of part (a).
(c) Where does the Taylor series in part (a) converge to $f(z)$?

8. Let $f(z)$ be analytic at $z = 0$ with $f(0) = 1$. Assume that f satisfies $[f(z)]^2 = f(2z)$ in some neighborhood of $z = 0$ and that $f(z) \not\equiv 1$.
(a) Write the Taylor series $f(z) = \sum_{n=0}^{\infty} a_n z^n$, and show that $a_0 = 1$, $a_2 = \frac{1}{2}a_1^2$, and $a_3 = \frac{1}{6}a_1^3$.
(b) With the Taylor series in part (a), show that the Taylor coefficients $\{a_n\}$ satisfy $2^n a_n = \sum_{k=0}^{n} a_k a_{n-k}$, $n \geq 0$. [*Hint:* Use the Cauchy product.]
(c) Show that $a_n = a_1^n/n!$ is a solution of the equation in part (b). [*Hint:* From the binomial series, we have that $(1 + 1)^n = \sum_{k=0}^{n} n!/k!(n-k)!$.]
(d) From part (c), what is a function $f(z)$ satisfying the given conditions?

9.* Let $f(z)$ be analytic in the strip $|\mathrm{Im}(z)| < \epsilon$ and be real on the real axis. Prove that $f(\bar{z}) = \overline{f(z)}$ in this strip. [*Hint:* Pick any point $\hat{z}$ in the strip and expand $f(z)$ in a Taylor series about the real point $x_0 = \mathrm{Re}(\hat{z})$. Use the fact that $f(x_0)$ is real to conclude something about its Taylor coefficients.]

10.* Here, we will lead the reader through a series of steps which will end with a proof of Theorem (6.3.19). Let $f(z)$ be analytic at z_0 with Taylor series $\sum_{n=0}^{\infty} [f^{(n)}(z_0)/n!]\,(z - z_0)^n$ valid in some disk D centered at z_0 and of radius $\hat{R}$.
(a) Let z be a point in D and C be a circle of radius ρ in the ζ-plane centered at z_0, lying inside D and containing z inside (see Figure 6.3.5). Use the Cauchy integral formulas to show that

$$f(z) - \sum_{n=0}^{N} \frac{f^{(n)}(z_0)}{n!}(z - z_0)^n = \frac{1}{2\pi i}\oint_C f(\zeta)\left[\frac{1}{\zeta - z} - \sum_{n=0}^{N}\frac{(z - z_0)^n}{(\zeta - z_0)^{n+1}}\right] d\zeta$$

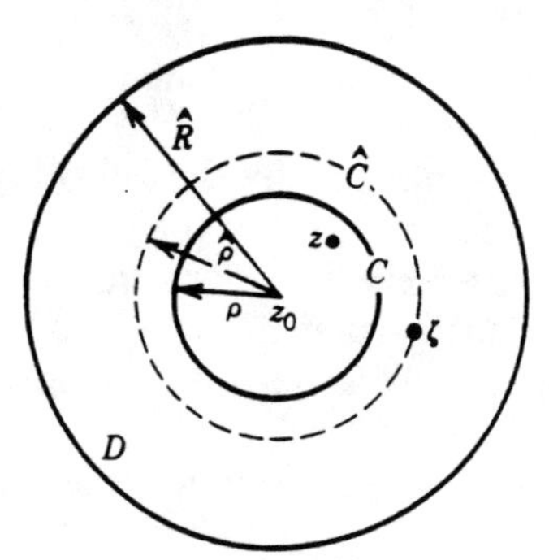

Figure 6.3.5

(b) Sum the partial geometric series in the integral and show that the above is of the form of the result in Theorem (6.3.19), where the remainder $g_{N+1}(z)$ is given by

$$g_{N+1}(z) = \frac{1}{2\pi i} \oint_C \frac{f(\zeta)\, d\zeta}{(\zeta - z)(\zeta - z_0)^{N+1}}$$

(c) Use the above to show that

$$g_{N+1}(z_0) = \frac{f^{N+1}(z_0)}{(N+1)!}$$

(d) Let $\hat{C}$ be a circle slightly larger than C and of radius $\hat{\rho}$ (see the figure). Why is

$$g_{N+1}(z) = \frac{1}{2\pi i} \oint_{\hat{C}} \frac{f(\zeta)\, d\zeta}{(\zeta - z)(\zeta - z_0)^{N+1}}$$

(e) Show that for ζ on $\hat{C}$, $|(\zeta - z)(\zeta - z_0)^{N+1}| \geq \hat{\rho}^{N+1}(\hat{\zeta} - \rho)$. Then, use the boundedness of $f(z)$ to show that the integrand in part (b) is bounded.

(f) Finally, appeal to Problem 10 of Section 5.3 to show that $g_{N+1}(z)$ is analytic inside C. Since C is arbitrary, the result if proved for all of D.

11. (a) Use Theorem (6.3.19) to prove **L'Hospital's rule.** That is, if $f(z)$ and $g(z)$ are analytic at $z = z_0$, and if the first N derivatives ($N \geq 0$) of *both* f and g vanish at z_0, while the $(N+1)$st derivative of g does not, then,

$$\lim_{z \to z_0} \left[\frac{f(z)}{g(z)}\right] = \frac{f^{(N+1)}(z_0)}{g^{(N+1)}(z_0)}$$

(b) Compute the following limits:

(i) $\displaystyle\lim_{z \to \pi} \left[\frac{\sin^2 z}{z - \pi}\right]$ (ii) $\displaystyle\lim_{z \to 0} \left[\frac{\cos z - 1}{e^z - 1}\right]$

(iii) $\displaystyle\lim_{z \to 0} \left[(1 - e^z)^{-1} \oint_0^z \frac{d\zeta}{1 + \zeta^2}\right]$, where C is the line joining $\zeta = 0$ to $\zeta = z$.

12. Show that for $0 < |z| < 2\pi$, there is a function $h(z)$, with $h(0) \neq 0$, such that $(e^z - 1)^{-1} = h(z)/z$.
13. Use Theorem (6.3.19) to explain why the following results seem reasonable. Let $w = f(z)$ be analytic at z_0, with $f(z_0) = w_0$.
    (a) If $f'(z_0) \neq 0$, then for each w "close enough" to w_0, there is a unique z "close enough" to z_0 satisfying $w = f(z)$. Hence, with $f'(z_0) \neq 0$, we would expect there to be a single-valued inverse function $z = g(w)$ (see Section 4.2).
    (b) If $f'(z_0) = 0$, then, for w close to w_0, there are *at least* two values of z close to z_0 satisfying $w = f(z)$. Hence, the inverse function should be multiple-valued.
14. Show that $(1 - e^z)^{1/2}$ has a branch point at $z = 0$. [*Hint:* Use Theorem (6.3.19) and the definition of branch point in Section 3.2.]
15. Where are the zeros in the finite plane and what are their orders for the following functions?
    (a) $z^4(e^z - 1)$ (b) $\cos^3 z$ (c) $z \operatorname{Log}(z + 1)$
    (d) $e^{e^z} - 1$ (e) $\sin z \cdot (\cos^2 z - 1)$
16.* Prove the following:
    (a) If $f(z)$ has a zero of order m at $z = z_0$, then $f'(z)$ has a zero of order $m - 1$ at $z = z_0$.
    (b) If $f(z)$ is entire and has a zero of order m at z_0, then $\int_{z_0}^{z} f(\zeta)\, d\zeta$ has a zero of order $m + 1$ at $z = z_0$.
    (c) If $f_1(z)$ has a zero of order m_1 at z_0, and if $f_2(z)$ has a zero of order m_2 at z_0, then $f_1(z) \cdot f_2(z)$ has a zero of order $m_1 + m_2$ at z_0. Also, if $m_1 \geq m_2$, then $f_1(z)/f_2(z)$ can be defined so as to be analytic at z_0 and the resulting function will have a zero of order $m_1 - m_2$ there.
17. What is the order of the zero $z = 0$ of the following functions defined by integrals?
    (a) $\int_0^z e^{\zeta^2}\, d\zeta$ (b) $\int_0^z (e^{\zeta} - 1) \sin(\zeta)\, d\zeta$
    (c) $\int_0^z \zeta \sin(e^{\zeta} - 1)\, d\zeta$ (d) $\int_0^z \zeta \operatorname{Log}(e^{\zeta} + 1)\, d\zeta$,
    where in (d) the contour of integration does not cross any branch cut of the logarithm.
18. (a)* Let $w = f(z)$ be analytic at $z = z_0$ with $f'(z_0) = 0$. If N is the order of the zero of $f'(z)$ at $z = z_0$, show that if $\Delta\theta$ is the angle between two curves intersecting at z_0 in the z-plane, then the angle between the images of these curves at the point $w_0 = f(z_0)$ is $(N + 1)\Delta\theta$.
    (b) Where are the following functions *not* conformal? In a neighborhood of these points, indicate how local angles are changed.
    (i) e^{z^2} (ii) $\cos z$
    (iii) $z^2(1 + z^2)^{-1}$ (iv) $(e^z - 1)^3$
19. Let $w = \sin^2 z$.
    (a) Where is w not conformal?
    (b) How do the local angles change in a neighborhood of these points?
    (c) Verify the results in part (*b*) by computing the image of the strip $0 < \operatorname{Re}(z) < \pi/2$, $0 < \operatorname{Im}(z) < \infty$ under this mapping.

6.4 LAURENT SERIES

We know from the previous sections that a function possesses a power series expansion $\sum_{n=0}^{\infty} a_n(z - z_0)^n$ in some neighborhood of z_0 if and only if it is analytic at z_0. The question we pose here is whether a similar type of series representation in powers of $(z - z_0)$ holds when $f(z)$ is not analytic at z_0. To see what might be in store for us, consider

$$f(z) = \frac{1}{z(1 - z)}$$

Clearly, f has a singularity at $z = 0$. However, since $1/(1 - z) = \sum_{n=0}^{\infty} z^n$, for $|z| < 1$, we have

$$f(z) = \frac{1}{z} \sum_{n=0}^{\infty} z^n = \sum_{n=-1}^{\infty} z^n \qquad \text{for } 0 < |z| < 1$$

Thus, f has an expansion in positive and negative powers of z. That this is true under more general circumstances will be proved below. But first we offer the following definition.

(6.4.1) ***Definition.*** A **Laurent series about $z = z_0$** is an infinite series of the form

$$\sum_{n=-\infty}^{\infty} c_n(z - z_0)^n$$

which, in general, involves positive and negative powers of $z - z_0$.

■

(6.4.2) ***Theorem.*** Let $f(z)$ be analytic in the **annular region** (see Figure 6.4.1a)

(6.4.3) $$D\colon\ R_1 < |z - z_0| < R_2$$

where we admit the possibility that $R_1 = 0$ or $R_2 = \infty$. Then in D, f

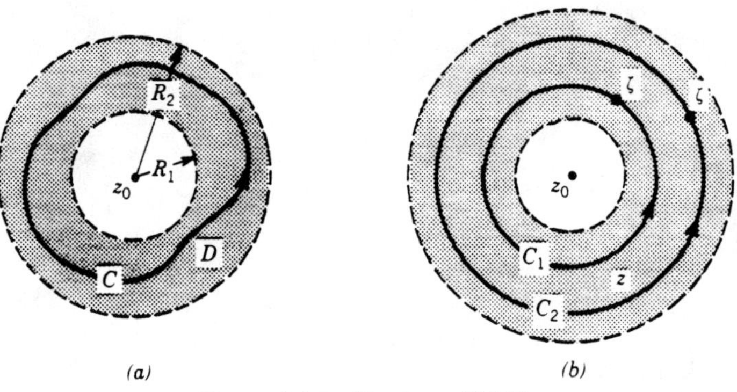

Figure 6.4.1 Theorem (6.4.2).

possesses a Laurent series expansion

$$f(z) = \sum_{n=-\infty}^{\infty} c_n(z - z_0)^n \qquad R_1 < |z - z_0| < R_2 \tag{6.4.4}$$

which converges to $f(z)$ and where the **Laurent coefficients** c_n are given by

$$c_n = \frac{1}{2\pi i} \oint_C \frac{f(\zeta)\, d\zeta}{(\zeta - z_0)^{n+1}} \qquad n = 0, \pm 1, \pm 2, \ldots \tag{6.4.5}$$

Here, C is any simple, closed, piecewise smooth curve lying in the annular region and encircling the disk $|z - z_0| < R_1$ (see the figure). Furthermore, the series converges uniformly in any smaller annular region $R_1 + \delta \leq |z - z_0| \leq R_2 - \delta$, where $\delta > 0$.

Proof. First pick any point z inside the annulus (see Figure 6.4.1b). Now, in the proof of Theorem (6.3.2) for Taylor series, we first used the Cauchy integral formula to represent $f(z)$ by $(2\pi i)^{-1} \oint_C f(\zeta)(\zeta - z)^{-1}\, d\zeta$. Here, we cannot do this since f may not be analytic in a whole neighborhood of z_0 (it is assumed analytic *only* in the annular region D). Instead, we use the result of Theorem (5.3.25) in Section 5.3 (the reader is urged to read the proof of that result). There, we learn that if the circles C_1 and C_2 are chosen as in Figure 6.4.1b, then

$$f(z) = \frac{1}{2\pi i}\left[\oint_{C_2} \frac{f(\zeta)}{\zeta - z}\, d\zeta - \oint_{C_1} \frac{f(\zeta)}{\zeta - z}\, d\zeta\right] \tag{6.4.6}$$

From the figure,

$$\text{(a)} \quad \left|\frac{z - z_0}{\zeta - z_0}\right| < 1 \qquad \text{for } \zeta \text{ on } C_2$$

$$\text{(b)} \quad \left|\frac{\zeta - z_0}{z - z_0}\right| < 1 \qquad \text{for } \zeta \text{ on } C_1 \tag{6.4.7}$$

Hence, as in the proof for Taylor series, we first write

$$\frac{1}{\zeta - z} = \begin{cases} \dfrac{1}{\zeta - z_0}\left[\dfrac{1}{1 - \left(\dfrac{z - z_0}{\zeta - z_0}\right)}\right] & \text{for } \zeta \text{ on } C_2 \\[2ex] \dfrac{-1}{z - z_0}\left[\dfrac{1}{1 - \left(\dfrac{\zeta - z_0}{z - z_0}\right)}\right] & \text{for } \zeta \text{ on } C_1 \end{cases} \tag{6.4.8}$$

To each term in brackets above, we add and subtract the first N terms of its geometric series. For example, for ζ on C_1, we would add and subtract

$$\sum_{k=0}^{N} \left(\frac{\zeta - z_0}{z - z_0}\right)^k = \frac{1 - \left(\dfrac{\zeta - z_0}{z - z_0}\right)^{N+1}}{1 - \left(\dfrac{\zeta - z_0}{z - z_0}\right)}$$

and get, from (6.4.8),

$$\frac{1}{\zeta - z} = \frac{-1}{z - z_0}\left[\frac{1}{1 - \left(\frac{\zeta - z_0}{z - z_0}\right)} - \frac{1 - \left(\frac{\zeta - z_0}{z - z_0}\right)^{N+1}}{1 - \left(\frac{\zeta - z_0}{z - z_0}\right)} + \sum_{k=0}^{N}\left(\frac{\zeta - z_0}{z - z_0}\right)^k\right]$$

(6.4.9)
$$= \frac{1}{\zeta - z}\left(\frac{\zeta - z_0}{z - z_0}\right)^{N+1} - \sum_{k=0}^{N}\frac{(\zeta - z_0)^k}{(z - z_0)^{k+1}} \qquad (\text{let } n = -k - 1)$$

$$= \frac{1}{\zeta - z}\left(\frac{\zeta - z_0}{z - z_0}\right)^{N+1} - \sum_{n=-N-1}^{-1}\frac{(z - z_0)^n}{(\zeta - z_0)^{n+1}} \qquad (\text{for } \zeta \text{ on } C_1)$$

Similarly, for ζ on C_2, on (6.4.8) we add and subtract

$$\sum_{n=0}^{N}\left(\frac{z - z_0}{\zeta - z_0}\right)^n = \frac{1 - \left(\frac{z - z_0}{\zeta - z_0}\right)^{N+1}}{1 - \left(\frac{z - z_0}{\zeta - z_0}\right)}$$

to get

(6.4.10)
$$\frac{1}{\zeta - z} = \frac{1}{\zeta - z}\left(\frac{z - z_0}{\zeta - z_0}\right)^{N+1} + \sum_{n=0}^{N}\frac{(z - z_0)^n}{(\zeta - z_0)^{n+1}} \qquad (\text{for } \zeta \text{ on } C_2)$$

We now substitute (6.4.9) and (6.4.10) into the integrals in (6.4.6) and arrive at

(6.4.11)
$$f(z) - \sum_{n=-N-1}^{-1}(z - z_0)^n\left[\frac{1}{2\pi i}\oint_{C_1}\frac{f(\zeta)\,d\zeta}{(\zeta - z_0)^{n+1}}\right] - \sum_{n=0}^{N}(z - z_0)^n\left[\frac{1}{2\pi i}\oint_{C_2}\frac{f(\zeta)\,d\zeta}{(\zeta - z_0)^{n+1}}\right]$$
$$= \frac{1}{2\pi i}\left[\oint_{C_2}\left(\frac{f(\zeta)}{\zeta - z}\right)\left(\frac{z - z_0}{\zeta - z_0}\right)^{N+1} d\zeta - \oint_{C_1}\left(\frac{f(\zeta)}{\zeta - z}\right)\left(\frac{\zeta - z_0}{z - z_0}\right)^{N+1} d\zeta\right]$$

In the integrals on the left-hand side above, each integrand is analytic in ζ for ζ in the annulus D, and thus the integrals do not depend upon the curves C_1 and C_2. Hence, choose these curves to be the same and denote the common curve by C. Also, following exactly the same reasoning as in the proof of Taylor series [see Equation (6.3.6)], we can show, using (6.4.7), that the terms on the right-hand side of the equality above approach zero as $N \to \infty$. Thus, the limit of the left side is just the result (6.4.4)–(6.4.5). In Problem 6*a*, the reader is asked to show that the convergence is uniform on any smaller annulus.

■

Note that *if* $f(z)$ is analytic at z_0, and hence analytic in some deleted neighborhood $0 < |z - z_0| < R_2$, then here, the *Laurent series reduces to the Taylor series.* To see this, consider the Laurent coefficients (6.4.5). If C lies in this deleted neighborhood, then for $n \leq -1$ the integrand is analytic within and on C and, by the Cauchy–Goursat Theorem, $c_n = 0$ (for $n \leq -1$). Thus, there are no negative powers in (6.4.4). Now, since $f(\zeta)$ is assumed analytic inside C, the Cauchy integral formula applied to (6.4.5) yields $c_n = [f^{(n)}(z_0)]/n!$, for $n \geq 0$. There are just the Taylor coefficients.

As the next example shows, **a function might have different Laurent series representations valid in different annular regions.**

(6.4.12) ***Example.*** Find all of the possible Laurent series about $z = 0$ for $f(z) = 1/(1 - z)$.

Solution. First, note from Figure 6.4.2*a* that the z-plane can be split into two annular regions

$$\text{I:}\quad 0 < |z| < 1$$

$$\text{II:}\quad 1 < |z| < \infty$$

with f analytic inside each. (Note that f is singular on $|z| = 1$.) We will evaluate the Laurent coefficients directly using (6.4.5). For region I, we choose a curve C_{I} as shown in Figure 6.4.2*b*. If we designate the Laurent coefficients for *this* annular region by c_n^{I}, we get, with $f(\zeta) = (1 - \zeta)^{-1}$ in (6.4.5),

$$c_n^{\text{I}} = \frac{1}{2\pi i} \oint_{C_{\text{I}}} \frac{d\zeta}{(1 - \zeta)\zeta^{n+1}} \tag{6.4.13}$$

which by the Cauchy integral formulas yields

$$c_n^{\text{I}} = \begin{cases} 0 & n \leq -1 \\ \left(\dfrac{1}{n!}\right)\left(\dfrac{d}{d\zeta}\right)^n \left(\dfrac{1}{1 - \zeta}\right)\Bigg|_{\zeta = 0} & n \geq 0 \end{cases}$$

$$= \begin{cases} 0 & n \leq 1 \\ 1 & n \geq 0 \end{cases}$$

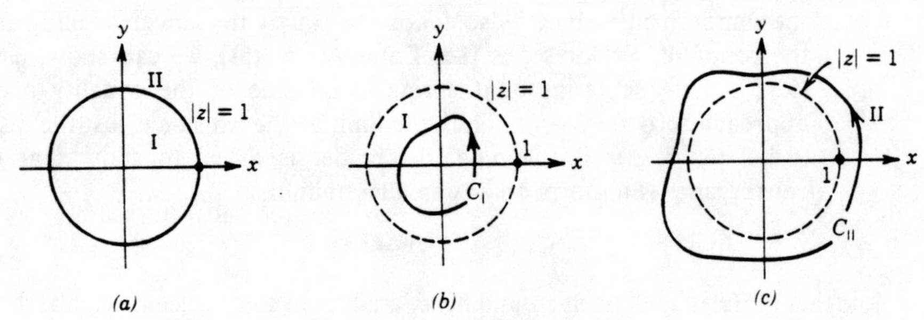

Figure 6.4.2 Example (6.4.12).

Hence, the Laurent series valid in region I is the geometric (Taylor) series:

(6.4.14)
$$\frac{1}{1-z} = \sum_{n=0}^{\infty} z^n \qquad |z| < 1$$

Now, if c_n^{II} designates the Laurent coefficients for region II, we find

$$c_n^{\mathrm{II}} = \frac{1}{2\pi i} \oint_{C_{\mathrm{II}}} \frac{d\zeta}{(1-\zeta)\zeta^{n+1}}$$

where C_{II} is shown in Figure 6.4.2*c*. The only difference between the above and (6.4.13) is that now inside C_{II} there is a singularity at $\zeta = 1$ which was not present inside C_{I}. Hence, again by the Cauchy integral formula,

$$c_n^{\mathrm{II}} = \begin{cases} -1 & n \le -1 \\ 0, & n \ge 0 \end{cases}$$

Thus, in region II we have the (*different*) Laurent series

$$\frac{1}{1-z} = -\sum_{n=-\infty}^{-1} z^n = -\sum_{n=1}^{\infty} z^{-1} \qquad 1 < |z| < \infty$$

■

As with Taylor series, one does not have to proceed with a direct evaluation of the Laurent coefficients in order to derive the Laurent series for a function, so long as it is known that in any given annular region of analyticity there is only one Laurent series representation for a function. That this is true is given by the next result, which is proved in Problem 6b.

(6.4.15) ***Theorem.*** Let $f(z)$ be analytic in the annulus $R_1 < |z - z_0| < R_2$. Then, if the series $\sum_{n=-\infty}^{\infty} c_n(z - z_0)^n$ converges uniformly to $f(z)$ for all z in this annulus, the coefficients c_n must be the Laurent coefficients (6.4.5).

■

This tells us that in **any annular region, Laurent series are unique**. The next few examples show how this can be used to easily generate Laurent series representations.

(6.4.16) ***Example.*** Find all of the possible Laurent series about $z = 1$ for $f(z) = 1/[(z + 1)(z + 2)^2]$.

Solution. From Figure 6.4.3, because of the number of singularities of f, there are three annular regions of interest centered at $z = 1$, namely,

$$\text{I:} \quad 0 < |z - 1| < 2$$

$$\text{II:} \quad 2 < |z - 1| < 3$$

$$\text{III:} \quad 3 < |z - 1| < \infty$$

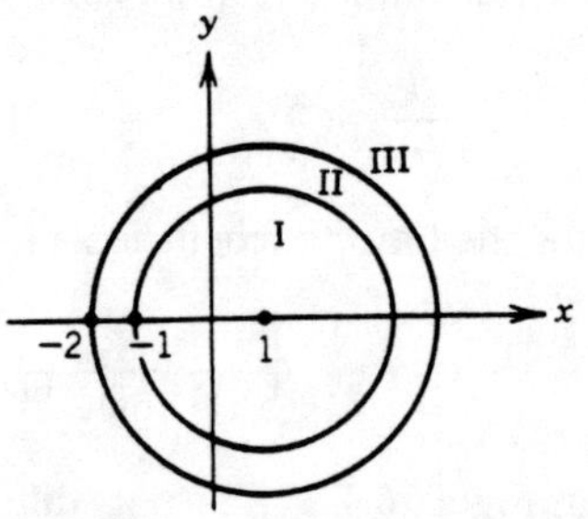

Figure 6.4.3 Example (6.4.16).

First, we perform the partial fraction expansion

$$f(z) = \frac{1}{z+1} - \frac{1}{z+2} - \frac{1}{(z+2)^2}$$

which we then express in terms of $z - 1$, *the powers of which will be in the Laurent series*, as

$$f(z) = \frac{1}{(z-1)+2} - \frac{1}{(z-1)+3} + \frac{d}{dz}\left[\frac{1}{(z-1)+3}\right] \tag{6.4.17}$$

Now in region I, $|z-1| < 2$ and $|z-1| < 3$. Thus, we first rewrite the above as

$$f(z) = \frac{1}{2}\left[\frac{1}{1+\left(\frac{z-1}{2}\right)}\right] - \frac{1}{3}\left[\frac{1}{1+\left(\frac{z-1}{3}\right)}\right] + \frac{1}{3}\frac{d}{dz}\left[\frac{1}{1+\left(\frac{z-1}{3}\right)}\right]$$

then expand each term in a geometric series, and finally differentiate the last series (see Problem 9) to get

$$f(z) = \sum_{n=0}^{\infty} (-1)^n[2^{-n-1} - (n+4)3^{-n-2}](z-1)^n \qquad \text{(in region I)}$$

In region II, where $2 < |z-1| < 3$, we rewrite (6.4.17) as

$$f(z) = (z-1)^{-1}\left[\frac{1}{1+\left(\frac{2}{z-1}\right)}\right] - \frac{1}{3}\left[\frac{1}{1+\left(\frac{z-1}{3}\right)}\right] + \frac{1}{3}\frac{d}{dz}\left[\frac{1}{1+\left(\frac{z-1}{3}\right)}\right]$$

Again, each term in brackets has a geometric series expansion, and we get

$$\begin{aligned} f(z) = &-\frac{1}{2}\sum_{n=-\infty}^{-1} (-1)^n 2^{-n}(z-1)^n \\ &- \sum_{n=0}^{\infty} (-1)^n (n+4)3^{-n-2}(z-1)^n \qquad \text{(in region II)} \end{aligned}$$

Finally, for region III, the reader is asked to verify that

$$f(z) = \sum_{n=-\infty}^{-2} (-1)^n[-2^{-n-1} + (2-n)3^{-n-2}](z-1)^n$$

(6.4.18) ***Example.*** Compute the Laurent series about $z = 1$ for $e^z/(z-1)$.

Solution. What we do here is to first expand e^z in a Taylor series about $z = 1$:

$$e^z = e \sum_{n=0}^{\infty} \frac{(z-1)^n}{n!} \qquad |z-1| < \infty$$

and then multiply by $(z-1)^{-1}$ to get the Laurent series

$$\frac{e^z}{z-1} = e \sum_{n=-1}^{\infty} \frac{(z-1)^n}{(n+1)!} \qquad 0 < |z-1| < \infty$$

■

(6.4.19) ***Example.*** Find those terms with negative powers and the first two terms with nonnegative powers in the Laurent series expansion of $1/(e^z - 1)$ about $z = 0$ which holds in the annulus $0 < |z| < 2\pi$.

Solution. We must first expand the denominator in a Taylor series about $z = 0$. However, since we do not want the full Laurent series (indeed, we would find it rather difficult to compute), we will not have need for the full Taylor series. To see where to stop, note that

$$e^z - 1 = \sum_{k=1}^{\infty} \frac{z^k}{k!} = z \sum_{n=0}^{\infty} \frac{z^n}{(n+1)!}$$

Hence,

$$\frac{1}{e^z - 1} = \frac{1}{z}\left(\frac{1}{1 + \frac{z}{2} + \frac{z^2}{3!} + \frac{z^3}{4!} + \cdots}\right) \tag{6.4.20}$$

The infinite series in the denominator must represent an entire function which vanishes only at $z = 2\pi i n$, $n = \pm 1, \pm 2, \ldots$. (Why?) Hence, the term in brackets must possess a Taylor series in the annulus $0 < |z| < 2\pi$. Due to the $1/z$ term, and because we wish to include terms only up to the positive powers z^0 and z in our Laurent series, we need only find the Taylor series through the z^2 term. Hence, by Theorem (6.3.19), we have

$$\frac{1}{e^z - 1} = \frac{1}{z}\left\{\frac{1}{1 + \left[\frac{z}{2} + \frac{z^2}{6} + z^2 g(z)\right]}\right\}$$

where $g(z)$ is analytic in the annulus and at $z = 0$. Since the term in square brackets vanishes at $z = 0$, we can write the geometric series

$$\frac{1}{1+\left[\frac{z}{2}+\frac{z^2}{6}+z^3 g\right]} = 1 - \left[\frac{z}{2}+\frac{z^2}{6}+z^3 g\right] + \left[\frac{z}{2}+\frac{z^2}{6}+z^3 g\right]^2 + \cdots$$

where "..." signifies higher powers of $(z/2) + (z^2/6) + z^3 g$. Thus, these terms will involve powers of z which are at least cubic. Substituting the above into (6.4.20) and combining terms yields

$$\frac{1}{e^z - 1} = \frac{1}{z}\left[1 - \frac{z}{2} + \frac{z^2}{12} + \cdots\right] = \frac{1}{z} - \frac{1}{2} + \frac{z}{12} + \cdots$$

where "..." denotes higher powers of z. This is now that part of the Laurent series for which we are looking.

■

(6.4.21) ***Example.*** Show that the function

$$f(z) = \operatorname{Log}\left(\frac{z}{z-1}\right)$$

is analytic for $|z| > 1$ and compute its Laurent series about $z = 0$ which is valid in this region.

Solution. From the properties of the principal logarithm, $f(z)$ is analytic except where

$$\operatorname{Im}\left(\frac{z}{z-1}\right) = 0 \quad \text{and} \quad \operatorname{Re}\left(\frac{z}{z-1}\right) \leq 0$$

These conditions are satisfied only for z on the real interval $0 \leq x \leq 1$. Hence, $f(z)$ is analytic in $|z| > 1$ and must possess a Laurent series in powers of z here. To compute it, note that if

$$f(z) = \sum_{k=-\infty}^{\infty} c_k z^k$$

then

$$f'(z) = \sum_{k=-\infty}^{\infty} k c_k z^{k-1} = \sum_{n=-\infty}^{\infty} (n+1) c_{n+1} z^n \tag{6.4.22}$$

(See Problem 9 for a proof of this.) Now, from the definition of f,

$$f'(z) = -\frac{1}{z(z-1)} = -\frac{1}{z^2}\left(\frac{1}{1-\frac{1}{z}}\right)$$

$$= -\frac{1}{z^2}\sum_{k=0}^{\infty}\left(\frac{1}{z}\right)^k \qquad (\text{for } |z| > 1)$$

$$= -\sum_{n=-\infty}^{-2} z^n$$

From this and (6.4.22), we find that

$$(n+1)c_{n+1} = \begin{cases} 0 & n \geq -1 \\ -1 & n \leq -2 \end{cases}$$

From this, we can compute the Laurent coefficients $\{c_n\}$ to find

$$f(z) = -\sum_{n=-\infty}^{-1} \frac{z^n}{n} \qquad |z| > 1$$

■

(6.4.23) ***Example.*** The function $e^{1/z}$ has $z = 0$ as its only singular point. To determine its Laurent series in the region $|z| > 0$, we first note that e^w is entire and has the Taylor series

$$e^w = \sum_{n=0}^{\infty} \frac{w^n}{n!} \qquad |w| < \infty$$

Hence, we simply replace w by $1/z$ to get a Laurent series convergent for $|1/z| < \infty$, that is, for $|z| > 0$. Thus,

$$e^{1/z} = \sum_{n=0}^{\infty} \frac{z^{-n}}{n!} \qquad 0 < |z| < \infty$$

■

PROBLEMS (The answers are given on page 439.)

(Problems with an asterisk are more theoretical.)

1. Compute all of the possible Laurent series for the following functions about the indicated point z_0. Indicate where each series converges to the function.

(a) $\dfrac{1}{z^2 - 1}$ $\quad z_0 = 1$ (b) $\dfrac{e^z}{z+1}$ $\quad z_0 = -1$

(c) $\dfrac{1}{z^2 - iz}$ $\quad z_0 = 1$ (d) $\dfrac{z}{(z+i)(z-i)^2}$ $\quad z_0 = 0$

(e) $\dfrac{e^{z^2} - 1}{z^4}$ $\quad z_0 = 0$ (f) $\dfrac{\mathrm{Log}(1+2z)}{z}$ $\quad z_0 = 0$

(g) $\dfrac{\sin z^2}{z}$ $\quad z_0 = 0$ (h) $\sin\left(\dfrac{1}{z}\right)$ $\quad z_0 = 0$

(i) $1/e^{1/z}$ $\quad z_0 = 0$

2. Find all of the terms with negative powers and the first two nonzero terms with nonnegative powers in the Laurent series of the following functions about the

indicated points z_0 and which converge in an annular region of the form $0 < |z - z_0| < R$ (there is only one such Laurent series). What is the largest value of R for each series?

(a) $\dfrac{e^z}{z^2 - 1}$ $z_0 = 1$ (b) $(\sin z)^{-1}$ $z_0 = n\pi$ [*Hint:* See Example (6.4.19)]

3. (a) Find that Laurent series expansion for $f(z) = 1/(2z^2 + iz)$ about $z = 0$ which converges at $z = i$.
   (b) Find that Laurent series expansion for $g(z) = 1/(z^2 + 2iz + 3)$ about $z = 0$ which converges at $z = 4i$.
   (c) For the function $g(z)$ in part (*b*), how many Laurent series are there about $z = 0$ which converge, for some value of z, to the number $g(2i)$?
4. Compute the Laurent series about $z = 0$ for $f(z) = [\sin(1/z)]/(1 - z)$ which converges to $f(z)$ for $0 < |z| < 1$.
5. Let $f(z) = (e^z - 1)^{-1}$ possess the three Laurent series

(i) $\sum_{n=-\infty}^{\infty} c_n z^n$ for $0 < |z| < 2\pi$

(ii) $\sum_{n=-\infty}^{\infty} d_n(z - 2\pi i)^n$ for $0 < |z - 2\pi i| < 2\pi$

(iii) $\sum_{n=-\infty}^{\infty} e_n(z + 2\pi i)^n$ for $0 < |z + 2\pi i| < 2\pi$

In terms of c_n, d_n, and e_n, what is the Laurent series for $f(z)$ about $z = 0$ which is valid for $2\pi < |z| < 4\pi$?

6.* The Laurent series $\sum_{k=-\infty}^{\infty} c_k(z - z_0)^k$ is said to **converge uniformly** in some region to $f(z)$ if for any ϵ there is an $N(\epsilon)$, so that for all $n, m > N(\epsilon)$, $|f(z) - \sum_{k=-n}^{m} c_k(z - z_0)^k| < \epsilon$ for all z in the region.
   (a) Prove that the Laurent series (6.4.4) for $f(z)$ converges uniformly to $f(z)$ in any smaller annulus $R_1 + \delta \le |z - z_0| \le R_2 - \delta$ $(\delta > 0)$. [*Hint:* Use a result similar to (6.4.11) to bound $|f(z) - \sum_{k=-n}^{m} c_k(z - z_o)^k|$. Then use (6.4.7) to compute the limit $n, m \to \infty$.]
   (b) Prove that if $f(z)$ is analytic in an annulus with center at z_0, and if $\sum_{n=-\infty}^{\infty} c_n(z - z_0)^n$ converges uniformly to $f(z)$ here, then the c_n must be given by the Laurent coefficients (6.4.5). Thus, Laurent series are unique in any given region of analyticity. [*Hint:* With the definition of uniform convergence in part (*a*), consider $(z - z_0)^{-l-1}[f(z) - \sum_{k=-n}^{m} c_k(z - z_0)^k]$, $l = 0, \pm 1, \pm 2, \ldots$. Integrate the result over any circle in the annulus and use the triangle inequality.]
7.* Prove that if $f(z)$ is analytic in $R_1 < |z - z_0| < R_2$ and possesses the Laurent series $\sum_{n=-\infty}^{\infty} c_n(z - z_0)^n$ there, then on any circle $|z - z_0| = r$ in the annulus, we have $|c_n| \le (M_r/r^n)$, where M_r is the maximum value of $|f(z)|$ on the circle.
8.* Prove that if $f(z)$ is analytic in the annulus $R_1 < |z - z_0| < R_2$, then it can be decomposed into the sum $f(z) = f_1(z) + f_2(z)$, where $f_1(z)$ is analytic for

$|z - z_0| < R_2$ and $f_2(z)$ is analytic for $|z - z_0| > R_1$. [*Hint:* Separate the Laurent series for $f(z)$ into a series of positive powers and a series of negative powers. Then use Theorem (6.2.2a) to conclude that the series of positive powers converges for $|z - z_0| < R_2$. Use a similar result to conclude something similar about the series of negative powers.]

9.* Prove that if $f(z)$ is analytic and has the Laurent series $\sum_{n=-\infty}^{\infty} c_n(z - z_0)^n$ in some annular region, then $f'(z)$ possesses the Laurent series $\sum_{n=-\infty}^{\infty} nc_n(z - z_0)^{n-1}$ in the same annular region. [*Hint:* What are the Laurent coefficients for $f'(z)$?]

10. Consider the Laurent series $f(z) = \sum_{n=-\infty}^{\infty} c_n z^n$ for a function $f(z)$ which is analytic in some annular region $r_1 < |z| < r_2$. If $|z| = r_0$ is a circle inside this annulus, show that $f(z)$ has the **Fourier series expansion**

$$f(r_0 e^{i\theta}) = \sum_{n=-\infty}^{\infty} A_n e^{in\theta}, \qquad A_n = \frac{1}{2\pi}\int_0^{2\pi} e^{-in\phi} f(r_0 e^{i\phi})\, d\phi.$$

6.5 CLASSIFICATION OF ISOLATED SINGULAR POINTS; POLES; ESSENTIAL SINGULARITIES

The main result of this section deals with the behavior of an analytic function in the neighborhood of an isolated singular point.

(6.5.1) ***Definition.*** The function $f(z)$ is said to have an **isolated singularity at $z = z_0$** if $f(z)$ is not analytic at z_0 but *is* analytic at every point in some (possibly small) neighborhood of z_0.

■

For example, **a rational function has isolated singularities at each of the zeros of its denominator**. Also, the function

$$f(z) = \frac{1}{e^z - 1}$$

has isolated singularities at $z = 2\pi i n$ $(n = 0, \pm 1, \ldots)$ while the function

$$g(z) = \frac{1}{\sin(1/z)}$$

has isolated singularities at $z_n = 1/n\pi$, $n = \pm 1, \pm 2, \ldots$. The point $z = 0$ is also a singular point of $g(z)$, but it is not isolated since any neighborhood of it must contain an infinite number of the singularities z_n. Finally, the principal logarithm $\operatorname{Log} z$ has singular points at $z = 0$ and at all real negative z, but none of these are isolated. (Why?)

Now, consider an analytic function $f(z)$ with an isolated singularity at $z = z_0$. From the previous section, $f(z)$ can have many Laurent series expansions about $z = z_0$, each valid in different annular regions. In particular, since z_0 is isolated, there must be some **deleted neighborhood** of z_0, $0 < |z - z_0| < R$, inside of which $f(z)$ is analytic. For this annular region, we have the following.

(6.5.2) ***Definition.*** Let z_0 be an isolated singular point of an analytic function $f(z)$ and consider the **local Laurent series expansion**

$$f(z) = \sum_{n=-\infty}^{\infty} c_n(z - z_0)^n \qquad 0 < |z - z_0| < R \qquad (R \neq 0) \tag{6.5.3}$$

which holds in some deleted neighborhood of z_0. Then,

(a) The **principal part of** f **at** z_0 is that part of this Laurent series consisting of the negative powers, that is,

$$\sum_{n=-\infty}^{-1} c_n(z - z_0)^n \qquad \text{(principal part)} \tag{6.5.4}$$

(b) z_0 is a **removable singularity** if the principal part is zero, that is, if $c_n = 0$ for $n \leq -1$. Thus, *for a removable singularity*

$$f(z) = \sum_{n=0}^{\infty} c_n(z - z_0)^n$$

(c) z_0 is a **pole** if the principal part has a finite number of terms. Thus, *for a pole,*

$$f(z) = \sum_{n=-N}^{\infty} c_n(z - z_0)^n \qquad c_{-N} \neq 0 \text{ and } N \geq 1$$

The positive integer N is called the **order of the pole**. If $N = 1$, the pole is called a **simple pole**.

(d) z_0 is an **(isolated) essential singularity** if the principal part does not terminate, that is, if there are an infinite number of negative powers in (6.5.3).

■

We should emphasize that the principal part of a Laurent series determines the **dominant (most singular) behavior** near the singular point, since it consists of all of the negative powers. How singular the function is near this point is then determined by the **nature of the singularity**, that is, whether it is removable, a pole, or essential. Also note from definitions (b), (c), and (d) above that these three cases are mutually exclusive. Hence, let us begin with a removable singularity.

(6.5.5) ***Theorem.*** If z_0 is an isolated singular point of $f(z)$, then

(a) z_0 is removable if and only if $\lim_{z \to z_0} [f(z)]$ exists.

(b) If z_0 is removable, $f(z)$ can be defined at z_0 in such a manner that the resulting function is analytic there.

Proof. First, assume that z_0 is removable. Then, the local Laurent series for f is

$$f(z) = \sum_{n=0}^{\infty} c_n(z - z_0)^n \qquad 0 < |z - z_0| < R$$

Now, by Theorem (6.2.2), the power series on the right has some nonzero radius of convergence which is at least as large as R and, by Theorem (6.2.14), must converge to some continuous function $g(z)$. Hence, the limit of the power series above exists as $z \to z_0$ and must equal c_0. Thus, we have proved half of the statement in (a) and have shown that

$$\lim_{z \to z_0} [f(z)] = c_0 = g(z_0)$$

If we now define $f(z_0) = c_0$, then the resulting function must coincide with the analytic function $g(z)$ above for $0 \le |z - z_0| < R$. Hence, we have proved part (b).

To prove the rest of part (a), we must show that if $\lim_{z \to z_0} [f(z)] = A$, then z_0 must be removable. To do this, consider the Laurent coefficients c_n from (6.5.3). From the previous section, these are given by

(6.5.6)
$$c_n = \frac{1}{2\pi i} \oint_C \frac{f(\zeta)}{(\zeta - z_0)^{n+1}} \, d\zeta$$

where C is any simple, piecewise smooth curve encircling $\zeta = z_0$ and lying in $0 < |\zeta - z_0| < R$. Since we are assuming that $\lim_{\zeta \to z_0} [f(z)] = A$, we have that, for any $\epsilon > 0$, there is a δ so that

(6.5.7)
$$0 < |\zeta - z_0| < \delta \Rightarrow |f(\zeta) - A| < \epsilon$$

Now, choose the curve C to be the circle $|\zeta - z_0| = r_1$, where $r_1 < \delta$, and rewrite (6.5.6) as

$$c_n = \frac{1}{2\pi i} \oint_{|\zeta - z_0| = r_1} \frac{[f(\zeta) - A]}{(\zeta - z_0)^{n+1}} \, d\zeta + \frac{A}{2\pi i} \oint_{|\zeta - z_0| = r_1} \frac{d\zeta}{(\zeta - z_0)^{n+1}}$$

For $n \le -1$, the second integral vanishes since the integrand is analytic. Hence, this together with (6.5.7) and the triangle inequality for integrals applied to the first integral above yields

$$|c_n| \le \frac{\epsilon(2\pi r_1)}{2\pi r_1^{n+1}} = \frac{\epsilon}{r_1^n} \qquad \text{for } n \le -1$$

Thus, these c_n can be made arbitrarily small (just choose ϵ small) and therefore must be zero. Hence, z_0 is a removable singularity.

■

(6.5.8) ***Example.*** The function $f(z) = (\sin z)/z$ has an isolated singularity at $z = 0$. To determine its nature, we expand $\sin z$ in a Taylor series about $z = 0$ to get

$$\sin z = z + z^3 g(z)$$

where $g(z)$ is analytic at $z = 0$ [see Theorem (6.3.19)]. Thus,

$$f(z) = \frac{z + z^3 g(z)}{z} \xrightarrow[z \to 0]{} 1$$

Since this limit exists, by the above theorem $z = 0$ is removable and the function

(6.5.9)
$$f(z) = \begin{cases} \dfrac{\sin z}{z} & z \neq 0 \\ 1 & z = 0 \end{cases}$$

is analytic (in fact, entire).

■

Poles. Unlike the behavior near a removable singularity, a function is *really singular* near a pole.

(6.5.10) ***Theorem.*** Let z_0 be an isolated singularity of $f(z)$. Then,

(a) z_0 is a pole if and only if $\lim_{z \to z_0} [f(z)] = \infty$, while $\lim_{z \to z_0} [(z - z_0)^k f(z)]$ exists for some $k \geq 1$. *The smallest* k *which yields a nonzero limit is the order of the pole.*

(b) z_0 is a pole of order N if and only if

(6.5.11)
$$f(z) = \frac{g(z)}{(z - z_0)^N} \qquad g(z_0) \neq 0$$

where g is analytic at z_0.

Proof. For part (a), first assume z_0 is a pole of order N. Then, for $|z - z_0| < R$,

$$f(z) = \frac{c_{-N}}{(z - z_0)^N} + \cdots + \frac{c_{-1}}{z - z_0} + \sum_{n=0}^{\infty} c_n (z - z_0)^n \qquad c_{-N} \neq 0$$

Clearly, $\lim_{z \to z_0} f(z) = \infty$.

Now the power series above, since it converges in some deleted neighborhood of z_0, must define a function which is analytic at $z = z_0$; hence $\lim_{z \to z_0} [(z - z_0)^k f(z)]$ exists for $k \geq N$, with N, the order of the pole, being the smallest value of k which yields a nonzero limit. Thus, we have proved half of part (a). In Problem 6b, the reader is asked to show that if the hypotheses of part (a) are satisfied, then z_0 is a pole.

For part (b), first assume (6.5.11) holds. Then, since $g(z)$ is analytic at z_0, it has the Taylor series $\sum_{n=0}^{\infty} a_n (z - z_0)^n$, with $a_0 = g(z_0) \neq 0$. Hence, from (6.5.11), the Laurent series for $f(z)$ is

$$f(z) = \frac{a_0}{(z - z_0)^N} + \cdots + \frac{a_{N-1}}{(z - z_0)} + \sum_{n=N}^{\infty} a_n (z - z_0)^{n-N}$$

which shows that $z = z_0$ is a pole of order N. The converse of this result is proved in Problem 6a.

■

(6.5.12) ***Example.*** Classify the singular points of $f(z) = (z^2 + iz + 2)/[(z^2 + 1)^2(z + 3)]$ and compute the principal part at each of these.

Solution. Clearly, $z = -3, \pm i$ are the singular points of f. To determine the type of singularity, *we separate out the singular part of* f *and do a Taylor series expansion of the remaining "nice" part*. Thus, for z near -3, we write

(6.5.13)
$$f(z) = \frac{g_1(z)}{z + 3} \quad \text{where} \quad g_1(z) = \frac{z^2 + iz + 2}{(z^2 + 1)^2}$$

Since $g_1(z)$ is analytic at $z = -3$, we have from its Taylor series

$$g_1(z) = \frac{11 - 3i}{100} + \text{(higher powers of } z + 3)$$

Thus, in some deleted neighborhood of $z = -3$

$$f(z) = \frac{11 - 3i}{100(z + 3)} + \text{(higher powers of } z + 3)$$

Hence, $z = -3$ is a simple pole, and the principal part of f at $z = -3$ is just the first term on the right-hand side above.

For z near i, we write $f(z)$ as

(6.5.14)
$$f(z) = \frac{g_2(z)}{(z - i)^2} \quad \text{where} \quad g_2(z) = \frac{z^2 + iz + 2}{(z + i)^2(z + 3)}$$

Now $g_2(z)$ is analytic at $z = i$ with $g_2(i) = 0$. Hence, the Taylor series for $g_2(z)$ about $z = i$ is of the form

$$g_2(z) = g_2'(i)(z - i) + \text{(higher-order powers)}$$

Thus,

$$f(z) = \frac{g_2'(i)}{z - i} + \text{(higher powers of } z - i)$$

and $z = i$ is either a simple pole [if $g_2'(i) \neq 0$] or a removable singularity [if $g_2'(i) = 0$]. Since $g_2'(i) = -3i/[4(3 + i)]$, we have a simple pole with principal part $g_2'(i)/(z - i)$.

The reader is asked to show that $z = -i$ is a pole of order 2 for $f(z)$, with principal part $g_3(-i)(z + i)^{-2} + g_3'(-i)(z + i)^{-1}$, where $g_3(z) = (z^2 + iz + 2)/[(z - i)^2(z + 3)]$.

■

(6.5.15) ***Example.*** Classify the singular point $z = 0$ of $f(z) = e^z/(z - \sin z)$ and compute its principal part there.

Solution. Intuitively, we must determine how fast the denominator vanishes at $z = 0$. To do this, we note from the Taylor series

$$\sin z = \sum_{n=0}^{\infty} \frac{(-1)^n z^{2n+1}}{(2n+1)!}$$

that we can write $\sin z = z + z^3 g(z)$, where $g(z)$ has the property that it is analytic at $z = 0$ and does not vanish there. Hence, $z - \sin z = -z^3 g(z)$, and thus,

(6.5.16)
$$f(z) = \frac{-e^z}{z^3 g(z)} \qquad g(0) \neq 0$$

Thus, by Theorem (6.5.10b), f has a pole of order 3 at $z = 0$.

To compute the principal part at $z = 0$, we must go further in the Taylor series for $z - \sin z$ and, in addition, expand the (analytic) numerator e^z in a Taylor series about $z = 0$. To this end, we have that

$$z - \sin z = \frac{1}{6}\left[z^3 - \frac{z^5}{20} + (\text{higher powers of } z)\right]$$

and

$$e^z = 1 + z + \frac{z^2}{2} + (\text{higher powers of } z)$$

Hence,

$$f(z) = \frac{6[1 + z + (z^2/2) + \text{higher powers of } z)]}{z^3\{1 - [(z^2/20) + (\text{higher powers of } z)]\}}$$

$$= \frac{6[1 + z + (11z^2/20) + (\text{higher powers of } z)]}{z^3}$$

where in the above we computed the first few terms of the geometric series for $\{1 - [(z^2/20) + (\text{higher powers of } z)]\}^{-1}$ [see Example (6.3.21) for a similar computation]. Hence, the principal part is

$$6[(1/z^3) + (1/z^2) + (11/20z)]$$

■

The next result shows that the order of a pole is often determined by the order of the zero of some denominator.

(6.5.17) ***Theorem.*** Let $g(z)$ and $h(z)$ be analytic at $z = z_0$ and assume that, at z_0, g(z) *has a zero of order* N and h(z) *has a zero of order* M. Then, z_0 is an isolated singularity of $f(z) = g(z)/h(z)$ and is

(a) A pole of order $M - N$ if $N - M < 0$.

(b) A removable singularity if $N - M \geq 0$, and in fact a zero of order $N - M$ if $N - M \geq 1$.

Proof. Since N and M are the orders of the zeros of $g(z)$ and $h(z)$, respectively, we have

$$g(z) = (z - z_0)^N g_1(z) \qquad \text{and} \qquad h(z) = (z - z_0)^M h_1(z)$$

where $g_1(z)$ and $h_1(z)$ are analytic at $z = 0$ with $g_1(0) \neq 0$ and $h_1(0) \neq 0$. Then, from the definition $f(z) = g(z)/h(z)$,

$$f(z) = \frac{p(z)}{(z - z_0)^{M-N}} \qquad p(0) \neq 0$$

where $p(z) = g_1(z)/h_1(z)$ is analytic and nonzero at $z = 0$. The results (a) and (b) follow immediately from this.

■

Since a polynomial can be factored into a product of powers $(z - z_k)^{n_k}$, each of which has a finite order zero, the above result shows that a **rational function possesses at worst poles at the zeros of its denominator**. In Problem 18, the reader is asked to prove that rational functions possess partial fraction expansions.

(Isolated) Essential Singularities. The simplest example of an (isolated) essential singularity was provided by Example (6.4.23), where we developed the Laurent series

(6.5.18)
$$e^{1/z} = \sum_{n=0}^{\infty} \frac{z^{-n}}{n!} \qquad 0 < |z| < \infty$$

Note that $z = 0$ is an (isolated) essential singularity of $e^{1/z}$ since the principal part $\sum_{n=1}^{\infty} z^{-n}/n!$ has an infinite number of terms in it. We also note an interesting property of this function. Let $\alpha \neq 0$ be *any complex number* and consider the sequence of points $z_n = (\operatorname{Log} \alpha + 2\pi i n)^{-1}$. Note that $\lim_{n\to\infty} z_n = 0$, and

$$\lim_{n\to\infty} e^{1/z_n} = \lim_{n\to\infty} [e^{\operatorname{Log} \alpha} \cdot e^{2\pi i n}] = \alpha$$

Hence, in any neighborhood of the (isolated) essential singularity $z = 0$, there are values of z such that $e^{1/z}$ is as close as desired to any arbitrary nonzero number α. The amazing fact is that this same result is true for any function which has an (isolated) essential singularity (see Problem 14). A deeper extension of this result, called Picard's Theorem, is stated (but not proved) below.

(6.5.19) ***Theorem* (Picard).** Let z_0 be an (isolated) essential singularity of $f(z)$. Then, in every deleted neighborhood of z_0 in which $f(z)$ is analytic and not constant, $f(z)$ takes on every complex value with the exception of at most one value.

■

(6.5.20) ***Example.*** Show that $z = 0$ is an (isolated) essential singularity for $f(z) = e^{z+(1/z)}$ and compute the principal part.

Solution. **We first write $f(z) = e^z \cdot e^{1/z}$ and expand e^z in a Taylor series and $e^{1/z}$ in a Laurent series about $z = 0$. Thus,**

$$f(z) = \left[\sum_{k=0}^{\infty} \frac{z^k}{k!}\right] \cdot \left[\sum_{n=0}^{\infty} \frac{z^{-n}}{n!}\right]$$

and the result of term-by-term multiplication yields

$$f(z) = \sum_{k=0}^{\infty} \frac{1}{(k!)^2} + \sum_{n=1}^{\infty} (z^n + z^{-n}) \left[\sum_{k=0}^{\infty} \frac{1}{k!(n+k)!}\right]$$

and the principal part is

$$\sum_{n=1}^{\infty} z^{-n} \left[\sum_{k=0}^{\infty} \frac{1}{k!(n+k)!}\right]$$

(6.5.21) ***Example.*** **The function $f(z) = 1/\sin(1/z)$ is singular at $z = 0$ and at $z_n = 1/n\pi, (n = \pm 1, \pm 2, \ldots$. The reader is asked to show that the points z_n are all simple poles. Note that $z = 0$, which is not an isolated singularity, is a limit point of these poles. In fact, this is a singularity of an entirely different type and will not be discussed here.**

The Point at Infinity. As with analytic functions, we have the following.

(6.5.22) ***Definition.*** The **point of infinity is an isolated singularity for $f(z)$** if $z = 0$ is an isolated singularity for $f(1/z)$. The nature of the singular point at $z = \infty$ for $f(z)$ is the same as the nature of the point $z = 0$ for $f(1/z)$.

Example. Let $f(z) = (z^4 + 2z + 1)/(z^2 + 5z + 1)$. Then,

$$f\left(\frac{1}{z}\right) = \frac{1 + 2z^3 + z^4}{z^2 + 5z^3 + z^4} = \frac{1}{z^2}\left(\frac{1 + 2z^3 + z^4}{1 + 5z + z^2}\right)$$

Since this function has a pole of order 2 at $z = 0$, $f(z)$ has a pole of order 2 at $z = \infty$. Now, expanding $(1 + 2z^3 + z^4)/(1 + 5z + z^2)$ in a Taylor series about $z = 0$, we find

$$f\left(\frac{1}{z}\right) = \frac{1}{z^2} - \frac{5}{z} + \cdots$$

Hence, the principal part of $f(z)$ at $z = \infty$ is $z^2 - 5z$. (Why?)

(6.5.23) ***Example.*** ***The entire function*** e^z has an essential singularity at $z = \infty$ because $e^{1/z}$ has an essential singularity at $z = 0$.

We will end this section with some interesting results concerning singularities at infinity.

(6.5.24) *Theorem*

(a) Let $f(z)$ be entire. Then, either f is constant or has a pole or essential singularity at $z = \infty$.

(b) A polynomial of order N has a pole of order N at $z = \infty$.

(c) A rational function is either analytic at $z = \infty$ or has a pole there.

Proof. For (a), first use the fact that since $f(z)$ is entire, it has a Taylor series

$$f(z) = \sum_{n=0}^{\infty} a_n z^n \qquad 0 \le |z| < \infty$$

Hence,

$$f\left(\frac{1}{z}\right) = \sum_{n=0}^{\infty} a_n z^{-n} \qquad 0 < |z| < \infty$$

from which we see that the Laurent series for $f(1/z)$ about $z = 0$ contains only nonpositive powers of z. Thus, if $f(z)$ is not constant, $z = 0$ is either a pole or an essential singularity of $f(1/z)$, with the same result at $z = \infty$ for $f(z)$.

For part (b), let $p(z) = \sum_{n=0}^{N} a_n z^n$, $a_N \ne 0$. Then, $p(1/z) = \sum_{n=0}^{N} a_n z^{-n}$ has a pole of order N at $z = 0$.

For part (c), consider the rational function

$$g(z) = \frac{a_N z^N + a_{N-1} z^{N-1} + \cdots + a_0}{b_M z^M + b_{M-1} z^{M-1} + \cdots + b_0}$$

with $a_N \ne 0$ and $b_M \ne 0$. Then,

$$g\left(\frac{1}{z}\right) = z^{M-N}\left(\frac{a_N + a_{N-1} z + \cdots + a_0 z^N}{b_M + b_{M-1} z + \cdots + b_0 z^M}\right)$$

Clearly, if $M - N \ge 0$, $g(1/z)$ is analytic at $z = 0$, while if $M - N < 0$, $g(1/z)$ has a pole of order $N - M$ at $z = 0$.

■

PROBLEMS (The answers are given on page 440.)

(Problems with an asterisk are more theoretical.)

1. Find the principal part of the following functions at each isolated singularity:

(a) $\dfrac{z}{e^z - 1}$ (b) $\dfrac{z^2 + z + 1}{(z - i)(z^2 + 2z + 1)^2}$ (c) $\dfrac{\sin z}{z(z - 1)}$

(d) $\dfrac{\operatorname{Log} z}{(z - 1)^2(z^2 + 4)}$ (e) $\dfrac{1}{\sin^2(1/z)}$ (f) $\dfrac{1 - \cos z}{z^4}$

2.* (a) Prove that if z_0 is a removable singularity of an analytic function $f(z)$, then $f(z)$ is bounded in some deleted neighborhood of z_0.

(b) Prove that if z_0 is an isolated singularity of an analytic function $f(z)$, and if $|f(z)| \le K/|z - z_0|^\alpha$ in some deleted neighborhood of z_0, where $K > 0$ and $\alpha < 1$, then z_0 is a removable singularity. [*Hint:* If $f(z) = \sum_{n=-\infty}^{\infty} c_n(z - z_0)^n$, use the integral expression for the Laurent coefficients, where the integral is taken over the circle $|z - z_0| = \epsilon$. Then, show that $\lim_{\epsilon \to 0} c_n = 0$ for $n \le -1$ and conclude that $c_n = 0$ for $n \le -1$.]

3. (a) Show that $\lim_{z \to 0} [z(1 + \sqrt{z})] = 0$, where the principal branch is used. Based upon Theorem (6.5.5a), can we conclude that $z = 0$, and if $z = 0$ is a removable singularity?

(b) If $g(z)$ is analytic at $z = 0$, and if $z = 0$ is a removable singularity of $g(z)/z^n$, what can you say about $g(z)$?

4. Classify the isolated singular points of the following functions and give the order of any poles

(a) $\dfrac{z}{e^{z^2} - 1}$ (b) $\dfrac{\sin z}{e^z - 1}$ (c) $z^2 \cos\left(\dfrac{1}{z}\right)$ (d) $\dfrac{\cot z}{z}$

(e) $\dfrac{iz}{z - i} + \dfrac{2i}{z^2 + i}$ (f) $\dfrac{e^{(z-1)^{-2}} \sin z \cos^3 z}{z(z - \pi/2)}$

(g) $\dfrac{\sin z^2}{z(e^z - 1)}$ (h) $(z - \pi)e^{1/z} \cot z$ (i) $\dfrac{e^{z-1} \sin\left(\dfrac{1}{z - 1}\right)}{z^2(z^2 - 4)}$

(j) $\dfrac{1 - \cos z}{\sin^3 z}$ (k) $\dfrac{(1 - e^z)\cos\left(\dfrac{1}{z - 8}\right)}{z^2(z - 8)(z^2 - 16)}$

(l) $z^{-n} \operatorname{Log}(1 + z^m)$ $(n, m = 0, 1, 2, \ldots$ (m) $\dfrac{1}{\sin z - \cos z}$

(n) $(1 + e^{1/z})^{-1}$

5. Consider the function $f(z) = [z - (2 - z^2)^{1/2}]^{-1}$ where the principal branch of the square root is taken.

(a) Show that all real values $z = x$ which satisfy $x^2 \ge 2$ are singular points of $f(z)$.

(b) Show that $z = 1$ is the only isolated singular point of $f(z)$.

(c) What is the nature of the isolated singular point $z = 1$? Compute the principal part of $f(z)$ here.

6.* (a) Complete the proof of Theorem (6.5.10b) and show that if z_0 is a pole of order N for $f(z)$, then (6.5.11) holds. [*Hint:* Factor $(z - z_0)^{-N}$ from $f(z)$ and show that the power series defines an analytic function.]

(b) Prove that if z_0 is an isolated singularity of $f(z)$ and $\lim_{z \to z_0} f(z) = \infty$, then z_0 is a pole. [*Hint:* Let $F(z) = 1/f(z)$. Use Theorem (6.5.5a) to show that $F(z)$ has a removable singularity at $z = z_0$. Then, show that $F(z)$ can be defined at z_0 so that it is analytic and vanishes there. Now, use part (b) of Theorem (6.5.10) to show that $1/F(z)$ has a pole at z_0.]

7. Let $f(z)$ be analytic in a simply connected domain D except at a single point z_0. Let $g(z) = f'(z)$. Describe the nature of the singular point z_0 for $g(z)$ if z_0 is
(a) A removable singular point for $f(z)$.
(b) A pole of order N for $f(z)$.

8. (a) Let z_0 be an isolated singular point of $f(z)$ and let $g(z)$ be analytic at z_0. What is the nature of the point z_0 for the function $f(z) + g(z)$?
(b) If $f(z)$ has a pole of order N at z_0, what is the nature of the singularity at z_0 for the function $(z - z_0)^k f(z)$, where k can take on the values $k = 0, \pm 1, \pm 2, \ldots$?

9. Let $f(z)$ have a pole of order N at $z = z_0$. What is the nature of the singularity z_0 for the function $f'(z)/f(z)$? Find the principal part.

10. (a) Prove that if $f(z)$ is analytic at z_0 and has a zero of order $N \geq 1$ there, then $1/f(z)$ has a pole of order N at z_0.
(b) Prove that if $f(z)$ has a pole of order $N \geq 1$ at z_0, then $1/f(z)$ has a zero of order N at z_0.
(c) Prove that if $f(z) \neq 0$ is analytic at z_0, then $1/f(z)$ is either analytic at z_0 or has a pole there. [*Hint:* Use part (*a*).]

11. (a) Let $f(z)$ have a pole of order $N \geq 1$ at $z = z_0$ and $g(z)$ be a polynomial of order M. What is the nature of the singularity at z_0 of the composite function $g(f(z))$?
(b) If $g(z)$ in part (*a*) is an entire function (not necessarily a polynomial), does your answer there change? Provide an example.

12. Let $f(z)$ possess an isolated singularity at $z = z_0$ and $g(z)$ be analytic at $z = z_0$. What is the nature of the singularity at $z = z_0$ of the function $f(z) \cdot g(z)$ if
(a) z_0 is a removable singularity of f?
(b) z_0 is a pole of order N of f?
(c) z_0 is an essential singularity of f? [*Hint:* Use the result of Theorems (6.5.5) and (6.5.10).]

13. Construct a function having all of the following properties: a removable singularity at $z = 2$, a pole of order 5 at $z = i$, and an essential singularity at $z = 1 + i$.

14.* The following series of steps leads to a proof of the **Casorati–*Weierstrass* *Theorem*** which states:

> *If* z_0 *is an (isolated) essential singularity of* f(z), *there are values of* z *such that* f(z) *comes arbitrarily close to any nonzero complex number* α.

To start, assume that there is an α and some deleted neighborhood of z_0 so that $f(z)$ is bounded away from α, that is, $|f(z) - \alpha| \geq K$, for some $K > 0$.
(a) Let $g(z) = [f(z) - \alpha]^{-1}$. Show that $g(z)$ is bounded near z_0. Hence, what must be the nature of the singular point z_0 for g?
(b) Why does $\lim_{z \to z_0} g(z)$ exist?
(c) Show that if the limit in (*b*) is zero, then $f(z)$ has a pole at z_0.

(d) Show that if the limit in (*b*) is not zero, then $f(z)$ has a removable singularity at z_0.

(e) Now conclude that the theorem is true.

15. Classify the nature of the point at infinity for those functions in Problem 4 which have isolated singularities at $z = \infty$.

16.* Show that the only function which has a pole of order 2 at $z = 0$, a simple pole at $z = i$, a zero of order 2 at $z = 1$, a simple zero at $z = 2$ and which is otherwise analytic (including at $z = \infty$) is a constant multiple of the rational function $(z-1)^2(z-2)/z^2(z-i)$. [*Hint:* Write $f(z) = (z-1)^2(z-2)g(z)/z^2(z-i)$, and use Theorems (6.5.5), (6.5.10), and (6.5.24a) to show that g is constant.]

17.* (a) Prove that an entire function which is bounded at infinity is contant. [*Hint:* Use Theorem (6.5.24a).]

(b) Prove that an entire function which has a pole at infinity of order N is a polynomial of order N. [*Hint:* Let $f(z) = \sum_{n=0}^{\infty} a_n z^n$, $0 \le |z| < \infty$, and show that all but a finite number of the a_n vanish.]

18.* Here, we will prove that *a function which has a finite number of poles, and at worst a pole at $z = \infty$, is a rational function and possesses a partial fraction expansion.* Thus, let $z_1, \ldots, z_N$ be the poles in the finite plane of orders $n_1, \ldots, n_N$, respectively. Let the order of the pole at infinity be n_∞.

(a) In some deleted neighborhood of each $z = z_l (l = 1, \ldots, N)$ and $z = \infty$, $f(z)$ has the local Laurent expansions

$$f(z) = \sum_{k=1}^{n_l} \frac{A_k^{(l)}}{(z-z_l)^k} + g_l(z) \qquad (\text{near } z = z_l)$$

while

$$f(z) = \sum_{k=1}^{n_\infty} A_k^{(\infty)} z^k + g_\infty(z) \qquad (\text{near } z = \infty)$$

where each $g_l(z)$ is analytic at $z = z_l$ while $g_\infty(z)$ is analytic at $z = \infty$. Why is the function

$$h(z) = f(z) - \sum_{l=1}^{N} \sum_{k=1}^{n_l} \frac{A_k^{(l)}}{(z-z_l)^k} - \sum_{k=1}^{n_\infty} A_k^{\infty} z^k$$

entire and analytic at $z = \infty$?

(b) From (a), conclude that $h(z)$ must be constant [see Theorem (6.5.24a)] and hence $f(z)$ is a rational function which has a partial fraction expansion derived from part (a).

6.6 RESIDUES AND THE RESIDUE THEOREM

Here, our main result will be the important residue theorem. We should say at the outset that even though the result follows almost trivially, its importance and implications are profound. In a sense, it is the culmination of all that has taken place in this chapter.

(6.6.1) ***Definition.*** Consider the local Laurent series for an analytic function $f(z)$ having an isolated singularity at $z = z_0$,

$$f(z) = \sum_{n=-\infty}^{\infty} c_n(z - z_0)^n \qquad 0 < |z - z_0| < R$$

The **residue of f at z_0**, denoted by $\mathbf{Res}[f]_{z=z_0}$, is the *coefficient of the power* $(z - z_0)^{-1}$ in the local Laurent series. That is, from the above,

(6.6.2)
$$\operatorname{Res}[f]_{z=z_0} = c_{-1}$$

■

(6.6.3) ***Example.*** Compute the residues at the singular points for the functions $f(z) = (z + 1)/[(z - 1)^2]$ and $g(z) = \sin z/z^3$.

Solution. The point $z = 1$ is a pole of order 2 for $f(z)$. To compute the residue there, we rewrite $f(z)$ as

$$f(z) = \frac{(z-1)+2}{(z-1)^2} = \frac{2}{(z-1)^2} + \frac{1}{z-1}$$

Hence, $\operatorname{Res}[f]_{z=1} = 1$.

For $g(z)$, which has a singularity at $z = 0$, we first expand the numerator $\sin z$ in the Taylor series $z - \frac{1}{6}z^3 + \cdots$ and write

$$g(z) = \frac{1}{z^2} - \frac{1}{6} + \text{(higher powers of } z)$$

Thus, at $z = 0$, which is a pole or order 2 for $g(z)$, $\operatorname{Res}[f]_{z=0} = 0$ due to the absence of the $1/z$ term.

■

We should remark that **the residue at a removable singular point is zero.** (Why?) Also, as the next result shows, there is a simple formula for the residue at a pole in terms of the derivatives of the function.

(6.6.4) ***Theorem.*** Let z_0 be a pole of $f(z)$ *at worst* of order $p \geq 1$. Then,

(6.6.5)
$$\operatorname{Res}[f]_{z=z_0} = \lim_{z \to z_0} \left\{ \frac{1}{(p-1)!} \left(\frac{d}{dz}\right)^{p-1} [(z - z_0)^p f(z)] \right\}$$

Also, if $g(z)$ and $h(z)$ are analytic at z_0, with $g(z_0) \neq 0$, and *if* g/h *has a simple pole at* z_0, then

(6.6.6)
$$\operatorname{Res}\left[\frac{g}{h}\right]_{z=z_0} = \frac{g(z_0)}{h'(z_0)}$$

Proof. If $f(z)$ has at worst a pole of order p at z_0, we have

$$f(z) = \sum_{n=-p}^{-1} c_n(z - z_0)^n + g(z) \qquad 0 < |z - z_0| < R$$

where $g(z)$ is the power series $\sum_{n=0}^{\infty} c_n(z - z_0)^n$, and hence is analytic at z_0 (see Section 6.2). The object is to compute c_{-1} above. However, due to the "singular" negative powers, we must first remove the most singular term $(z - z_0)^{-p}$ by multiplying by $(z - z_0)^p$ to get

$$(z - z_0)^p f(z) = c_{-p} + (z - z_0)c_{-p+1} + \cdots + (z - z_0)^{p-1} c_{-1} + (z - z_0)^p g(z)$$

If we now take $(p - 1)$ derivatives of both sides and then evaluate the limit at z_0, the result is (6.6.5). In Problem 3a, the reader is asked to derive (6.6.6).

■

With these results we can easily evaluate residues if we can first deduce the nature of the singular point. This in turn is often accomplished with a judicious use of Theorem (6.5.10) of the previous section. Also, the above result (6.6.5) is very reminiscent of the type of computation needed to perform partial fraction expansions (see Section 3.1).

(6.6.7) ***Example.*** Compute the residues at the singular points of

$$f(z) = \frac{1 + z + z^2}{(z - i)^2(z + 2)}$$

Solution. The function has a simple pole at $z = -2$ and a pole of order 2 at $z = i$. (Why?) Hence, by (6.6.5) with $p = 1$ and 2, respectively, we get

$$\text{Res}[f]_{z=-2} = \lim_{z \to -2} \left[\frac{1 + z + z^2}{(z - i)^2}\right] = \frac{3}{(2 + i)^2}$$

$$\text{Res}[f]_{z=i} = \lim_{z \to i} \left[\frac{d}{dz}\left(\frac{1 + z + z^2}{z + 2}\right)\right] = \frac{4i}{(2 + i)^2}$$

■

(6.6.8) ***Example.*** What are the residues at the essential singularities of the functions $z \sin(1/z)$ and $z \cos(1/z)$.

Solution. Since $z = 0$ is not a pole we cannot apply the previous results, and must compute the Laurent series directly. Since $\sin(1/z) = \sum_{n=0}^{\infty} [(-1)^n z^{-(2n+1)}]/[(2n + 1)!] = (1/z) - (1/6z^3) + \cdots$ and $\cos(1/z) = 1 - (1/2z^2) + \cdots$, we have $z \sin(1/z) = 1 - (1/6z^2) + \cdots$ and $z \cos(1/z) = z - (1/2z) + \cdots$. Hence,

$$\text{Res}\left[z \sin\left(\frac{1}{z}\right)\right]_{z=0} = 0 \quad \text{and} \quad \text{Res}\left[z \cos\left(\frac{1}{z}\right)\right]_{z=0} = -\frac{1}{2}$$

■

The next result exhibits the remarkable fact that *the contour integral of a function over a closed curve depends only on the residues of the function at those singularities lying inside the contour of integration.*

(6.6.9) ***Theorem* (Residue Theorem).** Let C be a simple, closed, piecewise smooth curve and let $f(z)$ be analytic inside and on C, except at a finite number of isolated singularities $z_1, \ldots, z_N$ lying interior to C. Then,

$$\oint_C f(z)\,dz = 2\pi i \sum_{k=1}^{N} \text{Res}[f]_{z=z_k}$$

Proof. Since $z_1, z_2, \ldots, z_N$ are isolated, there is a set of nonintersecting circles C_k, $k = 1, 2, \ldots, N$, with centers at the z_k and lying within C. By the Cauchy–Goursat Theorem for multiply connected domains, we have

$$\oint_C f(z)\,dz = \sum_{k=1}^{N} \oint_{C_k} f(z)\,dz \tag{6.6.10}$$

In some neighborhood of each z_k, consider the local Laurent series expansion for $f(z)$:

$$f(z) = \sum_{n=-\infty}^{\infty} c_n^{(k)}(z - z_k)^n$$

where the superscript k in the Laurent coefficients $\{c_n^{(k)}\}$ indicates that this is the Laurent series local to $z = z_k$. Now, by definition,

$$\text{Res}[f]_{z=z_k} = c_{-1}^{(k)} = \frac{1}{2\pi i}\oint_{C_k} f(z)\,dz$$

where the last equality follows from the integral definition (6.4.5) of the Laurent coefficients. Substituting this result into (6.6.10) proves the theorem.

■

(6.6.11) ***Example.*** Use the residue theorem to evaluate

$$\oint_{|z|=5} \frac{(e^z - 1)\,dz}{z(z-1)(z-i)^2}$$

Solution. The integrand has a removable singularity at $z = 0$, a simple pole at $z = 1$, and a pole of second order at $z = i$, all lying inside $|z| = 5$. Hence, with $f(z) = (e^z - 1)/[z(z-1)(z-i)^2]$,

$$\text{Res}[f]_{z=0} = 0, \qquad \text{Res}[f]_{z=1} = \lim_{z\to 1}\left[\frac{(e^z - 1)}{z(z-i)^2}\right] = \frac{i(e-1)}{2},$$

$$\text{Res}[f]_{z=i} = \lim_{z\to i}\left[\frac{d}{dz}\left(\frac{e^z - 1}{z(z-1)}\right)\right] = -\frac{3}{2}e^i + 1 + \frac{i}{2}$$

Thus,

$$\oint_{|z|=5} \frac{(e^z - 1)\,dz}{z(z-1)(z-i)^2} = \pi i(ie - 3e^i + 2) \tag{6.6.12}$$

■

(6.6.13) ***Example.*** Evaluate the two integrals

$$\oint_{|z|=1} \frac{z}{e^z - 1}\,dz \quad \text{and} \quad \oint_{|z|=1} z e^{1/z}\,dz$$

Solution. Unlike the previous example, to which we might also have applied the Cauchy integral formula, the integrands here are not amenable to this approach. Hence, we must use the residue theorem. For the first integral, note that since $z = 0$ is the only singularity of the integrand lying inside the unit circle, and since $e^z - 1 = z + \frac{1}{2}z^2 + \cdots$, $z = 0$ is a removable singularity with zero residue. Hence,

$$\oint_{|z|=1} \frac{z\,dz}{e^z - 1} = 0$$

Now, the second integrand has an essential singularity at $z = 0$ with Laurent series

$$ze^{1/z} = z \sum_{n=0}^{\infty} \frac{z^{-n}}{n!} = z + 1 + \frac{1}{2z} + \frac{1}{6z^2} + \cdots$$

Thus,

$$\oint_{|z|=1} ze^{1/z}\,dz = \frac{2\pi i}{2} = \pi i$$

■

It is interesting at this stage to compare the Cauchy integral formula with the residue theorem, since both are quite useful in evaluating contour integrals. The integrals to which the Cauchy integral formula can be applied are of the type

(6.6.14)
$$I = \oint_C \frac{g(z)\,dz}{(z - z_0)^{N+1}}$$

where z_0 lies inside C and $g(z)$ is analytic inside and on C. Hence, with our present nomenclature, the above integrand possesses at worst a pole of order $N + 1$ at $z = z_0$. (This will depend upon whether $g(z)$ vanishes at z_0, and to what order.) Now by the residue theorem and (6.6.5) with $p = N + 1$, we get

$$I = \frac{2\pi i}{N!} g^{(N)}(z_0)$$

which is the same result as that of the Cauchy integral formula for derivatives. However, the residue theorem is also applicable to integrals whose integrands possess isolated singularities other than poles. Hence, **the residue theorem is an extension of the Cauchy integral formulas.**

PROBLEMS (The answers are given on page 442.)

(Problems with an asterisk are more theoretical.)

1. Compute the residues at each of the isolated singular points for the following functions:

(a) $\dfrac{z^2 - z}{z^3}$ (b) $\dfrac{z^2}{z^3 + z}$ (c) $\dfrac{\sin z}{z^3(z-1)}$ (d) $\dfrac{z}{e^{z^2} - 1}$

(e) $z^3 \cos\left(\dfrac{1}{z}\right)$ (f) $z \cot z$ (g) $\dfrac{1}{\operatorname{Log}^2 z}$

(h) $\dfrac{1}{\sin(e^z)}$ (i) $(z - \pi)\csc z$ (j) $1/(1 - e^{1/z})$

(k) $\sin(e^{1/z})$ [*Hint:* In the integral expression for c_{-1} make the substitution $w = 1/z$.]

2. Provide an example of a function with a nonremovable singularity at $z = 1$ whose residue there is zero.

3. (a) *Prove:* If $g(z)$ and $h(z)$ are analytic at z_0 with $g(z_0) \neq 0$, and if $g(z)/h(z)$ has a simple pole at z_0, then $\operatorname{Res}[g/h]_{z=z_0} = g(z_0)/h'(z_0)$. [*Hint:* Use (6.6.5) and the definition of derivative.]

   (b) If $g(z)$ and $h(z)$ are analytic at z_0 with $g(z_0) \neq 0$, and if g/h has a pole of order 2 at z_0, what is $\operatorname{Res}[g/h]_{z=z_0}$ in terms of g and h?

4. Let $f(z)$ and $g(z)$ both possess an isolated singularity at $z = z_0$, and let $f(z)$ be even $[f(-z) = f(z)]$ and $g(z)$ be odd $[g(-z) = -g(z)]$.

   (a) Show that $f(z)$ and $g(z)$ both possess an isolated singularity at $z = -z_0$ *of the same type as at* $z = z_0$.

   (b) Show that $\operatorname{Res}[f]_{z=z_0} = -\operatorname{Res}[f]_{z=-z_0}$ and $\operatorname{Res}[g]_{z=z_0} = \operatorname{Res}[g]_{z=-z_0}$.

5. Evaluate the following integrals using the residue theorem:

(a) $\displaystyle\oint_{|z|=1} \frac{dz}{e^z - 1}$ (b) $\displaystyle\oint_{|z|=1} (z + z^2)e^{1/z}\,dz$

(c) $\displaystyle\oint_{|z|=5} \frac{z\,dz}{\sin z}$ (d) $\displaystyle\oint_{|z-1|=2} (z-1)e^{1/z^2}\,dz$

(e) $\displaystyle\oint_{|z|=3} \frac{dz}{z^4 + 16}$ (f) $\displaystyle\oint_{|z|=1/2} \frac{z^2 e^z}{1 - e^{z^4}}\,dz$

(g) $\displaystyle\oint_{|z-\pi|=\pi} \frac{dz}{\cos^2 z}$ (h) $\displaystyle\oint_{|z|=1} e^{2/z}\sin\left(\frac{1}{z}\right)dz$

(i) $\displaystyle\oint_{|z-2|=1} \frac{(1 - e^{z^2})}{z^3}\,dz$ (j) $\displaystyle\oint_{|z|=1/2} \frac{dz}{\operatorname{Log}(z+1)}$

(k) $\displaystyle\oint_{|z+1/2|=1} \frac{\sin z e^{1/(z-1)^2}}{z^2(z^2-1)}\,dz$ (l) $\displaystyle\oint_{|z|=1} e^{1/z^n}\,dz$

6. Consider the integral

$$I = \oint_{|z|=1/2} \frac{\sin(1/z)}{z-1}\,dz$$

(a) Evaluate I using residues. Express the answer in the form of an infinite series.

(b) Let $w = 1/z$ in I, evaluate the resulting integral, and hence sum the series in (*a*).

7. Let $f(z)$ have zeros at $z = z_k^{(0)}$ of order $n_k^{(0)}$, respectively, $k = 1, 2, \ldots, N_0$, and poles at $z = z_k^{(p)}$ of order $n_k^{(p)}$, $k = 1, 2, \ldots, N_p$, but otherwise be analytic. Let C be a simple closed contour surrounding all of the zeros and poles of f.

(a) Show that

$$\oint_C \frac{f'(z)}{f(z)}\,dz = 2\pi i(Z - P)$$

where $Z = \sum_{k=1}^{N_0} n_k^{(0)}$ and $P = \sum_{k=1}^{N_p} n_k^{(p)}$

(b) If $g(z)$ is analytic on and inside a smooth curve C, generalize the result in (a) and evaluate the integral $\oint_C [g(z) f'(z)/[f(z)]\,dz$.

8. Here, we will define the **residue at infinity**. Let $f(z)$ have an isolated singularity at $z = \infty$ [see Definition (6.5.22)] but otherwise be analytic in $|z| > R$ for some R. Let C be any circle $|z| = \hat{R}$, with $\hat{R} > R$, and inside of which lie *all* of the singularities of f. *We can thus view C as encircling the singularity at ∞.* We define

$$\text{Res}[f]_{z=\infty} = -\frac{1}{2\pi i}\oint_C f(z)\,dz$$

(a) Why do you suppose the negative sign is inserted in the above definition? [*Hint:* Where does infinity lie relative to C?]

(b) Let $w = 1/z$ in the above integral and show that

$$\text{Res}[f(z)]_{z=\infty} = -\text{Res}\left[w^{-2} f\left(\frac{1}{w}\right)\right]_{w=0}$$

(c) If z_k, $k = 1, 2, \ldots, N$, are the isolated singularities of f in the finite plane with corresponding residues r_k, and if r_∞ is the residue at infinity, show that $r_\infty + \sum_{k=1}^{\infty} r_k = 0$.

9. Use the results of Problem 8 to compute the residues at infinity for the functions in Problem 1a, b, c, e, and k.

10. In Problems 19 and 20 of Section 5.2 we showed that if $\Phi(z)$ is the **complex potential of a fluid flow**, and if C is a simple, closed, piecewise smooth curve in the flow region, then

$$\Gamma + i\Omega = \oint_C \Phi'\,dz \qquad \text{and} \qquad F = \overline{\frac{i\rho}{2}\oint_C (\Phi')^2\,dz}$$

where Γ is the **circulation** around C, Ω is the **flow** across C, and F is the (complex) **force** acting on a rigid body D immersed in the flow and bounded by C. Here, we will show that F is related to the circulation. The problem we will consider is that of a uniform flow $V_0 e^{i\alpha}$ at $z = \infty$ impinging upon a bounded, rigid body D as shown in Figure 6.6.1

(a) Assume that the complex potential $\Phi(z)$ has the property that its derivative $\Phi'(z)$ is analytic outside the rigid body and at infinity. With this assumption, why can we write $\Phi'(z) = A + B/z + C/z^2 + \cdots$ for $|z|$ large enough?

(b) Use the relationship $V(z) = \overline{\Phi'(z)}$, where $V(z)$ is the complex velocity, to show that the quantity A in part (a) is given by $A = V_0 e^{-i\alpha}$.

(c) From Figure 6.6.1, why is it true that

$$\oint_C \Phi' \, dz = \oint_{C_R} \Phi' \, dz \qquad \text{and} \qquad \oint_C (\Phi')^2 \, dz = \oint_{C_R} (\Phi')^2 \, dz$$

where C_R is some large circle?

(d) Argue that the flow Ω is zero if there are no sources of fluid outside the rigid body. Then use parts (a), (b), and (c) to conclude that $B = \Gamma/2\pi i$.

(e) Finally, conclude that the net force on the body is $-i\rho\Gamma V_0 e^{i\alpha}$, and is thus orthogonal to the incoming flow and proportional to the circulation.

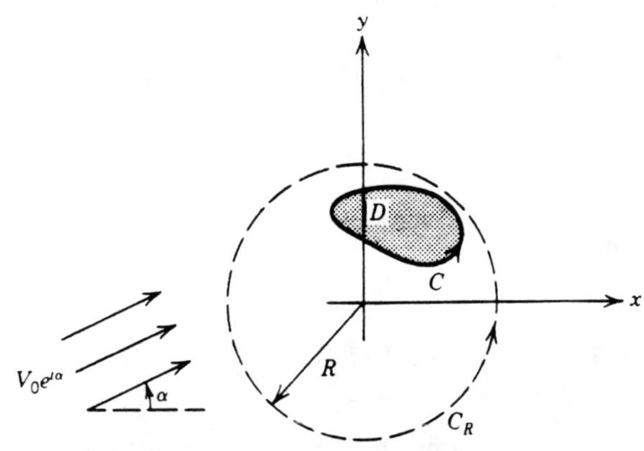

Figure 6.6.1

11.* Sometimes, residues and contour integrals can be used to sum series. Here, we will develop the theory needed to do this and in the next Problem apply it to some specific examples. Let $g(z)$ have simple poles at the integers $z = n$ $(n = 0, \pm 1, \pm 2, \ldots)$ with $\text{Res}[g]_{z=n} = 1$, but otherwise be analytic. Let $f(z)$ possess a *finite number* of simple poles at $z = z_k$, $k = 1, 2, \ldots, N$, with $\text{Res}[f]_{z=z_k} = \rho_k$. Assume that none of the z_k *is an integer*. Finally, let each C_n, $n = 0, 1, 2, \ldots$, be a square with vertices at $(1 \pm i)r_n$ and $(-1 \pm i)r_n$, where $n < r_n < n + 1$. Assume

(i) $|g(z)| \le K_1$ for z on any C_n

(ii) $|z^2 f(z)| \le K_2$ for $|z|$ large enough

show that

(a) For n large, $\oint_{C_n} g(z)\, f(z)\, dz = 2\pi i\left[\sum_{k=-n}^{n} f(k) + \sum_{k=1}^{N} \rho_k g(z_k)\right]$.

(b) $\lim_{n\to\infty}\left[\oint_{C_n} g(z)\, f(z)\, dz\right] = 0$. [*Hint:* Use the bounds on f and g.]

(c) $\sum_{k=-\infty}^{\infty} f(k)$ converges, and $\sum_{k=-\infty}^{\infty} f(k) = -\sum_{k=1}^{N} \rho_k g(z_k)$.

(d) The function $g(z) = \pi \cos(\pi z)/\sin(\pi z)$ satisfies all of the requirements for $g(z)$. [*Hint:* $|\pi \cos(\pi z)/\sin(\pi z)| = \pi\left|\dfrac{e^{2\pi i z}+1}{e^{2\pi i z}-1}\right|$. Now use the various triangle inequalities to bound this quantity on each part of C_n.]

12. In Problem 11, use $g(z) = \pi \cos(\pi z)/\sin(\pi z)$ to verify the following formulas:

(a) $\displaystyle\sum_{k=1}^{\infty} \frac{1}{k^2+\alpha^2} = \frac{\pi(e^{2\pi\alpha}+1)}{2\alpha(e^{2\pi\alpha}-1)} - \frac{1}{2\alpha^2}$ (α real and $\neq 0$). [*Hint:* Take $f(z) = 1/(z^2+\alpha^2)$ in Problem 11.]

(b) $\displaystyle\sum_{k=1}^{\infty} \frac{1}{k^2-\alpha^2} = \frac{1}{2\alpha^2} - \frac{\pi\cos(\pi\alpha)}{2\alpha\sin(\pi\alpha)}$ $\quad \alpha \neq \pm 1, \pm 2, \ldots$

(c) $\displaystyle\sum_{k=1}^{\infty} \frac{1}{k^2} = \frac{\pi^2}{6}$ [*Hint:* Take $f(z) = 1/z^2$ and redo Problem 11 to account for the double pole in $f(z)$ at $z = 0$.]

(d) $\displaystyle\sum_{k=1}^{\infty} \frac{1}{(2k-1)^4} = \frac{\pi^4}{96}$ [See the hint in (c).]

6.7 EVALUATION OF REAL INTEGRALS REVISITED; BRANCH POINT INTEGRALS

In Section 5.4, we showed how the Cauchy integral formulas could be used to evaluate some nontrivial real integrals. Here, we will reinterpret those results in terms of residue theory and use the residue theorem to extend the class of real integrals which can be evaluated. If these previous sections have already been studied, then the reader can proceed directly with what follows. If not, then we would strongly urge the reading of Examples (5.4.2), (5.4.15), and (5.4.30). These contain the rationale behind the examples to be done here. The only point of departure will be that instead of using the Cauchy integral formulas to evaluate contour integrals, as in the examples of Section 5.4, here we use the residue theorem.

To begin with, we consider real trigonometric integrals of the form, or which can be put into the form,

(6.7.1)
$$\int_0^{2\pi} F(\sin x, \cos x)\, dx$$

In Section 5.4, Example (5.4.2), we showed how to treat such integrals when F is a rational function of $\sin x$ and $\cos x$ using the Cauchy integral formula. The next example shows that the same approach can be used when the integrand is not a rational function. However, the residue theorem must now be invoked.

(6.7.2) ***Example.*** Evaluate

$$I = \int_0^{2\pi} e^{\cos x} \cos(\sin x)\, dx$$

Solution. Following the approach of Example (5.4.2) (again, we urge the reader to read, or reread, this) we first replace $\cos x$ by $\frac{1}{2}(e^{ix} + e^{-ix})$ and $\sin x$ by $(1/2i)(e^{ix} - e^{-ix})$ to get

$$I = \frac{1}{2}\int_0^{2\pi} e^{(1/2)(e^{ix}+e^{-ix})}\left[e^{(1/2)(e^{ix}-e^{-ix})} + e^{-(1/2)(e^{ix}-e^{-ix})}\right] dx$$

Now the complex substitution $w = e^{ix}$ will map the above real interval $[0, 2\pi]$ onto the unit circle in the w-plane and produce the following form for I:

$$I = \frac{1}{2}\oint_{|w|=1} \frac{(e^w + e^{1/w})}{iw}\, dw \tag{6.7.3}$$

Note that $w = 0$ is an essential singularity of the integrand, and thus the *Cauchy integral formula cannot be used here*. However, since

$$\operatorname{Res}\left[\frac{e^w + e^{1/w}}{w}\right]_{w=0} = 2$$

we find that

$$\int_0^{2\pi} e^{\cos x}\cos(\sin x)\, dx = 2\pi$$

■

The last example, and the results concerning trigonometric integrals in Section 5.4, are easily seen to be special cases of the following general result.

(6.7.4) ***Theorem.*** Let $F(\alpha, \beta)$ be a bounded, continuous function of α and β with the property that the complex function $f(w)$ defined by

$$f(w) = \frac{1}{iw} F\left(\frac{w - w^{-1}}{2i}, \frac{w + w^{-1}}{2}\right)$$

is analytic except for isolated singularities at the points $w = w_1, w_2, w_3, \ldots, w_N$. Then,

$$\int_0^{2\pi} F(\sin x, \cos x)\, dx = 2\pi i \sum_{k=1}^{N} \operatorname{Res}[f(w)]_{w=w_k} \tag{6.7.5}$$

where the *sum is taken over those singularities of* f(w) *which lie inside the unit circle* $|w| = 1$.

Improper Integrals Over the Real Axis. We now consider improper integrals of the form

(6.7.6)
$$I = \int_{-\infty}^{\infty} \frac{p(x)}{q(x)} \{\sin^n(ax) \text{ or } \cos^m(ax)\}\, dx$$

where $p(x)$ and $q(x)$ are polynomials, a is a real constant, and $q(x)$ has no real zeros. Such integrals were examined in detail in Examples (5.4.15) and (5.4.30) (again, the reader is strongly urged to read, or reread, these).

(6.7.7) ***Example.*** Evaluate

$$\int_{-\infty}^{\infty} \frac{\sin^2(3x)}{1+x^2}\, dx$$

Solution. To begin with, we note that

$$\sin^2(3x) = -\frac{1}{4}(e^{3ix} - e^{-3ix})^2$$
$$= -\frac{1}{4}(e^{6ix} - 2 + e^{-6ix})$$
$$= -\frac{1}{2}\text{Re}(e^{6ix} - 1)$$

Hence,

(6.7.8)
$$\int_{-\infty}^{\infty} \frac{\sin^2(3x)}{1+x^2}\, dx = -\frac{1}{2}\text{Re}\left[\int_{-\infty}^{\infty} \frac{(e^{6ix} - 1)}{1+x^2}\, dx\right]$$

We now consider the closed contour Γ_R shown in Figure 6.7.1. By the

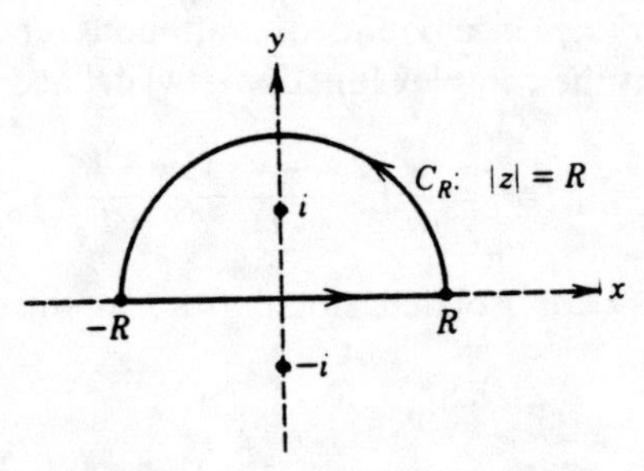

Figure 6.7.1 Contour Γ_R for Example (6.7.7).

residue theorem, we have

$$\oint_{\Gamma_R} \frac{e^{6iz} - 1}{1 + z^2}\, dz = 2\pi i \operatorname{Res}\left[\frac{e^{6iz} - 1}{1 + z^2}\right]_{z=i}$$
$$= \pi(e^{-6} - 1)$$

We now equate this to the integrals over both pieces of Γ_R in the figure to get

(6.7.9)
$$\int_{-R}^{R} \frac{e^{6ix} - 1}{1 + x^2}\, dx + \int_{C_R} \frac{e^{6iz} - 1}{1 + z^2}\, dz = \pi(e^{-6} - 1)$$

where C_R is the semicircle of radius R shown in the figure. Now on C_R, $z = Re^{i\theta}$, $0 \le \theta \le \pi$, and hence,

$$|e^{6iz}| = e^{-6R\sin\theta} \le 1$$

Since $\sin\theta \ge 0$ for this range of θ. Thus, for z on C_R,

$$\left|\frac{e^{6iz} - 1}{1 + z^3}\right| \le \frac{|e^{6iz}| + 1}{||z|^3 - 1|} \le \frac{2}{R^3 - 1}$$

By the triangle inequality for integrals, and since the length of C_R is πR, we have

$$\left|\int_{C_R} \frac{e^{6iz} - 1}{1 + z^3}\, dz\right| \le \left(\frac{2}{R^3 - 1}\right)(\pi R) \xrightarrow[R\to\infty]{} 0$$

Then, taking the limit as $R \to \infty$ in (6.7.9), we find

$$\int_{-\infty}^{\infty} \frac{e^{6ix} - 1}{1 + x^2}\, dx = \pi(e^{-6} - 1)$$

and hence, from (6.7.8),

$$\int_{-\infty}^{\infty} \frac{\sin^2(3x)}{1 + x^2}\, dx = \frac{\pi}{2}(1 - e^{-6})$$

■

The next result, which is proved in Problem 3, shows that integrals such as (6.7.6) depend only upon the residues of the integrand in the upper half-plane.

(6.7.10) ***Theorem.*** Let $p(x)$ and $q(x)$ be real polynomials with $q(x)$ having no real zeros, and let a be a nonnegative real number. Assume also that either

$$a > 0 \quad \text{and} \quad \text{degree}(q) - \text{degree}(p) \ge 1$$

or

$$a = 0 \quad \text{and} \quad \text{degree}(q) - \text{degree}(p) \ge 2$$

Then,

$$\text{(a)} \quad \int_{-\infty}^{\infty} \frac{p(x)}{q(x)} \cos(ax)\, dx = \text{Re}\left\{\sum_k \text{Res}\left[\frac{p(z)e^{iaz}}{q(z)}\right]_{z=z_k}\right\}$$

$$\text{(b)} \quad \int_{-\infty}^{\infty} \frac{p(x)\sin(ax)}{q(x)}\, dx = \text{Im}\left\{\sum_k \text{Res}\left[\frac{p(z)e^{iaz}}{q(z)}\right]_{z=z_k}\right\}$$

where the z_k, $k = 1, 2, \ldots, N$, are the zeros of $q(z)$ and *the sums are taken over those zeros lying in the upper half-plane.*

■

Improper Integrals Involving Branch Points. As a final application, we will show how to evaluate real integrals whose integrands, when extended into the complex plane, have branch point singularities. Even though the original integral itself does not depend upon the choice of the branch of the extended complex function, we will see that our calculations are sometimes simplified by an appropriate choice of branch cut.

(6.7.11) ***Example.*** Evaluate the real integral

$$I = \int_0^{\infty} \frac{x^a}{1 + x^2}\, dx \qquad -1 < a < 1$$

Solution. First note that for x very small, the integrand "behaves like" x^a, while for x large, it behaves like x^{a-2}. Hence, the conditions $-1 < a < 1$ guarantee that the improper integral exists. Also, when viewed as the complex function $z^a/(1 + z^2)$, the integrand has a branch point singularity at $z = 0$ and poles at $z = \pm i$. Therefore, as a first attempt at evaluation, we might try to use our old friend the closed contour consisting of the real interval $-R \leq x \leq R$ and the upper half of the circle $|z| = R$. However, to apply the residue theorem and pick up the contribution due to the pole at $z = i$, we must choose a branch of z^a whose branch cut does not intersect the upper half-plane. One such branch is defined by

$$z^a = |z|^a e^{ia\arg(z)} \qquad \frac{-\pi}{2} < \arg(z) \leq \frac{3\pi}{2} \tag{6.7.12}$$

and is depicted in Figure 6.7.2*a*. Now consider the closed contour Γ shown in part (*b*) of the figure. Note the small circle of radius ϵ indented around $z = 0$. The integral over this will be shown to be zero in the limit $\epsilon \to 0$, and it is technically needed in order to apply the residue theorem. Now, by the residue theorem, we have

$$\oint_{\Gamma} \frac{z^a}{1 + z^2}\, dz = \pi e^{\pi i a/2}$$

where we have used the fact that $i^a = e^{\pi i a/2}$ for our choice of branch of z^a.

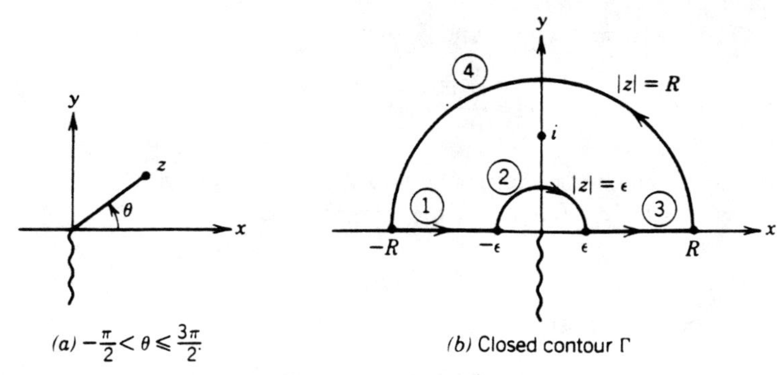

(a) $-\frac{\pi}{2} < \theta \leq \frac{3\pi}{2}$ (b) Closed contour Γ

Figure 6.7.2 Example (6.7.11).

Now, as usual, we equate this to the sum of the integrals over the four pieces of Γ designated by ①, ②, ③, and ④ in the figure. The following parametrizations of these four contours were chosen so that all arguments lie in the range specified in (6.7.12).

(6.7.13)

$$\begin{array}{llll} ①: & z = -te^{i\pi} & -R \leq t \leq -\epsilon & (\arg(z) = \pi) \\ ②: & z = \epsilon e^{i(\pi-\theta)} & 0 \leq \theta \leq \pi & (\arg(z) = \pi - \theta) \\ ③: & z = x & \epsilon \leq x \leq R & (\arg(z) = 0) \\ ④: & z = Re^{i\theta} & 0 \leq \theta \leq \pi & (\arg(z) = \theta) \end{array}$$

Hence, we get

(6.7.14)

$$\pi e^{\pi ia/2} = \int_{-R}^{-\epsilon} \frac{(-t)^a e^{\pi ia}}{1+t^2}\,dt + \int_{\epsilon}^{R} \frac{x^a}{1+x^2}\,dx$$
$$+ ie^{\pi ia}\epsilon^{a+1}\int_0^{\pi} \frac{e^{-i(a+1)\theta}}{1+\epsilon^2 e^{-2i\theta}}\,d\theta + i\int_0^{\pi} \frac{R^{a+1}e^{i(a+1)\theta}}{1+R^2e^{2i\theta}}\,d\theta$$

First, note that with the substitution $t = -x$, the first integral above is simply a multiple of the second (*this is how the branch of z^a manifests itself*). Also, since $a < 1$ and

$$\left|\frac{R^{a+1}e^{i(a+1)\theta}}{1+R^2e^{2i\theta}}\right| \leq \frac{R^{a+1}}{R^2-1} \qquad \text{for } R > 1$$

we can show in the usual way that the last integral is zero in the limit $R \to \infty$. Finally, the third integral is zero in the limit $\epsilon \to 0$ since $a > -1$ and

$$\left|\frac{\epsilon^{a+1}e^{-i(a+1)\theta}}{1+\epsilon^2 e^{-2ia\theta}}\right| \leq \frac{\epsilon^{a+1}}{1-\epsilon^2}$$

Therefore, with $\epsilon \to 0$ and $R \to \infty$ in (6.7.14), we get

$$\int_0^\infty \frac{x^a}{1+x^2}\,dx = \frac{\pi e^{\pi i a/2}}{1+e^{\pi i a}} = \frac{\pi}{2\cos(\pi a/2)}$$

Note that the last term above becomes infinite as $a \to \pm 1$. This simply reflects the divergence of the integral when $a = \pm 1$.

■

(6.7.15) ***Example.*** Evaluate $\int_0^\infty [x^a/(x+1)]\,dx$ for $-1 < a < 0$.

Solution. Unlike the previous example, the integrand here is singular on the real axis at $z = -1$. Hence, we cannot proceed as before and use a contour which includes the negative real axis. To evaluate this integral, we first choose the branch

$$z^a = |z|^a e^{ia\arg(z)} \qquad 0 < \arg(z) \le 2\pi \tag{6.7.16}$$

as shown in Figure 6.7.3*a*, and integrate around the contour Γ depicted in part (*b*) of the figure. There, the lines denoted by ① and ③ technically lie respectively a small distance above and below the branch cut on the positive real axis. In the limit as this small distance approaches zero, these two lines will approach the top and bottom of the branch cut. Hence, from the angle range in (6.7.16) we get, *in this limit*,

On ①: $\arg(z) = 0$

On ③: $\arg(z) = 2\pi$

Now the pieces ① → ④ of the contour Γ in Figure 6.7.3*b* can be parametrized by

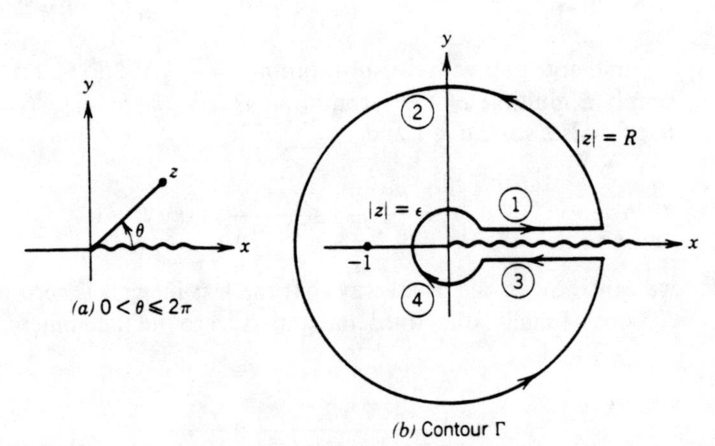

Figure 6.7.3 Example (6.7.15).

$$
\begin{array}{llll}
① & z = x & \epsilon \le x \le R & (\arg(z) = 0) \\
② & z = Re^{i\theta} & 0 \le \theta \le 2\pi & (\arg(z) = \theta) \\
③ & z = -te^{2\pi i} & -R \le t \le -\epsilon & (\arg(z) = 2\pi) \\
④ & z = \epsilon e^{i(2\pi - \theta)} & 0 \le \theta \le 2\pi & (\arg(z) = 2\pi - \theta)
\end{array} \tag{6.7.17}
$$

By the residue theorem,

$$\oint_{\Gamma} \frac{z^a}{1+z}\,dz = 2\pi i(-1)^a = 2\pi i e^{\pi i a}$$

If we equate this to the sum of the four integrals over the above pieces of Γ, we get

$$
\begin{aligned}
2\pi i e^{\pi i a} = & \int_{\epsilon}^{R} \frac{x^a}{x+1}\,dx + i\int_{0}^{2\pi} \frac{R^{a+1}e^{i(a+1)\theta}}{1+Re^{i\theta}}\,d\theta \\
& + \int_{-R}^{-\epsilon} \frac{(-t)^a e^{2\pi i a}}{1-t}(-dt) - ie^{2\pi i a}\int_{0}^{2\pi} \frac{\epsilon^{a+1}e^{-i(a+1)\theta}}{1+\epsilon e^{-i\theta}}\,d\theta
\end{aligned} \tag{6.7.18}
$$

Just as in the previous example, due to the range on a $(-1 < a < 0)$, we can easily show that the second and last integrals above approach zero as $R \to \infty$ and $\epsilon \to 0$, respectively. Also, the third integral with $x = -t$ is again proportional to the first, and in the limit we find

$$\int_{0}^{\infty} \frac{x^a}{1+x}\,dx = \frac{2\pi i e^{\pi i a}}{1-e^{2\pi i a}} = \frac{-\pi}{\sin(\pi a)} \qquad -1 < a < 0$$

■

In Problem 9, the reader is asked to prove the following general result.

(6.7.19) ***Theorem.*** Let $p(x)$ and $q(x)$ be real polynomials such that

(i) $q(x)$ has no *positive zeros.*

(ii) $p(0) \neq 0$ and $q(0) \neq 0$.

(iii) degree (q) − degree $(p) > 0$.

Then, for $-1 < a < [\deg(q) - \deg(p) - 1]$,

$$\int_{0}^{\infty} \frac{x^a p(x)}{q(x)}\,dx = \frac{2\pi i}{1-e^{2\pi i a}} \sum \operatorname{Res}\left[\frac{z^a p(z)}{q(z)}\right]$$

where the sum is taken over *all the poles* of $p(z)/q(z)$, and the branch of z^a is given as in Figure 6.7.3*a* by

$$z^a = |z|^a e^{ia \arg(z)} \qquad 0 < \arg(z) \le 2\pi$$

■

The next example shows that we can deal with integrals involving logarithms in much the same way as in the previous cases.

(6.7.20) ***Example.*** Evaluate $\int_0^\infty [\ln x/(x^2 + a^2)]\, dx$ for $a > 0$.

Solution. Here, we choose the branch of the logarithm to be

$$\log z = \ln|z| + i\arg(z) \qquad 0 < \arg(z) \leq 2\pi$$

and consider the closed contour Γ as depicted in Figure 6.7.4. (The reader, at the end of this example, is asked to show that if Γ were chosen as in

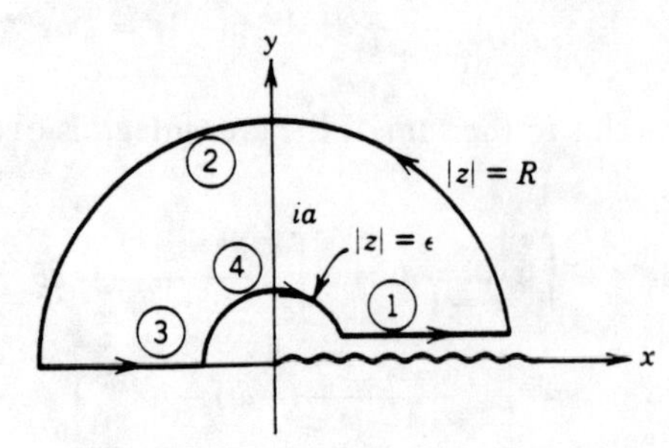

Figure 6.7.4 Contour Γ for Example (6.7.20).

Figure 6.7.3*b*, the procedure would not work.) Now, the four pieces of Γ can be parametrized by

$$\begin{array}{llll} ① & z = x & \epsilon \leq x \leq R & (\arg(z) = 0) \\ ② & z = Re^{i\theta} & 0 \leq \theta \leq \pi & (\arg(z) = \theta) \\ ③ & z = -xe^{\pi i} & -R \leq x \leq -\epsilon & (\arg(z) = \pi) \\ ④ & z = \epsilon e^{i(\pi - \theta)} & 0 \leq \theta \leq \pi & (\arg(z) = \pi - \theta) \end{array}$$

Since for our branch, $\log(ia) = \ln a + \pi i/2$, the residue theorem yields

$$\oint_\Gamma \frac{\log z}{z^2 + a^2}\, dz = \frac{\pi}{a}\left(\ln a + \frac{\pi i}{2}\right)$$

Setting this equal to the sum of the integrals over the four pieces of Γ leads to

$$\frac{\pi}{a}\left(\ln a + \frac{\pi i}{2}\right) = \int_\epsilon^R \frac{\ln x}{x^2 + a^2}\, dx + \int_0^\pi \frac{(\ln R + i\theta)iRe^{i\theta}}{R^2 e^{2i\theta} + a^2}\, d\theta$$

$$+ \int_{-R}^{-\epsilon} \frac{[\ln|x| + \pi i]}{x^2 + a^2}\, dx + \int_0^\pi \frac{[\ln \epsilon + i(\pi - \theta)]i\epsilon e^{i\theta}}{\epsilon^2 e^{-2i\theta} + a^2}\, d\theta$$

In the usual manner, we can show that the second and fourth integrals approach zero as $R \to \infty$ and $\epsilon \to 0$, respectively. (The occurrences of the singular terms $\ln R$ and $\ln \epsilon$ do not change this result, since they are very weakly singular.) Note, however, the *extra term* πi in the third integral. Now, taking the limit, we get

(6.7.21)
$$2\int_0^\infty \frac{\ln x}{x^2+a^2}\,dx + \pi i \int_0^\infty \frac{dx}{x^2+a^2} = \frac{\pi}{a}\left[\ln a + \frac{\pi i}{2}\right]$$

If we equate real and imaginary parts, we will not only get a value for the integral we want, namely,

$$\int_0^\infty \frac{\ln x}{x^2+a^2}\,dx = \frac{\pi \ln a}{2a} \qquad a > 0$$

but as an added bonus, we get the well-known result

$$\int_0^\infty \frac{dx}{x^2+a^2} = \frac{\pi}{2a}$$

■

The last example is a special case of the following.

(6.7.22) ***Theorem.*** Let $p(x)$ and $q(x)$ be real polynomials such that

(i) $p(x)$ and $q(x)$ are *even.*

(ii) $q(x)$ has *no real zeros.*

(iii) degree (q) − degree $(p) \geq 2$.

Then,

$$\int_0^\infty \frac{p(x)\ln x}{q(x)}\,dx = -\pi \operatorname{Im}\left\{\sum \operatorname{Res}\left[\frac{p(z)\log z}{q(z)}\right]\right\}$$

where the sum is taken over those poles of $p(z)/q(z)$ *which lie in the upper half-plane*, and the branch of $\log z$ is chosen to be

$$\log z = \ln|z| + i\arg(z) \qquad 0 < \arg(z) \leq 2\pi$$

■

Note that the above result cannot be used to evaluate an integral such as $\int_0^\infty [\ln x/x^3 + 1)]\,dx$, since the denominator has a real (negative) zero, and this violates condition (ii) of the theorem. On might expect, however, to be able to proceed as in Example (6.7.15) where a similar situation arose. There, we used the contour and branch depicted in Figure 6.7.3. You are urged to convince yourself that *this will not work*, since, as it turns out, the logarithmic term arising from the top and bottom of the cut (curves ① and ③ in Figure 6.7.3*b*) will cancel, and you will be left with just an integral of a rational function. To be able to evaluate an integral such as $\int_0^\infty [\ln x/(1 + x^3)]\,dx$, where the integrand may have poles on the negative real axis, it turns out that one should first consider $\int_0^\infty [(\ln x)^2/(x^3 + 1)]\,dx$. The general result is given below and in Problem 16*b*.

(6.7.23) ***Theorem.*** Let $p(x)$ and $q(x)$ be real polynomials such that

(i) $q(x)$ has no *positive zeros.*

(ii) degree (q) − degree $(p) \geq 2$.

Then,

$$\int_0^\infty \frac{p(x)\ln x}{q(x)}\,dx = -\frac{1}{2}\mathrm{Re}\left[\sum \mathrm{Res}\left(\frac{p(z)\log^2 z}{q(z)}\right)\right]$$

and

$$\int_0^\infty \frac{p(x)}{q(x)}\,dx = -\frac{1}{2\pi}\,\mathrm{Im}\left[\sum \mathrm{Res}\left(\frac{p(z)\log^2 z}{q(z)}\right)\right]$$

where the sum is taken over *all* the poles of $p(z)/q(z)$ and the branch of the logarithm is given by

$$\log z = \ln|z| + i\arg(z) \qquad 0 < \arg(z) \leq 2\pi$$

■

Note in the above that as a by-product we arrive at the value of the integral of rational functions such as $\int_0^\infty dx/(x^3 + x + 1)$. Such integrals could not be evaluated using any of the procedures previously developed, either in Section 5.4 or earlier in this section.

PROBLEMS (The answers are given on page 442.)

(Problems with an asterisk are more theoretical.)

1. Evaluate the integrals in Problem 1 of Section 5.4 using residues.
2. Evaluate the following integrals using residues:

(a) $\displaystyle\int_0^{2\pi} \frac{\cos^3(2x)}{5 + 4\cos(2x)}\,dx$ (compare with the use of the Cauchy integral formula)

(b) $\displaystyle\int_0^{2\pi} e^{\cos x}\cos(2x - \sin x)\,dx$

(c) $\displaystyle\int_0^{\pi} \frac{\cos(nx)\cos x}{1 - 2a\cos x + a^2}\,dx \qquad 0 < a < 1 \qquad n = 1, 2, \ldots$

(d) $\displaystyle\int_0^{\pi} \frac{[\sin(nx) - \alpha\sin(n-1)x]\sin(mx)}{1 - 2\alpha\cos x + \alpha^2}\,dx \qquad 0 < \alpha < 1$

3.* Prove Theorem (6.7.10). [*Hint:* Consider $\oint_{\Gamma_R} [p(z)/q(z)]\,e^{iaz}\,dz$, where Γ_R consists of the interval $-R \leq x \leq R$ and the semicircle C_R: $z = Re^{it}$, $0 \leq t \leq \pi$. Use Jordan's lemma (5.4.37) to show that

$$\lim_{R\to\infty} \left[\int_{C_R} (pe^{iaz}/q)\,dz\right] = 0]$$

4. Evaluate the integrals in Problems 7 and 10 of Section 5.4 using residues.

5. Evaluate the following integrals using residues:

(a) $\displaystyle\int_{-\infty}^{\infty} \frac{(a+bx)\cos(\alpha x)}{A+2Bx+x^2}\,dx \qquad A > B^2, \qquad \alpha > 0$

(b) $\displaystyle\int_{0}^{\infty} \frac{x^3\sin(ax)\,dx}{x^4+2b^2x^2\cos(2\lambda)+b^4} \qquad a > 0, \qquad b > 0, \qquad 0 < \lambda < \pi/2$

(c) $\displaystyle\int_{0}^{\infty} \frac{\sin(ax)\sin(bx)\,dx}{x^2+\alpha^2} \qquad 0 < a < b, \qquad \alpha > 0$

(d) $\displaystyle\int_{-\infty}^{\infty} \frac{x\cos(ax)\sin(bx)\,dx}{x^2+\alpha^2} \qquad 0 < a < b, \qquad \alpha > 0$

6. Consider the integral $\int_{-\infty}^{\infty} e^{ax}\,dx/(1+e^{bx})$, $0 < a < b$ (these inequalities are needed to ensure the existence of the integral). Note that the integrand has an infinite number of simple poles at $z_n = (i\pi/b)(2n+1)$, $n = 0, \pm 1, \ldots$.

(a) A first attempt at evaluation might be to consider a closed contour Γ_n consisting of the real interval $-R_n \le x \le R_n$ and the semicircle C_n: $z = R_n e^{i\theta}$, $0 \le \theta \le \pi$, where $R_n = 2\pi n/b$ (this avoids the poles of the integrand). We would then take the limit $R_n \to \infty$ and thus pick up the contributions from all of the poles. For this to work, we must have that the integral over the semicircle goes to zero with R_n. Show that in fact

$$\int_{C_n} \frac{e^{az}\,dz}{1+e^{bz}} \underset{n\to\infty}{\not\longrightarrow} 0$$

[*Hint:* Consider $\oint_{\Gamma_n} [e^{az}/(1+e^{bz})]\,dz$. Evaluate the residues at the z_n and then take the limit as $n \to \infty$. If the integral over the semicircle *did* approach zero, then you will have shown that a divergent series converges.]

(b) Evaluate the integral using the contour in Figure 6.7.5*a*. [*Hint:* Show that the integrals of $e^{az}/(1+e^{bz})$ over the two vertical lines approach

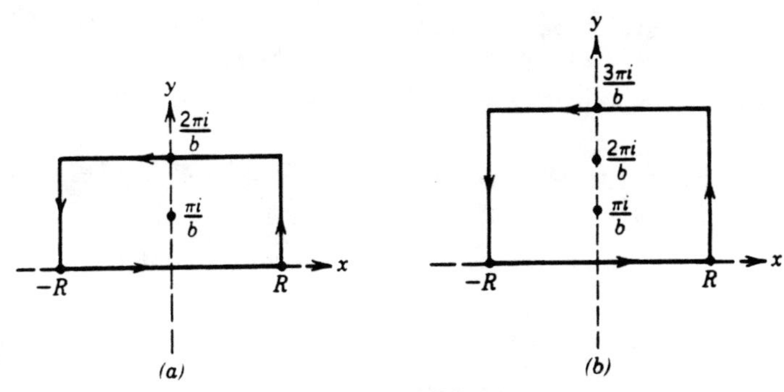

Figure 6.7.5

zero as $R \to \infty$, while the integrals over the two horizontal lines are proportional.]

(c) Evaluate the integral using the contour shown in Figure 6.7.5*b*. Can you generalize this?

7. Some real integrals can be evaluated using **indented contours** [see Example (5.4.40) of Section 5.4]. This problem shows that there is a relationship between the limit of a contour indented around a singularity and the residue there. Thus, consider the contour C_ϵ of Figure 6.7.6 which consists of an arc of a circle of radius ϵ with center at $z = z_0$.

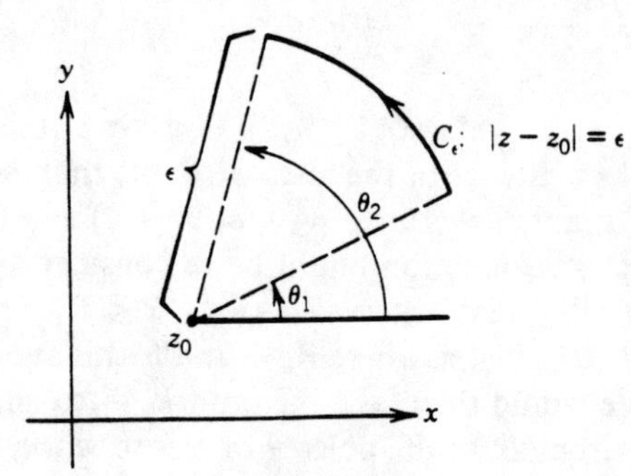

Figure 6.7.6

(a) Prove that if $f(z)$ has a *simple pole* at $z = z_0$, then

$$\lim_{\epsilon \to 0} \left[\int_{C_\epsilon} f(z)\, dz\right] = i(\theta_2 - \theta_1)\text{Res}[f]_{z=z_0}$$

[*Hint:* For ϵ small and z on C_ϵ, $f(z)$ can be represented by its Laurent series about $z = z_0$, where the series of nonnegative powers represents an analytic, and *hence bounded*, function $g(z)$. Show that $\lim_{\epsilon \to 0} [\int_{C_\epsilon} g(z)\, dz] = 0$].

(b) What is the limit in part (*a*) if $f(z)$ does not have a nonsimple pole at $z = z_0$?

8. (a) In Example (5.4.40) of Section 5.4, we showed how indented contours can be used to evaluate $\int_{-\infty}^{\infty} [(\sin x)/x]\, dx$. Reread this example and evaluate this same integral using the result of Problem 7*a*.

(b) Evaluate the following integrals.

(i) $\displaystyle\int_{-\infty}^{\infty} \frac{\sin^2 x}{x^2}\, dx$ [*Hint:* $\sin^2 x = -\frac{1}{2}\text{Re}(e^{2ix} - 1)$.]

(ii) $\displaystyle\int_{-\infty}^{\infty} \frac{(x - \sin x)}{x^3}\, dx$ [*Hint:* $x - \sin x = \text{Im}(ix - e^{ix} + 1)$. Why do you think the "1" is thrown in?]

(iii) $\displaystyle\int_{-\infty}^{\infty} \frac{[\sin(ax) - ax\cos(ax)]}{x^3}\, dx, \quad a > 0$ [*Hint:* See (ii).]

9.* Prove Theorem (6.7.19). [*Hint:* Consider the contour Γ of Figure 6.7.3*b*, and show that the conditions on a in the theorem ensure that the integrals over the two circles approach zero as $\epsilon \to 0$ and $R \to \infty$.]

10. Evaluate the following integrals.

(a) $\displaystyle\int_0^\infty \frac{x^a}{1 + x^{2n}}\,dx \qquad n = 1, 2, \ldots \text{ and } -1 < a < 2n - 1$

(b) $\displaystyle\int_0^\infty \frac{x^a}{(x + b)^n}\,dx \qquad b > 0, \qquad n = 1, 2, \ldots, \qquad \text{and } -1 < a < n - 1$

(c) $\displaystyle\int_0^\infty \frac{\sqrt{x}}{2x^3 - x + 1}\,dx$

(d) $\displaystyle\int_{-\infty}^\infty \frac{e^{ax}}{(1 + e^{bx})^n}\,dx \qquad 0 < a < nb, \qquad n = 1, 2, \ldots$

11. Consider the integral $I = \int_0^\infty [(x^a - 1)/(x^3 - 1)]\,dx$, $-1 < a < 2$.

(a) Use the branch $z^a = |z|^a e^{ia\theta}$, $-\pi/2 < \theta < 3\pi/2$, and the indented contour of Figure 6.7.7*a*, to show that this integral can be written in terms of the integral $\int_0^\infty (x^a e^{\pi i a} - 1)/(x^3 + 1)]\,dx$. You might find Problem 7 helpful.

(b) Evaluate the second integral in part (*a*) and find the value of I.

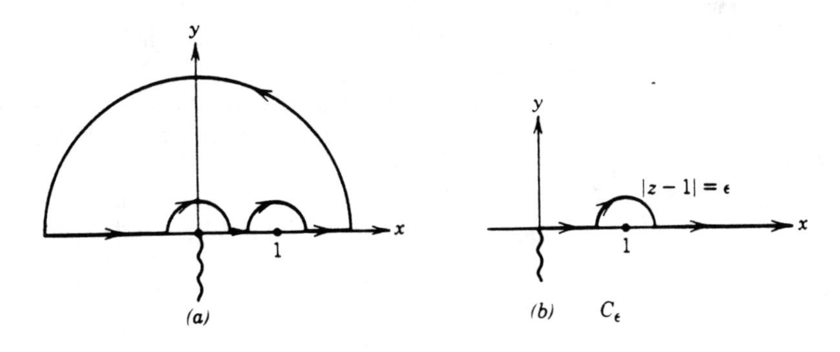

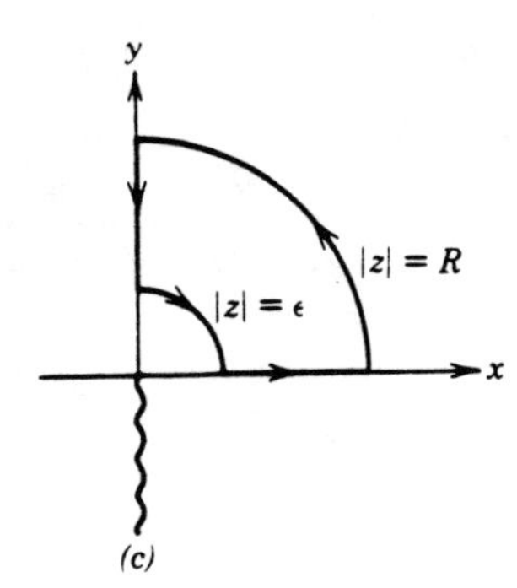

Figure 6.7.7

12. Consider the integral $I = \int_0^\infty [(x^a - 1)/(x^2 - 1)]\,dx$, $-1 < a < 1$.

(a) With the branch $z^a = |z|^a e^{ia\theta}$, $-\pi/2 < \theta < 3\pi/2$, show that

$$I = \lim_{\epsilon \to 0}\left[\int_{C_\epsilon} \frac{z^a - 1}{z^2 - 1}\,dz\right]$$

where C_ϵ is depicted in Figure 6.7.7*b*.

(b) Show that the contour C_ϵ in part (a) can be "deformed" onto the imaginary axis and that $I = \oint [(z^a - 1)/(z^2 - 1)]\,dz$, where C is the positive imaginary axis traversed from $z = 0$ to $z = \infty$.

(c) Parametrize the imaginary axis in a manner consistent with the branch of z^a and show that

$$I = \sin\left(\frac{\pi a}{2}\right)\int_0^\infty \frac{x^a}{x^2 + 1}\,dx$$

(d) Finally, evaluate I.

13. If $p(x)$ and $q(x)$ are real polynomials satisfying the conditions of Theorem (6.7.19), and if $-1 < \alpha < \deg(q) - \deg(p) - 1$, evaluate the integrals $\int_0^\infty [x^\alpha p(x)\cos(\beta \ln x)/q(x)]\,dx$ and $\int_0^\infty [x^\alpha p(x)\sin(\beta \ln x)/q(x)]\,dx$. [*Hint:* In Theorem (6.7.19), let $a = \alpha + i\beta$.]

14. (a) If $0 < a < 1$, evaluate the integrals $\int_0^\infty (\sin x/x^a)\,dx$ and $\int_0^\infty (\cos x/x^a)\,dx$. Express your answers in terms of $\int_0^\infty x^{-a}e^{-x}\,dx$. This integral is the **gamma function** $\Gamma(1 - a)$. [*Hint:* Consider the contour in Figure 6.7.7*c*.]

(b) Use the result of part (a) to evaluate $\int_0^\infty (\sin x/x^a)\,dx$, where $1 < a < 2$.

15. Consider $\int_{-1}^{1} \sqrt{[(1 - x)/(1 + x)]}\,dx$.

(a) Show that with the principal branch of the square root, the function $[(z - 1)/(z + 1)]^{1/2}$ is analytic everywhere except on the line joining $z = -1$ to $z = 1$.

(b) Show that $\int_C [(z - 1)/(z + 1)]^{1/2}\,dz \to -2i\int_{-1}^{1}\sqrt{(1 - x)/(1 + x)}\,dx$ in the limit as the radii of the small circles in Figure 6.7.8*a* tend to zero. Here,

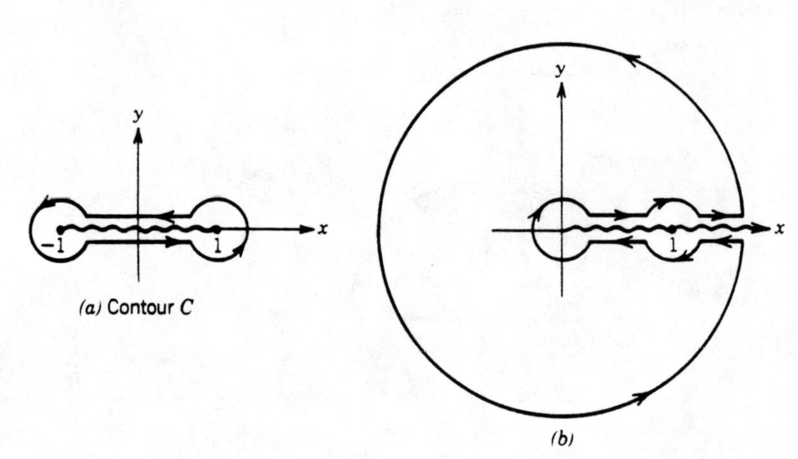

Figure 6.7.8

the principal branch of the square root is used, and C is the contour in the figure. [*Hint:* What are the principal arguments of $(z-1)/(z+1)$ above and below the cut?]

(c) Evaluate the integral on the left in part (a) using the **residue at infinity** (see Problem 8 of Section 6.6) and show that $\int_{-1}^{1} \sqrt{(1-x)/(1+x)}\,dx = \pi$.

16.* **(a)** Use the contour of Figure 6.7.4 to prove Theorem (6.7.22).

(b) Use the contour of Figure 6.7.3*b* to derive the results of Theorem (6.7.23). [*Hint:* Consider $\int [p(z)/q(z)](\log^2 z)\,dz$.]

17. Evaluate the following integrals:

(a) $\displaystyle\int_0^\infty \frac{(\ln x)\,dx}{(x^2+a^2)(x^2+1)}$ (b) $\displaystyle\int_0^\infty \frac{\ln x}{(x+a)^2}\,dx \qquad a > 0$

(c) $\displaystyle\int_0^\infty \frac{dx}{(x+2)^2(x^2+4x+3)}$

18. (a) Use the approach of Problem 16b to show that if $p(x)$ and $q(x)$ satisfy the conditions of Theorem (6.7.23), then $\int_0^\infty [p(x)(\ln x)^n/q(x)]\,dx$ $(n = 1, 2, \ldots)$ can be written in terms of the integrals $\int_0^\infty [p(x)(\ln x)^k/q(x)]\,dx$, where $k = 0, 1, \ldots, n-1$. [*Hint:* Integrate $p(z)(\log z)^{n+1}/q(z)$ over the contour in Figure 6.7.3*b*.]

(b) Use the approach in part (a) to evaluate $\int_0^\infty [(\ln x)^2/(1+x^2)]\,dx$

19. Evaluate $\int_0^\infty [\sqrt{x}\ln x/(x^2+1)]\,dx$.

20. Use the contour shown in Figure 6.7.8*b* to evaluate $\int_0^\infty [\ln x/(x^2-1)]\,dx$.

6.8 SOME CONSEQUENCES OF TAYLOR AND LAURENT SERIES

Here, as in Section 5.6, we will present some (mostly theoretical) results which exhibit some more of the remarkable properties of analytic functions. We begin by proving an interesting theorem known as the Schwarz lemma.

(6.8.1) ***Theorem*** **(Schwarz Lemma).** Let $f(z)$ be analytic in the unit disk $|z| < 1$ and satisfy $|f(z)| \le M$ there. Then, if $f(0) = 0$,

(a) $|f(z)| \le M|z| \qquad$ for $|z| < 1$.

(b) If there is a $\hat{z}$ in the unit disk such that $|f(\hat{z})| = M|\hat{z}|$, then $f(z) = kz$, with $|k| = M$.

Proof. Since $f(0) = 0$, we have from Theorem (6.3.19) that $f(z) = z\,g(z)$, where $g(z)$ is analytic in $|z| < 1$. From the bound on f, we have that $|g(z)| \le M/r$ for z on the circle $|z| = r$, and this holds for any $r < 1$. Thus, M/r is the maximum of $|g(z)|$ on the circle $|z| = r$, and by the maximum principle (5.6.14b), we have

$$|g(z)| \le M/r \qquad |z| \le r$$

If we now take the limit $r \to 1$, we get $|g(z)| \le M$ for $|z| < 1$. This proves (a).

If $|f(\hat{z})| = M|\hat{z}|$, then $|g(\hat{z})| = M$ at a point $\hat{z}$ which is interior to the unit disk. Hence, by the maximum principle (5.6.14a), we must have that $g(z) = k$, with the constant k satisfying $|k| = M$. Thus, we have proved (b).

■

(6.8.2) ***Example.*** Let $f(z)$ be analytic with an analytic inverse inside the unit disk and map the unit disk onto itself. If the origin is mapped into the origin $[f(0) = 0]$, show that the mapping $w = f(z)$ must be a rotation.

Solution. Let the mapping and its inverse be defined by

$$w = f(z) \qquad \text{and} \qquad z = g(w)$$

The functions f and g both map the unit disk onto itself and vanish at the origin, that is,

$$|f(z)| \leq 1 \qquad \text{for } |z| \leq 1 \qquad \text{and} \qquad f(0) = 0$$

$$|g(w)| \leq 1 \qquad \text{for } |w| \leq 1 \qquad \text{and} \qquad g(0) = 0$$

Hence, by the Schwarz lemma applied to both $f(z)$ and $g(w)$, we get (with $M = 1$)

$$|f(z)| \leq |z| \qquad \text{and} \qquad |g(w)| \leq |w|$$

Since $z = g(w)$ and $w = f(z)$, these lead to $|f| \leq |g|$ and $|g| \leq |f|$. Thus,

$$|f(z)| = |g(w)| = |z|$$

and hence $|f(z)| = |z|$ for all $|z| < 1$. Thus, by part (b) of the Schwarz lemma, we have $f(z) = kz$, with $|k| = 1$. This is just a rotation.

■

Argument Principle and Rouché's Theorem. Here, we will prove a very interesting result concerning the zeros and poles of an analytic function and show how it can be applied to some practical and theoretical problems.

(6.8.3) ***Theorem*** **(Argument Principle).** Let D be a simply connected domain and $f(z)$ (*not identically zero*) be analytic in D except possibly for a finite number of poles. Let C be a simple, closed contour in D not passing through any of the zeros or poles of $f(z)$. Define

Z = number of zeros of f inside C

P = number of poles of f inside C

$\Delta \arg(f)|_C$ = change in $\arg[f(z)]$ as C is traversed counterclockwise

In these definitions, we count the zeros and poles each according to its multiplicity. [For example, $z^3/(1 - z)^2$ has $Z = 3$ and $P = 2$ if C is a circle of radius 2.] Then,

$$Z - P = \frac{1}{2\pi} \Delta \arg(f)|_C = \frac{1}{2\pi i} \oint_C \frac{f'(z)}{f(z)} dz \tag{6.8.4}$$

Proof. From Problem 7*a* in Section 6.6, we learn that

$$Z - P = \frac{1}{2\pi i} \oint_C \frac{f'(z)}{f(z)}\, dz$$

which is half of our result. (The reader is urged to work out this problem.) Now let C_w be the image of C under the mapping $w = f(z)$. Note that C_w is also a closed contour, but it *need not be simple* since $f(z)$ need not be 1–1 on C. If we make the change of variables $w = f(z)$ in the above integral, then $dw = f'(z)\, dz$, and we get

$$Z - P = \frac{1}{2\pi i} \oint_{C_w} \frac{dw}{w} = N$$

where N is the number of times C_w encircles the origin $w = 0$. The theorem follows from the fact that

$$N = \frac{1}{2\pi} [\text{change in } \arg(w)]$$

$$= \frac{1}{2\pi} \Delta \arg(f)|_C$$

∎

(6.8.5) ***Example.*** Let $f(z) = z^2$ and C be the unit circle. The number of zeros of f (there are no poles) inside C is 2, while the image of the unit circle under the mapping $w = z^2$ is again the unit circle, *but traversed twice*. Hence, the change in argument of w as the unit circle is traversed is 4π, and $(1/2\pi)(4\pi) = 2$, thus verifying the theorem.

∎

Very often in the investigation of the **stability of dynamical systems**, it is important to determine when certain rational functions have *poles* in the right half-plane. This in turn depends upon whether a polynomial has *zeros* in the right half-plane. The next example shows how the argument principle can be used in such a determination. More general results of this nature fall under the category of **Nyquist criteria**.

(6.8.6) ***Example.*** How many zeros does the polynomial $p(z) = z^3 - z^2 + z + 1$ have in the right half-plane?

Solution. Consider the contour C_R shown in Figure 6.8.1*a*. Since $p(z)$ has no poles, by the argument principle,

$$Z = \frac{1}{2\pi} \Delta \arg[p(z)]|_{C_R}$$

where Z is the number of zeros of p inside C_R. Now, if R is large enough, the above will yield all of the zeros of p in the right half-plane. For z on the

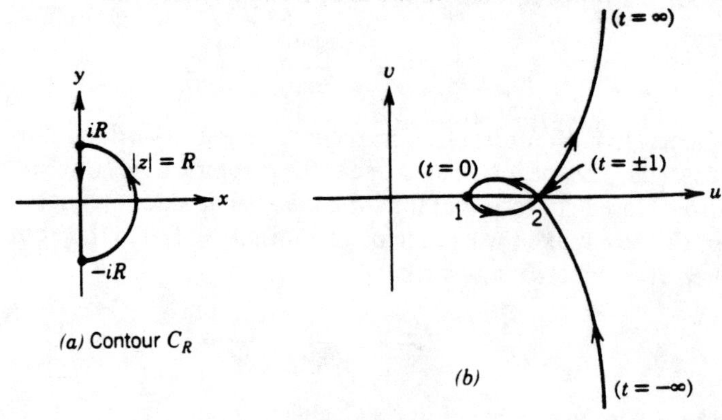

Figure 6.8.1

semicircle, we have $z = Re^{i\theta}$, $-\pi/2 \leq \theta \leq \pi/2$, and $p(z) = R^3e^{3i\theta} - R^2e^{2i\theta} + Re^{i\theta} + 1$, and for R large, $p(Re^{i\theta}) \approx R^3e^{3i\theta}$, and hence $\arg(p) \approx 3\theta$. Now, θ changes by $\pi/2 - (-\pi/2) = \pi$, and thus,

(6.8.7) $$\Delta \arg(p)\big|_{|z|=R} = 3\pi \qquad \text{(on the semicircle)}$$

Now, on the imaginary axis, $z = -it$, $-R \leq t \leq R$, and hence $p(z) = it^3 + t^2 - it + 1$. Thus, as the whole imaginary axis in the z-plane is traversed (let $R \to \infty$), the real and imaginary parts of $p(z)$ traverse the curve

(6.8.8) $$u = 1 + t^2, \qquad v = t^3 - t \qquad -\infty < t < \infty$$

The objective is to plot this curve in the uv-plane and to compute the argument change. Note from the above that

(a) $u \geq 1$.
(b) $v(-\infty) = -\infty$, $v(+\infty) = +\infty$, $u(\pm\infty) = +\infty$.
(c) $dv/du = (3t^2 - 1)/2t \xrightarrow[t \to \pm\infty]{} \pm\infty$. Thus, the curve is vertical at the extremities.
(d) For $t = -1, 0, 1$, we have $v = 0$ and $u = 2, 1, 2$, respectively.

With this information, we can plot the curve. It is shown in Figure 6.8.1*b*. From this, we find

$$\Delta \arg[p(z)] = \pi \qquad \text{(on the imaginary axis)}$$

Combining this with (6.8.7), we get

$$Z = \frac{1}{2\pi}(3\pi + \pi) = 2$$

Thus there are exactly two roots in the right half-plane.

■

(6.8.9) ***Theorem* (Rouché).** Let $f(z)$ and $g(z)$ be analytic in a domain D, and let C be a simple, closed contour in D not passing through any of the zeros of either f or $f + g$. Assume

$$|f(z)| > |g(z)| \qquad \text{for } z \text{ on } C$$

Then, *$f(z)$ and $f(z) + g(z)$ have the same number of zeros* (including multiplicities) inside C.

Proof. From the argument principle, since $f + g$ has no poles,

$$\frac{1}{2\pi i}\oint_C \frac{(f' + g')}{f + g}\,dz = \text{no. of zeros of } f + g \text{ in } C$$

Now,

$$\begin{aligned}\frac{f' + g'}{f + g} &= \frac{f'}{f} + \left[\frac{f' + g'}{f + g} - \frac{f'}{f}\right]\\ &= \frac{f'}{f} + \frac{fg' - gf'}{f(f + g)}\\ &= \frac{f'}{f} + \frac{(1 + g/f)'}{1 + g/f}\end{aligned}$$

Hence, since $(1/2\pi i)\oint_C (f'/f)\,dz$ = no. of zeros of f in C, we get

$$\text{no. of zeros of } f + g = (\text{no. of zeros of } f) + \frac{1}{2\pi i}\oint_C \frac{(1 + g/f)'}{1 + g/f}\,dz$$

Let us now show that the integral above is zero. Since we are assuming that $|f| > |g|$ on C, we get

$$\left|\left(1 + \frac{g}{f}\right) - 1\right| = \left|\frac{g}{f}\right| < 1$$

hence, for z on C, the quantity $1 + (g/f)$ lies inside the circle centered at 1 and of radius 1. Since this circle never loops around the origin, there is no change in the argument of $1 + (g/f)$, and thus,

$$\frac{1}{2\pi i}\oint_C \frac{(1 + g/f)'}{1 + g/f}\,dz = 0$$

■

(6.8.10) ***Example.*** Here, we will show how Rouché's Theorem can easily be used to prove the fundamental theorem of algebra that *every polynomial of degree n has n complex roots*. Let $p(z) = z^n + a_{n-1}z^{n-1} + \cdots + a_1 z + a_0$, and define $f(z) = z^n$ and $g(z) = a_{n-1}z^{n-1} + \cdots + a_1 z + a_0$. Note that $p(z) = f(z) + g(z)$. Also,

$$\lim_{z\to\infty}\left|\frac{g(z)}{f(z)}\right| = 0 \tag{6.8.11}$$

hence, *for z large enough*, $|g/f| < 1$. Thus,

$$|f(z)| > |g(z)| \quad \text{for } |z| = R \quad \text{and} \quad R \text{ large}$$

Thus, by Rouché, f and $f + g(=p)$ have the same number of zeros in $|z| < R$. But $f(z) = z^n$ has exactly n zeros at $z = 0$; and so, since R is arbitrary, $p(z)$ has n zeros in the plane.

■

Oftentimes, it is not enough just to know how many zeros a function possesses—we would also like to have some idea as to where they are located. This sometimes requires that we split the function into the sum of two parts so that we can use Rouché's Theorem.

(6.8.12) ***Example.*** Show that the three roots of

$$p(z) = z^3 + z + 1$$

all lie in the annulus $\hat{R}_0 \le |z| \le R_0$, where $\hat{R}_0$ and R_0 are the unique real roots of

$$\hat{R}_0^3 - \hat{R} = 1 \quad \text{and} \quad R_0^3 + R_0 = 1$$

Solution. We will apply Rouché's Theorem twice. First, let

$$f_1(z) = z^3 \quad \text{and} \quad g_1(z) = z + 1$$

On the circle $|z| = R$, we have

$$|f_1| = R^3 \quad \text{and} \quad |g_1| \le R + 1$$

We now ask for those R for which $R^3 > R + 1$, thus ensuring that $|f_1| > |g_1|$. The cubic function

$$h_1(R) = R^3 - R - 1$$

is plotted in Figure 6.8.2*a*. Note from h_1 that $R^3 > R + 1$ when $h_1(R) > 0$, and this is guaranteed when $R > R_0$, where R_0 is the real positive root of $h_1(R)$. Since $f_1(z) = z^3$ has three roots inside the circle $|z| = R$ and the conditions of Rouché's Theorem are satisfied on this circle if $R > R_0$, we can conclude that $f_1 + g_1$ also has three roots inside, that is,

(6.8.13) $$p(z) \text{ has three roots in } |z| \le R_0$$

Now, define

$$f_2(z) = z + 1 \quad \text{and} \quad g_2(z) = z^3$$

Then, on the circle $|z| = R$,

(6.8.14) $$|f_2| \ge |R - 1| \quad \text{and} \quad |g_2| = R^3$$

We now ask when $|R - 1| > R^3$, for this ensures that $|f_2| > |g_2|$. Here, however, we must consider the two cases $R \ge 1$ and $R < 1$. For $R \ge 1$,

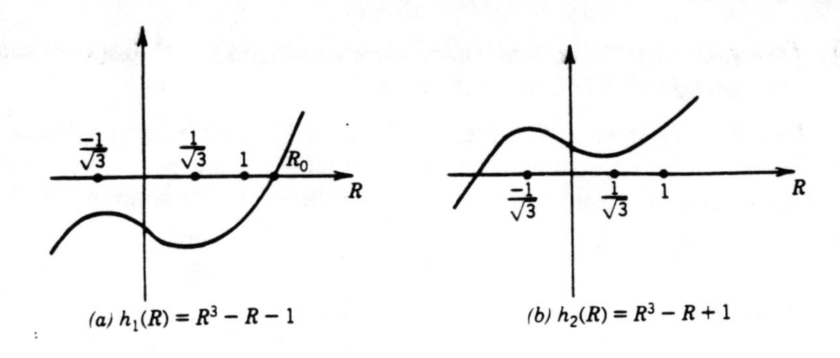

(a) $h_1(R) = R^3 - R - 1$ (b) $h_2(R) = R^3 - R + 1$

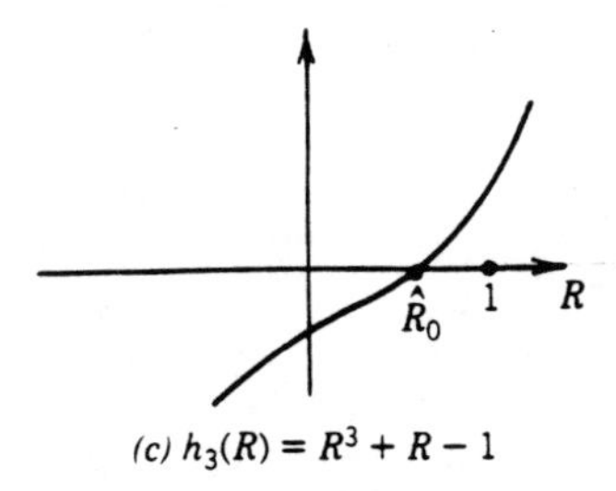

(c) $h_3(R) = R^3 + R - 1$

Figure 6.8.2

$|R - 1| = R - 1$, and our condition is that

$$h_2(R) = R^3 - R + 1 < 0 \qquad R \geq 1$$

In Figure 6.8.2*b*, we plot this cubic $h_2(R)$. It is clear that for $R \geq 1$, $h_2(R)$ is never negative. Thus, Rouché's Theorem is not applicable.

For $R < 1$, $|R - 1| = 1 - R$, and the condition $|f_2| > |g_2|$ yields, from (6.8.14),

$$h_3(R) = R^3 + R - 1 < 0 \qquad R < 1$$

The cubic h_3 is shown in Figure 6.8.2*c*, and it is clear that $h_3(R) < 0$ for $0 \leq R < \hat{R}_0$. Thus, f_2 and $p = f_2 + g_2$ have the same number of roots in $|z| < \hat{R}_0$. However, $f_2 = z + 1$ has only one root at $z = -1$, which is outside of this disk. Thus, $p(z)$ has no roots for $|z| < \hat{R}_0$. This, combined with (6.8.13), completes our result.

■

Analytic Continuation. In Chapter 2, the prime motivation used to define an elementary function such as e^z was that the resulting complex function should reduce to the usual real function e^x when z is real and be analytic in some neighborhood of the real axis. Here, we will show that such an analytic continuation of e^x from the real axis into the complex plane, if there is one, is unique. But first we must prove some results concerned with the vanishing of an analytic function.

(6.8.15) ***Theorem.*** Let $F(z)$ be analytic in a domain D and vanish inside of some disk contained in D. Then, $F(z) \equiv 0$ in D.

Proof. Let D be as depicted in Figure 6.8.3, with z_0 designating the center of the disk (shown shaded) in which $F(z)$ vanishes. If we assume that F does not vanish in all of D, then there must be some point z^* at which

$$F(z^*) \neq 0$$

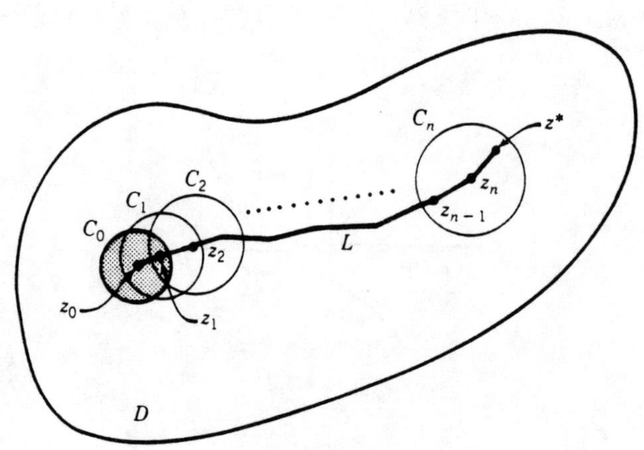

Figure 6.8.3

We will now show this to be impossible and hence conclude that $F(z) \equiv 0$ in D.

Since D is a domain, and hence connected, we can join z_0 to z^* by a broken line segment L lying entirely in D. Now, a theorem in real analysis tells us that there is a minimum distance δ between the points on L and the boundary of D. We now construct a sequence of points $z_1, z_2, \ldots, z_n$ on L so that the distance between z_k and z_{k+1}, for each $k = 1, 2, \ldots, n-1$, is

(6.8.16)
(a) $\delta/2$, if the radius of the shaded disk is greater than δ.
(b) Less than the radius of the shaded disk, if this radius is less than δ.

We now construct a sequence of circles $C_1, C_2, \ldots, C_n$, each of radius δ, with centers respectively at $z_1, z_2, \ldots, z_n$. Now, $f(z)$ is analytic inside each of these circles and thus possesses a Taylor series in powers of $(z - z_k)$, for circle C_k. Consider the first circle C_1. By construction, this must contain the center z_0 of the original circle C_0 inside of which $f = 0$. Likewise, by construction, the circle C_0 must contain the center z_1 of C_1. Now, since $F(z) \equiv 0$ in C_0, all of its derivatives vanish here. Hence, since z_1 is inside C_0, $F(z)$ and all its derivatives vanish at z_1. Thus, the Taylor series for F about $z = z_1$, which we know converges inside C_1, must be zero, and hence $F(z) \equiv 0$ inside C_1. We can now apply the same argument to each of the other circles and finally conclude that $F(z^*) = 0$.

■

With the above result, the reader is asked in the Problem 12 to prove the following.

(6.8.17) ***Theorem***

(a) If $F(z)$ is analytic in a domain D and vanishes on some curve in D, then $F(z) \equiv 0$ in D.

(b) If $f(z)$ and $g(z)$ are analytic in a domain D and $f(z) = g(z)$ on a curve in D, then $f(z) \equiv g(z)$ in D.

■

We now define the concept of analytic continuation and then prove a uniqueness theorem.

(6.8.18) ***Definition*** **(Analytic Continuation).** Let $f_1(z)$ be analytic in a domain D_1 and $f_2(z)$ analytic in a domain D_2. Assume that D_1 and D_2 intersect in a domain D, as shown in Figure 6.8.4. Then, if $f_1(z) = f_2(z)$ in D, we call f_2 an analytic continuation of f_1 into the domain D_2. Likewise, f_1 is an analytic continuation of f_2 into D_1.

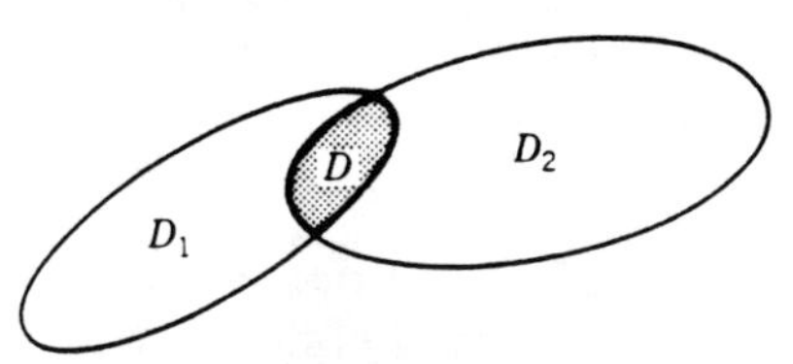

Figure 6.8.4

■

From the above, we easily see that, taken together, the function

(6.8.19)
$$F(z) = \begin{cases} f_1(z) & z \text{ in } D_1 \\ f_2(z) & z \text{ in } D_2 \end{cases}$$

is analytic in the union $D_1 \cup D_2$. The next result shows that an analytic continuation, if it exists, must be unique.

(6.8.20) ***Theorem.*** Let D_1, D_2, and D be as in Figure 6.8.4. Then, if $f_1(z)$ is analytic in D_1, there is only one function which is analytic in D_2 and which agrees with f_1 on D.

Proof. Assume there are two functions $f_2(z)$ and $f_3(z)$, both analytic in D_2 and both equal to $f_1(z)$ on D. Then, $f_2 - f_3 \equiv 0$ in D; and by Theorem (6.8.17b), $f_2 - f_3 \equiv 0$ in D_2.

■

In the Problem 14, the reader is asked to prove the following.

(6.8.21) ***Theorem.*** Let $f(x)$ be a real-valued function defined on some interval of the real axis. Then there is at most one function $f(z)$ which is analytic in a domain containing this interval and which reduces to $f(x)$ here.

■

What we learn from this result is that **a real-valued function can be continued into the complex plane as an analytic function in at most one way**. This is not to say that all real functions can be analytically continued into the complex plane.

(6.8.22) ***Example.*** The real function $f(x) = |x|$ defined on the interval $-1 < x < 1$ cannot be analytically extended. For if it could, its analytic extension would have to be differentiable at $z = 0$. However, $|x|$ is not differentiable at $x = 0$.

■

The final question to which we will briefly address ourselves pertains to how we might *construct* the analytic continuation of an analytic function into another domain. One such procedure is to **analytically continue a function by means of power series**. For a discussion of this, we refer the reader to any more advanced textbook on complex functions. Another procedure is discussed in the next theorem and extended in Problem 17.

(6.8.23) ***Theorem*** **(Schwarz Reflection Principle).** Let D be a domain part of whose boundary is an interval I of the real axis. Furthermore, each point on this interval is to have the property that if a small disk is drawn with the point as center, then half the disk lies in D and the other half outside of D (see Figure 6.8.5*a*). Let $f(z)$ be

(i) Analytic in D.
(ii) Continuous on I.
(iii) Real on I.

Then, the function

(6.8.24)
$$F(z) = \begin{cases} f(z) & z \text{ in } D \cup I \\ \overline{f(\bar{z})} & z \text{ in } \bar{D} \end{cases}$$

where $\bar{D}$ is the reflection of D across I, is analytic in the union $D \cup \bar{D} \cup I$. Hence, $\overline{f(\bar{z})}$ is the analytic continuation of $f(z)$ into $\bar{D}$.

Proof. That $F(z)$ is analytic in $\bar{D}$ follows from the Cauchy–Riemann equations. For if $f(z) = u(x, y) + i\,v(x, y)$, then

$$\overline{f(\bar{z})} = u(x, -y) - i\,v(x, -y)$$

The reader is asked to show that since $u(x, y)$ and $v(x, y)$ satisfy the Cauchy–Riemann equations, so do $u(x, -y)$ and $-v(x, -y)$. The difficult task is to show that $F(z)$ is analytic on the real interval I. To do this, we will

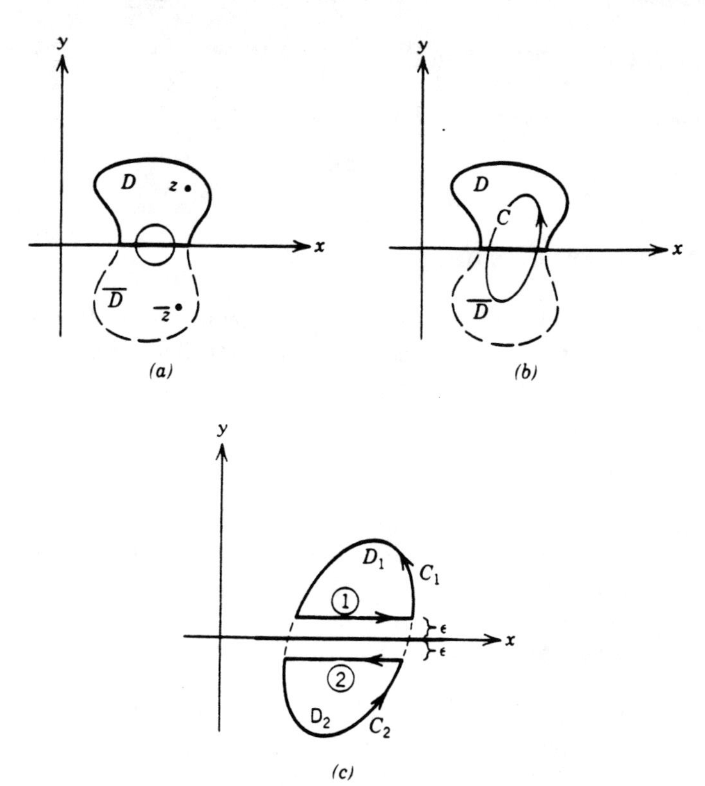

Figure 6.8.5

use Morea's Theorem (5.6.1) and show that if C is any simple closed contour in a domain which includes I, then $\oint_C F(z)\,dz = 0$. First, note from (6.8.24), and the fact that $f(z)$ is real on the interval I (that is, $\overline{f(x)} = f(x)$ for each x on I), that $F(z)$ is continuous *on* I. Now consider the closed curve C in Figure 6.8.5*b*. In part (*c*) of the figure, we show C broken up into curves C_1, C_2, and the dotted segments. We have also introduced the two lines ① and ② lying distances ϵ above and below the real axis, respectively. Now, since $F(z)$ is analytic in both domains D_1 and D_2 shown in the figure, we have, by the Cauchy–Goursat Theorem,

$$\int_{C_1} F(z)\,dz + \int_{C_2} F(z)\,dz + \int_{①} F(z)\,dz + \int_{②} F(z)\,dz = 0 \qquad \textbf{(6.8.25)}$$

and this holds for all values of ϵ. Now, as $\epsilon \to 0$, the lines ① and ② both merge onto the piece of the interval shown, however traversed in opposite directions. Hence, by the continuity of $F(z)$ on I, the last two integrals above cancel. Also in this limit, C_1 and C_2 join to form C. Hence,

$\oint_C F(z)\,dz = 0$. Since C is arbitrary, $F(z)$ is analytic, by Morea's Theorem, on the piece of the interval I contained inside of C. Since we can apply this argument to any piece of I, $F(z)$ is analytic on I.

■

PROBLEMS (The answers are given on page 444.)

1. Here, the reader is asked to extend the Schwarz lemma to the following: Let $f(z)$ be analytic in $|z| < 1$ and satisfy $|f| \le M$. If $\{z_1, z_2, \ldots, z_n\}$ are the zeros of f inside $|z| = 1$ (some may be repeated), then

$$|f(z)| \le M \prod_{k=1}^{n} \left| \frac{z - z_k}{1 - z\bar{z}_k} \right|$$

[*Hint:* Show that $f(z)/\prod_{k=1}^{n} |(z - z_k)/(1 - z\bar{z}_k)|$ can be defined at each z_k so as to be analytic in $|z| < 1$. Also show that for each k, $|(z - z_k)/(1 - z\bar{z}_k)| = 1$ for $|z| = 1$ and use this to show that $\prod_{k=1}^{n} |(z - z_k)/(1 - z\bar{z}_k)| \ge 1 - \epsilon (\epsilon > 0)$ on $|z| = 1 - \delta$, where ϵ and δ are small and $\epsilon \to 0$ as $\delta \to 0$. Now use the maximum principal to bound $|f|/\prod_{k=1}^{n} |(z - z_k)/(1 - z\bar{z}_k)|$ in $|z| < 1 - \delta$ and let $\delta \to 0$.]

2. Here, we extend the Schwarz lemma to the case when $f(0)$ does not necessarily vanish at $z = 0$. Thus, assume (i) $f(z)$ is analytic in $|z| \le 1$; (ii) $f(z)$ is not a constant function; (iii) $|f(z)| \le M$ in $|z| \le 1$.
   (a) Use the maximum principle (5.6.14) to show that $|f(0)| < M$, and hence the function $M^2 - \overline{f(0)}f(z)$ never vanishes in $|z| < 1$.
   (b) If $|w| \le 1$ and $|a| < 1$, show that $|(w - a)/(1 - \bar{a}w)| \le 1$. [*Hint:* $(w - a)/(1 - \bar{a}w)$ is a bilinear function of w.]
   (c) Use part (b) to show that $g(z) = M[f(z) - f(0)]/[M^2 - \overline{f(0)}f(z)]$ satisfies the hypotheses of the Schwarz lemma, and hence $|f(z) - f(0)| \le (1/M)|M^2 - \overline{f(0)}f(z)| \cdot |z|$.
   (d) From (c), use the triangle inequalities to show that

$$|f(z)| \le \frac{M[M|z| + |f(0)|]}{M - |z| \cdot |f(0)|} \qquad \text{for } |z| \le 1$$

3. Here, we derive the most general map of the unit disk onto itself (see Problem 14c of Section 2.5). Thus, let $w_1 = f(z)$ be the most general map of the unit disk onto itself, with $f(0) = \alpha, |\alpha| < 1$. Show that $w = (w_1 - \alpha)/(1 - \bar{\alpha}w_1)$ also maps the unit disk in the w_1-plane onto itself. What can you then say about $w = [f(z) - \alpha]/[1 - \bar{\alpha}f(z)]$. Now, use the result of Example (6.8.2) to show that $f(z) = e^{i\theta_0}(z - \beta)/(1 - \bar{\beta}z)$, where β is a complex number satisfying $|\beta| < 1$.

4. (a) Prove the following: Let $f(z)$ be analytic in the extended plane except for a finite number of poles, including possibly a pole at $z = \infty$. Then, $Z = P$, where Z and P are the *total* number of zeros and poles of f, with

multiplicities, and including those zeros and poles at $z = \infty$. [*Hint:* Take the contour C in the proof of the argument principal (6.8.3) to be a large enough circle which includes all of the finite zeros and poles inside. Then make the substitution $w = 1/z$ in the integral].

(b) Verify the result in (a) for rational functions.

5. Prove the following: Let $f(z)$ be analytic in a simply connected domain D and on its boundary C, and map C onto a piece of the real axis. Then $f(z)$ must be constant. [*Hint:* First show that $f(z)$ is real for z in D. Then, use this to conclude from the Cauchy–Riemann equations that f = constant. To show f is real on D, assume $f(z) = \alpha$, with $\text{Im}(\alpha) \neq 0$ and for some z in D. Then, show that $\text{Im}[f(z) - \alpha]$ is of one sign on C and use the argument principal to show that $f(z) \neq \alpha$.]

6. Use the **Nyquist criteria** approach of Example (6.8.6) to determine how many roots there are in the right half-plane for the following polynomials $p(z)$. [*Hint:* You must plot the curve $u = \text{Re}[p(-it)]$, $v = \text{Im}[p(-it)]$, $-\infty < t < \infty$, and determine the argument change as it is traversed. Perhaps the easiest way to do this is to determine where $u = 0$ and $v = 0$, and also where u and v are increasing or decreasing.]

(a) $p(z) = z^3 + z + 1$

(b) $p(z) = z^3 + z^2 + z + 4$

(c) $p(z) = z^5 + z^2 + z - 1$

(d) $p(z) = z^4 + z^3 + z^2 - z + 8$ (Be careful here, the curve "levels off" at $t = \pm\infty$.)

7. Use Rouché's Theorem to prove the following:

(a) If $h(z)$ is analytic and nonvanishing, and $|h(z)| < 1$, for $|z| \leq 1$, then $h(z) - z^n$ has n roots inside the unit circle.

(b) If $a > e^3/15$, then $e^z - a(z + 2z^2)$ has two roots inside $|z| < 3$.

8. Use Rouché's Theorem to verify the following:

(a) $16z^4 - 8iz^3 + 1 = 0$ has four roots in $|z| < 3/4$. [*Hint:* Take $f(z) = 16z^4 + 1$ and $g(z) = -8iz^3$.]

(b) $z^5 + z - 15 = 0$ has five roots in the annulus $\frac{3}{2} < |z| < 2$.

(c) $16z^4 - 12z + 1 = 0$ has four roots in $|z| < 1$ and one root in $|z| < \frac{3}{4}$.

9. The following steps, using Rouche's Theorem, leads to a proof of:

Theorem. If $f(z)$ is analytic at z_0 with $f'(z_0) \neq 0$, then f is locally invertible at z_0.

(a) Let $w_0 = f(z_0)$ and show that there is a circle $|z - z_0| = r_0$ on which $|f(z) - w_0| \geq m$ for some positive number m. [*Hint:* Use the fact that the zeros of $f(z) - w_0$ are isolated to conclude that there is a circle inside and on which $f(z) \neq w_0$.]

(b) Let $\hat{w}$ lie in the disk $|w - w_0| < m$, and define $F(z) = f(z) - w_0$, $G(z) = w_0 - \hat{w}$. Use Rouché's Theorem to show that $f(z) = \hat{w}$ has at least one root in $|z - z_0| < r_0$.

(c) Finally, to show that f is invertible, we must show that there is only *one* value of z in $|z - z_0| < r_0$ so that $f(z) = \hat{w}$ from (b). (Why?) Use the fact that $f'(z_0) \neq 0$ to show this.

10. In the previous problem, show that if $f'(z)$ has a zero of order $N - 1$ at z_0 $(N \geq 1)$, then there is a neighborhood of $z = z_0$ which is mapped onto a disk in the w-plane covered N times. [*Hint:* What is the multiplicity of the zero of $f(z) - w_0$ at $z = z_0$?]
11. (a) Prove that if $f(z)$ is analytic in a domain D, and if there is a sequence of points $\{z_n\}$ in D so that $f(z_n) = 0$ and $z_n \to z_0$ (z_0 in D), then $f(z) = 0$ in D. [*Hint:* Show that $f(z_0) = 0$, which contradicts the fact that the zeros of an analytic function are isolated. Then use Theorem (6.8.15).]
    (b) Prove that if $f(z)$ and $g(z)$ are analytic in a domain D, and $f(z_n) = g(z_n)$, with $z_n \to z_0$ (z_0 in D), then $f(z) = g(z)$ in D.
12. Use the results of Problem 11 to prove Theorem (6.8.17).
13. For each of the following, use the results of Problem 11b to construct analytic functions $f(z)$ having the properties given in each part. Why is there only one such function?
    (a) f is analytic at $z = 0$ and $f(1/n) = 1/n$, $n = 1, 2, \ldots$.
    (b) f is analytic at $z = 0$ and $f(1/n^2) = 1/n^4$.
    (c) f is analytic at $z = 1$ and $f(1 - 1/n) = 1/n$.
14. Use Theorem (6.8.17) to prove Theorem (6.8.21).
15. The following procedure uses analytic continuation to establish what is called **permanence of form**. That is, certain relationships which are known to hold in real calculus also must hold for the corresponding complex functions.
    (a) We know that $\sin(x + \pi/2) = \cos x$ for all real x. Argue that the complex function $f(z) = \sin(z + \pi/2) - \cos z$ vanishes on the real axis and hence must be identically zero off the real axis.
    (b) Use the approach of part (a) to prove that $\sin(z_1 + z_2) = \sin z_1 \cos z_2 + \cos z_1 \sin z_2$ for complex z_1 and z_2. [*Hint:* First show that $\sin(z + x_2) = \sin z \cos x_2 + \cos z \sin x_2$ for all complex z and all real x_2.]
    (The next two problems extend the Schwarz reflection principle.)
16. Use the same method of proof as in the Schwarz reflection principle to prove the following extension of Theorem (6.8.20): Let D_1 and D_2 be two domains separated by a smooth curve C as shown in Figure 6.8.6*a*. Let $f_1(z)$ be analytic in D_1 and continuous on $D_1 \cup C$ while $f_2(z)$ is analytic in D_2 and continuous on $D_2 \cup C$. If $f_1 = f_2$ on C, then f_1 and f_2 are analytic continuations of each other.
17. (a) Let a domain D_z have as part of its boundary a line segment L_z (see Figure 6.8.6*b*). Let $f(z)$ be analytic in D_z and continuous on $D_z \cup L_z$, and map D_z onto a domain D_w with *L_z mapped onto another line segment L_w* on the boundary of D_w. How can you analytically continue f across the line segment L_z? Prove your answer.
    (b) If $f(z)$ is entire and real on the real axis, and imaginary on the imaginary axis, then prove $f(z)$ is odd, that is, $f(-z) = -f(z)$. [*Hint:* Use part (a), where D_z is the first quadrant and there are two lines on the boundary.]

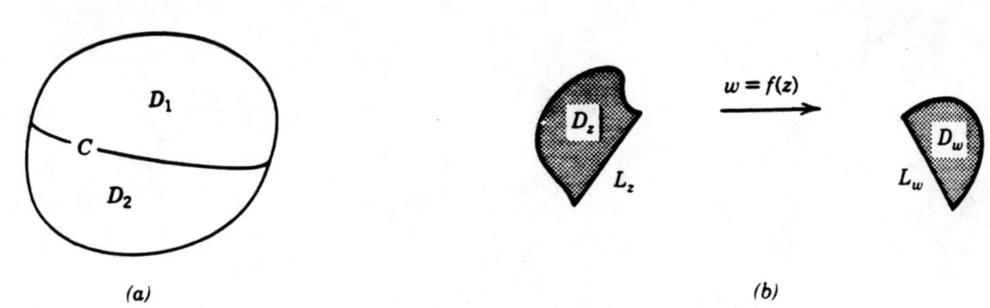

Figure 6.8.6

REFERENCES

A COMPLEX VARIABLES (ELEMENTARY AND INTERMEDIATE)

[A.1] Carrier, G., Krook, M., and Pearson, C., *Functions of a Complex Variable, Theory and Technique*. New York, McGraw-Hill, 1966.

[A.2] Churchill, R., Brown, J., and Verhey, R., *Complex Variables and Applications*. 3rd ed. New York, McGraw Hill, 1976.

[A.3] Dettman, J. W., *Applied Complex Variables*. New York, Macmillan, 1965.

[A.4] Eves, H. W., *Functions of a Complex Variable, Vols, I and II*. Boston, Mass., Prindle, Weber and Schmidt, 1966.

[A.5] Franklin, P., *Functions of Complex Variables*. Englewood Cliffs, N.J., Prentice-Hall, 1958.

[A.6] Greenleaf, F., *Introduction to Complex Variables*. Philadelphia, Saunders, 1972.

[A.7] Hamilton, H., *A Primer of Complex Variables with an Introduction to Advanced Techniques*. Belmont, Calif., Brooks Cole, 1966.

[A.8] Nehari, Z., *Introduction to Complex Variables*. Revised ed. Boston, Allyn and Bacon, 1968.

[A.9] Polya, G., and Latta, G., *Complex Variables*. New York, Wiley, 1974.

[A.10] Staff, E. B., and Snyder, A. D., *Fundamentals of Complex Analysis for*

Mathematics, Science and Engineering. Englewood Cliffs, N.J., Prentice-Hall, 1976.

[A.11] Wunsch, A. D., *Complex Variables with Applications.* Reading, Mass., Addison-Wesley, 1983.

B COMPLEX VARIABLES (ADVANCED)

[B.1] Ahlfors, L., *Complex Analysis, and Introduction to the Theory of Analytic Functions.* New York, McGraw-Hill, 1953.

[B.2] Bieberbach, L., *Conformal Mapping.* 4th ed. New York, Chelsea, 1964.

[B.3] Copson, E. T., *An Introduction to the Theory of a Complex Variable.* Oxford, Clarendon Press, 1962.

[B.4] Henrici, P., *Applied and Computational Complex Analysis, Vol. I.* New York, Wiley, 1974.

[B.5] Knopp, K., *Theory of Functions, Vols. I and II.* New York, Dover, 1947.

[B.6] Levinson, N., and Redheffer, R., *Complex Variables.* San Francisco, Holden-Day, 1970.

[B.7] Markushevich, A. I., *Theory of Functions of a Complex Variable.* Englewood Cliffs, N.J., Prentice-Hall, 1965.

[B.8] Markushevich, A. I., *Complex Numbers and Conformal Mappings.* New York, Pergamon, 1962.

[B.9] Nehari, Z., *Conformal Mapping.* New York, McGraw-Hill, 1952.

[B.10] Sansone, J., and Gerretsen, J., *Lectures on the Theory of Functions of a Complex Variable, Vol. I.* Groningen, Netherlands, Noordhoff-Groningen, 1960.

C ADVANCED CALCULUS AND ANALYSIS

[C.1] Amazigo, J., and Rubenfeld, L. A., *Advanced Calculus and its Applications to the Engineering and Physical Sciences.* New York, Wiley, 1980.

[C.2] Buck, R. C., *Advanced Calculus.* New York, McGraw-Hill, 1978.

[C.3] Kaplan, W., *Advanced Calculus.* Reading, Mass., Addison-Wesley, 1952.

D FLUID MECHANICS

[D.1] Milne-Thompson, L. M., *Theoretical Hydrodynamics.* 4th ed. New York, Macmillan, 1960.

[D.2] Prandtl, L., and Tietjens, O. G., *Fundamentals of Hydro- and Aeromechanics.* New York, Dover, 1957.

[D.3] von Mises, R., and Friedrichs, K. O., *Fluid Dynamics.* New York, Springer-Verlag, 1971.

E ELECTRICITY AND MAGNETISM

[E.1] Coulson, C. A., *Electricity*. London, Oliver and Boyd, 1961.

[E.2] Mason, M., and Weaver, W., *The Electromagnetic Field*. New York, Dover, 1929.

[E.3] Slater, J., and Frank, N., *Electromagnetism*. New York, McGraw-Hill, 1947.

[E.4] Sommerfeld, A., *Electrodynamics*. New York, Academic Press, 1964.

F ELASTICITY

[F.1] England, A. H., *Complex Variable Methods in Elasticity*. New York, Wiley, 1971.

[F.2] Sokolnikoff, I. S., *Methematical Theory of Elasticity*. McGraw-Hill, 1956.

G GENERAL APPLICATIONS

[G.1] Rothe, R., Ollendorff, F., and Pohlhausen, K., *Theory of Functions as Applied to Engineering Problems*. Cambridge, Mass., MIT Press, 1933.

[G.2] Walker, M., *Conjugate Functions for Engineers*. London, Oxford, 1933.

ANSWERS TO PROBLEMS

CHAPTER 1

Section 1.1, page 10

1. (a) $2^{1/2}, \frac{7}{5}, \frac{1}{5}$ (b) $\frac{1}{5}\sqrt{170}, -\frac{7}{5}, \frac{11}{5}$
(c) $1, 1, 0$ (d) $1, 0, 1$ (e) $\sqrt{27}, -5, \sqrt{2}$
(f) $10^{-2}, \frac{7}{2500}, -\frac{6}{625}$

2. (b) $2^{(n-1)/2}c_n$, where $c_{n+8} = c_n$ and $c_0 = \sqrt{2}, c_1 = 1 + i, \quad c_2 = i\sqrt{2}$, $c_3 = -1 + i, c_4 = -\sqrt{2}, c_5 = -1 - i, c_6 = -i\sqrt{2}, c_7 = 1 - i$.
(c) $2^{(n-1)/2}\bar{c}_n$, where c_n is given in (a).

6. (b) (i) $\operatorname{Im}(z_1\bar{z}_2) = 0$; (ii) $\operatorname{Re}(z_1\bar{z}_2) = 0$.
(c) $\operatorname{Im}[(z_2 - z_1)(\bar{z}_3 - \bar{z}_4)] = 0$ and $\operatorname{Im}[(z_3 - z_2)(\bar{z}_4 - \bar{z}_1)] = 0$, area = $|\operatorname{Im}[(z_2 - z_1)(\bar{z}_3 - \bar{z}_1)]|$.
(d) $\frac{1}{2}$ the area in (c).

7. (a) Any a, b so that $b = -(2 + 11i) - (3 + 4i)a$.
(b) $a = -\frac{11}{4}, b = \frac{25}{4}$.

10. (a) $k = b/2a$

11. (a) $\pm 2^{-1/2}(1 + i)$ (b) $\pm\left(\alpha + \dfrac{i}{2\alpha}\right), \alpha = [\frac{1}{2}(1 + \sqrt{2})]^{1/2}$
(c) $-i, \frac{1}{2}(\pm\sqrt{3} + i)$ (d) $n = 3$: $1, \frac{1}{2}(1 \pm i\sqrt{3})$; $n = 4$: $\pm 1, \pm i$

13. (a) Reflection in the origin.
(b) Reflection in the imaginary axis.

14. (a) Inside, on or outside the unit circle and on the ray joining z to the origin.
(b) Real parts have same signs; imaginary parts have opposite signs.

16. (a) Upper half-plane.
(b) On the side containing α, of the line through the origin and perpendicular to the line joining α and $-\alpha$.
(c) Outside of the disk $|z + (i/2)| = \frac{1}{2}$.
(d) z real or purely imaginary.
(e) z real or on the lines $\text{Im}(z) = \pm\sqrt{3}\,\text{Re}(z)$.
(f) On the line segment $x = \frac{3}{2}$, $-1 \leq y \leq 0$.
(g) On the parabola $x = \frac{1}{2}(y^2 - 1)$.

18. $z = \dfrac{1}{a + b}(az_1 + bz_2)$

19. (a) $|z - z_0| \leq R_0^2$, $\text{Im}(z - z_0) > 0$, $z_0 = x_0 + iy_0$
(b) $(1 - i)z + (1 + i)\bar{z} > 4$ (c) $|z| > 1$, $|z - (3 + i)| < 6$
(d) $5(z^2 + \bar{z}^2) + 6|z|^2 = 16$, $\text{Re}(z) > 0$
(e) $(a - b)(z^2 + \bar{z}^2) + 2(a + b)|z|^2 + 2(c - id)z + 2(c + id)\bar{z} + 4e = 0$
(f) $|z - z_0| = |z_0|$, $z_0 = x_0 + iy_0$

21. (a) $k = 1$: perpendicular bisector of line joining z_0 and z_1 (if $z_0 \neq z_1$); $k \neq 1$: a point if $z_0 = z_1$ or a circle if $z_0 \neq z_1$.
(b) No points if $|z_0| = 1$; unit disk if $|z_0| \neq 1$.
(c) $k < 1$: no points; $k = 1$: interval $[1, 2]$; $k > 1$: ellipse.
(d) Whole z-plane.

23. (e) 3

24. (a) 21a $k = 1$: unbounded, connected; $k \neq 1$: bounded, connected.
21b Open, bounded, connected, domain (if $|z_0| \neq 1$).
21c bounded, connected (if $k \geq 1$).
21d Unbounded, connected.
(b) Open, unbounded, disconnected.

Section 1.2, page 20

1. (a) $\dfrac{\pi}{4} + 2\pi n, \dfrac{\pi}{4}$ (b) $\dfrac{\pi}{2} + 2\pi n, \dfrac{\pi}{2}$ (c) $\dfrac{5\pi}{6} + 2\pi n, \dfrac{5\pi}{6}$

(d) $-\dfrac{3\pi}{2} + 2\pi n, \dfrac{\pi}{2}$ (e) $\pi + 2\pi n, \pi$ (f) $-\dfrac{\pi}{3} + 2\pi n, -\dfrac{\pi}{3}$

(g) $\tan^{-1}\frac{1}{2}$, $\text{Tan}^{-1}\frac{1}{2}$

7. (a) All nonnegative z. (b) z imaginary. (c) No z.

8. No z if $k < 2^{-1/2}$; one z if $k = 2^{-1/2}$ or $k > 1$; two z if $2^{-1/2} < k \leq 1$.

9. (a) No points if $k \leq -\pi$ or $k > \pi$. Half-ray from z_0 if $-\pi < k \leq \pi$.
(b) z real and $\neq 0$.

10. $z = 0$ $\text{Arg}(z_1) - \text{Arg}(z_2) + \pi$; $z = z_2$: $\text{Arg}(z_2) - \text{Arg}(z_1 - z_2) + \pi$

12. (b) (i) $\dfrac{\sin[\frac{1}{2}(N+1)\theta]}{\sin(\frac{1}{2}\theta)}\cos\left(\dfrac{N\theta}{2}\right)$ (ii) $\dfrac{\sin[\frac{1}{2}(N+1)\theta]\sin\left(\dfrac{N\theta}{2}\right)}{\sin(\frac{1}{2}\theta)}$

14. (b) $3-2i$

15. (a) Translate from z through the vector respresented by A.
(b) (i) At w, where $|w|=|\alpha|\cdot|z|$ and $\arg(w)=\arg(z)+\arg(\alpha)$; (ii) along half-ray through z and distance $|z|^{-1}$ from origin; (iii) at distance $|z|^n$ from origin and at argument $n.[\arg(z)]$.

16. (a) 2^{12} (b) -2^{-50} (c) $\pi/4$ (d) $2^n\left[\cos\left(\dfrac{n\pi}{6}\right)-i\sin\left(\dfrac{n\pi}{6}\right)\right]$

17. (a) $\pm 2^{5/8}(\cos\theta_0+i\sin\theta_0)$, $\pm i2^{5/8}(\cos\theta_0+i\sin\theta_0)$, $\theta_0=\pi/16$

(b) $\pm(1+i)$, $\pm(1-i)$ (c) $\pm 2^{-1/4}\left(\cos\dfrac{\pi}{8}+i\sin\dfrac{\pi}{8}\right)$

(d) $2^{-1/2}\cdot$[results of (a)]

(e) $2^{1/3}\left(\cos\dfrac{\pi}{9}-i\sin\dfrac{\pi}{9}\right)$, $2^{1/3}\left[\cos\left(\dfrac{5\pi}{9}\right)+i\sin\left(\dfrac{5\pi}{9}\right)\right]$,

$2^{1/3}\left[\cos\left(\dfrac{11\pi}{9}\right)+i\sin\left(\dfrac{11\pi}{9}\right)\right]$

18. (a) $\pm 2^{1/2}(1+i)$, $\pm 2^{1/2}(1-i)$ (b) $0, -2i, i\pm\sqrt{3}$

(c) $i, \dfrac{-i+\sqrt{3}}{2}$ (d) $i, -1$

19. (a) $-1, \frac{1}{2}(1\pm i\sqrt{3})$ (b) $1, \frac{1}{2}(-\sqrt{3}\pm i)$

20. (b) $\cos(n\theta)=\displaystyle\sum_{l=0}^{N_1}\frac{n!(\cos\theta)^{n-2l}(\sin\theta)^{2l}(-1)^l}{(2l)!(n-2l)!}$

$$\sin(n\theta)=\sum_{l=0}^{N_2}\frac{n!(\cos\theta)^{n-2l-1}(\sin\theta)^{2l+1}(-1)^l}{(2l+1)!(n-2l-1)!}$$

where n even: $N_1=\dfrac{n}{2}$, $N_2=\dfrac{n-2}{2}$; n odd: $N_1=N_2=\dfrac{n-1}{2}$

21. (b) $\cos^n\theta=2^{-n}\displaystyle\sum_{k=0}^{n}\frac{n!\cos(2k-n)\theta}{k!(n-K)!}$

$$\sin^{2n}\theta=2^{-2n}\sum_{k=0}^{2n}\frac{(2n)!(-1)^{n-k}}{k!(2n-k)!}\cos(2k-2n)\theta$$

$$\sin^{2n+1}\theta=2^{-(2n+1)}\sum_{k=0}^{2n+1}\frac{(-1)^{n-k}(2n+1)!}{k!(2n+1-k)!}\sin(2k-2n-1)\theta$$

22. (a) $n\pi,\ 2\pi n \pm \frac{\pi}{3},\ n = 0, \pm 1, \ldots$

(b) $\frac{\pi}{2} + 2\pi n,\ \theta_0 + 2\pi n,\ \tan\theta_0 = \frac{3}{4}$

(c) $k^2 \leq 5$

CHAPTER 2

Section, 2.1, page 34

1. (a) $z \neq i$: $\frac{1}{[x^2 + (y-1)^2]^2}\{x(x^2 + y^2 - 1) + i[(2 - y)(x^2 + y^2) - y]\}$

(b) $z \neq 2i$: $\frac{1}{x^2 + (y-2)^2}\{x(x^2 + y^2) - 4xy + x + i[y(x^2 + y^2) + 2(x^2 - y^2) + 2 - y]\}$

(c) $z \neq 0$: $(x^2 + y^2)^{-1}[x^2 - y^2 + x + iy(1 + 2x)]$

(d) all z: $2(x^2 - y^2)$

(e) $z \neq \pm\sqrt{2}(1 - i)$: $\frac{1}{(x^2 - y^2)^2 + 4(xy + 2)^2}\left\{\begin{matrix} x(x^2 + y^2) + 4y \\ +i[-y(x^2 + y^2) - 4x] \end{matrix}\right\}$

(f) $z \neq 0, -i$: $[x^2 + (y+1)^2]^{-1}[x - 1 - y - i(x + 1 + y)]$

(g) $z \neq 0$; $(x^2 + y^2)^{-1}[x(x^2 + y^2 + 1) + iy(x^2 + y^2 - 1)]$

(h) all z: $x^3 + iyx^2$

(i) $z \neq 0$: $(x^2 + y^4)^{-1}[x^2y^2 - xy + i(x^3 + y^3)]$

2. (a) Single-valued. (b) Single-valued. (c) Multiple-valued. (d) Single-valued. (e) Single-valued. (f) Multiple-valued.

3. (a) $\frac{i}{2}(\bar{z}^2 + |z|^2)$ (b) $\frac{i}{4}(4|z|^3 + \bar{z}^2 - z^2)$

(c) $\frac{z + \bar{z} + 2i}{(1 - i)z + (1 + i)\bar{z}}$ (d) $\frac{1}{2|z|^2}[z^2(1 + |z|) + \bar{z}^2(|z| - 1)]$

4. (a) $\frac{2}{5}(i + 2),\ 1 + i$ (b) $2i,\ \frac{1}{2}(1 + 3i)$ (c) $\frac{\pi}{2}, \frac{-\pi}{4}$ (d) $\sqrt{2},\ -i$

5. (a) With $z = x$, $u = \begin{cases} \pm\sqrt{x} & x \geq 0 \\ 0 & x < 0 \end{cases}$, $v = \begin{cases} 0 & x \geq 0 \\ \pm\sqrt{-x} & x < 0 \end{cases}$

6. (a) For z not real or imaginary, u and v are as in (2.1.8) with x replaced by $x^2 - y^2 + 1$ and y replaced by $2xy$. For

$$z = iy, \sqrt{z^2+1} = \begin{cases} \pm\sqrt{1-y^2} & y^2 \le 1 \\ \pm i\sqrt{y^2-1} & y^2 > 1 \end{cases}$$

$$z = x, \sqrt{z^2+1} = \pm\sqrt{1+x^2}$$

(b) $z, \frac{z}{2}(-1 \pm i\sqrt{3})$

7. (a) (i) Horizontal line segment $v = y_0^2$, $x_0 \le u \le x_1$; (ii) Vertical line segment $u = x_0$, $y_0^2 \le v \le y_1^2$.

8. (a) Reflection in the real axis.
(b) Reflection in the real axis, then translated to the right one unit.

9. (a) Region below the line $v = R^2$ and above parabola $v = u^2$.
(b) Line segment $v = R_0^2$, $-R_0 < u < R_0$.

10. (a) The half-line $u = x_0$, $v > x_0$. (b) Wedge $v > u$, $u > 0$.

11. (a) Unit circle. (b) Real axis. (c) Lower half-plane.
(d) Line $\text{Im}(w) = -\frac{1}{2}$.
(e) Region in fourth quadrant outside of unit circle.

12. Quarter-plane.

13. (a) Imaginary axis. (b) Line segment $v = 0$, $-2 < u < 2$.

14. (a) Parallelogram with vertices at $w = 2, 2i, -1+i, 1-i$.

15. (a) $|w| = |z|^2$, $\arg(w) = 2\arg(z)$.
(b) (i) Upper half-plane; (ii) wedge, bounded by positive real axis and ray $\arg(w) = 2\theta_0$; (iii) Unit disk covered twice;
(iv) annular sector $4 < |w| < 9$, $\pi/2 < \arg(w) < \pi$.

16. Quarter disk $|w| < 1$, $\text{Re}(w) > 0$, $\text{Im}(w) > 0$.

17. (a) 0 (b) i (c) 0 (d) ∞ (e) No limit.
(f) 0 (g) 0 (h) ∞ (i) 0
(j) ∞ if $m < n$, 1 if $m = n$, 0 if $m > n$.
(k) No limit.

18. (a) $z \ne -2 \pm i$ (b) Except at zeros of $q(z)$.
(c) $z \ne 0$ (d) Except line segment $x = 0$, $|y| \ge 1$.
(e) Everywhere. (f) Except line segment $x = 0$, $|y| \le 1$.
(g) Everywhere. (h) Everywhere. (i) $z \ne 0$

20. (a) $z^2 + 1$ (b) $\dfrac{z+i}{z(1+i)+i}$ (c) $(x^2 - y^2)^{-1} + 4ix^2y^2$
(d) $x^2 - 4y^2 + 8ixy$

Section 2.2, page 43

1. (a) nz^{n-1} (b) Nowhere. (c) $f' = 1$ at $\text{Im}(z) = 1/2$
(d) $f' = -z^{-2}, z \ne 0$ (e) Nowhere. (f) $f' = 0$ at $z = 0$

7. (a) $\text{Im}(z) = 1, f' = 2$ (b) $z = 0, f' = 0$ (c) $z = -i, f' = i$
(d) $f' = 3z2$ (e) z real, $f' = 2x$ (f) $f' = -z^{-2}, z \ne 0$

(g) $z = 0, f' = 0$ (h) $z = 0, f' = 0$ (i) Nowhere. (j) Nowhere.
(k) $z = 0, f' = 0$ (l) $z = 0, f' = 0$ (m) $z = \pm 1, f' = 1 - i$
(n) $f' = 2e^{x^2-y^2}\{x\cos(2xy) - y\sin(2xy) + i[x\sin(2xy) + y\cos(2xy)]\}$
(o) $z = 0, f' = 0$ (p) Everywhere, $f' = f$.
(q) Nowhere. (r) $z \neq 0$, z not negative real, $f' = z^{-1}$
(s) $z = 0, f' = 0$

8. $a = -\frac{1}{2}, b = -2, c = \frac{1}{2}$

12. (a) (i), (ii) (b) Degree (denominator) $\geq$ degree (numerator)

16. (a) $\dfrac{\partial \hat{f}}{\partial z} = \frac{1}{2}[u_x + v_y - i(u_y - v_x)], \dfrac{\partial \hat{f}}{\partial \bar{z}} = \frac{1}{2}[u_x - v_y + i(u_y + v_x)]$

Section 2.3, page 52

2. (a) Yes, $2\pi n$. (b) No.

3. (a) $\cos 1 + i \sin 1$ (b) $e[\cos 1 + i \sin 1]$
(c) ie^2 (d) $e^{\cos 1}[\cos(\sin 1) + i \sin(\sin 1)]$

5. (a) $\cos 2 + i \sin 2$ (b) does not exist. (c) Does not exist.
(d) $-i$ (e) $e^{\cos 1}[\cos(\sin 1) - i \sin(\sin 1)]$ (f) 1

6. (a) $i\left(\dfrac{\pi}{2} + 2\pi n\right)$

(b) $\pm\dfrac{(1+i)}{\sqrt{2}}\left(\dfrac{\pi}{4} + 2\pi n\right)^{1/2}, n = 0, 1, 2, \ldots;$

$\pm\dfrac{(1-i)}{\sqrt{2}}\left(\dfrac{-\pi}{4} - 2\pi n\right)^{1/2}, n = -1, -2, \ldots$

(c) $\dfrac{\ln 2^{1/2} - i[(\pi/4) + 2\pi n]}{\frac{1}{4}\ln^2 2 + [(\pi/4) + 2\pi n]^2}$ (d) $\ln 2 + i\left(\dfrac{\pi}{3} + 2\pi n\right)$

(e) $\dfrac{i}{3}\left(-\dfrac{\pi}{2} + 2\pi n\right)$ (f) $i\left(-\dfrac{\pi}{4} + \pi n\right)$ (g) $i\left(\pm\dfrac{2\pi}{3} + 2\pi n\right)$

(h) $\ln 2 + i\left(-\dfrac{\pi}{6} + 2\pi n\right)$ (i) $i\left(\dfrac{\pi}{2} + \pi n\right)$

(j) $\left|\dfrac{\pi}{2} + 2\pi k\right| + i\left(\pm\dfrac{\pi}{2} + 2\pi n\right), n, k = 0, \pm 1, \pm 2, \ldots$

7. (a) (i) On the hyperbolas $xy = (\pi/2)(n + \frac{1}{2}), n = 0, \pm 1, \ldots;$
(ii) On the hyperbolas $xy = n\pi/2, n = 0, \pm 1, \ldots.$
(b) (i) $\text{Re}(z) < 0$; (ii) $\text{Re}(z) = 0$; (iii) $\text{Re}(z) > 0$.

9. $z = z_0 + R_0 e^{it}$ for t real

10. (a) Entire, $2ze^{z^2}$. (b) Nowhere differentiable.
(c) Nowhere differentiable. (d) Entire, $e^z \cdot e^{e^z}$.
(e) $z \neq 2\pi i n, \dfrac{(e^z - 1 - ze^z)}{(e^z - 1)^2}$ (f) $z \neq 0, -z^{-2}e^{1/z}$

(g) $z \neq 1$, $-2(z-1)^{-3}e^{(z-1)^{-2}}$ (h) Nowhere differentiable.
(i) $z = 0$, $f' = 0$, nowhere analytic.

14. (a) Circular arc $|w| = e^{x_0}$, $y_0 < \arg(w) < y_1$.

15. (a) Quarter-disk $|w| < 1, 0 < \arg(w) < \pi/2$.
(b) Annular sector $1 < |w| < e^{x_0}, 0 < \arg(w) < y_0$.
(c) Extension of the unit disk covered an infinite number of times.
(d) Spiral $|w| = e^{\arg(w)}$ starting at $w = 1$.
(e) Annulus $e^{-\sqrt{2}} < |w| < 1$ covered an infinite number of times.

Section 2.4, page 60

1. (a) Ellipse $u^2 + \frac{1}{4}v^2 = k^2$.
(b) Rectangle $kx_0 < u < kx_1, ky_0 < v < ky_1$.
(c) Quarter-disk $|w| < 2k, u > k, v > k$.

2. (a) Upper half-plane.
(b) In lower half-plane, the exterior of disk centered at $w = 1$ of radius 1.
(c) Disk $|w - 2(i-1)| < 4$. (d) Left half-plane.
(e) In upper half-plane, the exterior of the unit disk.

3. Translate through $-\alpha$, rotate through θ_0, translate through α.

4. (a) $w = \dfrac{\rho_0}{R_0}(z - z_0)e^{i\theta_0} + w_0$, θ_0 arbitrary.

(b) $w = \dfrac{(w_1 - w_0)}{(z_1 - z_0)}(z - z_0) + w_0$ (it is unique).

(c) No linear map unless $\arg(w_2 - w_0) - \arg(w_1 - w_0) = \arg(z_2 - z_0) - \arg(z_1 - z_0)$. Then answer is as in (b).

5. $w = iz + 2$

6. If corresponding sides of the two rectangles are in the same ratio.

7. (a) Upper half of the disk $|w| < 16$.
(b) Annular sector $r_0^2 < |w| < r_1^2, 2\theta_0 < \arg(w) < 2\theta_1$.

8. That part of the upper half-plane lying to the right of the parabola $u = 1 - \frac{1}{4}v^2$.

9. That part of the upper half-plane lying to the left of the parabola $u = 1 - \frac{1}{4}v^2$.

10. (a) Upper half-plane. (b) Upper half-plane.
(c) Wedge $(3\pi/4) < \arg(w) < (9\pi/4)$.

11. (b) Half-plane $\mathrm{Im}(w) > 1$.

12. (a) If $b = a^2/3$, in which case

$$w = \left(z + \frac{a}{3}\right)^3 + c - \frac{a^3}{27}$$

(b) If

$$a_k = \frac{n!}{k!(n-k)!}\left(\frac{a_{n-1}}{n}\right)^{n-k}, k = 1, 2, \ldots, n-2$$

in which case

$$w = \left(z + \frac{a_{n-1}}{n}\right)^n + a_0 - \left(\frac{a_{n-1}}{n}\right)^n$$

Section 2.5, page 74

1. (b) $\dfrac{R_0}{\bar{z} - \bar{z}_0}$ (c) $z_0 + \dfrac{R_0^2}{\bar{z} - \bar{z}_0}$

2. (c) ∞
3. (c) $\bar{z}$ (d) Symmetric points.
4. (a) Half-disk $|w + \frac{1}{2}| < \frac{1}{2}$ and $\text{Re}(w) > -\frac{1}{2}$.
   (b) Half-disk $|w - \frac{1}{2}| < \frac{1}{2}$, $\text{Im}(w) > 0$.
   (c) Quarter-plane $\text{Re}(w) > 1$, $\text{Im}(w) < 0$.
   (d) Disk $\left|w + \dfrac{i}{2}\right| < \dfrac{1}{2}$.
   (e) Region inside disk $\left|w - \dfrac{1}{4}(1 - i)\right| < \dfrac{\sqrt{2}}{4}$ and outside disk, $\left|w - \dfrac{1}{4}(1 + i)\right| < \dfrac{\sqrt{2}}{4}$.
5. (a) Half-plane $\text{Re}(w) > 1$.
   (b) Wedge $u - v < 1$, $u > 1$.
   (c) Disk $|w - (1 + i)| \leq \sqrt{2}$.
   (d) Half-plane $\text{Re}(w) \leq \frac{1}{2}$.
   (e) Negative real axis.
   (f) Exterior of unit disk.
6. (a) Quarter-plane $\text{Re}(w) > \frac{1}{2}$, $\text{Im}(w) > 0$.
   (b) Half-plane $v < u/(1 + \sqrt{2})$ covered infinite times.
7. Region between two circles or region exterior to two nonintersecting disks or half-plane missing the region exterior to a disk.
8. (a) $\dfrac{(3 - i)z + 1 - i}{(i - 1)z - (1 + i)}$ (b) $\dfrac{(1 - i)}{2}\left[\dfrac{z - i}{-z + 1}\right]$ (c) $\dfrac{(1 + i)z + 2i}{z + (2i - 1)}$
9. (a) $\dfrac{(1 + 3i)z + 2(1 - i)}{(1 + i)(z - 2i)}$ (c) $-i\left(\dfrac{z + 2}{z - 2i}\right)$ (d) $\dfrac{-z + 2}{z - 2i}$
10. (a) $\dfrac{(1 + 3i)z + (1 - 11i)}{(i - 1)(z + i)}$ (b) $\dfrac{2z + (1 - 3i)}{2z - (1 + 3i)}$
11. (a) $2(z - i) + 1$ (b) $\dfrac{(3 + \sqrt{3})z - 6i}{(1 - \sqrt{3})z - 2i}$

12. (a) $\dfrac{z^2 - i}{z^2 + i}$ (b) $\dfrac{1}{4}\left(\dfrac{z+1}{z-1}\right)^2$ (c) $\dfrac{e^{iz} + i}{e^{iz} - 1}$

13. No, since one of the two circles must map onto another circle.

14. (b) $w = \dfrac{\bar{z}_0 z - e^{i\theta_0} z}{z - e^{i\theta_0}}$ $\quad \mathrm{Im}(z_0) > 0$

(c) $w = e^{i\psi_0}\left(\dfrac{z - \alpha}{\bar{\alpha} z - 1}\right)$ $\quad |\alpha| < 1$

Section 2.6, page 81

2. (a) $\cosh 1$ (b) $\cos\left(\sinh \dfrac{\pi}{2}\right) + i \sin\left(\sinh \dfrac{\pi}{2}\right)$

(c) $\cos(\cos 1)\cosh(\sin 1) - i \sin(\cos 1)\sinh(\sin 1)$

5. (a) $(n + \frac{1}{2})\pi i$ (b) $n\pi + i(-1)^{n+1}\sinh^{-1} 2$

(c) $\ln(\pi|n|) + 2\pi i k + i\begin{cases} 0, n > 0 \\ \pi, n < 0 \end{cases}$ $\quad (n = \pm 1, \pm 2, \ldots, k = 0, \pm 1, \ldots)$

(d) $\pi k + i \sinh^{-1}\left(\left|\dfrac{\pi}{2} + 2\pi n\right|\right)\begin{cases} (-1)^k, & n \geq 0 \\ (-1)^{k+1} & n < 0 \end{cases}$

$n = 0, \pm 1, \ldots, k = 0, \pm 1, \ldots$

(e) $\pi n + \frac{1}{2}(\pi - \theta_0), \theta_0 = \tan^{-1}(\frac{4}{3})$ (f) $[(-1)^n \sinh^{-1} 1 - in\pi]^{-1}$

6. (a) $\alpha \neq \pm i$

8. $z = n\pi$; nowhere; $f'(n\pi) = 0$.

11. (a) Real interval $(0, 1)$ if $y_0 = 0$. Part of ellipse $(u^2/\cosh^2 y_0) + (v^2/\sinh^2 y_0) = 1$ in first quadrant if $y_0 > 0$. Part of above ellipse in fourth quadrant if $y_0 < 0$.

12. (a) (i) Second quadrant; (ii) fourth quardant.
(b) (i) First quadrant; (ii) fourth quadrant.

14. $2\pi i n$

15. (a) $-(n + \frac{1}{2})\pi i$ (b) $n\pi i$ (c) $(-1)^n \sinh^{-1} 1 + i(n + \frac{1}{2})\pi$
(d) $\ln|\alpha| + i[\arg(\alpha) + 2\pi n]$

Section 2.7, page 91

2. (a) $\ln 5 + i\left(-\dfrac{7\pi}{4} + 2\pi n\right)$ (b) $1 + i\left(2 + \dfrac{\pi}{2} + 2\pi n\right)$

(c) $-\dfrac{\pi i}{2}$ (d) $\dfrac{\pi i}{2}$ (e) $\dfrac{\pi i}{2}$ (f) $1 + \dfrac{\pi i}{4}$

(g) $-\dfrac{1}{2}\ln 2 + i\left(-\dfrac{\pi}{12} + 2\pi n\right)$ (h) $\ln(\sinh e) + i\left(\dfrac{\pi}{2} + 2\pi n\right)$

(i) $-i \sinh \dfrac{\pi}{4}$

3. $0, \frac{1}{2}\ln 2 + \frac{3\pi i}{4}$

4. (a) $\frac{1}{2}(1 + i\sqrt{3})$ (b) No solutions. (c) $-\frac{\pi}{3} < \arg(z) < \frac{\pi}{3}$

(d) $-\infty < \text{Re}(z) < \infty, -\pi < \text{Im}(z) \le \pi$
(e) The unit circle except $z = -1$.
(f) The right half of unit circle, including $z = \pm i$.

5. z negative real, or $\text{Re}(z) > 0$ or $\text{Re}(z) = 0$ and $\text{Im}(z) > 0$.

6. (a) $\ln^2 \pi + \frac{\pi^2}{4} + i\pi \ln \pi$ (b) 0 (c) 0 (d) 0

(e) No limit.

7. $\theta_0 < \arg(z) \le \theta_0 + 2\pi$, where $-2\pi \le \theta_0 < 0$, $z = 1$.

8. (a) On the lines $x > 0$, $y = (2n + 1)\pi$.
(b) On those pieces of the hyperbola $x^2 - y^2 = 1$ lying in the second and fourth quadrants.
(c) On negative real axis and on the pieces of the hyperbola $y^2 - 3x^2 = 1$ in the first and fourth quadrants.
(d) Everywhere.

9. (a) Half-strip $-\infty < u < 0, -\pi < v \le \pi$.
(b) Strip $-\infty < u < \infty, 0 < v < \pi$.
(c) Half-strip $-\infty < u < 0, -\pi/2 < v < \pi/2$.
(d) First quadrant.
(e) Two disjoint strips $-\infty < u < \infty$, $0 < v \le \pi/2$ and $-\infty < u < \infty$, $-3\pi/2 < v < -\pi$.
(f) Strip $-\infty < u < \infty, 0 < v < \pi/2$.

10. (a) $e^{[(\pi/4) + 2\pi n]}\left[\cos\left(\frac{\ln 2}{2}\right) + i\sin\left(\frac{\ln 2}{2}\right)\right]$

(b) $-\frac{\pi}{2} + 2\pi n + 2\pi ik$ $(n = 0, \pm 1, \ldots, k = 0, \pm 1, \ldots)$

(c) $\pm\left|\frac{\pi}{4} + \pi n\right|^{1/2}\begin{cases}1 + i & n = 0, 1, 2, \ldots \\ 1 - i & n = -1, -2, \ldots\end{cases}$

(d) $e^{-(\pi/2)(\pi + 2\pi n)}\left[\cos\left(\frac{\pi \ln 2}{2}\right) + i\sin\left(\frac{\pi \ln 2}{2}\right)\right]$ (e) $e^{-\pi + 2\pi n}$

(f) $e^{-[(\pi/2) + 2\pi n]\sin 1}\left\{\cos\left[\left(\frac{\pi}{2} + 2\pi n\right)\cos 1\right] + i\sin\left[\left(\frac{\pi}{2} + 2\pi n\right)\cos 1\right]\right\}$

(g) $\frac{\pi i}{4}, \frac{-3\pi i}{4}$

11. (a) 1 (b) No solutions. (c) No solutions. (d) $-2^{-1/2}$

12. $-\pi/n < \text{ang}(z) \le \pi/n$.

14. Strips $-\infty < y < \infty, 2\pi n < x < (2n+1)\pi$ ($n = 0, \pm 1, \ldots$); half-lines $x = 2\pi k$, $y > 0$ or $x = (2k+1)\pi$, $y < 0$ ($k = 0, \pm 1, \ldots$).

16. (a) Quarter-disk $|w| < 1$, $\mathrm{Re}(w) > 0$, $\mathrm{Im}(w) > 0$.

(b) Two eighth-disks $|w| < |$with $-\frac{\pi}{2} < \arg(w) < -\frac{\pi}{4}$ and

$\frac{\pi}{4} < \arg(w) < \pi/2$.

(c) Strip $-\infty < u < \infty, 0 < v < \pi/4$.

17. (a) (prin.)$(z-i)^{1/2}$ (b) $e^{\pi i/8}$(prin.)$(z-i)^{1/2}$

(c) $\frac{1}{\pi}\,\mathrm{Log}\left[\frac{(1+i)}{\sqrt{2}}(z-i)\right]$

Section 2.8, page 98

2. (a) $(\pi/2) + 2\pi n$; $\pi/2$ (b) $n\pi$; 0

(c) $x \geq 1$: $2\pi n - i\ln(x \pm \sqrt{x^2-1})$; $-i\ln(x + \sqrt{x^2-1})$

$-1 < x < 1$: $2\pi n \pm \tan^{-1}\left(\frac{\sqrt{1-x^2}}{x}\right)$; $\tan^{-1}\left(\frac{\sqrt{1-x^2}}{x}\right)$,

where $0 < \tan^{-1}\lambda < \pi$

$x \leq -1$: $\pi(1+2n) - i\ln(-x \pm \sqrt{x^2-1})$; $\pi - i\ln(-x - \sqrt{x^2-1})$

(d) $2\pi n - i\ln(-y + \sqrt{1+y^2})$ and $\pi + 2\pi n - i\ln(y + \sqrt{1+y^2})$;
$-i\ln(-y + \sqrt{1+y^2})$

(e) $\frac{\pi}{2} + 2\pi n - i\ln(y + \sqrt{1+y^2})]$ and $-\frac{\pi}{2} - i\ln(-y + \sqrt{1+y^2}) + 2\pi n$;

$\frac{\pi}{2} - i\ln(y + \sqrt{1+y^2})$

(f) $\pi n + \frac{\pi}{4} - \frac{i}{2}\ln\left(\frac{\cos\theta_0}{1+\sin\theta_0}\right)$; $\frac{-i}{2}\ln\left(\frac{\cos\theta_0}{1+\sin\theta_0}\right) + \frac{\pi}{4}$ for $\cos\theta_0 > 0$

$\pi n - \frac{\pi}{4} - \frac{i}{2}\ln\left(\frac{-\cos\theta_0}{1+\sin\theta_0}\right)$; $\frac{-\pi}{4} - \frac{i}{2}\ln\left(\frac{-\cos\theta_0}{1+\sin\theta_0}\right)$ for $\cos\theta_0 < 0$

3. (b) $\frac{d}{dz}(\cos^{-1}z) = \frac{-i}{\sqrt{z^2-1}}$; $\frac{d}{dz}(\tan^{-1}z) = \frac{1}{1+z^2}$

4. (c) No, since whatever branch is chosen, $\mathrm{Re}[\mathrm{Sin}^{-1}(\sin z)]$ must lie in the range $(-\pi, \pi)$.

5. (a) $g_1 = g_2 + 2\pi k$ or $g_1 = 2\pi k - g_2$ $k = 0, \pm 1, \pm 2, \ldots$

(b) Yes.

6. (a) $h_1 = h_2 + \pi k$ $k = 0, \pm 1, \ldots$

(b) On imaginary axis where $|\mathrm{Im}(z)| \geq 1$.

7. (a) $-i\sqrt{z^2-1}$ (b) $\sqrt{1-z^2}$

CHAPTER 3

Section 3.1, page 104

1. (a) $z = -2$, isolated. (b) $z = -2, 1 \pm i\sqrt{3}$, isolated.
(c) $z = \pm i$, isolated. (d) $\text{Re}(z) = 0$ and $|\text{Im}(z)| \geq 1$, not isolated.
(e) $z = 0, -3, 9$, isolated. (f) $z = 0, \pm i\sqrt{3}$, isolated.
(g) $z = 0$, isolated; $\text{Im}(z) = 0$ and $\text{Re}(z) \leq -1$, not isolated.
(h) $\text{Im}(z) = 0$ and $\text{Re}(z) \leq 0$, not isolated.
(i) Imaginary axis, not isolated.

2. (a) $\dfrac{1}{(1+2i)^2}\left[\dfrac{1}{z-1} - \dfrac{1}{z+2i} + \dfrac{2(i-2)}{(z+2i)^2}\right]$

(b) $\dfrac{1}{12}\left[\dfrac{4i}{z-i} + \dfrac{3(3-2i)}{z-2i} + \dfrac{3+2i}{z+2i}\right]$ (c) $\dfrac{5}{(z+1)^2} - \dfrac{3}{(z+1)^3}$

5. (a) $z + 5 + \dfrac{1}{3}\left(\dfrac{77}{z-4} - \dfrac{5}{z-1}\right)$

(b) $1 + \dfrac{1}{2}\left[\dfrac{1}{z} + \dfrac{1}{z^2} + \dfrac{7}{z-2} + \dfrac{9}{(z-2)^2}\right]$ (c) $z + i - 1 - \dfrac{i}{z-i}$

Section 3.2, page 116

1. (a) $z = \pm i$ (b) $z = \pm 1$ (c) $z = 0, 1$ (d) $z = 0, 1, \infty$
(e) $z = \alpha_1, \alpha_2, \ldots, \alpha_n$, $z = \infty$ only if n is odd. (f) $z = \pm i, \infty$
(g) $z = 0, \infty$ and $z = 1$ if the branch of $\log z$ is such that $\log 1 = 0$.
(h) $z = 0, i$
(i) $z = 0, \infty$ and $z = 1$ if the branch of $\sqrt{z}$ is such that $\sqrt{1} = -1$.
(j) $z = \pm 1, \pm i, \infty$
(k) $z = 0, \infty$, and e^{-1} if the branch of $\log z$ is such that $\arg(e^{-1}) = 0$.
2. (b) $z^{1/n}$ is real only when z is real and positive (for n even) or z is real (n odd).
6. (a) $z = 0, \pm 1, \infty$
(c) $-\pi < \theta_1, \theta_2, \theta_3 \leq \pi$. $0 < x < 1$: $f = i[x(1-x^2)]^{1/2}$;
$-1 < x < 0$: $f = -[x(1-x^2)]^{1/2}$;
$x < -1$: $f = -i[x(x^2-1)]^{1/2}$.
7. (a) $z = \pm 1, \infty$
(b) Choose the principal logarithm and $\sqrt{z^2-1} = |z^2-1|^{1/2}$; $e^{(i/2)(\theta_1+\theta_2)}$,
$-\pi < \theta_1, \theta_2 \leq \pi$.
8. (a) When a branch of $\sqrt{z}$ is chosen so that $\sqrt{1} = -1$.

Section 3.3, page 122

1. (a) Harmonic. (b) 4 (c) Harmonic.
(d) $-4(x^2+y^2)^{-2}(x^2-y^2)$, $x^2+y^2 \neq 0$
(e) $-x^{-2}y^{-2}(x^2+y^2)$, $x \neq 0$, $y \neq 0$
(f) $(a^2-b^2)e^{ax}\sin(by)$, harmonic if $b = \pm a$.

3. (b) It is the real part of $\log f(z)$.
4. (b) (i) $\phi(\alpha) = a + b\alpha$; (ii) $\phi(\alpha) = a + b \ln \alpha$.

5. (d) (i) $\frac{1}{4}(x^2 + y^2)$; (ii) $\frac{1}{8}(x^2 + y^2)(ax + by)$; (iii) $\frac{xy}{12}(x^2 + y^2)$.

7. (a) $-\frac{1}{2}(x^2 - y^2) - \frac{1}{3}(x^3 - 3xy^2)$ (b) $e^{2y}\cos(2x)$
(c) $-2x + 2xy$ (d) None if $b \neq \pm a$; if $b = \pm a$, $v = e^{ax}\sin(ay)$.
(e) None. (f) $-x - 3x^2y + y^3$ (g) None.

(h) $\cos x \cosh y$ (i) $\dfrac{x}{x^2 + y^2}$ (j) None.

13. (b) Since $f'(0) = 0$.

Section 3.4, page 134

3. Pressure is greatest at that point on a streamline where the speed is lowest.
4. Streamlines are the spirals $r = ce^{-a_0\theta/\kappa_0}$, $c > 0$.
7. (d) $z = \pm R_0 e^{i\alpha}$. Both are on the boundary.
8. (b) $V = 0$ on the real axis right below the source.

(c) $\mathbf{F} = \dfrac{\rho}{y_0}\mathbf{j}$, where $y_0 = \mathrm{Im}(z_0)$.

10. (b) $V = a_0\left(1 - \dfrac{1}{\bar{z}^2}\right) - \dfrac{i\Gamma}{2\pi\bar{z}}$

(c) Speed $= 2a_0|(\Gamma/4\pi a_0) + \sin\theta|$, on $z = e^{i\theta}$.
Pressure $= \rho[\text{constant} - 2a_0^2[(\Gamma/4\pi a_0) + \sin\theta]^2$.
(d) There is a net vertical force. (e) $\mathbf{F} = (\rho\Gamma a_0)\mathbf{j}$
(f) $\Gamma < 4\pi a_0$: $V = 0$ at $z = (1/4\pi a_0)(-i\Gamma \pm \sqrt{16\pi^2 a_0^2 - \Gamma^2})$; both of these are on the boundary. $\Gamma = 4\pi a_0$: $V = 0$ at $z = -i$, which is also on the boundary. $\Gamma > 4\pi a_0$: $V = 0$ at $z = (-i/4\pi a_0)[\Gamma \pm \sqrt{\Gamma^2 - 16\pi^2 a_0^2}]$; one is outside the cylinder and one is inside.

11. (d) At $z = i$.

Section 3.5, page 146

9. (a) $-2q_0[\log(z - z_0) - \log(z - z_1)]$

10. (b) $\dfrac{2q_0 e^{\bar{z}}(e^{\bar{z}_0} - e^{z_0})}{(e^{\bar{z}} - e^{\bar{z}_0})(e^{\bar{z}} - e^{z_0})}$

11. (a) $x_0 = 0$; $(1 + k^2)y_0 = h_0(1 - k^2)$; $2ky_0 = r_0(1 - k^2)$
(b) $-2q_0 \ln k = V_0$

(c) $z_0 = i\sqrt{h_0^2 - r_0^2}$; $q_0 = \dfrac{-V_0}{2\ln\left[\dfrac{1}{r_0}(h_0 - \sqrt{h_0^2 - r_0^2})\right]}$;

$\Phi = -2q_0[\log(z - z_0) - \log(z - \bar{z}_0)]$

13. (a) $\Phi = -2q_0[\mathrm{Log}(z - z_0) - \log(z - \bar{z}_0) + \log(z + z_0) - \log(z + \bar{z}_0)]$
(b) Images at $z = z_0 e^{2\pi ik/n}, \bar{z}_0 e^{2\pi ik/n}$ $(k = 0, 1, \ldots, n - 1)$;
$\Phi = -2q_0 \sum_{k=0}^{n-1} [\log(z - z_0 e^{2\pi ik/n}) - \log(z - \bar{z}_0 e^{2\pi ik/n})]$
(c) There will be an infinite number of images, and the infinite series won't converge.

14. $z_1 = R_0^2/\bar{z}_0$; $\phi = -2q_0\left[\ln|z - z_0| - \ln\left|z - \frac{R_0^2}{\bar{z}_0}\right|\right] - 2q_0 \ln\left(\frac{R_0}{|z_0|}\right)$

CHAPTER 4

Section 4.1, page 160

1. (a) $z_0 = z_1 = 2$; closed, simple, smooth.
(b) $z_0 = z_1 = 0$; closed, piecewise smooth.
(c) $z_0 = z_1 = -1$; closed, simple, piecewise smooth.
(d) $z_1 = -1$ (no initial point); smooth.
(e) $z_0 = z_1 = 1 + i$; closed piecewise smooth.
(f) $z_0 = 0, z_1 = 3$; piecewise smooth.
(g) No initial and final points; piecewise smooth.
(h) $z_0 = -2 + 5i, z_1 = -2 - i$; piecewise smooth.
(i) $z_0 = z_1 = 0$; closed, simple, piecewise smooth.
(j) $z_0 = z_1 = 0$; closed; smooth.

3. (a) $z = 1 + e^{-2it}, -\pi \le t \le 0$
(b) $z = (1 + i)(1 - |t|), -1 \le t \le 1$
(c) $z = \begin{cases} e^{-i(1+t)} & -\pi - 1 \le t < -1 \\ -t & -1 \le t \le 1 \end{cases}$
(i) $z = (1 - \cos t)e^{-it}, -2\pi \le t \le 0$

7. (a) $\cos(2t) + i\sin(2t)$
(b) $e^{t\cos t}[\cos(t\sin t) - i\sin(t\sin t)]$
(c) $\begin{cases} \sin t & -1 \le t \le 1 \\ \sin 3\cosh t + i\cos 3\sinh t & 1 < t \le 3 \end{cases}$

8. (a) $z = i + 2e^{-it}, -2\pi \le t \le 0$
(b) $z = \begin{cases} t & 0 \le t \le 1 \\ 1 + i(t - 1) & 1 < t \le 2 \\ (1 + i)(3 - t) & 2 < t \le 3 \end{cases}$
(c) $z = z_0 + a\cos t - ib\sin t, -\pi \le t \le 0 (z_0 = x_0 + iy_0)$
(d) $z = \cosh t + 2i\sinh t, -\infty < t < \infty$
(e) $z = -\left(t + \frac{i}{t}\right), -\infty < t < 0$
(f) $z = \begin{cases} t & -1 \le t \le 1 \\ e^{i(t-1)} & 1 < t \le 1 + \pi \end{cases}$

(g) $z = \begin{cases} 1 + it & -1 \le t \le 1 \\ 2 - t + i & 1 < t \le 3 \\ -1 + i(4 - t) & 3 < t \le 5 \\ t - 6 - i & 5 < t \le 7 \end{cases}$

11. (a) $\phi(\tau) = 2\tau$ (b) $\phi(\tau) = \cosh^{-1} \tau$ (positive branch)

12. (a) $t + i - is$ (b) $t(1 + s) + i(1 - s)$ (c) $i + te^{-\pi is/2}$
(d) $(1 + s)e^{it}$ (e) $\cos t + i(3 - s) \sin t$

(f) $\left(1 - \frac{s}{2}\right)\cos t + i\left(3 - \frac{5s}{2}\right)\sin t$

13. (a) $(1 - s)\,z(t) + sA_0$

Section 4.2, page 167

2. In any region R having the property that
(a) If z is in R, then none of the points $z + 2\pi n$ or $(2n + 1)\pi - z$ is in R, $n = 0, \pm 1, \ldots$.
(b) If z is in R, then $z + 2\pi i n\,(n = 0, \pm 1, \ldots)$ is not in R.
(c) The whole z-plane.
(d) Whole z-plane.
(e) If z is in R, then $-z$ and $\pm iz$ are not in R.
(f) If z is in R, then $z + 2\pi i n$ and $(2n + 1)\pi i - z$ are not in R.
(g) If z is in R, then $-z$ is not in R.

3. (a) $z \neq (n + \frac{1}{2})\pi$ (b) Everywhere. (c) Everywhere.
(d) Everywhere. (e) $z \neq 0$ (f) $z \neq (n + \frac{1}{2})\pi i$ (g) $z \neq 0$

4. (a) f is onto.

5. (a) The left half-plane.
(b) Yes. $z = \log w$ where $0 < \arg(w) \le 2\pi$.

6. (b) $z = e^w$

7. Invertible everywhere with $z = g(w) = u + i(v - u)$.

8. (a) Invertible on any domain D_z having the property that if z is in D_z, then $\bar{z}$ is not.
(b) Invertible everywhere.

Section 4.3, page 177

2. All of these are at $z = 0$:
(a) $\mathbf{T} = \mathbf{i}$ (b) $\mathbf{T} = -\mathbf{i}$ (c) Does not exist.
(d) $\mathbf{T} = \alpha_1\mathbf{i} + \alpha_2\mathbf{j}$ where $\alpha = \alpha_1 + i\alpha_2$ (e) $\mathbf{T} = \mathbf{i}$ (f) $\mathbf{T} = \mathbf{j}$
(g) $\mathbf{T} = \mathbf{i}$ (h) $\mathbf{T} = -\mathbf{i}$ (i) $\mathbf{T} = \mathbf{j}$ (j) $\mathbf{T} = -\mathbf{j}$

3. (a) At $z = 1$, $\Delta\theta_{1,2} = \pi/2$; at $z = -1$, $\Delta\theta_{1,2} = 3\pi/2$
(b) At $z = 0$, $\Delta\theta_{1,2} = 0$
(c) At $z = 1$, $\Delta\theta_{1,2} = \pi$

4. Since $w' \neq 0$ anywhere.

5. (b) (i) $\pi/2$; (ii) $\Delta\phi_{1,2} = \pi/2n = (1/n)\Delta\theta_{1,2}$; (iii) $\Delta\phi_{1,2} = (\pi/2n)(4n - 3)$.

6. (b) Strip $\mathrm{Re}(w) > 0, 0 < \mathrm{Im}(w) < \pi/2$.
7. (b) $\Delta\phi_{1,2} = 3\pi/2$; The map is not analytic.

Section 4.4, page 188

2. (a) $\hat{U} = v^2 - u^2 + 6u - 9$ (b) $\hat{U} = e^{2e^u \cos v} \sin(2e^u \sin v)$

3. (b) $\hat{U} = \dfrac{4v^2 - (u^2 + v^2 - 1)^2}{[(u-1)^2 + v^2]^2}$ (c) $h = x^2$

Section 4.5, page 201

2. (a) $w = 2z/(1 - z)$

3. (a) $\dfrac{1}{\pi}\mathrm{Arg}(z^4 + 4)$ (b) $1 - \dfrac{1}{\pi}\mathrm{Arg}\left[-\left(\dfrac{z-1}{z+1}\right)^2\right]$

(c) $1 - \dfrac{1}{\pi}\mathrm{Arg}\left[2 - \left(\dfrac{2 - z^4}{2 + z^4}\right)^2\right]$ (d) $1 + \dfrac{1}{\pi}\mathrm{Arg}[i(z - 1 + i)^2]$

(e) $1 - \dfrac{1}{\pi}\mathrm{Arg}\left[\sin^2\left(\dfrac{iz}{2}\right)\right]$

4. (a) $\dfrac{2}{\pi}\mathrm{Arg}(z)$ (b) $\dfrac{1}{2} + \dfrac{1}{\pi}\mathrm{Re}[\mathrm{Sin}^{-1}(2e^{-z} + 1)]$

(c) $\dfrac{3}{2} - \dfrac{1}{\pi}\mathrm{Re}\left\{\mathrm{Sin}^{-1}\left[1 - 2\left(\dfrac{z+1}{z-1}\right)^2\right]\right\}$

5. (a) $1 - \dfrac{1}{\pi}\mathrm{Arg}\left\{-1 + \left[\dfrac{1 - (z - 1 + i)^4}{1 + (z - 1 + i)^4}\right]^2\right\}$

(b) $1 - \dfrac{1}{\pi}\mathrm{Arg}[e^{\pi(i-1)(z-1)/\sqrt{2}} - 1]$

(c) $1 - \dfrac{1}{\pi}\mathrm{Arg}\left[\dfrac{-\sin(\frac{1}{2}\theta_2)}{\sin(\frac{1}{2}\theta_1)} e^{(i/2)(\theta_2 - \theta_1)}\left(\dfrac{z - e^{i\theta_1}}{z - e^{i\theta_2}}\right)\right]$

(d) $\dfrac{1}{\pi}\mathrm{Arg}[\cos^2(i\,\mathrm{Log}\,z)]$

6. $\frac{1}{2}(x^2 + y^2)^{-1}[x^2 + (y - 1)^2 - 1]$

8. (a) $\alpha = \dfrac{1}{16}(19 - \sqrt{105})$, $R_1 = \left[\dfrac{2\alpha - 1}{\alpha(2 - \alpha)}\right]^{1/2}$

(b) $T = T_2 + \dfrac{1}{R_1}(T_1 - T_2)\ln\left|\dfrac{z - \alpha}{\alpha z - 1}\right|$

Section 4.6, page 212

1. (a) $\Phi = \sin^2\left(\dfrac{-\pi i z}{2A}\right)$ (b) $\Phi = (\mathrm{prin.})\, z^{\pi/\alpha}$

3. (f) Radius $= a + b$; $\gamma = \sqrt{a^2 - b^2}$

4. (a) $\Phi = \dfrac{1}{2}\left[\dfrac{(z + \sqrt{z^2 - 3})e^{-i\alpha}}{3} + \dfrac{3e^{i\alpha}}{z + \sqrt{z^2 - 3}}\right]$

(b) Add $\dfrac{i\Gamma}{2\pi}\log(z + \sqrt{z^2 - 3})$ to the result in (a).

7. (a) $\phi = -2q_0 \ln\left|\dfrac{z^{\pi/\alpha} - z_0^{\pi/\alpha}}{z^{\pi/\alpha} - \bar{z}_0^{\pi/\alpha}}\right|$ (use principal branch)

8. $-2q_0 \ln\left|\dfrac{\sin^2\left(\dfrac{\pi z}{2h_0}\right) - \sin^2\left(\dfrac{\pi z_0}{2h_0}\right)}{\sin^2\left(\dfrac{\pi z}{2h_0}\right) - \sin^2\left(\dfrac{\pi \bar{z}_0}{2h_0}\right)}\right|$

9. $-2q_0 \ln\left|\dfrac{e^{2\pi/z} - e^{2\pi/z_0}}{e^{2\pi/z} - e^{2\pi/\bar{z}_0}}\right|$

10. $-2q_0 \ln\left|\dfrac{(z - z_0)\left(z - \dfrac{1}{z_0}\right)}{(z - \bar{z}_0)\left(z - \dfrac{1}{\bar{z}_0}\right)}\right|$

11. $-2q_0 \ln\left|\dfrac{z + \sqrt{z^2 - 1} - z_0 - \sqrt{z_0 - 1}}{z + \sqrt{z^2 - 1} - \bar{z}_0 - \sqrt{\bar{z}_0^2 - 1}}\right|$

12. (a) $-2q_0 \ln\left|\dfrac{(z - z_0)(z + \bar{z}_0)}{(z - \bar{z}_0)(z + z_0)}\right|$

(b) $-2q_0 \ln\left|\dfrac{(\sqrt{z} - \sqrt{z_0})(\sqrt{z} + \sqrt{\bar{z}_0}}{(\sqrt{z} - \sqrt{\bar{z}_0})(\sqrt{z} + \sqrt{z_0})}\right|$ (use principal branch)

(c) $-2q_0 \ln\left|\dfrac{(e^{z/2} - e^{z_0/2})(e^{z/2} + e^{\bar{z}_0/2})}{(e^{z/2} - e^{\bar{z}_0/2})(e^{z/2} + e^{z_0/2})}\right|$

CHAPTER 5

Section 5.1, page 230

1. (a) $\frac{1}{2}\ln 2 - \pi i/4$ (b) $-\frac{1}{2}(1 + i)$ (c) $\frac{1}{2}(1 + i)$
(d) $\pi i - 2$ (e) $2^{5/4}e^{\pi i/8} - 2^{1/2}(1 + i)$

(f) $\begin{cases} 2\pi & n = 0 \\ 0 & n \neq 0 \end{cases}$

2. $\dfrac{\pi}{4}\sqrt{2}$

4. (b) (i) $(s-a)^{-1}$, $\text{Re}(s) > \text{Re}(a)$; (ii) $n!s^{-n-1}$, $\text{Re}(s) > 0$;
(iii) $s(s^2+a^2)^{-1}$, $\text{Re}(s) > |\text{Im}(a)|$; (iv) $a(s^2+a^2)^{-1}$, $\text{Re}(s) > |\text{Im}(a)|$;
(v) $s(s^2-a^2)^{-1}$, $\text{Re}(s) > |\text{Re}(a)|$; (vi) $a(s^2-a^2)^{-1}$, $\text{Re}(s) > |\text{Re}(a)|$;

(vii) $\dfrac{a}{(s-b)^2+a^2}$, $\text{Re}(s) > \text{Re}(b) + |\text{Im}(a)|$.

5. (a) $\dfrac{4}{27}[(10)^{3/2}-1]$ (b) $\pi/\sqrt{2}$ (c) 8

6. (a) $2\pi i$ (b) $\dfrac{21}{2}+\dfrac{i}{2}(e^6+e^2-2e)$ (c) $i[e^{i-1}-1]$

(d) $-\frac{1}{6}+i$ (e) $\frac{1}{4}(1-i)$ (f) 0 (g) 0

7. (a) $\dfrac{1}{2}\ln 2 - \dfrac{3\pi i}{4}$ (b) $-\dfrac{1}{2}\ln 2 - \dfrac{3\pi i}{4}$ (c) 0

(d) $\dfrac{1}{2}\left[\dfrac{\sinh(2+i)}{2+i} - \dfrac{\sinh(2-i)}{2-i}\right]$ (e) $\dfrac{2i}{i+1}\cosh(\pi/2)$

(f) 0 (g) $\dfrac{2i}{3}$ (h) 0

8. (b) $-\dfrac{4i}{3}$ **9.** $2\pi i$

10. (a) $\begin{cases} 0 & z_0 \text{ outside square} \\ 2\pi i & z_0 \text{ inside square} \end{cases}$ (b) $\frac{1}{4}(\pi i - \ln 3)$ (c) 0

16. (a) $-\pi/4$ (b) 0

Section 5.2, page 244

2. (a) Any region not containing the origin, $-\pi i$.
(b) Everywhere, $-\frac{1}{2}[\cos(-3+4i)-1]$.
(c) Any region not containing the origin or any part of the negative real axis, 0.
(d) Same as in (c), $ie^{-\pi/2}/(1+i)$.

6. (a) 0 (b) $2\pi i \begin{cases} 0 & k > n \\ \dfrac{n(n-1)\cdots(n-k+1)}{(k-1)!} & k \le n \end{cases}$ (c) 0

(d) 0 (e) $-\dfrac{2\pi}{3}$ (f) 0 (g) 0

8. $K_3 = 0$, $K_4 = -1$, $K_5 = 2$
9. $I_3 = -3$, $I_4 = 7$, $I_5 = 4$, $I_6 = 0$
13. (b) $\deg(q) - \deg(p) \ge 2$
19. (c) When there are no sources.

Section 5.3, page 259

1. (a) 0 (b) 0 (c) 0 (d) 0

(e) $0, n \le 0$ or $\dfrac{2\pi i}{(n-1)!}\begin{cases} 0 & n \ge 1 \text{ and odd} \\ (-1)^{(n/2)-1} & n \ge 1 \text{ and even} \end{cases}$

(f) $-\pi i$ (g) 0 (h) $\dfrac{-\pi i}{8}$ (i) 0 (j) $\dfrac{3\pi}{2}$

(k) 0 (l) $\dfrac{-\pi i}{2}$ (m) $\pi\left[1 + \dfrac{\pi}{2} + i(1 - \ln 2)\right]$

(n) 0 (o) $\dfrac{2\pi}{25}(4 + 3i)$ (p) $2\pi i$ (q) $\dfrac{2\pi i \bar{z}_0}{1 - |z_0|^2}\begin{cases} 1 & |z_0| > 1 \\ -1 & |z_0| < 1 \end{cases}$

2. (a) 0 (b) $\dfrac{\pi i}{60}$ (c) $\begin{cases} 0 & |z| < 1 \\ \dfrac{-2\pi i\, g(z)}{z} & |z| > 1 \end{cases}$

4. $|z| \ne 1,\ 2\pi i\begin{cases} 0 & |z| > 1 \\ \dfrac{1}{2z}(e^z - e^{-z}) & |z| < 1 \text{ and } z \ne 0 \\ 1 & z = 0 \end{cases}$

5. $2 + \pi(e^{-i} + ie)$ **6.** $2\pi i, 0; 2\pi i e^i(1 - i), 0$

13. (a) $\Gamma + i\Omega = 2\pi i A;\ F = 0$ (b) $\Gamma = \Omega = 0;\ F = 2\pi\rho \overline{Bg''(0)}$
(c) $\Gamma + i\Omega = 2\pi i B;\ F = -2\pi\rho a_0 \bar{B}$

Section 5.4, page 279

1. (a) $2\pi/\sqrt{3}$ (b) $2\pi/\sqrt{5}$ (c) $2\pi/3\sqrt{3}$ (d) 0 (e) 0

(f) $\dfrac{\pi}{2}$ (g) $\dfrac{2\pi}{\sqrt{a^2 - 1}}\begin{cases} 1 & a > 1 \\ -1 & a < -1 \end{cases}$ (h) $\dfrac{2\pi}{\sqrt{a(a+1)}}$

(i) $\dfrac{2\pi}{|1 - a^2|}$

4. (a) $n = 2k$:

$$-\frac{\cos\theta}{2k}\left[(\sin\theta)^{2k-1} + \sum_{j=1}^{k-1}\frac{(2k-1)(2k-3)\cdots(2k-2j+1)(\sin\theta)^{2k-2j-1}}{2^j(k-1)(k-2)\cdots(k-j)}\right]$$

$$+\frac{1\cdot 3\cdot 5\cdots(2k-1)}{2^k k!}\theta$$

$n = 2k + 1$:

$$-\frac{\cos\theta}{2k+1}\left[(\sin\theta)^{2k} + \sum_{j=0}^{k-1}\frac{2^{j+1}k(k-1)\cdots(k-j)(\sin\theta)^{2k-2j-2}}{(2k-1)(2k-3)\cdots(2k-2j-1)}\right]$$

(b) $n=2k$:

$$\frac{\sin\theta}{2k}\left[(\cos\theta)^{2k-1}+\sum_{j=0}^{k-1}\frac{(2k-1)(2k-3)\cdots(2k-2j+1)(\cos\theta)^{2k-2j-1}}{2^j(k-1)(k-2)\cdots(k-j)}\right]$$

$$+\frac{1\cdot3\cdot5\cdots(2k-1)\theta}{2^k k!}$$

$n=2k+1$:

$$\frac{\sin\theta}{2k+1}\left[(\cos\theta)^{2k}+\sum_{j=0}^{k-1}\frac{2^{j+1}k(k-1)\cdots(k-j)(\cos\theta)^{2k-2j-2}}{(2k-1)(2k-3)\cdots(2k-2j-1)}\right]$$

6. (a) (ii) $2\pi/\sqrt{3}$ (b) $\pi\sqrt{3}/2$

7. (a) $\pi/\sqrt{2}$ (b) $-\pi/27$ (c) $\pi/3$ (d) $\pi/32$
(e) $\pi/3$ (f) $\pi/4$

8. (b) $\dfrac{\pi}{n\sin[(m+1)\pi/n]}$ (c) $1/3$

10. (a) $\pi e^{-5}/2$ (b) $\dfrac{\pi}{e}(\cos 2+2\sin 2)$ (c) $\dfrac{\pi}{4}(3e^{-2}+1)$

11. (b) $\dfrac{\pi}{2(b^2-a^2)}\left[\dfrac{e^{-2b}-1}{b}-\dfrac{e^{-2a}-1}{a}\right]$

(c) $\dfrac{\pi}{8(b^2-a^2)}\left[\dfrac{e^{-4b}-4e^{-2b}+3}{b}-\dfrac{e^{-4a}-e^{-2a}+3}{a}\right]$

12. (a) π (b) $\dfrac{1\cdot3\cdot5\cdots(2n-3)\pi}{1\cdot3\cdot5\cdots(2n-2)}$

(c) $$\frac{\pi\sin(n\pi/m)}{m\sin\left[\dfrac{(p+1)\pi}{m}\right]\sin\left[\dfrac{(p+1+n)\pi}{m}\right]}$$

13. (a) $\dfrac{\sqrt{\pi}}{2\sqrt{2}}$ (b) $\dfrac{\sqrt{\pi}}{2\sqrt{2}}$

Section 5.5, page 297

3. (a) $$\frac{x_0 e^{i\alpha_1}\displaystyle\int_{u_1}^{w}(w-u_1)^{-\alpha_1/\pi}(w-u_2)^{-\alpha_1/\pi}\,dw}{\left[\frac{1}{2}(u_2-u_1)\right]^{1-(2\alpha_1/\pi)}\displaystyle\int_0^1(1-t^2)^{-\alpha_1/\pi}\,dt}-x_0$$

6. (e) $A=ih/\pi$ **9.** (b) $A=2h/\pi$, $B=0$

10. (a) $w = \sqrt{1 + \frac{1}{z^2}}$, where $\sqrt{\zeta} = |\zeta|^{1/2}e^{i\psi/2}, 0 < \psi \le 2\pi$

(b) $\frac{1}{\pi}\,\text{Arg}\left[\frac{1 + \sqrt{z^{-2}+1}}{1 - \sqrt{z^{-2}+1}}\right]$

Section 5.6, page 310

11. $u(x, y) = \text{Re}[f(z)]$ in $|z| < 1$

14. (a) Max $= \sqrt{5}$ at $z = 1 + i$ (b) Max $= e$ at $z = 1$

CHAPTER 6

Section 6.1, page 324

1. (a) $-i$ (b) 0 (c) ∞ (d) $\begin{cases} 0 & \text{Re}(\alpha) > 0 \\ \text{Div.} & \text{Re}(\alpha) \le 0 \end{cases}$

(e) $\begin{cases} 0 & |\alpha/\beta| < 1 \\ 1 & |\alpha/\beta| > 1 \\ 1/2 & \alpha = \beta \\ \text{Div.} & \alpha \ne \beta \text{ and } |\alpha/\beta| = 1 \end{cases}$ (f) 0 (g) $\begin{cases} 0 & \alpha \ne 0 \\ 1 & \alpha = 0 \end{cases}$

(h) $\begin{cases} 0 & \text{Im}(\alpha) > 0 \\ \text{Div.} & \text{Im}(\alpha) \le 0 \end{cases}$ (i) $\begin{cases} \text{Log}\,\alpha & \alpha \ne 0 \\ \text{Div.} & \alpha = 0 \end{cases}$ (j) 0

(k) 0 (l) $\begin{cases} 0 & \text{Re}(\alpha) < 0 \\ 1 & \alpha = 0 \\ \text{Div.} & \text{Re}(\alpha) > 0 \text{ on } \text{Re}(\alpha) = 0 \text{ and } \text{Im}(\alpha) \ne 0 \end{cases}$

3. (a) 0 (b) $\pi/2$

4. (b) No; $z_n = e^{in}$ (d) No; $z_n = -1 + \frac{i}{n}(-1)^n$

7. (a) $\begin{cases} \text{Conv.} & \text{Re}(\alpha) \le 0 \\ \text{Div.} & \text{Re}(\alpha) > 0 \end{cases}$ (b) Conv. (c) Conv.

(d) $\begin{cases} \text{Conv.} & |\alpha| > 1 \\ \text{Div.} & |\alpha| < 1 \end{cases}$

10. (a) $(1/2i)(z + i)$ for $\text{Im}(z) > 0$; otherwise diverge.
(b) $(1 - e^{1/z})^{-1}$ for $\text{Re}(z) < 0$; otherwise diverge.
(c) $z/(z - 1)$ for $|z| > 1$; otherwise diverge.

14. (a) $u(\theta) = f(e^{i\theta})$

Section 6.2, page 335

1. (a) $|z-i|=1$ (b) All z. (c) $|z|<1/2$
   (d) $|z|<|a|^{1/2}$ (e) All z. (f) $|z|<e$

5. (a) $|z|<\dfrac{1}{\max\{a,b\}}-\delta,\ \delta>0$ (b) $\left|z+\dfrac{1}{2}\right|<\dfrac{1}{2}-\delta$

   (c) $|z-i|<e-\delta$ (d) Everywhere. (e) $|z|<1-\delta$

9. (a) $-\displaystyle\int_0^1 \frac{\ln(1+x)}{x}\,dx-\oint_{-1}^{z}\frac{\operatorname{Log}(1-\zeta)}{\zeta}\,d\zeta$, where C is inside the unit circle and does not pass through $\zeta=0$.

   (b) $1+\dfrac{(1-z)\operatorname{Log}(1-z)}{z}$

Section 6.3, page 349

2. (a) $(-1)^k\displaystyle\sum_{n=0}^{\infty}\frac{(-1)^n(z-\pi k)^{2n+1}}{(2n+1)!}$, all z

   (b) $(-1)^k\displaystyle\sum_{n=0}^{\infty}\frac{(-1)^n[z-(k+\frac{1}{2})\pi]^{2n}}{(2n)!}$, all z

   (c) $\displaystyle\sum_{n=0}^{\infty}\frac{(-1)^n(z-2)^n}{(2-i)^{n+1}}$, $|z-2|<\sqrt{5}$

   (d) $(1+3i)^{-1}\displaystyle\sum_{n=0}^{\infty}[i^{n+1}-3^{n+1}]z^n$, $|z|<1/3$

   (e) $\displaystyle\sum_{n=0}^{\infty}\frac{(-1)^n z^{2(n+1)}}{n+1}$, $|z|<1$

   (f) $\dfrac{1}{2}\displaystyle\sum_{n=0}^{\infty}(-1)^n(1-3^{-n-1})(z-1)^n$, $|z-1|<1$

   (g) $\displaystyle\sum_{n=0}^{\infty}(-1)^n(z-1)^n\left[1+\frac{i}{(1-i)^{n+1}}+\frac{n+1}{(1-i)^{n+2}}\frac{-i(n+1)(n+2)}{2(1-i)^{n+3}}\right]$,
   $|z-1|<1$

   (h) $\displaystyle\sum_{n=0}^{\infty}\frac{\beta(\beta-1)\cdots(\beta-n+1)}{n!}\alpha^{\beta-n}z^n$, $|z|<\begin{cases}|\alpha| & \text{for } \operatorname{Re}(\alpha)\ge 0\\ |\operatorname{Im}(\alpha)| & \text{for } \operatorname{Re}(\alpha)<0\end{cases}$

   (i) $\dfrac{1}{2}\displaystyle\sum_{n=0}^{\infty}\frac{z^n}{n!}[(1+i)^n+(1-i)^n]$, all z

3. (a) $\displaystyle\sum_{n=0}^{\infty}\frac{(-1)^n z^{2n+1}}{(2n+1)n!}$ (b) $\displaystyle\sum_{n=0}^{\infty}\frac{(-1)^n z^{2n+1}}{n!(n+1)}$

(c) $z + \sum_{n=1}^{\infty} \frac{(-1)^n(2n-1)!z^{2n+1}}{2^{2n-1}(2n+1)(n-1)!}$

4. (b) $\ln 2 - \frac{z}{4} - \frac{3}{32}z^2 - \frac{5}{96}z^3, |z| < 1$

6. (b) $\sum_{n=0}^{\infty} \frac{(-1)^n z^{2n}}{(2n+1)!}$

7. (a) $\frac{1}{2} + \frac{z}{4} + 0.z^2$ (c) $|z| < \pi$

8. (d) $e^{kz}, k \neq 0$

11. (b) (i) 0; (ii) 0; (iii) -1.

15. (a) $z = 0$, order 5, and $z = 2\pi i n$ $(n = \pm 1, \pm 2, \ldots)$, order 1

(b) $z = \left(n + \frac{1}{2}\right)\pi$, order 3 (c) $z = 0$, order 2

(d) $z = \ln(2\pi|k|) + 2\pi i n + \frac{\pi i}{2}\begin{cases} 1 & k > 0, \text{ order } 1 \\ -1 & k < 0, \text{ order } 1 \end{cases}$

$(k = \pm 1, \pm 2, \ldots; n = 0, \pm 1, \pm 2, \ldots)$

(e) $z = n\pi$ $(n = 0, \pm 1, \pm 2, \ldots)$, order 3

17. (a) 1 (b) 3 (c) 3 (d) 2

18. (b) (i) $z = 0$, doubled; (ii) $z = n\pi$; doubled; (iii) $z = 0$, doubled; (iv) $z = 2\pi i n$ $(n = 0. \pm 1, \ldots)$, tripled.

19. (a) $z = n\pi/2$ (b) Doubled.

Section 6.4, page 361

1. (a) $\sum_{n=-1}^{\infty} \frac{(-1)^{n+1}(z-1)^n}{2^{n+2}} (0 < |z-1| < 2); \sum_{n=-2}^{-\infty} (-1)^n 2^{-n-2}(z-1)^n$,

$|z-1| > 2$

(b) $e^{-1} \sum_{n=-1}^{\infty} \frac{(z+1)^n}{(n+1)!}, 0 < |z+1| < \infty$

(c) $i \sum_{n=0}^{\infty} (-1)^n[1 - (1-i)^{-n-1}](z-1)^n, 0 \leq |z-1| < 1$

$i\left[\sum_{n=-\infty}^{-1} (-1)^{n+1}(z-1)^n - \sum_{n=0}^{\infty} (-1)^n(1-i)^{-n-1}(z-1)^n\right]$,

$1 < |z-1| < \sqrt{2}$

$i \sum_{n=-\infty}^{-1} (-1)^{n+1}[1 - (1-i)^{n-1}](z-1)^n, |z-1| > \sqrt{2}$

(d) $\sum_{n=0}^{\infty} \frac{(-1)^n}{2} z^{2n} - \frac{i}{2} \sum_{n=1}^{\infty} (n+1) i^{-n} z^n, 0 \le |z| < 1$

$\frac{1}{2} \sum_{n=-\infty}^{-1} [(-1)^{n+1} z^{2n} - i^{n+1} z^n], 1 < |z-1|$

(e) $z^{-2} + \sum_{n=0}^{\infty} \frac{z^{2n}}{(n+2)!}, 0 < |z|$

(f) $\sum_{n=0}^{\infty} \frac{(-1)^n 2^n z^n}{(n+1)!}, 0 < |z|$

(g) $\sum_{n=0}^{\infty} (-1)^n \frac{z^{4n+1}}{(2n+1)!}, 0 < |z|$

(h) $\sum_{n=-\infty}^{0} (-1)^n \frac{z^{2n-1}}{(1-2n)!}, 0 < |z|$

(i) $\sum_{n=-\infty}^{0} (-1)^n \frac{z^n}{(-n)!}, 0 < |z|$

2. (a) $\frac{e}{2}[(z-1)^{-1} + \frac{1}{2} + \frac{1}{4}(z-1)], R = 2$

(b) $(-1)^n \left[(z - n\pi)^{-1} + \frac{1}{6}(z - n\pi) + \frac{7}{360}(z - n\pi)^3 \right], R = \pi$

3. (a) $\sum_{n=-\infty}^{-2} (-1)^n i^{-n-2} 2^{n+1} z^n$

(b) $-\frac{1}{4} \sum_{n=-\infty}^{-1} i^{-n}[1 + (-1)^n 3^{-n-1}] z^n$

(c) Two, since $g(z) = g(2i) \Rightarrow z = 2i, -4i$.

4. $\frac{1}{2i} \sum_{n=-\infty}^{\infty} c_n z^n, c_n = \begin{cases} \sin 1 & n \ge 0 \\ \sum_{k=-n}^{\infty} \frac{i^k}{k!} [1 - (-1)^k] & n < 0 \end{cases}$

5. $\sum_{n=-\infty}^{\infty} (c_n + d_n + e_n) z^n$

Section 6.5, page 371

1. (a) Prin. part $= 0$ at $z = 0$.

Prin. part $= \frac{2\pi i n}{z - 2\pi i n}$, at $z = 2\pi i n$ $(n = \pm 1, \pm 2, \ldots)$.

(b) Prin. part $= \frac{-i}{4(z-i)}$, at $z = i$.

$$\text{Prin. part} = \frac{1}{(i-1)(z+1)^2} + \frac{i}{2(z+1)} \text{ at } z = -1.$$

(c) Prin. part = 0 at $z = 0$; prin. part $= \dfrac{\sin 1}{z-1}$ at $z = 1$.

(d) $\dfrac{1}{5(z-1)}$, at $z = 1$; $\dfrac{\ln 2 + (\pi i/2)}{4i(2i-1)^2(z-2i)}$, at $z = 2i$; $\dfrac{-\ln 2 + (\pi i/2)}{4i(2i+1)^2(z+2i)}$, at $z = -2i$.

(e) $\dfrac{1}{n^4\pi^4}\left[\dfrac{1}{(z-1/n\pi)^2} + \dfrac{2\pi n}{(z-1/n\pi)}\right]$, at $z = \dfrac{1}{n\pi}$, $n = \pm 1, \pm 2, \ldots$.

(f) $\dfrac{1}{2!z^2}$, at $z = 0$.

3. (b) $\left.\dfrac{d^k g}{dz^k}\right|_{z=0} = 0$, $k = 0, 1, \ldots, n-1$

4. (a) $z_n = \pm\pi\sqrt{|n|}\begin{cases} 1+i & n \geq 0 \\ 1-i & n < 0 \end{cases}$ are simple poles.

(b) $z = 0$ is removable; $z = 2\pi i n$ $(n = \pm 1, \pm 2, \ldots)$, simple poles.
(c) $z = 0$ is essential.
(d) $z = 0$ is pole of order 2; $z = n\pi (n = \pm 1, \pm 2, \ldots)$, simple poles.

(e) $z = i$, simple pole; $z = \pm\dfrac{1-i}{\sqrt{2}}$, simple poles.

(f) $z = 0$, removable; $z = 1$, essential; $z = \pi/2$, removable.
(g) $z = 0$, removable; $z = 2\pi i n$ $(n = \pm 1, \pm 2, \ldots)$, simple poles.
(h) $z = 0$, essential; $z = \pi$, removable; $z = n\pi$ $(n \neq 0, 1)$, simple poles.
(i) $z = 0$, pole order 2; $z = \pm 2$, simple poles; $z = 1$, essential.
(j) $z = 0$, simple pole; $z = n\pi$ $(n \neq 0)$, pole order 3.
(k) $z = 0, \pm 4$, simple pole; $z = 8$, essential.
(l) $z = 0$ is removable if $m \geq n$; pole order $n - m$ if $m < n$.
(m) $z = (\pi/4) + \pi n$ $(n = 0, \pm 1, \ldots)$ are simple poles.
(n) $z = 0$ is essential.

5. (c) Simple pole; prin. part $= 1/2(z-1)$.

7. (a) Removable. (b) Pole order $N + 1$.

8. (a) Same as for $f(z)$.
(b) Pole of order $N - k$ if $k \leq N - 1$; removable if $k \geq N$.

9. Simple pole with principal part $-N/(z - z_0)$.

11. (a) Removable if $M = 0$; pole of order $M \cdot N$ if $M \neq 0$.
(b) Yes. Let $f(z) = 1/z$ and $g(z) = e^z$.

12. (a) Removable. (b) Removable, or pole, at worst, order N.
(c) Essential.

15. (a) Nonisolated. (b) Nonisolated. (c) Pole order 3.
(d) Nonisolated. (e) Removable. (f) Essential.
(g) Nonisolated. (h) Nonisolated. (i) Essential.
(j) Nonisolated. (k) Essential. (l) $m = 0 \Rightarrow$ removable.
(m) Nonisolated. (n) Removable.

Section 6.6, page 379

1. (a) $\operatorname{Res}|_{z=0} = 1$ (b) $\operatorname{Res}|_{z=0} = 0$, $\operatorname{Res}|_{z=\pm i} = 1/2$
(c) $\operatorname{Res}|_{z=0} = 1$, $\operatorname{Res}|_{z=1} = \sin 1$.
(d) $\operatorname{Res}|_{z=0} = 1$, $\operatorname{Res}|_{z=\pm\sqrt{2\pi i n}} = \frac{1}{2}$, $n = \pm 1, \pm 2, \ldots$
(e) $\operatorname{Res}|_{z=0} = \frac{1}{24}$ (f) $\operatorname{Res}|_{z=0} = 0$; $\operatorname{Res}|_{z=n\pi} = n\pi$ $\operatorname{Res}|_{z=1} = 1$

(h) $\operatorname{Res}|_{z=\ln(\pi|n|)+i\theta_{k,n}} = \dfrac{(-1)^n}{n\pi}$, $n = \pm 1, \pm 2, \ldots, \theta_{k,n} = 2\pi k + \begin{cases} 0 & n \geq 1 \\ \pi & n \leq -1 \end{cases}$

(i) $\operatorname{Res}|_{z=n\pi} = \pi(-1)^n(n-1)$

(j) $\operatorname{Res}|_{z=1/2\pi i n} = \dfrac{-1}{4\pi^2 n^2}$, $n = \pm 1, \pm 2, \ldots$

(k) $\operatorname{Res}|_{z=0} = \cos 1$

2. $1/z^2$

3. (b) $\dfrac{2}{3[h''(z_0)]^2}[3h''(z_0)g'(z_0) - g(z_0)h'''(z_0)]$

5. (a) $2\pi i$ (b) $4\pi i/3$ (c) 0 (d) $2\pi i$
(e) 0 (f) $-2\pi i$ (g) 0 (h) $2\pi i$ (i) 0

(j) $2\pi i$ (k) $2\pi i\left[\dfrac{e^{1/4}\sin 1}{2} - e\right]$ (l) $\begin{cases} 0 & n \leq 0 \text{ or } n \geq 2 \\ 2\pi i & n = 1 \end{cases}$

6. (a) $-2\pi i \displaystyle\sum_{k=0}^{\infty} \frac{(-1)^k}{(2k+1)!}$ (b) $-2\pi i \sin 1$

7. (b) $2\pi i\left[\displaystyle\sum_{k=1}^{N_0} n_k^{(0)} g(z_k^{(0)}) - \sum_{k=1}^{N_p} n_k^{(p)} g(z_k^{(p)})\right]$

9. (a) -1 (b) -1 (c) $-1 - \sin 1$
(e) $-1/24$ (k) $-\cos 1$

Section 6.7, page 392

2. (a) $\dfrac{-13\pi}{48}$ (b) π (c) $\dfrac{\pi a^{n-1}}{2}\left(\dfrac{1+a^2}{1-a^2}\right)$

(d) $\begin{cases} 0 & m < n \\ \dfrac{\pi}{2}\alpha^{m-1} & m \geq n \end{cases}$

5. (a) $\pi e^{-\alpha\sqrt{A-B^2}}\left[\dfrac{a-bB}{\sqrt{A-B^2}}\cos(\alpha B)+b\sin(\alpha B)\right]$

(b) $\dfrac{\pi}{2}e^{-ab\cos\lambda}\left[\dfrac{\sin(2\lambda-ab\sin\lambda)}{\sin(2\lambda)}\right]$

(c) $\dfrac{\pi}{4\alpha}\left[e^{-\alpha(b-a)}-e^{-\alpha(a+b)}\right]$ (d) $-\pi e^{-a\alpha}\sinh(b\alpha)$

6. (b) $\dfrac{\pi}{b}\sin\dfrac{\pi a}{b}$

7. (b) Does not exist.

8. (b) (i) π; (ii) $\pi/2$; (iii) $\pi a^2/4$.

10. (a) $\dfrac{\pi}{2n\sin\left(\dfrac{\pi(a+1)}{2n}\right)}$ (b) $\dfrac{\pi(-1)^n(a+1)a(a-1)\cdots(a-n+2)b^{a-n+1}}{(a+1)(n-1)!\sin(\pi a)}$

(c) $\dfrac{\pi}{2}\left[\dfrac{2^{3/4}}{10}\left(6\cos\left(\dfrac{\pi}{8}\right)+2\sin\left(\dfrac{\pi}{8}\right)\right)-\dfrac{2}{5}\right]$

(d) $\dfrac{\pi(-1)^n(a-b)(a-2b)\cdots(a-(n-1)b)}{b^n(n-1)!\sin\left(\dfrac{\pi(a-b)}{b}\right)}$

11. (a) $\dfrac{\pi(e^{2\pi ia/3}-1)}{(e^{2\pi i/3}-1)\sin\left(\dfrac{2\pi}{3}\right)}+\displaystyle\int_0^\infty\frac{(x^a e^{\pi ia}-1)}{x^3+1}\,dx$

(b) $\dfrac{\pi\sin(\pi a/3)}{3\sin\left(\dfrac{\pi}{3}\right)\sin\left(\dfrac{\pi(a+1)}{3}\right)}$

12. (d) $\dfrac{\pi\sin(\pi a/2)}{2\sin\left(\dfrac{\pi(a+1)}{2}\right)}$

14. (a) $\displaystyle\int_0^\infty\frac{\sin x}{x^a}\,dx=\cos\left(\frac{\pi a}{2}\right)\Gamma(1-a)$; $\displaystyle\int_0^\infty\frac{\cos x}{x^a}\,dx=\sin\left(\frac{\pi a}{2}\right)\Gamma(1-a)$

(b) $\dfrac{\sin\left[\dfrac{\pi}{2}(a-1)\right]\Gamma(2-a)}{a-1}$

17. (a) $\dfrac{\pi\ln a}{2a(1-a^2)}$ (b) $\dfrac{\ln a}{a}$ (c) $\dfrac{1}{2}(\ln 3-1)$

18. (a)

$$-\frac{1}{n+1}\sum \operatorname{Res}\left[\frac{p(z)(\log z)^{n+1}}{q(z)}\right] - \frac{1}{n+1}\sum_{k=0}^{n-1}\frac{(n+1)!(2\pi i)^{n-k}}{k!(n+1-k)!}\int_0^\infty \frac{p(x)(\ln x)^k}{q(x)}\,dx$$

(b) $\pi^3/8$ 19. $\pi^2/2\sqrt{2}$ 20. $\pi^2/4$

Section 6.8, page 408

6. (a) 2 (b) 2 (c) 3 (d) 2
13. (a) $f(z) = z$ (b) $f(z) = z^2$ (c) $f(z) = 1 - z$
17. Let z be in D_z and z^* be the image of z in L_z. Let $w = f(z)$ and w^* be the image of w in L_w. Then define $f(z^*) = w^*$.

A

SOME CONFORMAL MAPPINGS

Here, we collect some of the results concerned with the elementary functions and qualitatively illustrate the geometry of their mapping properties.

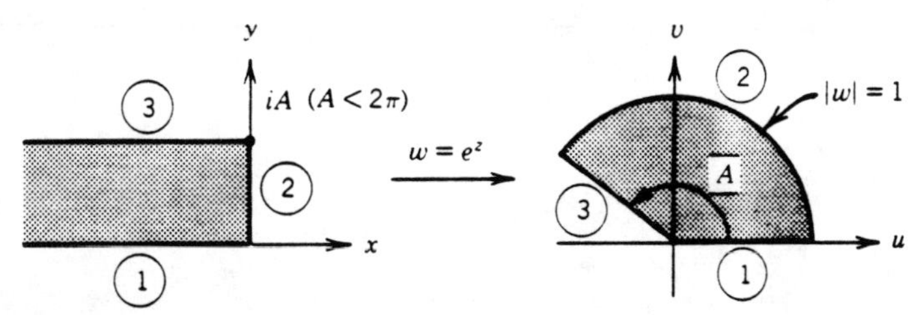

Figure A.1

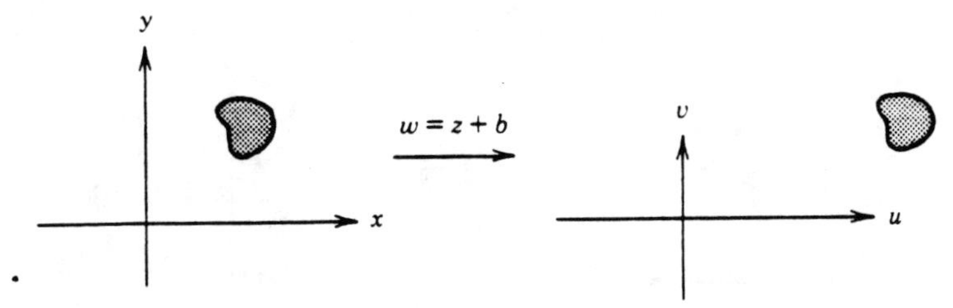

Figure A.2

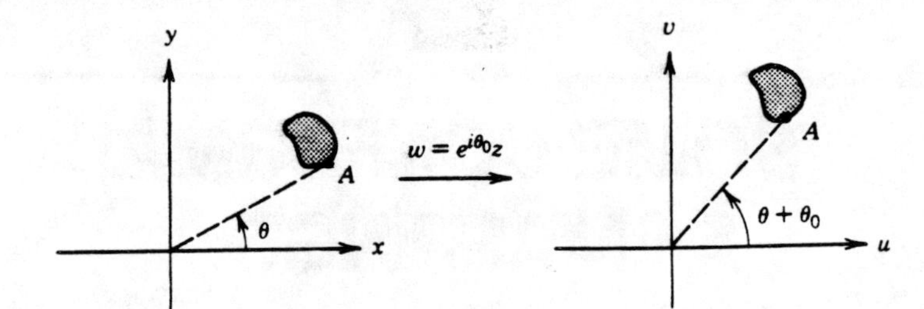

Figure A.3

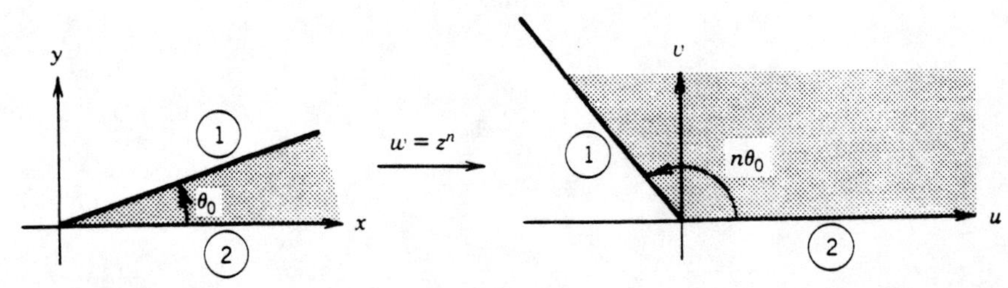

Figure A.4

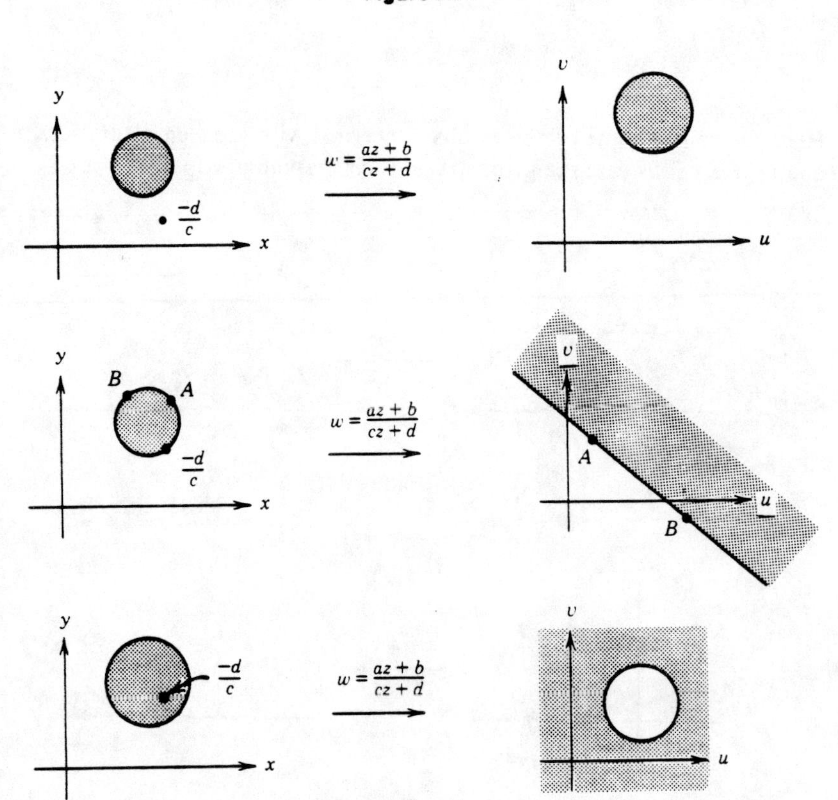

Figure A.5

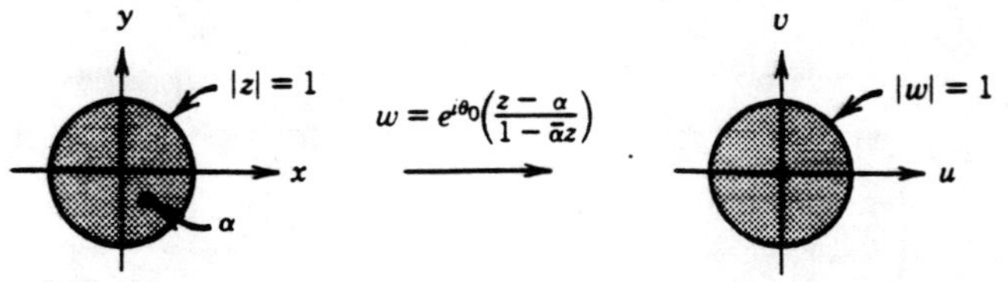

y $|z| = 1$ α x

$$w = e^{i\theta_0}\left(\frac{z - \alpha}{1 - \bar{\alpha}z}\right)$$

v $|w| = 1$ u

Figure A.6

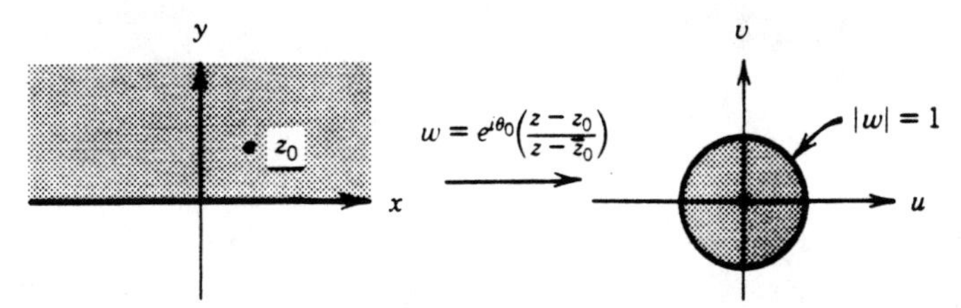

Figure A.7

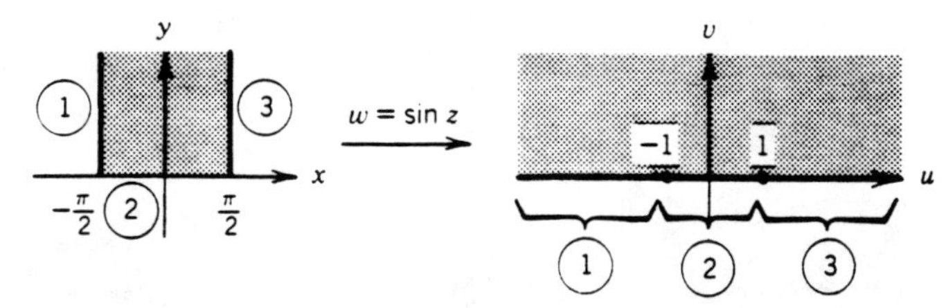

Figure A.8

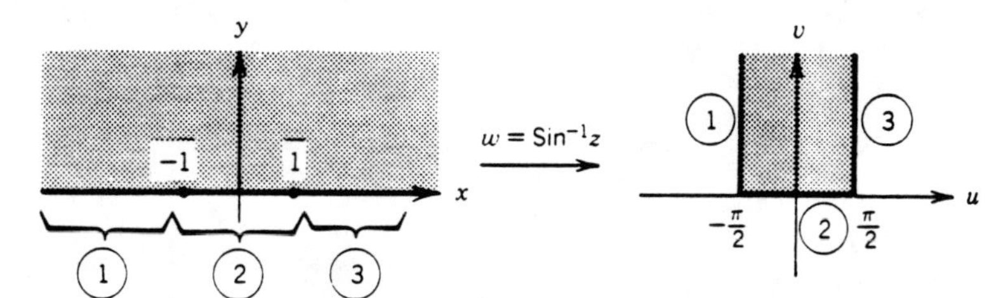

Figure A.9

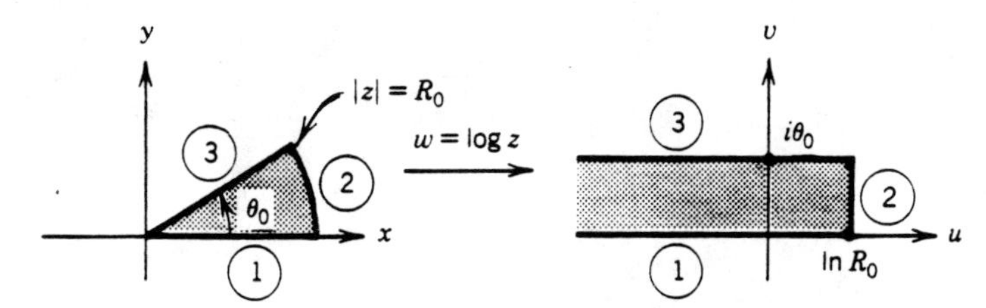

Figure A.10

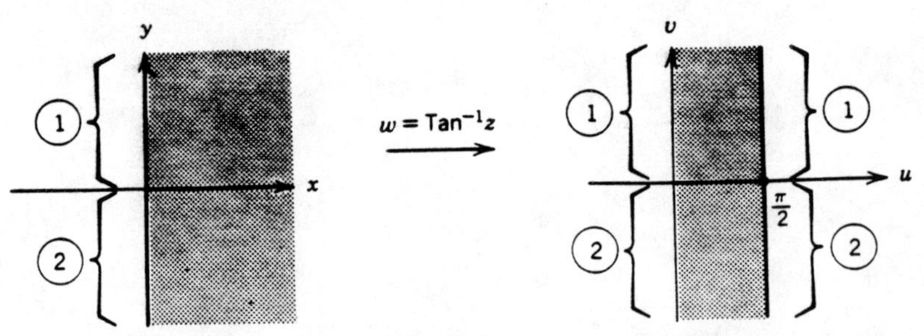

Figure A.11

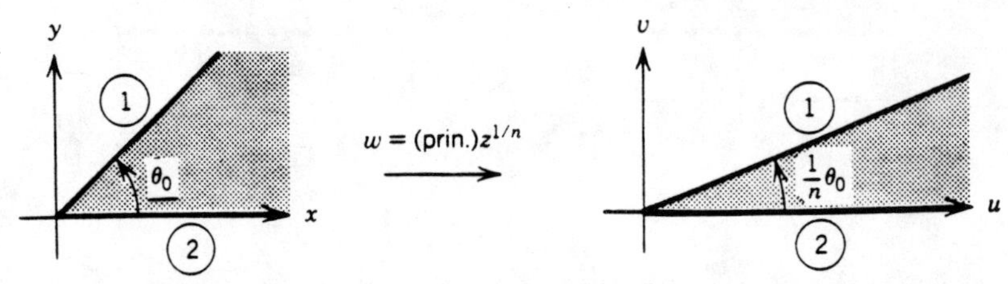

Figure A.12

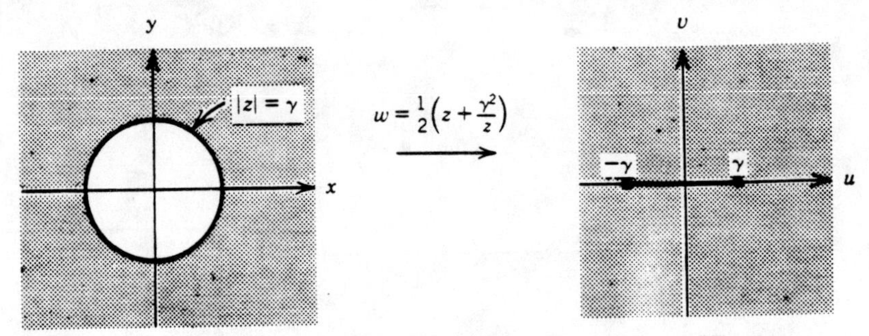

Figure A.13

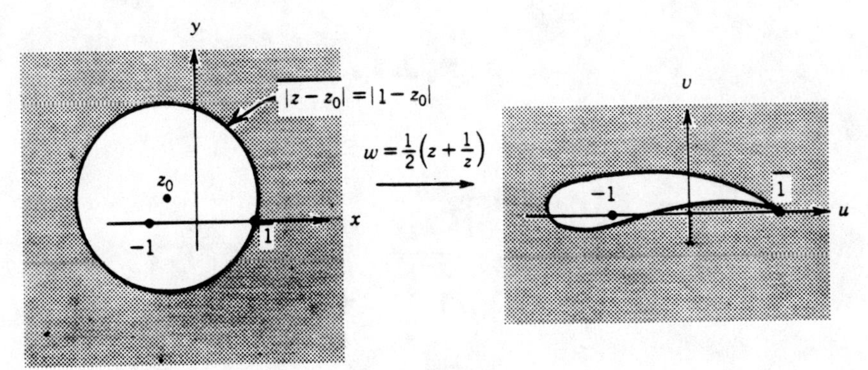

Figure A.14

B

STEADY-STATE TEMPERATURE DISTRIBUTIONS

Here, we derive, in as self-contained a manner as possible, the equations governing two-dimensional, steady-state temperature distributions. The physical formulation is not meant to be complete, and the interested reader is directed to other references on the subject.

To begin with, consider a region in space occupied by a **thermally conducting material**, with (x, y, z) representing a point in space and t representing time. We define

(B.1)

$T(x, y, z; t)$ = the **temperature** at (x, y, z) at time t

$\rho(x, y, z)$ = **density** (mass per unit volume) at (x, y, z) of the material

$Q(x, y, z; t)$ = amount of **heat** at (x, y, z) at time t

$s(x, y, z; t)$ = the rate per unit volume at which heat is produced at time t due to **heat sources** at (x, y, z)

What follows are some experimental observations from which we are able to formulate our **mathematical model.**

(B.2) **Experimental Facts**

(a) In a thermally conducting body, heat flows from a point of higher temperature to one of lower temperature.

(b) The amount of heat per unit time crossing a small surface element at a point is proportional to the surface area and to the normal component of the temperature gradient at that point. The (positive) factor of proportionality, κ, is called the **coefficient of thermal conductivity**. It may vary from point to point and hence, in general, is a function $\kappa(x, y, z)$ of position.

(c) The total amount of heat in a small bit of material is proportional to the mass and temperature of the material. This (positive) factor of proportionality, $c(x, y, z)$, is called the **specific heat** of the body.

Now let V be any region in the thermally conducting material (see Figure B.1), S its bounding surface, and $\mathbf{n}$ the **unit outer normal field** to S. If $\Delta\sigma$ is a small bit of surface area on S, then, from (a) and (b) above, we have

$$\text{Rate at time } t \text{ at which heat enters } V \text{ through } \Delta\sigma = \kappa(\mathbf{n} \cdot \nabla T)\, \Delta\sigma$$

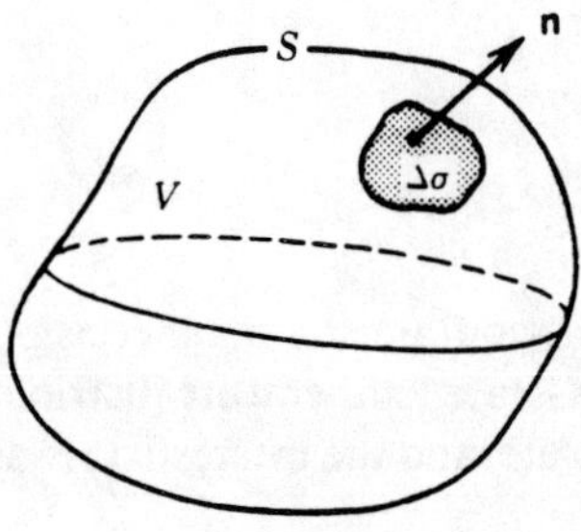

Figure B.1

Hence,

$$\text{Rate at time } t \text{ at which heat enters } V \text{ through its whole boundary } S = \iint_S \mathbf{n} \cdot (\kappa \nabla T)\, d\sigma$$

(B.3)

$$= \iiint_V \nabla \cdot (\kappa \nabla T)\, dV$$

where ∇ is the **gradient operator**, and the second equality follows from the **divergence theorem**[1] if we assume enough smoothness for κ and T.

Now, if ΔV represents a small amount of volume in V, then, from experimental fact (c),

$$\text{Amount of heat in } \Delta V \text{ at time } t = cT(\rho\, \Delta V)$$

[1] Vector integral theorems are discussed in most Advanced Calculus books.

and hence,

(B.4)
$$\text{Rate of increase at time } t \text{ of heat in } V = \frac{d}{dt}\iiint_V c\rho T\, dV$$

$$= \iiint_V c\rho \frac{\partial T}{\partial t}\, dV$$

The amount of heat in V can increase in time due to (i) heat entering through its surface and (ii) sources in V producing heat. Hence, from the above, (B.3), and the definition of s in (B.1), we have the **balance law**

(B.5)
$$\iiint_V c\rho \frac{\partial T}{\partial t}\, dV = \iiint_V \nabla \cdot (\kappa \nabla T)\, dV + \iiint_V s\, dV$$

Now, by the **mean value theorem for integrals**, we can replace each term of the above by the average value of the integrands at some point in V multiplied by the volume of V. Hence, if we divide by this volume and let V shrink to a point, we get the following **inhomogeneous heat equation** at each point in V:

(B.6)
$$c\rho \frac{\partial T}{\partial t} = \nabla \cdot (\kappa \nabla T) + s$$

In order to completely determine the temperature distribution in V, we must specify (i) **initial conditions** at $t = 0$ and (ii) **boundary conditions** on the surface S. At this point, we make the following assumptions.

(B.7) **Assumptions**

(a) The coefficient of thermal conductivity κ is a constant.
(b) There are no sources [$s = 0$ in (B.6)].

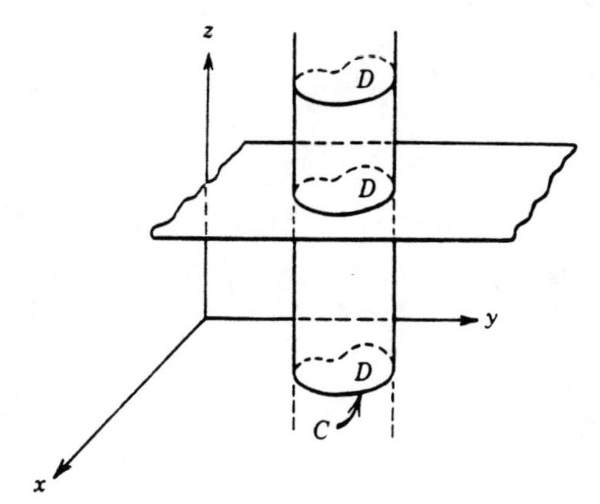

Figure B.2

(c) All initial effects decay to zero in time.

(d) The region V is two-dimensional (**cylindrical**), as shown in Figure B.2. By this we mean that every horizontal plane cuts it in the same cross section D bounded by a curve C.

(e) The boundary conditions attain, in time, a time-independent **steady state** which is the same on each horizontal cross section (that is, does not depend on the vertical variable z).

■

With these assumptions, we can *expect* the following:

Conclusions

(a) Since all (time-dependent) boundary terms ultimately reach a steady state and since the effects of the initial conditions die out, the temperature function T should also attain a steady state, and hence ultimately be independent of t.

(b) Since the geometry of the region is independent of the vertical coordinate and since the boundary conditions attain a steady state having the same properties, we expect that the steady-state temperature distribution T should depend only on x and y, that is, $T = T(x, y)$.

■

With the above assumptions and conclusions, we see that the heat equation (B.6) reduces to Laplace's equation. $T_{xx} + T_{yy} = 0$ in the two-dimensional domain D bounded by a curve C, as shown in Figure B.3*a*. Now, on the surface of a thermally conducting body, we can

(B.8)
(a) Prescribe the temperature distribution on one part of the boundary.
(b) Insulate the rest of the boundary.

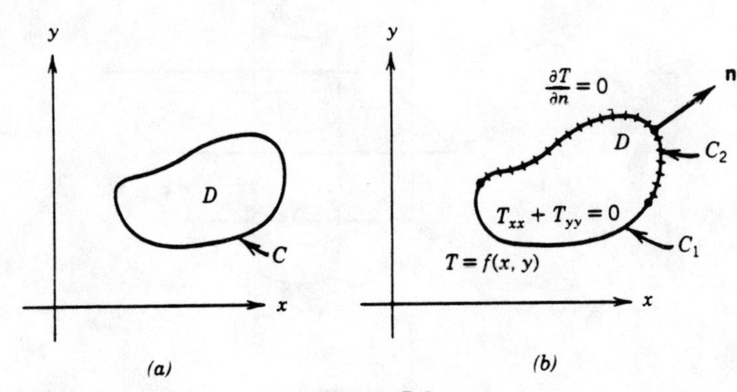

Figure B.3

A surface is said to be **insulated** if no heat can flow through it. From Experimental Fact (B.2b), the heat flow is proportional to $\mathbf{n} \cdot \nabla T$; and hence, on an insulated surface, we have

(B.9)
$$\mathbf{n} \cdot \nabla T \equiv \frac{\partial T}{\partial n} = 0 \qquad \text{(insulated boundary)}$$

where $\partial T/\partial n$ denotes the **normal derivative** of T in the direction of the outer normal to the boundary. Combining this with the boundary conditions (B.8), we arrive at the following **boundary value problem for a two-dimensional steady-state temperature distribution $T(x, y)$ in a region D bounded by a curve C** (see Figure B.3*b*).

(B.10)
$$\text{(a)} \quad T_{xx} + T_{yy} = 0 \qquad (x, y) \text{ in } D$$
$$\text{(b)} \quad T(x, y) = f(x, y) \qquad (x, y) \text{ on } C_1$$
$$\text{(c)} \quad \frac{\partial T}{\partial n} = 0 \qquad (x, y) \text{ on } C_2$$

Boundary condition (b), in which the temperature is prescribed on part of the boundary C_1, is called a **Dirichlet condition**, while (c), in which the normal derivative is prescribed on the remaining part of the boundary, is called a **Newmann condition.** Finally, if the region D is unbounded, we also prescribe that $T(x, y)$ be bounded at infinity.[2]

[2] We will assume that we have uniqueness, that is, there is only one sufficiently differentiable, bounded solution to our problem. The reader is referred to books on partial differential equations where these matters are discussed in detail.

C

FLUID FLOW

Here, we will derive some of the basic equations describing the steady flow of an ideal, irrotational, incompressible fluid. We refer the reader to any elementary textbook for a more thorough derivation.

If a fluid is flowing in some region of space, then at each instant of time, and to each point of this region, we can assign a vector (the **velocity of a fluid particle** passing through that point). By a **steady fluid flow** we mean that this vector will be the same even at different times when a different particle might be arriving at that same point. Thus, there will be a **fluid velocity field** defined (and which may be different) at each point of the fluid and which does not change with time. By an **ideal fluid** we mean that all thermal conductivity and **viscosity** effects have been ignored, while by a **two-dimensional fluid flow** we mean that the velocity field is parallel to a plane (which we will take to be the xy-plane) and is independent of the coordinate defining distance above the plane. This is of course an idealization which is met only approximately in practice. It might be set up as follows. A very long pipe, with many holes in it, is perpendicular to the xy-plane (see Figure C.1). If we neglect outside forces, then, if fluid is emitted from the pipe at an eventual steady rate, it will enter the surrounding space in such a manner that after **transient (initial) effects** decay, the resulting steady fluid motion will be approximately two-dimensional.

By the above, a two-dimensional, steady fluid can be described by a fluid velocity vector field $\mathbf{v}$ which depends only on the x- and y-coordinates and has no $\mathbf{k}$ component, that is,

(C.1) $$\mathbf{v}(x, y) = v_1(x, y)\mathbf{i} + v_2(x, y)\mathbf{j}$$

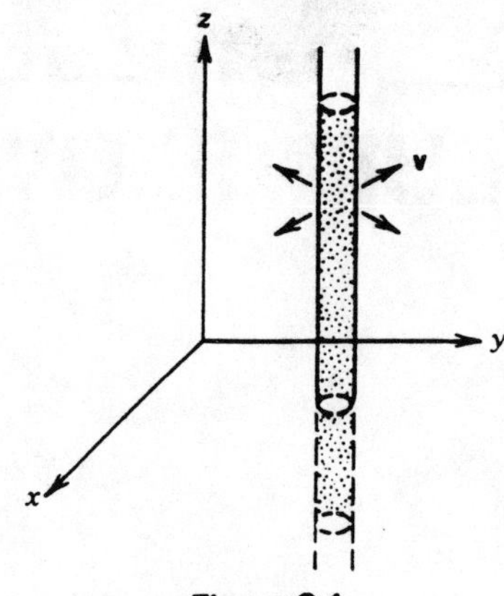

Figure C.1

where v_1 and v_2, the **i** and **j** components of **v**, are scalar functions of position. Since **v** is independent of the z-coordinate and is parallel to the xy-plane, we need only investigate the fluid flow in the xy-plane itself. Hence, as shown in Figure C.2, at each point (x, y) in the plane, we assign a vector $\mathbf{v}(x, y)$. To derive the mathematical equations describing our fluid flow, we must take for granted certain physical laws. For our purposes, these will be as follows.

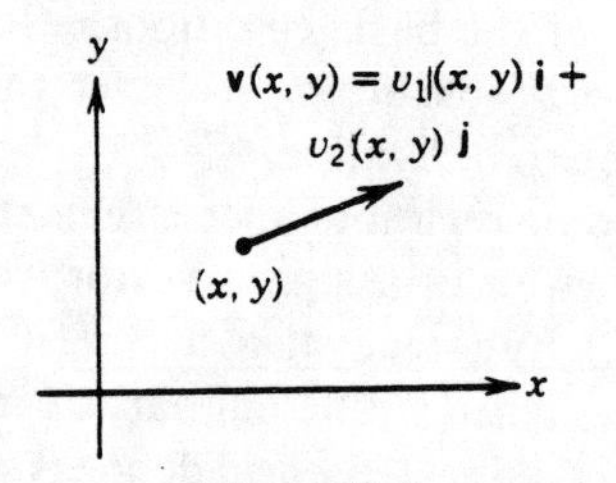

Figure C.2

(C.2) **Physical Laws**

(a) **Mass conservation**, which states that the time rate of increase of fluid in any region is due to fluid entering through the boundary of the region and to fluid produced by sources in the region.

(b) A consequence of the *full* fluid dynamical equations which states that in an ideal fluid, the **circulation** along any closed curve vanishes.[1] If C is any closed curve with unit tangent vector field **t**, then we define the circulation by

(C.3)
$$\text{circulation} = \oint_C (\mathbf{v} \cdot \mathbf{t})\, ds$$

[1] This can be derived from the basic equations of fluid dynamics, which are in any elementary textbook on the subject.

where the integral is a line interval with respect to arc length over C.[2] Such a fluid is called **irrotational**.

■

Now, consider an arbitrary region D in a two-dimensional steady fluid flow and bounded by a curve C with unit outer normal field $\mathbf{n}$ and tangent field $\mathbf{t}$ (see Figure C.3). Let ρ be the **density distribution** of the fluid (mass/unit area). By an

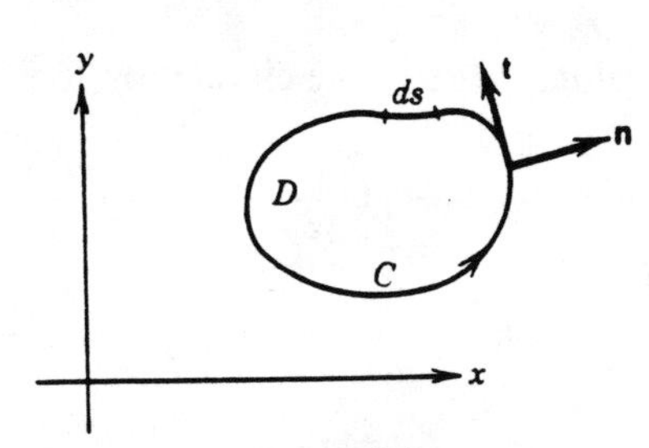

Figure C.3

incompressible fluid, we will mean one in which ρ is constant. Since $\mathbf{n}$ is the *outer* normal to C, the quantity $-\rho(\mathbf{n} \cdot \mathbf{v})\, ds$ represents the amount of fluid per unit time entering D through a *little bit* of its boundary ds. Hence, since ρ is assumed constant,

(C.4) Fluid entering D through its whole boundary C in unit time

$$= -\rho \oint_C \mathbf{n} \cdot \mathbf{v}\, ds = -\rho \iint_D \left(\frac{\partial v_1}{\partial x} + \frac{\partial v_2}{\partial y} \right) dx\, dy$$

where the second equality follows from a form of **Green's Theorem**.[3]

If $s(x, y)$ denotes a steady, two-dimensional **source density distribution** which produces fluid (mass per unit time per unit area), then

(C.5) $$\text{Fluid produced in } D \text{ per unit time due to sources} = \iint_D s(x, y)\, dx\, dy$$

Now the law of mass conservation (C.2**a**), *applied to a steady-state flow*, leads to the sum of the right-hand sides of (C.4) and (C.5) being zero, that is,

(C.6) $$\iint_D \left[-\rho \left(\frac{\partial v_1}{\partial x} + \frac{\partial v_2}{\partial y} \right) + s \right] dx\, dy = 0$$

If we assume that the integrand above is continuous, then, since we can take D to be

[2] Line integrals are discussed in any elementary calculus book.

[3] The reader is referred to any advanced calculus text for a discussion of vector integral theorems.

an arbitrarily small region, we can conclude that the integrand is zero. That is,

(C.7)
$$\frac{\partial v_1}{\partial x} + \frac{\partial v_2}{\partial y} = \frac{s}{\rho}$$

Note that, in the absence of sources,

(C.8)
$$\frac{\partial v_1}{\partial x} + \frac{\partial v_2}{\partial y} = 0 \qquad \text{(in a source-free region)}$$

This is called the **continuity equation.**

To derive a second equation, we go to the circulation law (C.2b) and write

$$0 = \oint_C \mathbf{v} \cdot \mathbf{t}\, ds = \iint_D \left(\frac{\partial v_1}{\partial y} - \frac{\partial v_2}{\partial x} \right) dx\, dy$$

where the second equality also follows from a form of Green's Theorem.[3] Hence, as above, we get

(C.9)
$$\frac{\partial v_1}{\partial y} - \frac{\partial v_2}{\partial x} = 0$$

Finally, a word about boundary conditions and the inviscid assumption. A fluid cannot enter a stationary, rigid body. Hence, if $\mathbf{n} = n_1 \mathbf{i} + n_2 \mathbf{j}$ represents the unit normal field at the boundary of such a body, then we have the following **boundary condition**:

(C.10)
$$\mathbf{v} \cdot \mathbf{n} = v_1 n_1 + v_2 n_2 = 0 \qquad \text{(on the boundary of a rigid body)}$$

If a fluid is **inviscid**, it cannot cling to a rigid body and hence fluid particles are permitted to slide along its surface. Thus, the above is the only boundary condition we can impose for an inviscid fluid (if the fluid were viscous, then $\mathbf{v}$, and hence both v_1 and v_2, would vanish on the boundary). Finally, if the region of fluid flow is unbounded, we will have need to impose **boundary conditions at infinity**.

For ease of reference, we collect our results below.

(C.11) **Basic Equations**

(a) $\mathbf{v} = v_1 \mathbf{i} + v_2 \mathbf{j}$ (velocity field)

(b) $\dfrac{\partial v_1}{\partial x} + \dfrac{\partial v_2}{\partial y} = 0$ (in a source-free region)

(c) $\dfrac{\partial v_1}{\partial y} - \dfrac{\partial v_2}{\partial x} = 0$ (irrotational fluid)

(d) $\mathbf{v} \cdot \mathbf{n} = 0$ (on a rigid body)

(e) $\rho \oint_C \mathbf{v} \cdot \mathbf{n}\, ds =$ fluid produced per unit time due to sources in the region bounded by the closed curve C.

P.131

[3] The reader is referred to any advanced calculus text for a discussion of vector integral theorems.

D

ELECTROSTATICS

Here, we will briefly develop the equations governing two-dimensional electrostatics. The reader is referred to any elementary book on electricity and magnetism for a more thorough development. To begin with, we have the following experimental observation.

(D.1) **Coulomb's Force Law**

> The electrostatic field between two small charged bodies is directed along the line joining them and is inversely proportional to the square of the distance between them.[1]

Mathematically, as depicted in Figure D.1, Coulomb's law states that the **force F** on a **charge of strength q_0** located at position (x, y, z) in space due to a charge of strength q_1 at (x_1, y_1, z_1) is given by

$$\mathbf{F} = \frac{q_0 q_1 \mathbf{r}_1}{|\mathbf{r}_1|^3} \tag{D.2}$$

[1] The reader will recognize this as being the same as **Newton's Law of Gravitation**.

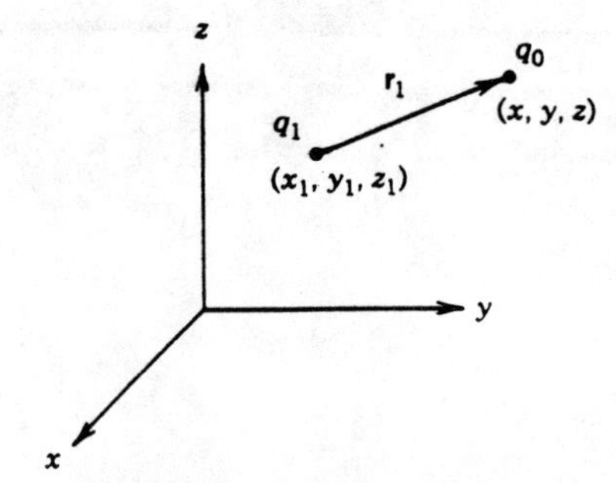

Figure D.1 Coulomb's law.

where[2]

(D.3)

$$\text{(a)}\quad \mathbf{r}_1 = (x - x_1)\mathbf{i} + (y - y_1)\mathbf{j} + (z - z_1)\mathbf{k}$$

$$\text{(b)}\quad |\mathbf{r}_1| = [(x - x_1)^2 + (y - y_1)^2 + (z - z_1)^2]^{1/2}$$

Note that if the charges q_0 and q_1 are of the same sign, the force is directed away from q_1 (**like charges repel**), while the opposite is true if q_0 and q_1 have different signs (**opposite charges attract**).

(D.4) ***Definition.*** The **electrostatic field intensity $\mathbf{E}_1$** at the point (x, y, z) due to charge q_1 is the vector function of position defined by

$$\mathbf{E}_1 = \frac{1}{q_0}\mathbf{F} = \frac{q_1\mathbf{r}_1}{|\mathbf{r}_1|^3}$$

and hence is the **force per unit charge** at (x, y, z) due to a charge q_1 at (x_1, y_1, z_1).

■

From (D.3), we can easily show that, for $(x, y, z) \neq (x_1, y_1, z_1)$,

(D.5)

$$\nabla\left(\frac{1}{|\mathbf{r}_1|}\right) = -\frac{\mathbf{r}_1}{|\mathbf{r}_1|^3} \quad \text{and} \quad \nabla^2\left(\frac{1}{|\mathbf{r}_1|}\right) = 0$$

where ∇ is the **gradient operator** $\mathbf{i}(\partial/\partial x) + \mathbf{j}(\partial/\partial y) + \mathbf{k}(\partial/\partial z)$ and ∇^2 is the **Laplacian** $(\partial^2/\partial x^2) + (\partial^2/\partial y^2) + (\partial^2/\partial z^2)$.[3] Using these results in the above definition, we can write

(D.6)

$$\mathbf{E}_1(x, y, z) = -\nabla\left(\frac{q_1}{|\mathbf{r}_1|}\right) \quad \text{with} \quad \nabla^2\left(\frac{q_1}{|\mathbf{r}_1|}\right) = 0$$

[2] There is usually a physical constant K in (D.2) whose value depends upon the units of charge. We will assume units of charge for which the constant is 1.

[3] The reader is referred to any advanced calculus text where these vector differential operators are discussed in detail.

Now consider a continuous, time independent **volume distribution of charge** distributed uniformly over a region R of space. If $\rho(\xi, \eta, \zeta)$ is the **charge density** (charge per unit volume) at a point (ξ, η, ζ) in R (see Figure D.2), then $\rho\, d\xi\, d\eta\, d\xi$ is the total amount of charge in a small volume $d\xi\, d\eta\, d\zeta$. Hence, the electric field intensity $\Delta\mathbf{E}(x, y, z)$ at (x, y, z) due to this amount of charge is, by (D.6),

(D.7)
$$\Delta\mathbf{E}(x, y, z) = -\nabla\left\{\frac{\rho(\xi, \eta, \zeta)\, d\xi\, d\eta\, d\zeta}{[(x-\xi)^2 + (y-\eta)^2 + (z-\zeta)^2]^{1/2}}\right\}$$

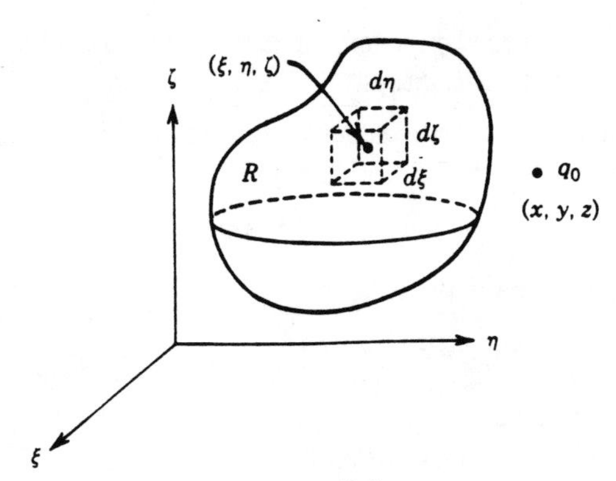

Figure D.2

Thus, "adding up" (integrating) all of these contributions and using the second of the equations in (D.6), we find that the electric field $\mathbf{E}$, and force $\mathbf{F}$ on a charge q_0, at a point (x, y, z) *outside* of R is given by

$$\text{(a)}\quad \mathbf{E}(x, y, z) = -\nabla\phi(x, y, z) \qquad \text{and} \qquad \mathbf{F} = q_0\mathbf{E}$$

where

(D.8)
$$\text{(b)}\quad \nabla^2\phi(x, y, z) = 0 \qquad (x, y, z) \text{ not in } R$$

The *scalar* **electric potential function (voltage)** $\phi(x, y, z)$ is given in terms of the charge density ρ by

(D.9)
$$\phi(x, y, z) = \iiint\limits_R \frac{\rho(\xi, \eta, \zeta)\, d\xi\, d\eta\, d\zeta}{[(x-\xi)^2 + (y-\eta)^2 + (z-\zeta)^2]^{1/2}}$$

Hence, we have that

> Outside of a charged region, the electric field intensity is the negative gradient of an harmonic function.

In addition, it is shown in any elementary textbook on electricity and magnetism, *that in a region possessing a charge density* ρ

(D.10)
$$\nabla^2\phi = -4\pi\rho$$

Using the relationship (D.8a) between force and the electric field, we can derive the following basic result.

(D.11) ***Theorem.*** The **work** W done in moving a charge q_0 from a point $P_1(x_1, y_1, z_1)$ to point $P_2(x_2, y_2, z_2)$ through an electric field, is q_0 multiplied by the **potential difference**, that is,

$$W = q_0[\phi(P_2) - \phi(P_1)]$$

■

Conductors. The types of problems in electrostatics which will be of interest to us will be ones involving conductors.

(D.12) ***Definition.*** A **perfect conductor** is a body in which or on which there is no resistance to the motion of charge.

■

With this definition, we have the following.

(D.13) ***Theorem.*** Inside a perfect conductor,

(a) $\mathbf{E} = \mathbf{0}$.

(b) ϕ = constant.

(c) There is no charge at rest.

Proof. Part (a) follows from the fact that by Definition (D.12), there is no force field inside a conductor. Hence, $\mathbf{E} = \mathbf{F}/q_0 = 0$. Thus, from (D.8a), $\nabla\phi = \mathbf{0} \Rightarrow \phi$ = constant in a conductor. Finally, since ϕ is constant, from (D.10) the static charge density ρ vanishes; thus, there are no charges at rest.

■

Now, if a battery is connected to a conductor, charge will flow freely between the conductor and the battery. When the battery is removed, the charge imparted to the conductor will attempt to arrange itself in an equilibrium configuration. Since in the final **steady state** no charge can be inside [(c) above], it will all have flowed to the surface and will remain as a steady-state **surface charge density**. In fact, on the surface of a perfect conductor we have the following.

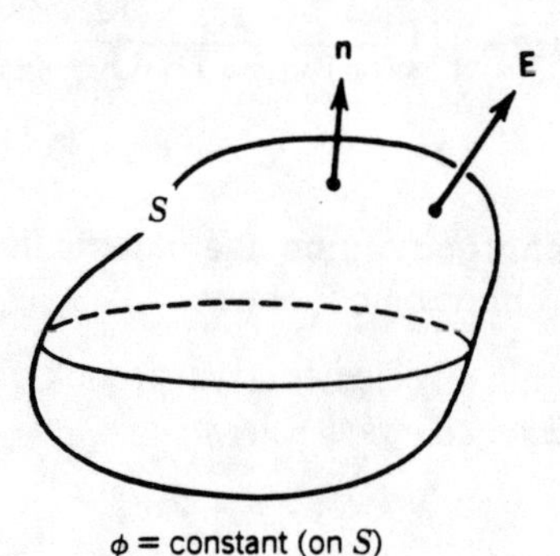

ϕ = constant (on S)

Figure D.3 Conductor.

(D.14) ***Theorem.*** Let S be the surface bounding a perfect conductor in equilibrium, with $\mathbf{n}$ the **unit outer normal field** to S (see Figure D.3). Then,

(a) At each point on S, $\mathbf{E}$ is normal to S.

(b) ϕ is constant on S (hence S is an **equipotential**).

(c) $\iint\limits_S \frac{\partial \phi}{\partial n}\, d\sigma = -4\pi Q$, where Q is the **total charge** on the surface of the conductor. The **normal derivative** $\partial\phi/\partial n$ is defined by $\mathbf{n} \cdot \nabla\phi$.

■

Let us consider a vertical cylindrical conductor R in space (see Figure D.4*a*). Then, R has the property that every horizontal plane cuts it in the same cross section. If we also assume that any surface charge densities on the surface S of R are independent of the vertical coordinate z, then we might expect that the field $\mathbf{E}$

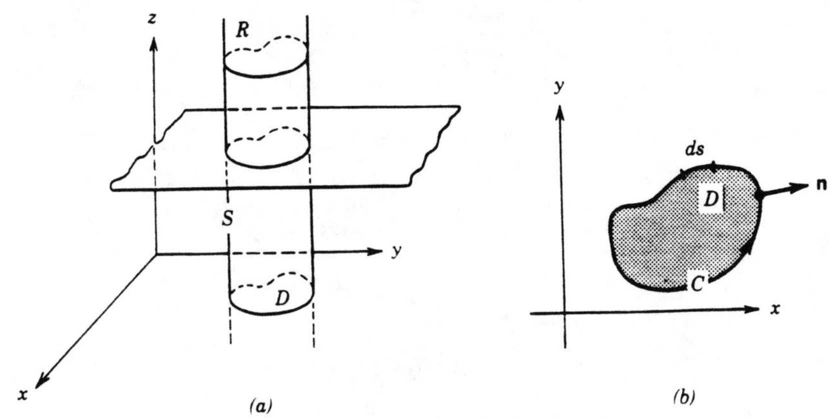

Figure D.4 Two-dimensional conductor.

and potential ϕ will also be independent of z. Thus, they will be functions $\mathbf{E}(x, y)$ and $\phi(x, y)$ of only x and y and will have the same distributions over any horizontal plane. Hence, if we consider the two-dimensional cross section D lying in the xy-plane and bounded by a curve C (see Figure D.4*b*), then, by (D.8b) and (D.14), the potential $\phi(x, y)$ is a solution of

(D.15)

(a) $\phi_{xx} + \phi_{yy} = 0$ (outside D).

(b) $\phi = \text{constant}$ (inside D and on C).

(c) If q is the total charge on C *per unit vertical length* on S, then

$$\oint_C \frac{\partial \phi}{\partial n}\, ds = -4\pi q$$

where the term on the left is a line integral over C.[4]

[4] In fact, this result is true even if C is not the boundary of a conductor. This follows from applying **Green's Theorem** to both sides of (D.10). Then, q is the charge inside and on C.

Note that if we can find a potential ϕ satisfying (D.15), then from (D.8) the electric field is given by

$$\mathbf{E} = -\left(\frac{\partial \phi}{\partial x}\mathbf{i} + \frac{\partial \phi}{\partial y}\mathbf{j}\right)$$

Finally, we list the types of problems which are often of interest.

(D.16) **Problems**

I. The surfaces of conductors $D_0, D_1, \ldots, D_N$ are raised to constant potentials $V_0, V_1, \ldots, V_N$, respectively. Find the potential ϕ (and hence **E**) in the region between them (see Figure D.5*a*).

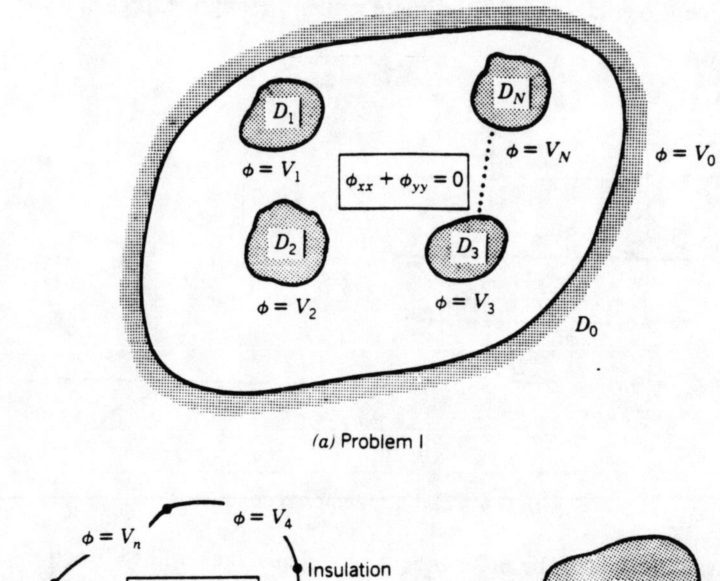

(a) Problem I

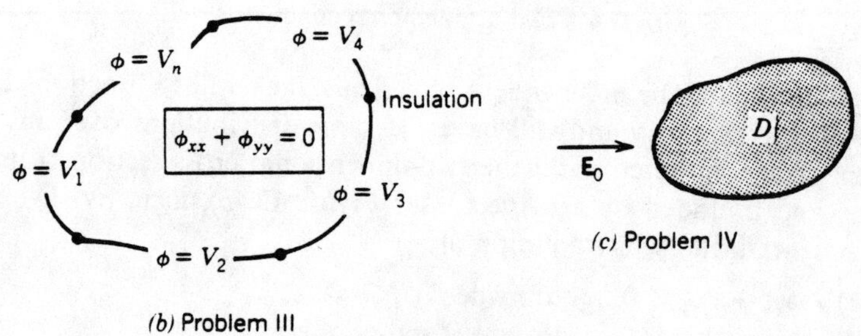

(b) Problem III

(c) Problem IV

Figure D.5

II. In a variation of Problem I, charges $q_0, q_1, \ldots, q_N$ are put on the surfaces of the conductors $D_0, D_1, \ldots, D_N$. Find the electric field between them.

III. N pieces of a very thin conductor in the shape of a planar curve are insulated from each other and then raised to (possibly different) potentials $V_1, \ldots, V_N$. Find the field in the region inside the curve (see Figure D.5*b*).

IV. A known electric field $\mathbf{E}_0$, with corresponding potential ϕ_0, is set up in the whole plane (for example, a constant, uniform field), and then a conductor D is placed in this field. How is the field $\mathbf{E}_0$ disturbed by the presence of the conductor? (See Figure D.5*c*).

INDEX